Linear Algebra to Differential Equations

Linear Algebra to Differential Equations

J. Vasundhara Devi
Sadashiv G. Deo
Ramakrishna Khandeparkar

CRC Press
Taylor & Francis Group
Boca Raton London New York

CRC Press is an imprint of the
Taylor & Francis Group, an **informa** business
A CHAPMAN & HALL BOOK

First edition published 2022
by CRC Press
6000 Broken Sound Parkway NW, Suite 300, Boca Raton, FL 33487-2742

and by CRC Press
2 Park Square, Milton Park, Abingdon, Oxon, OX14 4RN

CRC Press is an imprint of Taylor & Francis Group, LLC

Library of Congress Cataloging-in-Publication Data

Names: Vasundhara Devi, J., author. | Deo, Sadashiv G., 1933- author. |
Khandeparkar, Ramakrishna, author.
Title: Linear algebra to differential equations / J. Vasundhara Devi,
Sadashiv G. Deo, Ramakrishna Khandeparkar.
Description: First edition. | Boca Raton, FL : Chapman & Hall, CRC Press,
2021. | Includes bibliographical references and index.
Identifiers: LCCN 2021019504 (print) | LCCN 2021019505 (ebook) | ISBN
9780815361466 (hardback) | ISBN 9781032067988 (paperback) | ISBN
9781351014953 (ebook)
Subjects: LCSH: Algebras, Linear. | Differential equations.
Classification: LCC QA184.2 .V388 2021 (print) | LCC QA184.2 (ebook) |
DDC 512/.5--dc23
LC record available at https://lccn.loc.gov/2021019504
LC ebook record available at https://lccn.loc.gov/2021019505

ISBN: 978-0-815-36146-6(hbk)
ISBN: 978-1-032-06798-8(pbk)
ISBN: 978-1-351-01495-3(ebk)

DOI: 10.1201/9781351014953

Typeset in Computer Modern
by KnowledgeWorks Global Ltd.

To our teacher, guide and philosopher
Professor V. Lakshmikantham,
world citizen and teacher beyond comparison.

Contents

Preface

The authors feel the necessity to answer an important question: what is the need for another book on matrices, linear algebra and differential equations when there are hundreds of books available and some of them are excellent?

An answer in a nutshell is: racing time, enhanced computational literacy and facilities, expanding knowledge on all frontiers, similarity among various disciplines, complexities evolving due to changes in social, physical, technical and engineering domains giving rise to new problems–all call for a new book.

A book, in fact, a textbook that

(i) is complete in the basic knowledge of the subject,

(ii) has a lucid exposition of the necessary topics,

(iii) introduces the abstraction and at the same time is down to earth,

(iv) highlights the methods and approaches that are more useful,

(v) brings out the essential techniques,

(vi) points towards the similarities in various disciplines,

(vii) has illustrative examples,

(viii) introduces applications both classical and novel,

(ix) gives an overview of techniques that may be useful in the future and could go a long way in serving the needs of the students and applied scientists alike.

With this view the authors conceived this book.

This book has many new features. The fundamental knowledge essential for a student of linear algebra is given in the first two chapters. The authors quickly go through the basics and also include concepts like pseudo inverse and congruent inverse which are useful in applications. Eigen-value decomposition theorem and the introduction of singular values and singular vectors in the second chapter allow the time needed for a student to assimilate these ideas to be used later.

Chapter 3 begins with abstract concepts but quickly shifts to down-to-earth topics: change of basis matrix, linear transformations and matrices of linear transformations. These are some of the topics frequently used by applied scientists. The singular value decomposition theorem and the Gram–Schmidt orthogonalization process are given since these are essential tools for applied scientists, engineers and technocrats.

In Chapter 4, after highlighting the natural existence of errors in computation, the basic methods for solving a linear system of equations are given using both direct and iterative methods. As orthogonal transformations play a vital role in iterative techniques, the Householder method and the plane rotation method are given. Bounds for the eigen-values, the power method for finding the largest eigen-value and QR method are also described. A brief description of Krylov subspace methods introduce a broad spectrum of solvers to be discovered by the reader. As problems in nearly all the topics involving matrices can be solved by just giving a command in a computer, much stress has not been given in solving problems. In fact, the opinion of the authors is that the content of this chapter, is more tuned to assignments than to examinations. Further, in this chapter the basic techniques have been given and the refinements can be learnt by the reader whenever required.

Chapter 5 brings forth the use of linear algebra in applied areas. An attempt has been made to introduce and showcase some of them–both from classical applications and the latest topics. The reader will recognize that these applications are only the tip of an iceberg of the wide range and scope of applications of linear algebra.

A section has been included in Chapter 3 linking linear algebra to differential equations. It can be considered as a prologue to Chapters 6 to 9.

In Chapter 6, Kronecker products and sums are introduced. These are important tools to obtain eigen-values and eigen-vectors of higher-order matrices.

In Chapter 7 various types of derivatives available in literature have been given though most of the derivatives are seldom used. There is a lot of potential for both mathematicians and applied researchers to develop new types of differential equations and obtain new results.

Chapter 8 contains the study of a linear system of differential equations. This chapter is essential for engineering students using control systems and mechanical systems among others.

In Chapter 9, Linear matrix differential equations, an extension of the systems studied in Chapter 8 have been described.

Throughout the book the concepts have been illustrated through examples while routine mathematical proofs have been avoided. While a number of examples and problems are illustrative, nevertheless, the readers are encouraged to create their own problems as the process of creating a textbook problem is a backward process starting from the solution.

The authors would like to place on record their gratitude to all the scientists and mathematicians who have taken the subject to the level where applications of the subject have become innumerable.

Next the authors would like to record their regards to the management of Gayatri Vidya Parishad College of Engineering (Autonomous) (GVPCE(A)),

Visakhapatnam, India, for all the support rendered. Special mention must be made of Dr A. B. Koteswara Rao, Principal, GVPCE(A) for giving all the necessary support when the writing of the book was in progress including the comments and suggestions in the topic of Robotics. In developing the sections relating to Cryptography, Computer Graphics, Big Data and Bioinformatics, the authors wish to acknowledge the guidance provided by Professor R. G. Marshall, (Retired) Professor of Computer Science, Plymouth State University, New Hampshire, USA. Dr G. Sudheer and Dr K. L. Sai Prasad of Dept. of Mathematics, GVPCE (Women) have given timely suggestions for the betterment of the book and the authors acknowledge it.

The Dept. of Mathematics of GVPCE(A) has given its unstinted support. The authors feel it a privilege to express their appreciation to the department in general and in particular to Dr R.V.G. Ravi Kumar, Associate Professor and Head, Department of Mathematics, Dr A.R.J. Srikanth, Dr. Ch. V. Sreedhar, Dr Ch. Appala Naidu, Dr J. Satish, Dr D. Ravi Kumar, Mr. I.S.N.R.G. Bharat and Mr. S. Srinivasa Rao, all Assistant Professors of the department and Dr P. Sarada Varma, WOS(A), SERB, DST. Dr K.V. Varalakshmi, Assistant Professor, Dept. of Mechanical Engineering GVPCE(A) drew the figures in the book and Dr D. N. D. Harini, Associate Professor, Dept. of Computer Science and Engineering GVPCE(A) and Mr S. Kanthi Kiran, Assistant Professor, Dept. of Information Technology, GVPCE(A) helped problems related to Computer Graphics and PCA. Professor D. Varadaraju, Professor, Dept. of Mechanical Engineering, GVPCE(A) has given immense support in reading of the proofs. They all are sincerely acknowledged. Dr N. Deepika Rani, Dept. of Electronics and Communications Engineering, GVPCE(A) has given timely help with her team of students: Ms P Mayuri, Ms B. Prema and Ms V. Shyamala – all M. Tech, ECE students, in typesetting the book in LATEX and the authors record their appreciation. The contribution of Dr Lymuel McRae, Consultant, McLean, Virginia, USA, Dr F. A. McRae of Catholic University of America, USA, Dr Z. Drici of Illinois Wesleyan University, USA and Mr J. V. S. Dattatreya of Affine Tech Systems Private Limited India in ensuing correctness of final version of the book is sincerely acknowledged. The authors extend heartfelt thanks to the Janvi Enterprises at Dombivili(E) Maharashtra State, India for providing computer facilities for quick communication between the authors while the book writing was in progress.

Dr Aastha Sharma and Ms Shikha Garg of Taylor & Francis group have demonstrated utmost patience and gave complete support in a very turbulent phase of the writing of the book. The authors express their appreciation.

Last but not the least the authors record their deep sense of gratitude to their respective families, teachers and friends for being with them throughout.

J. Vasundhara Devi
Sadashiv G. Deo
Ramakrishna Khandeparkar

Condolences

It is with a heavy heart that we express our sorrow at the loss of our co-author, Dr Ramakrishna Khandeparkar, while the book is still in progress. His contribution to the book is sincerely acknowledged.

J. Vasundhara Devi
Sadashiv G. Deo

Chapter 1

Vectors and Matrices

1.1 Introduction

This is the foundation chapter for the book. The building blocks of all the topics covered in the book constitute this chapter. Section 1.2 begins with the definitions of a scalar and vector and proceeds with the study of their properties. These properties provide a base for the theory developed in later chapters. Matrices are introduced in terms of vectors in Section 1.3. This definition is useful as nearly in all applications a matrix consisting of linearly independent vectors or ortho-normal vectors is utilized to transform a vector into a new vector in a given situation. Various types of matrices which will be useful later are introduced in Section 1.4, while Section 1.5 deals with elementary operations on matrices. These elementary operations are the only operations that can be done within a matrix and their importance is in developing algorithms for numerical techniques. Elementary matrices resulting from elementary operations are introduced in this section. Determinants, which attribute a single numerical value to a matrix, are given along with their properties in Section 1.6. The contents of Section 1.7 consist of the inverse of a matrix together with its properties. Section 1.8 deals with the partitioning of a matrix, wherein basic algebraic properties involving the partition are introduced. Section 1.9 introduces the concepts of pseudo inverse and congruent inverse.

1.2 Scalars and Vectors

The concepts of a scalar and a vector are very essential and are basic fundamental notions describing the physical world.

A scalar is used to denote quantification or measurement and to express magnitude. For example, mass, temperature, length, volume of a body and time are all scalars.

DOI: 10.1201/9781351014953-1

Definition 1.2.1 *A scalar is a quantity completely specified by a real or a complex number.*

A scalar is denoted by small letters a, b, c, x, y, z for real numbers and $a + ib$, $x + iy$ and so on for complex numbers.

Velocity and acceleration are typical examples of a vector. In physical terms, a vector has both magnitude and direction.

On the real line $\mathbb{R}$, a vector describes the distance to the left or to the right of zero. Here, zero is the point of reference.

In the plane $\mathbb{R}^2$, the vector $(a \;\; b)$ is the position vector of the point $(a \;\; b)$. It gives the distance of the point from the origin $(0 \;\; 0)$ as well as the direction.

In the space $\mathbb{R}^3$, corresponding to any point $(x \;\; y \;\; z)$, a position vector is associated to it. This vector gives both the distance and direction from $(0 \;\; 0 \;\; 0)$, the origin, which is the frame of reference in $\mathbb{R}^3$.

Definition 1.2.2 *An n-dimensional vector* **a** *is an ordered set of 'n' scalars written in ordinary parentheses as* $\mathbf{a} = (a_1 \;\; a_2 \cdots \;\; a_i \;\; \cdots \;\; a_n)$. *The scalar '$a_i$' is called the ith component of the vector.*

A vector having 'n' components is called an 'n'-tuple.

Example 1.2.1 *In the 3-dimensional space, a point* $(x \;\; y \;\; z)$ *can be considered as a vector, called the position vector of the point* $(x \;\; y \;\; z)$ *and is represented as in Fig 1.1.*

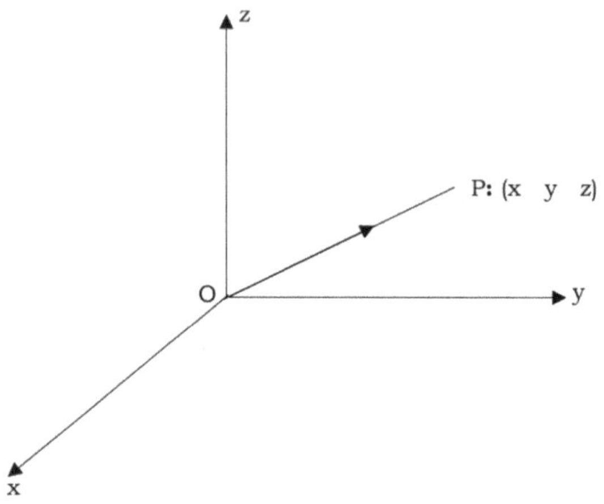

Fig. 1.1 A vector in 3D space

Observation. It can be observed that in a 2-dimensional space or in a 3-dimensional space, a vector is a directed line segment, having both magnitude and direction.

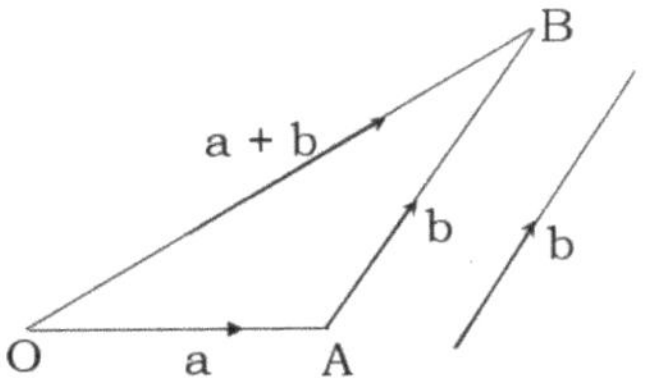

Fig. 1.2 Triangle law of addition

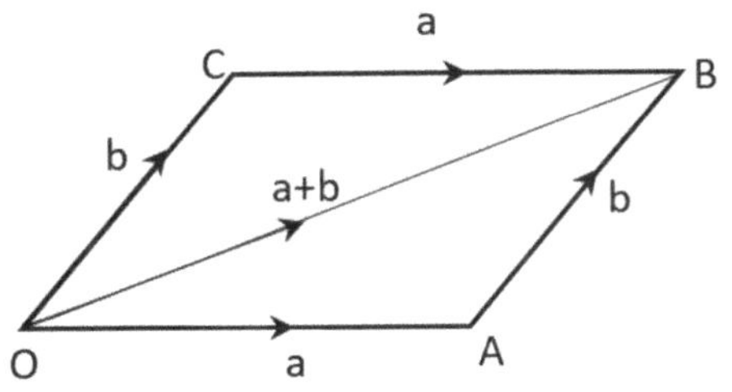

Fig. 1.3 Parallelogram law of addition

Definition 1.2.3 *Two vectors* $\mathbf{a} = (a_1 \ \ a_2 \ \ \cdots \ \ a_n)$ *and* $\mathbf{b} = (b_1 \ \ b_2 \ \ \cdots \ \ b_n)$ *are equal if and only if* $a_i = b_i$, $i = 1, 2, \cdots, n$.

Definition 1.2.4 *A zero vector is a vector all of whose components are zero,* $\mathbf{a} = (0 \ \ 0 \ \ \cdots, 0)$ *and is designated as* $\mathbf{0}$.

Definition 1.2.5 *A vector* $\mathbf{w} = (w_1 \ \ w_2 \ \ ... \ \ w_n)$ *is the negative vector of* $\mathbf{v} = (v_1 \ \ v_2 \ \ ... \ \ v_n)$ *if and only if* $\mathbf{w} = (-v_1 \ \ -v_2 \ \ ... \ \ -v_n) = -\mathbf{v}$.

Definition 1.2.6 *The sum of two vectors* $\mathbf{a} = (a_1 \ \ a_2 \ \ \cdots \ \ a_n)$ *and* $\mathbf{b} = (b_1 \ \ b_2 \ \ \cdots \ \ b_n)$ *is denoted by* $\mathbf{a} + \mathbf{b}$ *and is defined as* $\mathbf{a} + \mathbf{b} = (a_1 + b_1 \ \ a_2 + b_2 \ \ ... \ \ a_n + b_n)$. *That is, the sum of the two vectors* $\mathbf{a}$ *and* $\mathbf{b}$ *is a vector whose* i*th component is the sum* $a_i + b_i$.

Geometrically, if $\mathbf{a}$ and $\mathbf{b}$ are any two vectors in a plane, then the sum $\mathbf{a} + \mathbf{b}$ is obtained from the triangle formed by vectors $\mathbf{a}$ and $\mathbf{b}$ or from the parallelogram with $\mathbf{a}$ and $\mathbf{b}$ as adjacent sides. The following diagrams, Fig 1.2 and Fig 1.3 are self-explanatory.

Properties of vector addition in a plane are given in the following figures. Fig. 1.4 shows that vector addition is commutative, that is, $\mathbf{a} + \mathbf{b} = \mathbf{b} + \mathbf{a}$.

Fig. 1.5 shows that vector addition is associative, that is,
$(\mathbf{a} + \mathbf{b}) + \mathbf{c} = \mathbf{a} + (\mathbf{b} + \mathbf{c})$.

Fig. 1.6 explains the concept of difference of two vectors $\mathbf{a}$ and $\mathbf{b}$.

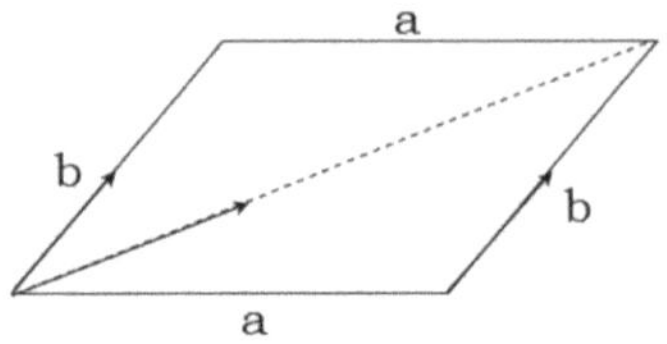

Fig. 1.4 Vector addition is commutative

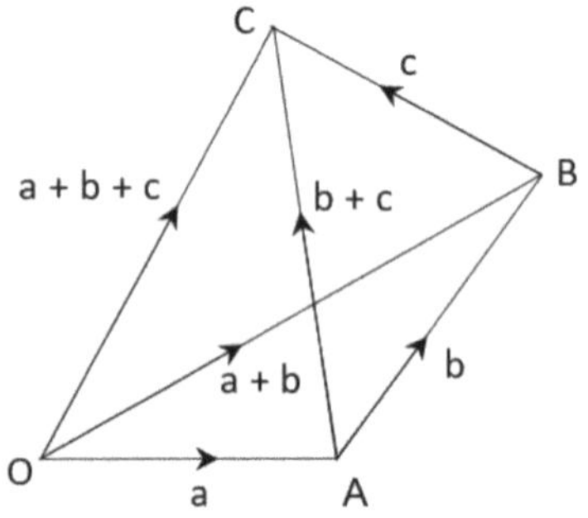

Fig. 1.5 Vector addition is associative

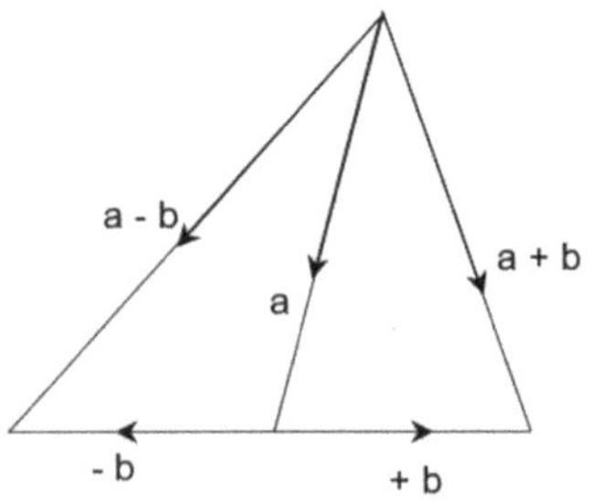

Fig. 1.6 Difference of two vectors

Note. The properties of associativity and commutativity of addition hold good for n-tuples, $n \geq 1$.

Definition 1.2.7 *Multiplication of a vector* $\mathbf{a}$ *by a scalar* $\lambda \in \mathbb{R}$ *is defined as a vector* $\lambda\mathbf{a} = (\lambda a_1 \ \ \lambda a_2 \ \ \cdots \ \ \lambda a_n)$, *that is, the vector obtained by multiplying each component* a_i *of* $\mathbf{a}$ *with* λ.

Observation. If $\lambda > 0$, then the vector $\lambda\mathbf{a}$ is in the direction of $\mathbf{a}$. If $\lambda < 0$, then the vector $\lambda\mathbf{a}$ is in the opposite direction of $\mathbf{a}$. If $\lambda = 0$, then the vector $\lambda\mathbf{a}$ is the zero vector.

If $|\lambda| > 1$, then $\lambda\mathbf{a}$ is said to be a magnification or dilation of $\mathbf{a}$ and if $|\lambda| < 1$, then $\lambda\mathbf{a}$ is called a contraction of $\mathbf{a}$.

A combination of the above definitions gives rise to an important notion that plays a pivotal role in linear algebra.

Definition 1.2.8 *If* $\mathbf{v}_1, \mathbf{v}_2, \cdots, \mathbf{v}_n$ *are* n *vectors and* $\alpha_1, \alpha_2, ..., \alpha_n$ *are* n *scalars, then the linear combination of the* n *vectors is defined as a vector* $\mathbf{v} = \alpha_1\mathbf{v}_1 + \alpha_2\mathbf{v}_2 + ... + \alpha_n\mathbf{v}_n$.

Example 1.2.2 *If*

$$\mathbf{v_1} = \begin{bmatrix} 1 \\ 2 \\ 3 \end{bmatrix}, \quad \mathbf{v_2} = \begin{bmatrix} 3 \\ 2 \\ 4 \end{bmatrix} \text{ and } \mathbf{v_3} = \begin{bmatrix} 2 \\ 1 \\ 8 \end{bmatrix},$$

$$\text{then } 3\mathbf{v_1} + 2\mathbf{v_2} + \mathbf{v_3} = \begin{bmatrix} 11 \\ 11 \\ 25 \end{bmatrix}.$$

Example 1.2.3 *In* $\mathbb{R}^2$, *let* $\mathbf{i} = (1 \ \ 0)$ *and* $\mathbf{j} = (0 \ \ 1)$. *Then any position vector* $\mathbf{r} = (x \ \ y)$ *can be written as* $\mathbf{r} = x(1 \ \ 0) + y(0 \ \ 1)$.

Example 1.2.4 *In* $\mathbb{R}^3$, *let* $\mathbf{i} = (1 \ \ 0 \ \ 0)$, $\mathbf{j} = (0 \ \ 1 \ \ 0)$ *and* $\mathbf{k} = (0 \ \ 0 \ \ 1)$. *Then any position vector* $\mathbf{r} = (x \ \ y \ \ z)$ *can be written as* $\mathbf{r} = x(1 \ \ 0 \ \ 0) + y(0 \ \ 1 \ \ 0) + z(0 \ \ 0 \ \ 1) = x\mathbf{i} + y\mathbf{j} + z\mathbf{k}$.

Result 1.2.1 To extend this idea to n-tuples, denote $\mathbf{e}_1 = (1 \ \ 0 \ \ ... \ \ 0)$, $\mathbf{e}_2 = (0 \ \ 1 \ \ 0 \ \ ..., 0), ..., \mathbf{e}_n = (0 \ \ 0 \ \ ... \ \ 0 \ \ 1)$, where

$$\mathbf{e}_i = \begin{cases} 1 \ \textit{in the ith component} \\ 0 \ \textit{elsewhere.} \end{cases}$$

Then any vector, $\mathbf{v} = (x_1 \ \ x_2 \ \ ... \ \ x_n) = x_1\mathbf{e}_1 + x_2\mathbf{e}_2 + ... + x_n\mathbf{e}_n$.

Definition 1.2.9 *The vectors* $\mathbf{v}_1, \mathbf{v}_2, ..., \mathbf{v}_n$ *are said to be linearly independent if and only if a linear combination*

$$\alpha_1 \mathbf{v}_1 + \alpha_2 \mathbf{v}_2 + ... + \alpha_n \mathbf{v}_n = \mathbf{0}, \quad \alpha_i \in \mathbb{R},$$

implies $\alpha_i = 0$ *for all i=1,2,...,n.*

Example 1.2.5 *Show that* $\mathbf{v}_1 = (1 \ \ 1 \ \ 0)$, $\mathbf{v}_2 = (1 \ \ 0 \ \ 1)$ *and* $\mathbf{v}_3 = (0 \ \ 1 \ \ 1)$ *are linearly independent.*
Consider $\alpha_1(1 \ \ 1 \ \ 0) + \alpha_2(1 \ \ 0 \ \ 1) + \alpha_3(0 \ \ 1 \ \ 1) = 0$
then $\alpha_1 + \alpha_2 = 0$, $\alpha_1 + \alpha_3 = 0$ *and* $\alpha_2 + \alpha_3 = 0$, *which gives* $\alpha_1 = \alpha_2 = \alpha_3 = 0$.

Definition 1.2.10 *The vectors* $\mathbf{v}_1, \mathbf{v}_2, ..., \mathbf{v}_n$ *are said to be linearly dependent if they are not linearly independent.*

Alternatively, the vectors $\mathbf{v}_1, \mathbf{v}_2, ..., \mathbf{v}_n$ are linearly dependent if there exist scalars $\alpha_1, \alpha_2, ..., \alpha_n$, not all zero, such that the linear combination

$$\alpha_1 \mathbf{v}_1 + \alpha_2 \mathbf{v}_2 + ... + \alpha_n \mathbf{v}_n = \mathbf{0}.$$

Example 1.2.6 *Show that* $\mathbf{v}_1 = (1 \ \ 3 \ \ 4), \mathbf{v}_2 = (2 \ \ 9 \ \ 5)$ *and* $\mathbf{v}_3 = (4 \ \ 15 \ \ 13)$ *are linearly dependent.*

Consider $\alpha_1(1 \ \ 3 \ \ 4) + \alpha_2(2 \ \ 9 \ \ 5) + \alpha_3(4 \ \ 15 \ \ 13) = 0$. Then simplifying gives $\alpha_1 + 2\alpha_2 + 4\alpha_3 = 0$, $3\alpha_1 + 9\alpha_2 + 15\alpha_3 = 0$ and $4\alpha_1 + 5\alpha_2 + 13\alpha_3 = 0$ which yields upon solving $\alpha_1 = 2$, $\alpha_2 = 1$ and $\alpha_3 = -1$. Hence the result.

The length or magnitude of a vector is a very important notion and is as follows.

Definition 1.2.11 *The magnitude or the length of a vector* $\mathbf{v} = (v_1 \ \ v_2 \ \ ... \ \ v_n)$ *is given by the Euclidean distance as*

$$|\mathbf{v}| = l(\mathbf{v}) = \sqrt{v_1^2 + v_2^2 + ... + v_n^2}$$

Example 1.2.7 *If* $\mathbf{v} = (2, 5, 7)$ *then the Euclidean distance is*

$$|\mathbf{v}| = l(\mathbf{v}) = \sqrt{2^2 + 5^2 + 7^2} = \sqrt{78}.$$

Definition 1.2.12 *A unit vector is a vector whose magnitude or length is 1.*

Example 1.2.8 *Show that* $(1 \ \ 0), (0 \ \ 1), (\frac{1}{\sqrt{2}} \ \ \frac{1}{\sqrt{2}})$ *are all unit vectors in a plane.*

$$|(1,0)| = \sqrt{1} = 1, \quad |(0,1)| = \sqrt{1} = 1$$

and $|(\frac{1}{\sqrt{2}}, \frac{1}{\sqrt{2}})| = \sqrt{\frac{1}{2} + \frac{1}{2}} = 1$.

Normalization of a vector is a process of obtaining a unit vector by dividing a given vector by its length.

Example 1.2.9 *If* $\mathbf{v} = (10 \;\; 5 \;\; 4)$, *then,*

$$\mathbf{u} = \frac{\mathbf{v}}{|\mathbf{v}|} = \frac{1}{\sqrt{141}} (10 \;\; 5 \;\; 4)$$

is a unit vector.

The product of vectors can be defined in many ways. The dot product (also called scalar product or inner product) and vector product, which are widely used in science and engineering are considered below.

Definition 1.2.13 *Dot product of two vectors* $\mathbf{a} = (a_1 \;\; a_2 \;\; \ldots \;\; a_n)$ *and* $\mathbf{b} = (b_1 \;\; b_2 \;\; \ldots \;\; b_n)$ *is defined as*

$$\mathbf{a} \cdot \mathbf{b} = \sum_{i=1}^{n} a_i b_i = a_1 b_1 + a_2 b_2 + \ldots + a_n b_n.$$

The dot product in a plane is defined as follows.

Let $OA = \mathbf{a}$ and $OB = \mathbf{b}$ be vectors in the XY-plane making angles α and β with the X-axis. Let θ be the angle between $\mathbf{a}$ and $\mathbf{b}$ measured in the anti-clockwise direction. Then $\theta = \alpha - \beta$. Then from the Fig 1.7,

$$\cos\theta = \cos(\alpha - \beta) = \cos\alpha \;\; \cos\beta + \sin\alpha \;\; \sin\beta$$

$$= \frac{a_1 b_1}{|\mathbf{a}||\mathbf{b}|} + \frac{a_2 b_2}{|\mathbf{a}||\mathbf{b}|} = \frac{a_1 b_1 + a_2 b_2}{|\mathbf{a}||\mathbf{b}|},$$

we get, $|\mathbf{a}||\mathbf{b}| cos\theta = a_1 b_1 + a_2 b_2 = \mathbf{a} \cdot \mathbf{b}$,
i.e.

$$\mathbf{a} \cdot \mathbf{b} = |\mathbf{a}||\mathbf{b}| cos\theta.$$

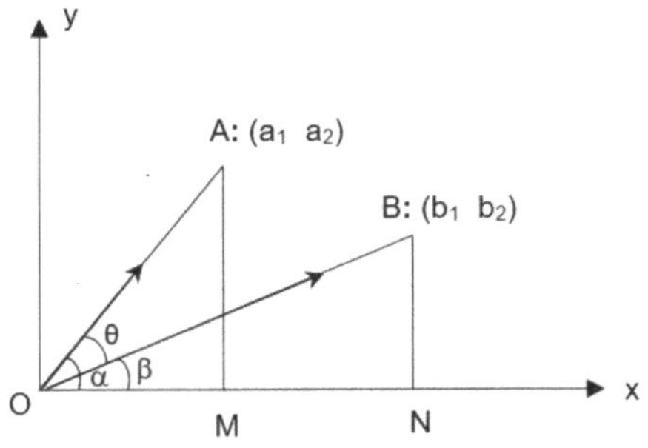

Fig. 1.7 Scalar product of two vectors

Thus in a plane or in the space, $\mathbf{a} \cdot \mathbf{b} = |\mathbf{a}||\mathbf{b}| cos\theta$, where θ is the angle between $\mathbf{a}$ and $\mathbf{b}$.

Some of the important properties of the dot product are:

1. The dot product of two nonzero vectors $\mathbf{a}$ and $\mathbf{b}$ is positive, zero or negative according to whether the angle θ between the vectors is less than, equal to or greater than $90°(=\frac{\pi}{2})$, respectively.

2. The dot product is commutative,

$$\mathbf{a}\cdot\mathbf{b} = \mathbf{b}\cdot\mathbf{a} = |\mathbf{a}||\mathbf{b}|cos\theta.$$

3. If $\mathbf{a}\cdot\mathbf{b} = \mathbf{0}$ then either $\mathbf{a} = \mathbf{0}$ or $\mathbf{b} = \mathbf{0}$; if $\mathbf{a} \neq \mathbf{0}$ and $\mathbf{b} \neq \mathbf{0}$ then $\mathbf{a}$ is perpendicular to $\mathbf{b}$.

Example 1.2.10 *(a) If* $\mathbf{a} = (2 \;\; 3 \;\; 4)$, $\mathbf{b} = (0 \;\; 1 \;\; 5)$ *then* $\mathbf{a}\cdot\mathbf{b} = 23$.
(b) if $\mathbf{e_1} = (1 \;\; 0)$ *and* $\mathbf{e_2} = (0 \;\; 1)$ *then* $\mathbf{e_1}\cdot\mathbf{e_2} = 0$, *that is,* $\mathbf{e_1}$ *is perpendicular to* $\mathbf{e_2}$ *and* $\mathbf{e_1}\cdot\mathbf{e_1} = 1$ *and* $\mathbf{e_2}\cdot\mathbf{e_2} = 1$.

Definition 1.2.14 *Projection of vector* $\mathbf{b}$ *on to a vector* $\mathbf{a}$ *is a scalar and is defined as*

$$\frac{\mathbf{b}\cdot\mathbf{a}}{|\mathbf{a}|}.$$

Definition 1.2.15 *The projection vector of* $\mathbf{b}$ *on to a vector* $\mathbf{a}$ *is given by* $\frac{\mathbf{b}\cdot\mathbf{a}}{|\mathbf{a}|^2}\mathbf{a}$.

Another kind of definition of a product between two vectors yields a vector and the definition is as follows.

Definition 1.2.16 *The cross product of any two vectors* $\mathbf{a} = (a_1 \; a_2 \; a_3)$ *and* $\mathbf{b} = (b_1 \; b_2 \; b_3)$ *is defined by*

$$\mathbf{a}\times\mathbf{b} = (a_2b_3 - a_3b_2 \quad -(a_1b_3 - a_3b_1) \;\; a_1b_2 - a_2b_1).$$

The following definition is in terms of angle θ.

Aliter definition

The cross product or the vector product of two vectors $\mathbf{a}$ and $\mathbf{b}$ is defined as

$$\mathbf{a}\times\mathbf{b} = |\mathbf{a}||\mathbf{b}|\sin\theta\,\mathbf{e},$$

where θ is the angle between $\mathbf{a}$ and $\mathbf{b}$ such that $0 \leq \theta \leq \pi$ and $\mathbf{e}$ is a unit vector perpendicular to both $\mathbf{a}$ and $\mathbf{b}$ and in a sense as per the right-hand rule.
The vector product is explained in Fig. 1.8.

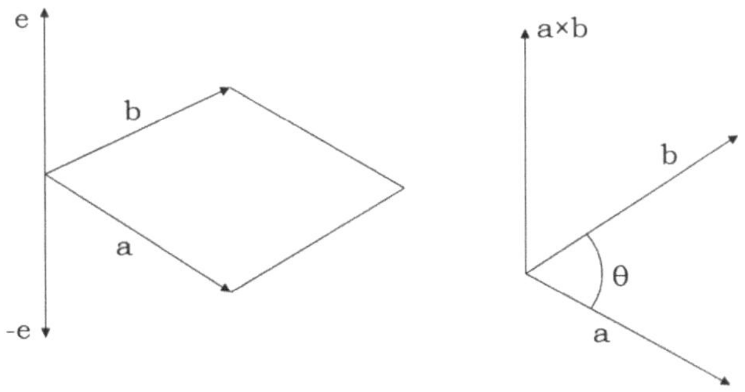

Fig. 1.8 Vector product of two vectors

Important properties of the cross product

1. $\mathbf{a} \times \mathbf{b}$ is a vector in the direction perpendicular to both $\mathbf{a}$ and $\mathbf{b}$.
2. The cross product of two vectors $\mathbf{a}$ and $\mathbf{b}$ gives the area of the parallelogram enclosed by the two vectors $\mathbf{a}$ and $\mathbf{b}$.
3. The area of the triangle whose sides are $\mathbf{a}$ and $\mathbf{b}$ is given by $\frac{1}{2}(\mathbf{a} \times \mathbf{b})$.

Note. Problems on vector products can be easily done once the concept of determinants is introduced in Section 1.6.

Example 1.2.11 *If* $\mathbf{i}$, $\mathbf{j}$, $\mathbf{k}$ *are mutually perpendicular unit vectors in* $\mathbb{R}^3$ *along the coordinate axes, then the following relations hold:*

$$\mathbf{i} \times \mathbf{j} = \mathbf{k}, \qquad \mathbf{j} \times \mathbf{k} = \mathbf{i}, \qquad \mathbf{k} \times \mathbf{i} = \mathbf{j};$$

$$\mathbf{j} \times \mathbf{i} = -\mathbf{k}, \qquad \mathbf{k} \times \mathbf{j} = -\mathbf{i}, \qquad \mathbf{i} \times \mathbf{k} = -\mathbf{j};$$

$$\mathbf{i} \times \mathbf{i} = \mathbf{j} \times \mathbf{j} = \mathbf{k} \times \mathbf{k} = \mathbf{0}.$$

EXERCISE 1.2

1. If $\mathbf{a} = (3 \quad -4 \quad 1)$ and $\mathbf{b} = (1 \quad 0 \quad 2)$ find

   (i) $\mathbf{a} + \mathbf{b}$ (ii) $5\mathbf{b}$

   (iii) $3\mathbf{a} + 2\mathbf{b}$ (iv) $3\mathbf{a} - 4$

2. Find $\mathbf{a} + \mathbf{b}$ and $\mathbf{a} \cdot \mathbf{b}$ for the following vectors:

(i) $a = (2\ \ 1\ \ 3), b = (-3\ \ 0\ \ 5)$

(ii) $a = (2\ \ 7\ \ 0), b = (-1\ \ 4\ \ 3)$

3. Find the cosine of the angle between the vectors $\mathbf{a} = (-4\ \ 3\ \ 4)$ and $\mathbf{b} = (-4\ \ 5\ \ -8)$.

4. Find a unit vector in the direction of the vector $(4\ \ 5\ \ 2)$.

5. Using vectors, prove that a parallelogram whose diagonals are equal is a rectangle.

6. Using vectors, prove that the median to the base of an isosceles triangle is perpendicular to the base.

7. If $\mathbf{a} = 2\mathbf{i} - 3\mathbf{j} + 6\mathbf{k},\ \mathbf{b} = 2\mathbf{i} - \mathbf{j} - 4\mathbf{k}$ then verify that

$$|\mathbf{a}|^2 - |\mathbf{b}|^2 = (\mathbf{a} + \mathbf{b}) \cdot (\mathbf{a} - \mathbf{b}).$$

8. If $|\mathbf{a}+\mathbf{b}| = |\mathbf{a}-\mathbf{b}|$ then prove that the vectors $\mathbf{a}$ and $\mathbf{b}$ are perpendicular.

9. Prove the identity $\mathbf{a} = (\mathbf{a} \cdot \mathbf{i})\mathbf{i} + (\mathbf{a} \cdot \mathbf{j})\mathbf{j} + (\mathbf{a} \cdot \mathbf{k})\mathbf{k}$, where $\mathbf{a}$ is any vector in the space.

10. Find a unit vector perpendicular to each of the vectors $4\mathbf{i} + 3\mathbf{j} + \mathbf{k}$ and $2\mathbf{i} - \mathbf{j} + 2\mathbf{k}$. Also, find the cosine of the angle between the given vectors.

11. If $\mathbf{a} + \mathbf{b} + \mathbf{c} = \mathbf{0}$, prove that the angle θ between the vectors $\mathbf{a}$ and $\mathbf{b}$ is given by

$$\cos\theta = \frac{\mathbf{c}^2 - \mathbf{a}^2 - \mathbf{b}^2}{2|\mathbf{a}||\mathbf{b}|}.$$

12. Find the work done in moving an object acted upon by a force $(4\ \ -3\ \ 2)$ along a vector $(-1\ \ -3\ \ 5)$.

1.3 Introduction to Matrices

In this section, following the approach used to define a vector, a new concept called a matrix will be introduced. A few types of matrices will be defined and basic operations of algebra will be given.

Definition 1.3.1 *A matrix of order $m \times n$ is an arrangement of n number of m-vectors each written in a column.*

If $\mathbf{x_1} = (a_{11}\ a_{21}\ ...\ a_{m1})$, $\mathbf{x_2} = (a_{12}\ a_{22}\ ...\ a_{m2})$, ..., $\mathbf{x_n} = (a_{1n}\ a_{2n}\ ...\ a_{mn})$. Then

$$\mathbf{A} = (\mathbf{x_1}\ \mathbf{x_2}\ ...\ \mathbf{x_n}) = \begin{bmatrix} a_{11} & a_{12} & \cdots & a_{1n} \\ a_{21} & a_{22} & \cdots & a_{2n} \\ \vdots & \vdots & \vdots & \vdots \\ a_{m1} & a_{m2} & \cdots & a_{mn} \end{bmatrix}_{m\times n}$$

is an $m \times n$ matrix.

Notation. For the sake of brevity and clarity, a matrix of order $m\times n$, $\mathbf{A}_{m\times n}$, is written as

$$\mathbf{A} = [a_{ij}]_{m\times n},$$

where $i = 1, 2, ..., m,\ j = 1, 2, \cdots, n$ and a_{ij} is the ith row, jth column entry of the matrix.

Special cases. (i) If $m = n = 1$ then $\mathbf{A} = [a_{11}]$ is a single entry matrix and is considered as a scalar.

(ii) If $n = 1$, $\mathbf{A} = \begin{bmatrix} a_{11} \\ a_{21} \\ \vdots \\ a_{m1} \end{bmatrix} = [a_{i1}],\quad i = 1, 2, \cdots, m$

is a column matrix and is called a column vector. Thus, a column vector is represented using a square bracket or a parenthesis.

(iii) If $m = 1$, then $\mathbf{A} = \begin{bmatrix} a_{11} & a_{12} & \ldots & a_{1n} \end{bmatrix} = [a_{1j}],\quad j = 1, 2, \cdots n$
is a row matrix and is known as a row vector. Thus, a row vector is represented using a square bracket or a parenthesis.

(iv) If $m = n$ then $\mathbf{A} = [a_{ij}]_{n\times n}$ is a square matrix of order n.

(v) If $m \neq n$, then $\mathbf{A} = [a_{ij}]_{m\times n}$ is a rectangular matrix.

Example 1.3.1

(i) $\mathbf{A} = [3]$ *is a single entry matrix.*

(ii) $\mathbf{A} = \begin{bmatrix} 1 \\ 5 \\ 3 \\ 2 \end{bmatrix}_{4\times 1}$ *is a column vector or a column matrix.*

(iii) $\mathbf{A} = \begin{bmatrix} 2 & 3 & 6 & 4 & 8 \end{bmatrix}_{1\times 5}$ *is a row matrix or a row vector.*

(iv) $\mathbf{A} = \begin{bmatrix} 1 & 3 \\ 4 & 5 \end{bmatrix}_{2\times 2}$ *is a square matrix.*

(v) $\mathbf{A} = \begin{bmatrix} 2 & 1 & 4 & 8 \\ 5 & 2 & 1 & 3 \end{bmatrix}_{2\times 4}$ *is a rectangular matrix.*

Some important matrices are defined below.

Definition 1.3.2 *Zero matrix, denoted by* $\mathbf{O}$ *or by* $\mathbf{0}$ *is a matrix of the form* $\mathbf{A} = [a_{ij}]_{m\times n} = [0]_{m\times n}$*, that is,* $a_{ij} = 0$ *for all* $i = 1, 2, \cdots, m$ *and* $j = 1, 2, \cdots, n$.

Definition 1.3.3 *Elements of the leading diagonal of a square matrix of order* n *are the entries* a_{ii}, $i = 1, 2, \cdots, n$.

Example 1.3.2 *Let* $\mathbf{A} = \begin{bmatrix} a & 3 & 4 & 1 \\ 1 & b & 5 & 2 \\ 2 & 3 & c & 1 \\ 0 & 1 & 2 & d \end{bmatrix}$*; then the leading diagonal consists of the entries* a, b, c *and* d.

Definition 1.3.4 *Diagonal matrix is a square matrix of order* n*, all of whose off-diagonal elements are zero, that is,*
$a_{ij} = 0$ *for* $i \neq j$ *in* $\mathbf{A} = [a_{ij}]_{n\times n}$.

Definition 1.3.5 *Trace of* $\mathbf{A}$ *is the sum of the diagonal elements of* $\mathbf{A} = [a_{ij}]_{n\times n}$ *and is written as*

$$tr(\mathbf{A}) = \sum_{i=1}^{n} a_{ii}.$$

Definition 1.3.6 *Identity matrix is a diagonal matrix whose diagonal elements* $a_{ii} = 1$*, for* $i = 1, 2, \cdots, n$*. It is denoted by* $\mathbf{I}_n$.

Example 1.3.3 *(i)* $\mathbf{A} = \begin{bmatrix} 2 & 0 & 0 \\ 0 & 8 & 0 \\ 0 & 0 & 5 \end{bmatrix}$ *is a diagonal matrix of order 3.*

(ii) $tr\,\mathbf{A} = 15$. *(iii)* $\begin{bmatrix} 1 & 0 & 0 & 0 \\ 0 & 1 & 0 & 0 \\ 0 & 0 & 1 & 0 \\ 0 & 0 & 0 & 1 \end{bmatrix}$ *is an identity matrix of order 4.*

Definition 1.3.7 *Lower triangular matrix is a square matrix whose entries above the leading diagonal are all zero, that is,*
$a_{ij} = 0$*, for all* $i < j$, $i, j = 1, 2, \cdots, n$.

Definition 1.3.8 *Upper triangular matrix is a square matrix whose entries below the leading diagonal elements are all zero, that is,*
$a_{ij} = 0$*, for all* $i > j$, $i, j = 1, 2, \cdots, n$.

Example 1.3.4 *(i)* $\begin{bmatrix} 1 & 0 & 0 & 0 \\ -2 & 3 & 0 & 0 \\ 1 & -1 & 1 & 0 \\ 2 & 0 & 1 & -1 \end{bmatrix}$ *is a lower triangular matrix of order 4.*

(ii) $\begin{bmatrix} 5 & 4 & -1 & 2 \\ 0 & 3 & 2 & -1 \\ 0 & 0 & 4 & 1 \\ 0 & 0 & 0 & 2 \end{bmatrix}$ *is an upper triangular matrix of order 4.*

With these few definitions of matrices, it is essential to introduce the algebra of matrices which involves defining the sum, scalar product and product of matrices. In order to define the sum, difference and equality of two matrices, it is necessary to take matrices of the same order.

Let $\mathbf{A} = [a_{ij}]_{m\times n}$ and $\mathbf{B} = [b_{ij}]_{m\times n}$.

Definition 1.3.9 *The sum of two matrices* $\mathbf{A}$ *and* $\mathbf{B}$ *is a matrix* $\mathbf{C} = [c_{ij}]_{m\times n}$, *where* $c_{ij} = a_{ij} + b_{ij}$ *for* $i = 1, 2, \cdots, m$ *and* $j = 1, 2, \cdots, n$.

Definition 1.3.10 *Scalar multiplication of a matrix* $\mathbf{A}$ *with* $\lambda \in \mathbb{R}$ *or* $\lambda \in \mathbb{C}$ *is the matrix* $\lambda\mathbf{A} = [\lambda a_{ij}]_{m\times n}$.

Definition 1.3.11 *The difference of two matrices* $\mathbf{A}$ *and* $\mathbf{B}$ *is a matrix* $\mathbf{C} = [c_{ij}]_{m\times n}$ *defined by* $\mathbf{C} = \mathbf{A} - \mathbf{B} = \mathbf{A} + (-1)(\mathbf{B}) = [a_{ij} - b_{ij}]_{m\times n}$.

Definition 1.3.12 *Two matrices* $\mathbf{A}$ *and* $\mathbf{B}$ *are said to be equal if and only if* $a_{ij} = b_{ij}$ *for all* $i = 1, 2, \cdots, m$ *and* $j = 1, 2, \cdots, n$.

Example 1.3.5 *(i) If* $\mathbf{A} = \begin{bmatrix} a & b & c \\ d & e & f \end{bmatrix}$ *and* $\mathbf{B} = \begin{bmatrix} p & q & r \\ x & y & z \end{bmatrix}$ *then*

$$\mathbf{A} + \mathbf{B} = \begin{bmatrix} a+p & b+q & c+r \\ d+x & e+y & f+z \end{bmatrix}.$$

Note that the orders of $\mathbf{A}, \mathbf{B}$ *and* $\mathbf{A} + \mathbf{B}$ *are same.*
(ii) The following matrices are not compatible for matrix addition or difference (subtraction).

$$\mathbf{A} = \begin{bmatrix} 2 \\ -3 \\ 4 \end{bmatrix}, \mathbf{B} = \begin{bmatrix} 4 & 1 \\ -7 & 0 \\ 2 & 2 \end{bmatrix}, \mathbf{C} = \begin{bmatrix} 5 & 7 & 1 \\ 2 & -1 & 3 \end{bmatrix}, \mathbf{D} = \begin{bmatrix} 1 & 5 & 1 \\ 2 & 4 & 3 \\ 3 & 3 & 2 \end{bmatrix}.$$

(iii) If $\mathbf{A} = \begin{bmatrix} 1 & 4 & 5 \\ 6 & -3 & 2 \end{bmatrix}$, $\mathbf{B} = \begin{bmatrix} 3 & -1 & 3 \\ 0 & -2 & 4 \end{bmatrix}$

then

$$\mathbf{A} - \mathbf{B} = \begin{bmatrix} -2 & 5 & 2 \\ 6 & -1 & -2 \end{bmatrix}.$$

.

(iv) If $\mathbf{A} = \begin{bmatrix} 4 & -5 & -3 \\ 3 & 2 & 1 \end{bmatrix}$ then $4\mathbf{A} = \begin{bmatrix} 16 & -20 & -12 \\ 12 & 8 & 4 \end{bmatrix}$.

Definition 1.3.13 *Multiplication of two matrices* $\mathbf{A} = [a_{ij}]_{m\times p}$ *and* $\mathbf{B} = [b_{ij}]_{p\times n}$ *is a product matrix* $\mathbf{C} = \mathbf{AB}$, *where* $\mathbf{C} = [c_{ij}]_{m\times n}$ *and* c_{ij} *is defined as*

$$c_{ij} = \sum_{k=1}^{p} a_{ik}b_{kj} = a_{i1}b_{1j} + a_{i2}b_{2j} + \cdots + a_{ip}b_{pj}.$$

Remark 1.3.1 *(i) The product of two matrices* **A** *and* **B** *is defined if and only if the number of columns of matrix* **A** *is equal to the number of rows of matrix* **B**.
(ii) Unlike scalars and vectors, the product **AB** *may not be equal to the product* **BA**.
(iii) The product **AB** *may exist but the product* **BA** *need not exist.*
(iv) If **AB** = **BA**, *then* **A** *and* **B** *are said to commute.*

Example 1.3.6 *Find the product* **AB** *and* **BA** *for*

(i) $\mathbf{A} = \begin{bmatrix} 2 & 4 \end{bmatrix}$, $\mathbf{B} = \begin{bmatrix} 2 \\ 3 \end{bmatrix}$

(ii) $\mathbf{A} = \begin{bmatrix} 1 & 2 & 3 \end{bmatrix}$, $\mathbf{B} = \begin{bmatrix} -2 \\ 1 \\ 5 \end{bmatrix}$

(iii) $\mathbf{A} = \begin{bmatrix} 1 & 2 & 0 \\ 2 & 3 & 1 \\ 3 & 4 & 2 \end{bmatrix}$, $\mathbf{B} = \begin{bmatrix} 0 & 2 & 1 \\ 2 & 2 & 0 \\ 2 & 0 & 1 \end{bmatrix}$

(iv) $\mathbf{A} = \begin{bmatrix} 2 & 4 & 8 \end{bmatrix}$, $\mathbf{B} = \begin{bmatrix} 1 & 2 \\ 0 & 1 \\ 2 & 3 \end{bmatrix}$.

Observe that not only **A** and **B** are compatible but **B** and **A** are also compatible for cases (i) to (iii). Hence the product **AB** and **BA** are defined for all cases except case (iv).

(i) $\mathbf{AB} = \begin{bmatrix} 2 & 4 \end{bmatrix} \begin{bmatrix} 2 \\ 3 \end{bmatrix} = [2 \times 2 + 4 \times 3] = [16],$

$$\mathbf{BA} = \begin{bmatrix} 2 \\ 3 \end{bmatrix} \begin{bmatrix} 2 & 4 \end{bmatrix} = \begin{bmatrix} 4 & 8 \\ 6 & 12 \end{bmatrix}.$$

(ii) $\mathbf{AB} = \begin{bmatrix} 1 & 2 & 3 \end{bmatrix} \begin{bmatrix} -2 \\ 1 \\ 5 \end{bmatrix} = [1 \times (-2) + 2 \times 1 + 3 \times 5] = [15]$

and $\mathbf{BA} = \begin{bmatrix} -2 \\ 1 \\ 5 \end{bmatrix} \begin{bmatrix} 1 & 2 & 3 \end{bmatrix} = \begin{bmatrix} -2 & -4 & -6 \\ 1 & 2 & 3 \\ 5 & 10 & 15 \end{bmatrix}.$

(iii) $\mathbf{AB} = \begin{bmatrix} 1 & 2 & 0 \\ 2 & 3 & 1 \\ 3 & 4 & 2 \end{bmatrix} \begin{bmatrix} 0 & 2 & 1 \\ 2 & 2 & 0 \\ 2 & 0 & 1 \end{bmatrix} = \begin{bmatrix} 4 & 6 & 1 \\ 8 & 10 & 3 \\ 12 & 14 & 5 \end{bmatrix}$

and $\mathbf{BA} = \begin{bmatrix} 0 & 2 & 1 \\ 2 & 2 & 0 \\ 2 & 0 & 1 \end{bmatrix} \begin{bmatrix} 1 & 2 & 0 \\ 2 & 3 & 1 \\ 3 & 4 & 2 \end{bmatrix} = \begin{bmatrix} 7 & 10 & 4 \\ 6 & 10 & 2 \\ 5 & 8 & 2 \end{bmatrix}.$

(iv) $\mathbf{AB} = \begin{bmatrix} 2 & 4 & 8 \end{bmatrix} \begin{bmatrix} 1 & 2 \\ 0 & 1 \\ 2 & 3 \end{bmatrix} = \begin{bmatrix} 18 & 32 \end{bmatrix},$

but $\mathbf{BA}$ is not defined.

As is the case with real numbers and vectors, addition of matrices is associative and commutative. Below are given results relating to algebra of matrices in terms of addition, scalar multiplication and product of matrices. The readers are advised to prove them. The proofs are simple and can be obtained by using the properties of real numbers for each entry of the matrix.

1. Properties of addition and scalar multiplication. Let $\mathbf{A}, \mathbf{B}, \mathbf{C}$ be matrices each of order $m \times n$ and λ and μ be any scalars. Then

   (a) $\mathbf{A} + \mathbf{B} = \mathbf{B} + \mathbf{A}$.

   (b) $\mathbf{A} + (\mathbf{B} + \mathbf{C}) = (\mathbf{A} + \mathbf{B}) + \mathbf{C}$.

   (c) $(\lambda + \mu)\mathbf{A} = \lambda\mathbf{A} + \mu\mathbf{A}$.

   (d) $\lambda(\mathbf{A} + \mathbf{B}) = \lambda\mathbf{A} + \lambda\mathbf{B}$.

   (e) $0.\mathbf{A} = \mathbf{O}_{m\times n}$.

   (f) $\lambda\mathbf{O}_{m\times n} = \mathbf{O}_{m\times n}$.

   (g) $\lambda(\mu\mathbf{A}) = (\lambda\mu)\mathbf{A} = \mu(\lambda\mathbf{A})$.

2. Properties of matrix multiplication. Let $\mathbf{A} = [a_{ij}]_{m\times p}$, $\mathbf{B} = [b_{ij}]_{p\times q}$, $\mathbf{C} = [c_{ij}]_{p\times q}$ and $\mathbf{D} = [d_{ij}]_{q\times n}$, then

   (a) $\mathbf{A}(\mathbf{BD}) = (\mathbf{AB})\mathbf{D}$.

(b) $\mathbf{A}(\mathbf{B} \pm \mathbf{C}) = \mathbf{AB} \pm \mathbf{AC}$.

(c) $(\mathbf{B} \pm \mathbf{C})\mathbf{D} = \mathbf{BD} \pm \mathbf{CD}$.

(d) $\mathbf{I}_{m\times m}\mathbf{A}_{m\times p} = \mathbf{A}_{m\times p}$.

(e) $\mathbf{A}_{m\times p}\mathbf{I}_{p\times p} = \mathbf{A}_{m\times p}$.

(f) $\mathbf{A}_{m\times p}\mathbf{O}_{p\times p} = \mathbf{O}_{m\times p}$.

(g) $\mathbf{O}_{k\times m}\mathbf{A}_{m\times p} = \mathbf{O}_{k\times p}$.

Example 1.3.7 *Suppose* $\mathbf{A} = \begin{bmatrix} 1 & 4 & -1 \\ 3 & 0 & 2 \end{bmatrix}$, $\mathbf{B} = \begin{bmatrix} 3 & 0 & 1 \\ 2 & 1 & 3 \\ 4 & 1 & 2 \end{bmatrix}$ *and* $\mathbf{C} = \begin{bmatrix} 2 & 1 \\ 0 & 2 \\ -3 & 1 \end{bmatrix}$. *Then show that* $(\mathbf{AB})\mathbf{C} = \mathbf{A}(\mathbf{BC})$.

Consider $(\mathbf{AB})\mathbf{C} = \begin{bmatrix} 7 & 3 & 11 \\ 17 & 2 & 7 \end{bmatrix} \begin{bmatrix} 2 & 1 \\ 0 & 2 \\ -3 & 1 \end{bmatrix} = \begin{bmatrix} -19 & 24 \\ 13 & 28 \end{bmatrix}$.

Similarly $\mathbf{A}(\mathbf{BC}) = \begin{bmatrix} 1 & 4 & -1 \\ 3 & 0 & 2 \end{bmatrix} \begin{bmatrix} 3 & 4 \\ -5 & 7 \\ 2 & 8 \end{bmatrix} = \begin{bmatrix} -19 & 24 \\ 13 & 28 \end{bmatrix}$.

Hence $(\mathbf{AB})\mathbf{C} = \mathbf{A}(\mathbf{BC}) = \mathbf{ABC}$.

Example 1.3.8 *Consider* $\mathbf{A} = \begin{bmatrix} 0 & 2 \\ 0 & 0 \end{bmatrix}$, $\mathbf{B} = \begin{bmatrix} 2 & 0 \\ 0 & 0 \end{bmatrix}$; *then* $\mathbf{AB} = \begin{bmatrix} 0 & 0 \\ 0 & 0 \end{bmatrix}$. *Thus* $\mathbf{A} \neq \mathbf{0}$, $\mathbf{B} \neq \mathbf{0}$ *but* $\mathbf{AB} = \mathbf{0}$.

Example 1.3.9 *Show that* $\mathbf{A}^3 = 4\mathbf{A}$, *where* $\mathbf{A} = \begin{bmatrix} 1 & -1 \\ -1 & 1 \end{bmatrix}$.

Consider $\mathbf{A}^2 = \begin{bmatrix} 1 & -1 \\ -1 & 1 \end{bmatrix} \begin{bmatrix} 1 & -1 \\ -1 & 1 \end{bmatrix} = \begin{bmatrix} 2 & -2 \\ -2 & 2 \end{bmatrix}$ *and hence*

$$\mathbf{A}^3 = \mathbf{A}^2\mathbf{A} = \begin{bmatrix} 2 & -2 \\ -2 & 2 \end{bmatrix} \begin{bmatrix} 1 & -1 \\ -1 & 1 \end{bmatrix} = \begin{bmatrix} 4 & -4 \\ -4 & 4 \end{bmatrix} = 4\mathbf{A}.$$

Example 1.3.10 *Show that* $(\mathbf{A} + k\mathbf{B}^2)^3 = k\mathbf{I}_3$, *where*

$\mathbf{A} = \begin{bmatrix} 0 & 1 & 0 \\ 0 & 0 & 1 \\ 0 & 0 & 0 \end{bmatrix}$ *and* $\mathbf{B} = \begin{bmatrix} 0 & 0 & 0 \\ 1 & 0 & 0 \\ 0 & 1 & 0 \end{bmatrix}$ *and* k *is any scalar.*

On multiplication, $\mathbf{B}^2 = \begin{bmatrix} 0 & 0 & 0 \\ 0 & 0 & 0 \\ 1 & 0 & 0 \end{bmatrix}$, $\mathbf{A} + k\mathbf{B}^2 = \begin{bmatrix} 0 & 1 & 0 \\ 0 & 0 & 1 \\ k & 0 & 0 \end{bmatrix}$ and

$$(\mathbf{A} + k\mathbf{B}^2)^2 = \begin{bmatrix} 0 & 0 & 1 \\ k & 0 & 0 \\ 0 & k & 0 \end{bmatrix}.$$

Further,

$$(\mathbf{A} + k\mathbf{B}^2)^3 = (\mathbf{A} + k\mathbf{B}^2)^2(\mathbf{A} + k\mathbf{B}^2) = \begin{bmatrix} k & 0 & 0 \\ 0 & k & 0 \\ 0 & 0 & k \end{bmatrix} = k\mathbf{I}_3,$$

hence the result.

Example 1.3.11 *Prove that* $\mathbf{A}^3 - 23\mathbf{A} - 40\mathbf{I}_3 = \mathbf{0}$, *where*

$$\mathbf{A} = \begin{bmatrix} 1 & 2 & 3 \\ 3 & -2 & 1 \\ 4 & 2 & 1 \end{bmatrix}.$$

On multiplication, $\mathbf{A}^2 = \begin{bmatrix} 19 & 4 & 8 \\ 1 & 12 & 8 \\ 14 & 6 & 15 \end{bmatrix}$, $\mathbf{A}^3 = \begin{bmatrix} 63 & 46 & 69 \\ 69 & -6 & 23 \\ 92 & 46 & 63 \end{bmatrix}$ and

$$\mathbf{A}^3 - 23\mathbf{A} - 40\mathbf{I}_3 = \begin{bmatrix} 63 & 46 & 69 \\ 69 & -6 & 23 \\ 92 & 46 & 63 \end{bmatrix} - 23\begin{bmatrix} 1 & 2 & 3 \\ 3 & -2 & 1 \\ 4 & 2 & 1 \end{bmatrix} - 40\begin{bmatrix} 1 & 0 & 0 \\ 0 & 1 & 0 \\ 0 & 0 & 1 \end{bmatrix},$$

$$= \begin{bmatrix} 0 & 0 & 0 \\ 0 & 0 & 0 \\ 0 & 0 & 0 \end{bmatrix},$$

thus proving the result.

Example 1.3.12 *If* $\mathbf{A} = \begin{bmatrix} 3 & -4 \\ 1 & -1 \end{bmatrix}$, *prove that* $\mathbf{A}^n = \begin{bmatrix} 1+2n & -4n \\ n & 1-2n \end{bmatrix}$.

The proof is given by the method of induction. Let $n = 2$, then

L.H.S. $= \mathbf{A}^2 = \begin{bmatrix} 5 & -8 \\ 2 & -3 \end{bmatrix}$ and R.H.S. is $\begin{bmatrix} 5 & -8 \\ 2 & -3 \end{bmatrix}$.

Hence the result is true for $n = 2$.
Assume that the result holds for $n = k$.

$$\mathbf{A}^k = \begin{bmatrix} 1+2k & -4k \\ k & 1-2k \end{bmatrix}.$$

To prove the result for $n = k + 1$,

$$\begin{aligned}
\text{L.H.S.} = \mathbf{A}^k\mathbf{A} &= \begin{bmatrix} 1+2k & -4k \\ k & 1-2k \end{bmatrix}\begin{bmatrix} 3 & -4 \\ 1 & -1 \end{bmatrix} \\
&= \begin{bmatrix} 1+2k+2 & -4k-4 \\ k+1 & 1-2k-2 \end{bmatrix} \\
&= \begin{bmatrix} 1+2(k+1) & -4(k+1) \\ k+1 & 1-2(k+1) \end{bmatrix} \\
&= \mathbf{A}^{k+1}.
\end{aligned}$$

As the result is true for $n = k+1$, by the principle of mathematical induction, the result holds good for all natural numbers.

Example 1.3.13 *Suppose that* $\mathbf{P}$ *and* $\mathbf{Q}$ *are square matrices of order* n*, show that* $\mathbf{P}$ *and* $\mathbf{Q}$ *commute if and only if* $\mathbf{P} - k\mathbf{I}$ *and* $\mathbf{Q} - k\mathbf{I}$ *commute for every real number* k*.*

If $\mathbf{P}$ and $\mathbf{Q}$ commute, i.e.

$$\mathbf{PQ} = \mathbf{QP} \tag{1.1}$$

$$\begin{aligned}
\text{Then } (\mathbf{P} - k\mathbf{I})(\mathbf{Q} - k\mathbf{I}) &= \mathbf{PQ} - \mathbf{P}k\mathbf{I} - k\mathbf{IQ} + k^2\mathbf{I} \\
&= \mathbf{PQ} - k\mathbf{P} - k\mathbf{Q} + k^2\mathbf{I};
\end{aligned}$$

$$(\mathbf{P} - k\mathbf{I})(\mathbf{Q} - k\mathbf{I}) = \mathbf{PQ} - k\mathbf{P} - k\mathbf{Q} + k^2\mathbf{I}; \tag{1.2}$$

$$\begin{aligned}
(\mathbf{Q} - k\mathbf{I})(\mathbf{P} - k\mathbf{I}) &= \mathbf{QP} - \mathbf{Q}k\mathbf{I} - k\mathbf{IP} + k^2\mathbf{I} \\
&= \mathbf{QP} - k\mathbf{Q} - k\mathbf{P} + k^2\mathbf{I};
\end{aligned}$$

$$(\mathbf{Q} - k\mathbf{I})(\mathbf{P} - k\mathbf{I}) = \mathbf{QP} - k\mathbf{Q} - k\mathbf{P} + k^2\mathbf{I}; \tag{1.3}$$

Since $\mathbf{PQ} = \mathbf{QP}$, expressions (1.2) and (1.3) are equal, and hence $\mathbf{P} - k\mathbf{I}$ and $\mathbf{Q} - k\mathbf{I}$ commute. Conversely, expressions (1.2) and (1.3) are equal only if (1.1) holds.

EXERCISE 1.3

1. State the orders of the following matrices:

(i) $\begin{bmatrix} 1 & 3 & 0 & -2 \\ 4 & 2 & 3 & 4 \\ 7 & 6 & -5 & 8 \\ -9 & 7 & 1 & 9 \end{bmatrix}$ (ii) $\begin{bmatrix} -4 \\ 3 \\ 6 \\ 0 \end{bmatrix}$

(iii) $\begin{bmatrix} -1 & 2 & 0 \\ 5 & 3 & -7 \end{bmatrix}$ (iv) $\begin{bmatrix} 2 & -4 & 0 & 11 \\ 6 & 6 & 3 & -5 \\ -1 & 7 & 8 & 2 \end{bmatrix}$

(v) $\begin{bmatrix} 0 & 0 & 0 & 0 & 0 \\ 0 & 0 & 0 & 0 & 0 \end{bmatrix}$.

2. A matrix has nine elements. Find the possible number of matrices that can be formed. State the possible orders if matrix has 11 elements.

3. Construct 4×3 matrices $[a_{ij}]$ whose elements are

   (i) $a_{ij} = i + 2j$ (ii) $a_{ij} = 3i - jj$ (iii) $a_{ij} = \frac{(i-j)^2}{3}$.

4. If $\begin{bmatrix} x - 2z & 3x - z \\ y + z & x + z \end{bmatrix} = \begin{bmatrix} -5 & 0 \\ 5 & 4 \end{bmatrix}$, find the values of x, y, z.

5. Find the values of $\alpha, \beta, \gamma, \delta$ from the matrix equation

$$\begin{bmatrix} 2\alpha - \delta & \alpha + 2\delta \\ 5\beta - \gamma & 4\beta + 3\gamma \end{bmatrix} = \begin{bmatrix} 3 & 4 \\ 11 & 24 \end{bmatrix}.$$

6. If $\mathbf{A} = \begin{bmatrix} 1 & 2 & -3 \\ 5 & 0 & 2 \\ 1 & -1 & 1 \end{bmatrix}$, $\mathbf{B} = \begin{bmatrix} 4 & 0 & 1 \\ 1 & 3 & -2 \\ 2 & 2 & 3 \end{bmatrix}$ and $\mathbf{C} = \begin{bmatrix} 3 & 4 & 2 \\ -1 & 2 & 0 \\ 2 & 0 & 3 \end{bmatrix}$.

   Then compute

   (i) $-3\mathbf{B}, 4\mathbf{C}$

   (ii) $\mathbf{A}^2 - 2\mathbf{B}^2 + \mathbf{C}^2$

   (iii) $\mathbf{A} + (\mathbf{B} - \mathbf{C})$

   Also verify $\mathbf{A} + (\mathbf{B} - \mathbf{C}) = (\mathbf{A} + \mathbf{B}) - \mathbf{C}$.

7. If $\mathbf{A} = \begin{bmatrix} 1 & 2 & 3 \\ 1 & 0 & -1 \\ -1 & 3 & 2 \end{bmatrix}$, $\mathbf{B} = \begin{bmatrix} 3 & 1 \\ 2 & 0 \\ 4 & -1 \end{bmatrix}$ and $\mathbf{C} = \begin{bmatrix} 2 & 0 & -2 & 1 \\ 1 & 2 & 3 & -4 \end{bmatrix}$.
   Verify that $\mathbf{A}(\mathbf{BC}) = (\mathbf{AB})\mathbf{C}$.

8. Show that for all p, q, r, s the matrices $\mathbf{A} = \begin{bmatrix} p & q \\ -q & p \end{bmatrix}$ and $\mathbf{B} = \begin{bmatrix} r & s \\ -s & r \end{bmatrix}$ commute.

9. If $\mathbf{A} = \begin{bmatrix} 1 & 2 \\ -3 & 0 \end{bmatrix}$, find $\mathbf{A^2} + \mathbf{3A} + \mathbf{5I}$ where $\mathbf{I}$ is second-order identity matrix.

10. If $\mathbf{A_x} = \begin{bmatrix} \cos x & \sin x \\ -\sin x & \cos x \end{bmatrix}$, then show that $(\mathbf{A}_x)^n = \begin{bmatrix} \cos nx & \sin nx \\ -\sin nx & \cos nx \end{bmatrix}$, where n is a positive integer.

11. If $\mathbf{A} = \begin{bmatrix} 1 & -3 & 2 \\ 2 & 1 & -3 \\ 4 & -3 & -1 \end{bmatrix}$, $\mathbf{B} = \begin{bmatrix} 2 & 1 & -2 \\ 3 & -2 & -1 \\ 2 & -5 & 0 \end{bmatrix}$, $\mathbf{C} = \begin{bmatrix} 1 & 4 & 0 \\ 2 & 1 & 1 \\ 1 & -2 & 2 \end{bmatrix}$.
Verify that $\mathbf{AB} = \mathbf{AC}$. (Note that the cancellation law in general does not hold good for matrix multiplication.)

12. If $\mathbf{A} = \begin{bmatrix} p & q \\ -r & s \end{bmatrix}$, prove that $\mathbf{A}^2 - (p+s)\mathbf{A} + (ps - qr)\mathbf{I} = 0$ where $\mathbf{I}$ is second-order-identity matrix.

13. If $\mathbf{A} = \begin{bmatrix} 3 & 5 & -1 \\ 1 & 2 & 5 \end{bmatrix}$, $\mathbf{B} = \begin{bmatrix} 4 & 2 & -5 \\ -3 & 5 & 3 \end{bmatrix}$, then solve the matrix equations for $\mathbf{X}$ (i) $\mathbf{3A} + \mathbf{5X} = \mathbf{2B}$, (ii) $\alpha\mathbf{A} + \beta\mathbf{B} = \gamma X$, where α, β and γ are scalars.

14. Prove that $\mathbf{A} = \begin{bmatrix} \cos^2 x & \sin x \cos x \\ \sin x \cos x & \sin^2 x \end{bmatrix} \begin{bmatrix} \cos^2 y & \sin y \cos y \\ \sin y \cos y & \sin^2 y \end{bmatrix}$ is a zero matrix, where x and y differ by an odd multiple of $\frac{\pi}{2}$.

15. Let $\mathbf{A}$ be a square matrix. $\mathbf{A}^k$ is defined by the relation $\mathbf{A}^k = \mathbf{A}\mathbf{A}^{k-1}$, where k is a positive integer. Show that

    (i) $\mathbf{A}^m\mathbf{A}^n = \mathbf{A}^{m+n}$

    (ii) $[\mathbf{A}^m]^n = \mathbf{A}^{mn}$.

1.4 Types of Matrices

In this section, various types of matrices are introduced. These matrices are important and are useful in studying physical phenomena.

Definition 1.4.1 *Transpose of a matrix* $\mathbf{A} = [a_{ij}]_{m\times n}$ *is a matrix* $\mathbf{A}^T = [a'_{ij}]_{n\times m}$ *where* $a'_{ij} = a_{ji}$, $i = 1, 2, \ldots, m$ *and* $j = 1, 2, \ldots, n$.

Example 1.4.1 *If* $\mathbf{A} = \begin{bmatrix} 1 & -1 \\ 1 & 2 \\ 4 & 3 \end{bmatrix}$ *then* $\mathbf{A}^T = \begin{bmatrix} 1 & 1 & 4 \\ -1 & 2 & 3 \end{bmatrix}$.

The following properties are satisfied by the transpose of a matrix.

(i) $(\mathbf{A}^T)^T = \mathbf{A}$

(ii) $(\mathbf{A} + \mathbf{B})^T = \mathbf{A}^T + \mathbf{B}^T$.

(iii) $(k\mathbf{A})^T = k\mathbf{A}^T$, where k is any scalar.

(iv) $(\mathbf{AB})^T = \mathbf{B}^T\mathbf{A}^T$, where the product $\mathbf{AB}$ is well defined.

Example 1.4.2 *If* $\mathbf{A} = \begin{bmatrix} 3 & 2 \\ -4 & 0 \\ -1 & 5 \end{bmatrix}$, $\mathbf{B} = \begin{bmatrix} -2 & -1 & 0 \\ 5 & -3 & -2 \end{bmatrix}$ *then*

$$\mathbf{AB} = \begin{bmatrix} 4 & -9 & -4 \\ 8 & 4 & 0 \\ 27 & -14 & -10 \end{bmatrix} \text{ and } (\mathbf{AB})^T = \begin{bmatrix} 4 & 8 & 27 \\ -9 & 4 & -14 \\ -4 & 0 & -10 \end{bmatrix}.$$

Further, $\mathbf{B}^T\mathbf{A}^T = \begin{bmatrix} -2 & 5 \\ -1 & -3 \\ 0 & -2 \end{bmatrix} \begin{bmatrix} 3 & -4 & -1 \\ 2 & 0 & 5 \end{bmatrix} = \begin{bmatrix} 4 & 8 & 27 \\ -9 & 4 & -14 \\ -4 & 0 & -10 \end{bmatrix}.$

Thus, $(\mathbf{AB})^T = \mathbf{B}^T\mathbf{A}^T$.

Definition 1.4.2 *The conjugate matrix,* $\bar{\mathbf{A}}$, *of a matrix* $\mathbf{A} = [a_{ij}]_{m\times n}$ *is a matrix* $\bar{\mathbf{A}} = [b_{ij}]_{m\times n}$ *where* $b_{ij} = \bar{a_{ij}}$ *that is* b_{ij}= *conjugate of* a_{ij} *for* i=1,2,...m, j=1,2,...,n.

Example 1.4.3 *Let* $\mathbf{A} = \begin{bmatrix} 1+2i & 1 & 1+3i \\ i & -2+i & 5+3i \end{bmatrix}$,

then $\bar{\mathbf{A}} = \begin{bmatrix} 1-2i & 1 & 1-3i \\ -i & -2-i & 5-3i \end{bmatrix}$.

The following properties are satisfied by conjugate matrices. Let $\mathbf{A}_{m\times n}$, $\mathbf{B}_{m\times n}$ be two matrices. Then

1. $\overline{(\overline{\mathbf{A}})} = \mathbf{A}$.
2. $\overline{\mathbf{A}+\mathbf{B}} = \overline{\mathbf{A}} + \overline{\mathbf{B}}$.
3. $\overline{\mathbf{AB}} = \overline{\mathbf{A}}\,\overline{\mathbf{B}}$, here $m = n$.

Combining the above two definitions gives rise to a new definition.

Definition 1.4.3 *The conjugate transpose of a matrix,* $\mathbf{A}^c$, *of a given matrix* $\mathbf{A}_{m\times n}$ *is defined as* $\mathbf{A}^c = \overline{\mathbf{A}^T}$.

The conjugate transpose of matrix satisfies

(i) $(\mathbf{A}^c)^c = \mathbf{A}$ (ii) $\mathbf{A}^c = \overline{\mathbf{A}^T} = \overline{\mathbf{A}}^{\mathbf{T}}$.

Example 1.4.4 *Verify the above statements for the matrix*

$$\mathbf{A} = \begin{bmatrix} 2+i & 3+2i & 4+3i \\ 5+4i & 6+5i & 7+6i \\ 1 & 2 & 2+3i \end{bmatrix}.$$

Then $\mathbf{A}^T = \begin{bmatrix} 2+i & 5+4i & 1 \\ 3+2i & 6+5i & 2 \\ 4+3i & 7+6i & 2+3i \end{bmatrix}$ and $\mathbf{A}^c = \begin{bmatrix} 2-i & 5-4i & 1 \\ 3-2i & 6-5i & 2 \\ 4-3i & 7-6i & 2-3i \end{bmatrix}$.

The conclusion is left to the reader.

Definition 1.4.4 *A symmetric matrix* $\mathbf{A}_{n\times n}$ *is a square matrix satisfying the criterion* $\mathbf{A} = \mathbf{A}^T$.

Example 1.4.5 $\mathbf{A} = \begin{bmatrix} a & h & g \\ h & b & f \\ g & f & c \end{bmatrix}$ and $\begin{bmatrix} 3 & -2 & 4 \\ -2 & 5 & 6 \\ 4 & 6 & -1 \end{bmatrix}$

are symmetric matrices.

Aliter definition

$\mathbf{A}_{m\times n}$ is a symmetric matrix if and only if $a_{ij} = a_{ji}, \quad i,j = 1,2,3,\ldots,n$.

Definition 1.4.5 *A skew-symmetric matrix* $\mathbf{A}_{m\times n}$ *is a square matrix satisfying* $a_{ij} = -a_{ji}, \quad i,j = 1,2,3,\ldots,n$. *In other words* $\mathbf{A} = -\mathbf{A}^T$.

Example 1.4.6 $\mathbf{A} = \begin{bmatrix} 0 & h & g \\ -h & 0 & f \\ -g & -f & 0 \end{bmatrix}$ and $\mathbf{B} = \begin{bmatrix} 0 & -2 & 4 \\ 2 & 0 & 5 \\ -4 & -5 & 0 \end{bmatrix}$

are skew-symmetric matrices.

Observation. For skew-symmetric matrices, the diagonal elements are zero.

The following are some important properties of symmetric and skew-symmetric matrices.

1. If $\mathbf{A}$ and $\mathbf{B}$ are symmetric matrices (skew-symmetric) then $\mathbf{A}+\mathbf{B}$, $\mathbf{A}-\mathbf{B}$ are also symmetric matrices (skew-symmetric) matrices.

2. For any square matrix $\mathbf{A}$, $\mathbf{A}+\mathbf{A}^T$ is symmetric and $\mathbf{A}-\mathbf{A}^T$ is skew-symmetric.

Result 1.4.1 A square matrix $\mathbf{A}$ can be expressed as a sum of a symmetric matrix and a skew-symmetric matrix in a unique way.

Example 1.4.7 *Express the matrix* $\mathbf{A} = \begin{bmatrix} 2 & -3 & -1 \\ 3 & 6 & 4 \\ 5 & 1 & 0 \end{bmatrix}$ *as the sum of symmetric and skew-symmetric matrices.*

Solution. Consider $\mathbf{A} + \mathbf{A}^T = \begin{bmatrix} 4 & 0 & 4 \\ 0 & 12 & 5 \\ 4 & 5 & 0 \end{bmatrix}$ and $\mathbf{A} - \mathbf{A}^T = \begin{bmatrix} 0 & -6 & -6 \\ 6 & 0 & 3 \\ 6 & -3 & 0 \end{bmatrix}$

Set $\mathbf{P} = \frac{1}{2}(\mathbf{A} + \mathbf{A}^T)$ and $\mathbf{Q} = \frac{1}{2}(\mathbf{A} - \mathbf{A}^T)$;

hence $\mathbf{A} = \begin{bmatrix} 2 & 0 & 2 \\ 0 & 6 & \frac{5}{2} \\ 2 & \frac{5}{2} & 0 \end{bmatrix} + \begin{bmatrix} 0 & -3 & -3 \\ 3 & 0 & \frac{3}{2} \\ 3 & \frac{-3}{2} & 0 \end{bmatrix}$ is in the required fashion.

Definition 1.4.6 *A Hermitian matrix is a square matrix* $\mathbf{A} = [a_{ij}]_{n\times n}$ *satisfying the condition* $a_{ij} = \overline{a_{ji}}$.

Observation: By the definition, $a_{ij} = \overline{a_{ji}}$ means that the diagonal elements of a Hermitian matrix are real.

Example 1.4.8 $\mathbf{A} = \begin{bmatrix} 3 & 5-2i \\ 5+2i & 8 \end{bmatrix}$.

Definition 1.4.7 *A skew-Hermitian matrix is a square matrix* $\mathbf{A} = [a_{ij}]_{n\times n}$ *satisfying the condition* $a_{ij} = -\overline{a_{ji}}$.

Observation: The definition implies that $a_{ij} = -\overline{a_{ji}}$, which means the diagonal elements of a skew-Hermitian matrix are either zero or purely imaginary.

Example 1.4.9 $\mathbf{A} = \begin{bmatrix} 0 & -3-2i \\ 3-2i & 0 \end{bmatrix}$, $\mathbf{B} = \begin{bmatrix} 2i & 4-4i \\ -4-4i & -i \end{bmatrix}$.

Result 1.4.2 If $\mathbf{A}$ is a Hermitian matrix, then $\mathbf{B} = \mathbf{iA}$ is a skew-Hermitian matrix.

Proof. Let $\mathbf{A}$ be a Hermitian matrix. Then $a_{ij} = \overline{a_{ji}}$. Let $a_{ji} = a + ib$. Then $a_{ij} = a - ib$.
Set $\mathbf{B} = i\mathbf{A} \implies b_{ij} = ia_{ij} = i(a - ib) = ai + b$,
and $b_{ji} = ia_{ji} = i(a + ib) = ai - b = -(b - ia) = -\overline{(b + ai)} = -\overline{b_{ij}}$.

Hence $\mathbf{B}$ is a skew-Hermitian matrix.

Definition 1.4.8 *An orthogonal matrix is a real square matrix satisfying* $\mathbf{AA}^T = I$.

Example 1.4.10 *Let* $\mathbf{A} = \begin{bmatrix} \cos x & -\sin x \\ \sin x & \cos x \end{bmatrix}$. *Then* $\mathbf{A}$ *is an orthogonal matrix.*

Solution. Since $\mathbf{AA}^T = \begin{bmatrix} \cos x & -\sin x \\ \sin x & \cos x \end{bmatrix} \begin{bmatrix} \cos x & \sin x \\ -\sin x & \cos x \end{bmatrix} = \begin{bmatrix} 1 & 0 \\ 0 & 1 \end{bmatrix}$.

Result 1.4.3 The product of any two orthogonal matrices of same order is orthogonal.

Proof. Let $\mathbf{A}_{m\times m}$ and $\mathbf{B}_{m\times m}$ be orthogonal. Thus, $\mathbf{AA}^T = I$ and $\mathbf{BB}^T = I$. Consider $(\mathbf{AB})(\mathbf{AB})^T = (\mathbf{AB})(\mathbf{B}^T\mathbf{A}^T) = \mathbf{I}$.

Definition 1.4.9 *An Idempotent matrix is a square matrix satisfying* $A^2 = A$.

Example 1.4.11 *Show that* A *is an Idempotent matrix where*

$$\mathbf{A} = \begin{bmatrix} -1 & 3 & 5 \\ 1 & -3 & -5 \\ -1 & 3 & 5 \end{bmatrix}.$$

Solution. Left as an exercise.

Definition 1.4.10 *A Nilpotent matrix is a square matrix for which* $\mathbf{A}^n = 0$, *where* n *is a positive integer.*

Example 1.4.12 *Show that* $\mathbf{A} = \begin{bmatrix} 1 & -3 & -4 \\ -1 & 3 & 4 \\ 1 & -3 & -4 \end{bmatrix}$ *is a Nilpotent matrix. Give the value of* n.

Solution. $\mathbf{A^2} = \mathbf{O}$. Thus $n = 2$.

Definition 1.4.11 *A periodic matrix of order* n *is a square matrix satisfying the relation* $\mathbf{A}^{n+1} = \mathbf{A}$, *for some least positive integer* n.

Example 1.4.13 *Show that* $\mathbf{A} = \begin{bmatrix} 1 & -2 & -6 \\ -3 & 2 & 9 \\ 2 & 0 & -3 \end{bmatrix}$ *is a periodic matrix of period 2.*

Solution. Left as an exercise.

Definition 1.4.12 *An involutory matrix is a square matrix satisfying the relation* $\mathbf{A}^2 = \mathbf{I}$.

Example 1.4.14 *Show that* $\mathbf{A} = \begin{bmatrix} 0 & 1 & -1 \\ 4 & -3 & 4 \\ 3 & -3 & 4 \end{bmatrix}$ *is an involutory matrix.*

Solution. Left as an exercise.

Definition 1.4.13 *An Unitary matrix is a square matrix satisfying the relation* $\mathbf{AA}^c = \mathbf{I}$.

EXERCISE 1.4

1. Let $\mathbf{A}$ and $\mathbf{B}$ be two matrices that are compatible. Then prove that
   (i) $(\mathbf{A}+\mathbf{B})^T = \mathbf{A}^T + \mathbf{B}^T$
   (ii) $(k\mathbf{A})^T = k\mathbf{A}^T$
   (iii) $(\mathbf{AB})^T = \mathbf{B}^T\mathbf{A}^T$.
2. If $\mathbf{A}$ is a square matrix, show that
   (i) $\mathbf{A}+\mathbf{A}^T$ is symmetric; (ii) $\mathbf{A}-\mathbf{A}^T$ is skew-symmetric.
3. Prove that, if $\mathbf{A}$ is a square matrix, then $\mathbf{AA}^T$ and $\mathbf{A}^T\mathbf{A}$ are both symmetric.
4. If $\overline{\mathbf{A}}$, $\overline{\mathbf{B}}$ are the conjugates of the matrices $\mathbf{A}$ and $\mathbf{B}$, respectively, and are compatible then show that
   (i) $\overline{(\mathbf{A}+\mathbf{B})} = \overline{\mathbf{A}} + \overline{\mathbf{B}}$
   (ii) $\overline{(k\mathbf{A})} = \overline{k\mathbf{A}}$, if k is complex
   (iii) $\overline{\mathbf{AB}} = \overline{\mathbf{A}}\,\overline{\mathbf{B}}$.
5. Prove that every Hermitian matrix $\mathbf{H}$ can be expressed as $\mathbf{H} = \mathbf{P}+\mathbf{Q}$, where $\mathbf{P}$ is real and skew-symmetric and $\mathbf{Q}$ is real and symmetric.
6. Show that the following matrix is a Nilpotent matrix of the order 2.
$$\mathbf{A} = \begin{bmatrix} 1 & -1 & -2 \\ -1 & 1 & 2 \\ 1 & -1 & -2 \end{bmatrix}.$$
7. Find a necessary and sufficient condition for the matrix
$$\mathbf{A} = \begin{bmatrix} a & -b & -c \\ -a & b & c \\ a & -b & -c \end{bmatrix}$$
   (i) to be Nilpotent of order 2;
   (ii) to be Idempotent.
8. Complete Example 1.4.11.
9. Do Examples 1.4.13 and 1.4.14.

1.5 Elementary Operations and Elementary Matrices

Sometimes it is convenient to transform a given matrix into a diagonal matrix or a triangular matrix. This can be done by applying certain operations on the given matrix $\mathbf{A}$. It is important to note that these operations are done within a matrix and these are the only operations that can be performed within a matrix.

Definition 1.5.1 *The elementary row (column) operations on a given matrix are as follows.*

Let R_i and R_j be the ith row and jth row, respectively of the matrix $\mathbf{A}$ and C_k, C_l be the kth column and lth column, respectively of the matrix $\mathbf{A}$.

(i) Interchange of two rows (columns), that is $R_i \leftrightarrow R_j$ $(C_k \leftrightarrow C_l)$.

(ii) Multiplying a row (column) with a constant, that is $R_i \to \alpha R_i$ $(C_k \to \alpha C_k)$, where α is a constant.

(iii) Multiplying a row (column) with a constant and adding it to another row (column), that is $R_j \to R_j + \alpha R_i$ $(C_k \to C_k + \alpha C_l)$.

Example 1.5.1 *Consider* $\mathbf{A}_{4\times 4} = \begin{bmatrix} 1 & 3 & 8 & 4 \\ 3 & 2 & 1 & 1 \\ 4 & 12 & 32 & 16 \\ 8 & 5 & 4 & 3 \end{bmatrix}$

(i) Interchanging of two rows: $R_3 \leftrightarrow R_2$ gives

$$\sim \begin{bmatrix} 1 & 3 & 8 & 4 \\ 4 & 12 & 32 & 16 \\ 3 & 2 & 1 & 1 \\ 8 & 5 & 4 & 3 \end{bmatrix}$$

(ii) Multiplying a row with a constant: multiply R_2 with $\frac{1}{4}$, that is, $R_2 \to \frac{1}{4}R_2$

$$\sim \begin{bmatrix} 1 & 3 & 8 & 4 \\ 1 & 3 & 8 & 4 \\ 3 & 2 & 1 & 1 \\ 8 & 5 & 4 & 3 \end{bmatrix}$$

(iii) Multiplying a row with a constant and adding it to another row: $R_2 \to R_2 + (-1)R_1$ (multiplying R_1 with (-1) and adding it to R_2).

$$\sim \begin{bmatrix} 1 & 3 & 8 & 4 \\ 0 & 0 & 0 & 0 \\ 3 & 2 & 1 & 1 \\ 8 & 5 & 4 & 3 \end{bmatrix}.$$

Note that similar operations can be done on columns. It is important to note that these row (column) operations do not change the properties of the matrix.

Definition 1.5.2 *Elementary matrix is obtained by doing a single row (column) operation on an identity matrix.*

Example 1.5.2 *(i)* $\begin{bmatrix} 1 & 0 & 0 \\ 0 & 0 & 1 \\ 0 & 1 & 0 \end{bmatrix}$ *interchanging* $R_2 \leftrightarrow R_3$.

(ii) $\begin{bmatrix} 3 & 0 \\ 0 & 1 \end{bmatrix}$, $R_1 \to 3R_1$.

(ii) $\begin{bmatrix} 1 & 0 \\ 3 & 1 \end{bmatrix}$, $R_2 \to R_2 + 3R_1$.

Result 1.5.1 A row operation done on matrix $\mathbf{A}$ is equivalent to premultiplying $\mathbf{A}$ with the corresponding elementary matrix.

Example 1.5.3 $\mathbf{A} = \begin{bmatrix} 3 & 4 & 2 \\ -1 & 0 & 1 \\ 4 & 2 & 2 \end{bmatrix}$

$R_1 \leftrightarrow R_2$ *gives* $\begin{bmatrix} -1 & 0 & 1 \\ 3 & 4 & 2 \\ 4 & 2 & 2 \end{bmatrix}$.

Doing the same operation on Identity matrix $\mathbf{I}_3$ and premultiplying with $\mathbf{A}$ yields

$$\begin{bmatrix} 0 & 1 & 0 \\ 1 & 0 & 0 \\ 0 & 0 & 1 \end{bmatrix} \begin{bmatrix} 3 & 4 & 2 \\ -1 & 0 & 1 \\ 4 & 2 & 2 \end{bmatrix} = \begin{bmatrix} -1 & 0 & 1 \\ 3 & 4 & 2 \\ 4 & 2 & 2 \end{bmatrix}.$$

Similarly, other operations can be verified.

Result 1.5.2 A column operation performed on matrix $\mathbf{A}$ is equivalent to post-multiplying $\mathbf{A}$ with the corresponding elementary matrix.

Example 1.5.4 *Let* $\mathbf{A} = \begin{bmatrix} 6 & 3 & 2 \\ 2 & 1 & 0 \\ 4 & 2 & 2 \end{bmatrix}$

Then $C_1 \to \frac{1}{2}C_1$ *gives* $\mathbf{A} = \begin{bmatrix} 3 & 3 & 2 \\ 1 & 1 & 0 \\ 2 & 2 & 2 \end{bmatrix}$.

Now consider $\begin{bmatrix} 6 & 3 & 2 \\ 2 & 1 & 0 \\ 4 & 2 & 2 \end{bmatrix} \begin{bmatrix} \frac{1}{2} & 0 & 0 \\ 0 & 1 & 0 \\ 0 & 0 & 1 \end{bmatrix}$

$$= \begin{bmatrix} 3 & 3 & 2 \\ 1 & 1 & 0 \\ 2 & 2 & 2 \end{bmatrix}.$$

Definition 1.5.3 *Equivalent Matrices. Two matrices* **A** *and* **B** *are said to be equivalent if one can be obtained from the other by a sequence of elementary row and/or column operations.*

Also, two matrices are said to be equivalent if one is obtained from the other by pre-multiplying and/or post-multiplying with a sequence of elementary matrices.

Notation. $\mathbf{A} \sim \mathbf{B}$ means that **A** is equivalent to **B**.

EXERCISE 1.5

1. Verify Result 1.5.1 for the following matrices by (i) $R_2 \leftrightarrow R_3$ (ii) $R_2 \to 2R_2$ (iii) $R_3 \to R_3 - 2R_3$.

   (i) $\mathbf{A} = \begin{bmatrix} 2 & 3 & 1 & 0 \\ -1 & 2 & 1 & 1 \\ -2 & 4 & 2 & 2 \end{bmatrix}$ (ii) $\begin{bmatrix} 8 & 6 & 3 \\ 9 & 2 & 1 \\ 2 & 4 & 5 \end{bmatrix}$

   (iii) $\begin{bmatrix} 1 & 3 \\ 2 & 4 \\ 4 & 4 \\ 2 & 1 \end{bmatrix}$ (iv) $\begin{bmatrix} 4 & 1 & 2 & 3 \\ -2 & 5 & 6 & 4 \\ 3 & 2 & 2 & 3 \end{bmatrix}$.

2. Verify Result 1.5.2 for the following matrices by (i) $C_1 \leftrightarrow C_3$ (ii) $C_2 \to 2C_2$ (iii) $C_3 \to C_3 + 2C_2$.

   (i) $\mathbf{B} = \begin{bmatrix} 1 & 0 & 2 & 1 \\ 3 & 1 & 2 & 1 \\ 1 & 2 & 4 & 2 \\ 1 & 1 & -2 & -1 \end{bmatrix}$ (ii) $\begin{bmatrix} 4 & 3 & 11 \\ 2 & 5 & 10 \end{bmatrix}$

(iii) $\begin{bmatrix} 2 & 3 & 1 \\ -1 & 2 & -2 \\ 3 & -2 & 5 \end{bmatrix}$ (iv) $\begin{bmatrix} -3 & 8 & 5 & 4 \\ 2 & 1 & 3 & 2 \\ 8 & 6 & -1 & -2 \end{bmatrix}$.

1.6 Determinants

In this section, the notion of a determinant is introduced and its properties are discussed.

Definition 1.6.1 *A determinant is a numerical value associated with every square matrix* $\mathbf{A}$ *and is denoted by* $det\mathbf{A}$ *or* $|\mathbf{A}|$.

The determinant of a matrix $\mathbf{A} = [a]_{1\times 1}$ is defined as $det\mathbf{A} = a$.

The determinant of $\mathbf{A} = \begin{bmatrix} a & b \\ c & d \end{bmatrix}_{2\times 2}$ is defined as $det\mathbf{A} = ad - bc$.

In order to define the determinant of higher-order matrices, the following concepts are needed.

Definition 1.6.2 *A submatrix* $\mathbf{B}$ *of a given matrix* $\mathbf{A}_{m\times n}$ *is obtained by deleting one or more rows and (or) one or more columns of* $\mathbf{A}$.

Definition 1.6.3 *The* (i,j)*th minor* M_{ij} *of a matrix* $\mathbf{A}_{n\times n}$ *is the determinant of the matrix obtained by removing the ith row and the jth column of* $\mathbf{A}$.

Definition 1.6.4 *The* (i,j)*th cofactor of a matrix* $\mathbf{A}_{n\times n}$ *is given by* $A_{ij} = (-1)^{i+j}M_{ij}$.

To find the determinant of a $\mathbf{A}_{n\times n}$ matrix, the following procedure given by Laplace is useful.

Laplace's rule to find the determinant of a matrix of order n

Consider the elements of the *ith* row, a_{ij}, $1 \leq j \leq n$, where i is fixed. Corresponding to each element $a_{i1}, a_{i2}, \cdots, a_{in}$ find the cofactors $A_{i1}, A_{i2}, \cdots, A_{in}$; then the determinant of $\mathbf{A}$ is given by $|\mathbf{A}| = det\mathbf{A} = a_{i1}A_{i1} + a_{i2}A_{i2} + \cdots + a_{in}A_{in}$.

Observation

(i) The $det\mathbf{A}$ can be found using any row i, $i = 1, 2, 3, \cdots, n$, as fixed.

(ii) The $det\mathbf{A}$ can be found similarly using any fixed column $j = 1, 2, 3, \cdots, n$.

Geometrical representation

Consider two position vectors $\mathbf{x} = (a_{11}\ \ a_{21})$ and $\mathbf{y} = (a_{12}\ \ a_{22})$.
To find the area of the parallelogram formed by these vectors, write

$$\mathbf{A} = [x\ y] = \begin{bmatrix} a_{11} & a_{12} \\ a_{21} & a_{22} \end{bmatrix},$$

then $|\mathbf{A}| = a_{11}a_{22} - a_{12}a_{21}$ gives the required area.

If $\mathbf{x} = (a_{11}\ \ a_{21}\ \ a_{31})$, $\mathbf{y} = (a_{12}\ \ a_{22}\ \ a_{32})$ and $\mathbf{z} = (a_{13}\ \ a_{23}\ \ a_{33})$,

then $|\mathbf{A}| = \begin{vmatrix} a_{11} & a_{12} & a_{13} \\ a_{21} & a_{22} & a_{23} \\ a_{31} & a_{32} & a_{33} \end{vmatrix}$ is the volume of the parallelepiped formed by the three vectors $\mathbf{x}, \mathbf{y}$ and $\mathbf{z}$.

Similarly, the determinant of $\mathbf{A}_{n\times n}$ matrix gives the volume of an n-parallelepiped formed by n-vectors.

Example 1.6.1 *Using the second row find the det of the matrix,*

$$\mathbf{A} = \begin{bmatrix} 3 & 5 & -7 \\ 4 & 1 & -12 \\ 2 & 9 & -3 \end{bmatrix}.$$

Then

$$|\mathbf{A}| = 4(-1)^{2+1}\begin{vmatrix} 5 & -7 \\ 9 & -3 \end{vmatrix} + 1(-1)^{2+2}\begin{vmatrix} 3 & -7 \\ 2 & -3 \end{vmatrix} - 12(-1)^{2+3}\begin{vmatrix} 3 & 5 \\ 2 & 9 \end{vmatrix} = 17.$$

Properties of determinants

Some useful properties of the determinant are stated below without proof. An example is given to illustrate the property. Let $\mathbf{A}$ be an $n \times n$ matrix. Then

1. $|\mathbf{A}| = |\mathbf{A}^T|$.

Example 1.6.2 *Let* $\mathbf{A} = \begin{bmatrix} 2 & -1 & 1 \\ 3 & 2 & 4 \\ -1 & 0 & 3 \end{bmatrix}$; *Then* $\mathbf{A}^T = \begin{bmatrix} 2 & 3 & -1 \\ -1 & 2 & 0 \\ 1 & 4 & 3 \end{bmatrix}$

and $|\mathbf{A}| = |\mathbf{A}^T| = 27$.

2. Let $\mathbf{B}$ be the matrix obtained by interchanging two rows (columns) of $\mathbf{A}$. Then

$$|\mathbf{B}| = -|\mathbf{A}|.$$

Example 1.6.3 *Let* $\mathbf{A} = \begin{bmatrix} 3 & -2 & 2 \\ 1 & 2 & -3 \\ 4 & 1 & 2 \end{bmatrix}$ *and* $\mathbf{B} = \begin{bmatrix} 1 & 2 & -3 \\ 3 & -2 & 2 \\ 4 & 1 & 2 \end{bmatrix}$;

then $|\mathbf{A}| = 35$ *and* $|\mathbf{B}| = -35 = -|\mathbf{A}|$.

3. If any two rows (columns) of the matrix $\mathbf{A}$ are identical or proportional (that is, $R_j = aR_j$ *or* $C_k = bC_l$), then $|\mathbf{A}| = 0$.

Example 1.6.4 *(i) Let* $A = \begin{bmatrix} 2 & -1 & -2 \\ -3 & 2 & 3 \\ -1 & 0 & 1 \end{bmatrix}$, *then* $|\mathbf{A}| = 0$.

(ii) Consider $\mathbf{A} = \begin{bmatrix} 5 & 1 & 2 \\ 2 & 4 & 6 \\ 25 & 5 & 10 \end{bmatrix}$, *then* $|\mathbf{A}| = 0$.

4. Let $\mathbf{B}_{n \times n}$ be the matrix obtained by multiplying a row (column) of the matrix $\mathbf{A}$ by a scalar (i.e.) $R_i \to aR_i[C_k \to aC_k]$ then $|\mathbf{B}| = a|\mathbf{A}|$.

Example 1.6.5 *(i) Let* $\mathbf{A} = \begin{bmatrix} 1 & -2 & 3 \\ 3 & -1 & 1 \\ 4 & 2 & -3 \end{bmatrix}$ *and* $\mathbf{B} = \begin{bmatrix} 1 & -2 & 6 \\ 3 & -1 & 2 \\ 4 & 2 & -6 \end{bmatrix}$,

then $|\mathbf{A}| = 5$ *and* $|\mathbf{B}| = 10 = 2|\mathbf{A}|$.

5. Let $\mathbf{B}_{n \times n}$ be the matrix obtained by doing the third elementary row operation on $\mathbf{A}$. Then $|\mathbf{B}| = |\mathbf{A}|$.

Example 1.6.6 *Let* $\mathbf{A} = \begin{bmatrix} 2 & 1 & 1 \\ 3 & 2 & 4 \\ 1 & 0 & 3 \end{bmatrix}$ *then* $|\mathbf{A}| = 5$.

Let $R_1 \to R_1 + 5R_2$ and write $\mathbf{B}$ as

$$\mathbf{B} = \begin{bmatrix} 2+15 & 1+10 & 1+20 \\ 3 & 2 & 4 \\ 1 & 0 & 3 \end{bmatrix}. \text{ Then } |\mathbf{B}| = 5.$$

6. If in any row (column) of a matrix, every entry is a sum of two numbers then the matrix can be expressed as a sum of two matrices of the same order and the det of the sum equals the det of the original matrix.

Example 1.6.7 *Let*

$$\mathbf{A} = \begin{bmatrix} 2+b & 1+q & 4+r \\ 4 & 1 & 3 \\ 7 & 2 & 1 \end{bmatrix}$$
$$= \begin{bmatrix} 2 & 1 & 4 \\ 4 & 1 & 3 \\ 7 & 2 & 1 \end{bmatrix} + \begin{bmatrix} b & q & r \\ 4 & 1 & 3 \\ 7 & 2 & 1 \end{bmatrix}$$
$$= \mathbf{B} + \mathbf{C} \quad (say)$$

then $|\mathbf{A}| = |\mathbf{B}| + |\mathbf{C}|$ *(Prove!)*

Example 1.6.8 *If* $\mathbf{A} = \begin{bmatrix} 1 & a & a^2 \\ 1 & b & b^2 \\ 1 & c & c^2 \end{bmatrix}$

then $|\mathbf{A}| = (a-b)(b-c)(c-a)$.

Applying row operations $R_1 \to R_1 - R_2$ and $R_2 \to R_2 - R_3$

$$\text{then } |\mathbf{A}| = \begin{vmatrix} 0 & a-b & a^2-b^2 \\ 0 & b-c & b^2-c^2 \\ 1 & c & c^2 \end{vmatrix}$$

$$= (a-b)(b-c) \begin{vmatrix} 0 & 1 & a+b \\ 0 & 1 & b+c \\ 1 & c & c^2 \end{vmatrix}$$
$$= (a-b)(b-c)(c-a).$$

Example 1.6.9 *Evaluate the determinant of* $\mathbf{A} = \begin{bmatrix} 1 & \omega & \omega^2 \\ \omega & \omega^2 & 1 \\ \omega^2 & 1 & \omega \end{bmatrix}$ *where* ω *is complex and is one of the cube roots of unity.*

Applying the column operations $C_1 \to C_1 + C_2 + C_3$

$$\mathbf{A} = \begin{bmatrix} 1+\omega+\omega^2 & \omega & \omega^2 \\ 1+\omega+\omega^2 & \omega^2 & 1 \\ 1+\omega+\omega^2 & 1 & \omega \end{bmatrix},$$

since $1+\omega+\omega^2 = 0$, $det\mathbf{A} = 0$.

Example 1.6.10 *Show that the det of* $\mathbf{A} = 4a^2b^2c^2$ *for*

$$\mathbf{A} = \begin{bmatrix} a^2 & bc & ac+c^2 \\ a^2+ab & b^2 & ac \\ ab & b^2+bc & c^2 \end{bmatrix}.$$

Applying row operations $R_1 \to R_1 + R_2 + R_3$

$$\mathbf{A} = \begin{bmatrix} 2a(a+b) & 2b(b+c) & 2c(c+a) \\ a^2+ab & b^2 & ac \\ ab & b^2+bc & c^2 \end{bmatrix}$$

$$|\mathbf{A}| = 2abc \begin{vmatrix} a+b & b+c & c+a \\ a+b & b & a \\ b & b+c & c \end{vmatrix}$$

Applying row operations again on the matrix obtained $R_2 \to R_2 - R_1$ gives

$$|\mathbf{A}| = 2abc \begin{vmatrix} a+b & b+c & c+a \\ 0 & -c & -c \\ b & b+c & c \end{vmatrix}$$

$$= 2abc[(a+b)(-c^2 + c(b+c)) + b[(b+c)(-c) + c(c+a)]] = 4a^2b^2c^2.$$

Result 1.6.1 If $\mathbf{A}$ is an orthogonal matrix, then $|\mathbf{A}| = \pm 1$. Since $|\mathbf{A}| = |\mathbf{A}^T|$, $|\mathbf{A}\mathbf{A}^T| = |\mathbf{A}|^2 = |\mathbf{I}| = 1$. Hence the result $|\mathbf{A}| = \pm 1$.

EXERCISE 1.6

1. If A_{ij} is the cofactor of a_{ij} in the matrix $\mathbf{A} = [a_{ij}]_{3\times 3}$ and $|\mathbf{A}| = D$, then show that

$$\begin{vmatrix} a_{11} & a_{12} & a_{13} \\ a_{21} & a_{22} & a_{23} \\ a_{31} & a_{32} & a_{33} \end{vmatrix} \begin{vmatrix} A_{11} & A_{12} & A_{13} \\ A_{21} & A_{22} & A_{23} \\ A_{31} & A_{32} & A_{33} \end{vmatrix} = D^3.$$

2. Find the product of the determinants of the matrices,

$$\begin{bmatrix} 1 & 2 & 3 \\ 2 & 3 & 1 \\ 3 & 1 & 2 \end{bmatrix} \text{ and } \begin{bmatrix} 2 & 3 & 4 \\ 3 & 4 & 2 \\ 4 & 2 & 3 \end{bmatrix}.$$

3. If a, b, c have all different values and $\mathbf{A} = \begin{bmatrix} a & a^2 & a^3-1 \\ b & b^2 & b^3-1 \\ c & c^2 & c^3-1 \end{bmatrix}$, then prove that $abc = 1$ using $|\mathbf{A}| = 0$.

4. Let $\mathbf{A} = \begin{bmatrix} 1+p^2-q^2 & 2pq & -2q \\ 2pq & 1-p^2+q^2 & 2p \\ 2q & -2p & 1-p^2-q^2 \end{bmatrix}$, then show that $|\mathbf{A}| = (1+p^2+q^2)^3$.

5. Find the square of the determinant in a determinant form

$$\begin{vmatrix} 0 & x & y & 0 \\ x & 0 & 0 & a \\ y & 0 & 0 & b \\ 0 & a & b & 0 \end{vmatrix}.$$

6. Prove that

$$\begin{vmatrix} \sin(x+\frac{\pi}{4}) & \sin x & \cos x \\ \sin(x+\frac{\pi}{4}) & \cos x & \sin x \\ 1 & a & 1-a \end{vmatrix} = \left[\frac{1}{\sqrt{2}}(\sqrt{2}-1)\right] \cos 2x$$

for any value of a.

1.7 Inverse of a Matrix

In Section 1.2, the sum and scalar multiplication of vectors and in Section 1.3, the sum, scalar multiplication and product of matrices were introduced parallel to the definitions of real numbers. Now comes the turn of the reciprocal or the inverse of a matrix.

Observing that a reciprocal of a real number, $a \neq 0$ satisfies the relation

$$a.\frac{1}{a} = 1 = \frac{1}{a}.a;$$

the question that arises is whether there is a similar concept for matrices. The problem is, given a matrix $\mathbf{A}_{m\times n}$, is it possible to find matrices $\mathbf{B}_{n\times m}$ and $\mathbf{C}_{n\times m}$ such that the matrix product is compatible and $\mathbf{A}_{m\times n}\ \mathbf{B}_{n\times m} = \mathbf{I}_m$ and $\mathbf{C}_{n\times m}\ \mathbf{A}_{m\times n} = \mathbf{I}_n$.

The answer leads to the following notion.

Definition 1.7.1 *Let a matrix $\mathbf{A}_{m\times n}$ be given.*

(i) *A matrix $\mathbf{B}_{n\times m}$ is called the left inverse of $\mathbf{A}$, if $\mathbf{B}_{n\times m}\ \mathbf{A}_{m\times n} = \mathbf{I}_n$.*

(ii) *A matrix $\mathbf{C}_{n\times m}$ is called the right inverse of $\mathbf{A}$, if $\mathbf{A}_{m\times n}\ \mathbf{C}_{n\times m} = \mathbf{I}_m$.*

These inverses may or may not exist and this fact is illustrated in the following example.

Example 1.7.1 *Consider* $\mathbf{A} = \begin{bmatrix} 1 & -1 & 2 \\ 1 & 0 & 1 \end{bmatrix}$.

Find the left inverse and right inverse for **A**, if they exist.
Solution: The left inverse $\mathbf{B}_{3\times 2}$, if it exists satisfies the equation,

$$\mathbf{B}_{3\times 2}\mathbf{A}_{2\times 3} = \mathbf{I}_3.$$

This means that to find the six entries in $\mathbf{B}_{3\times 2}$; there are nine equations. Hence it is obvious that **B** cannot be found.

So consider $\mathbf{C}_{3\times 2} = \begin{bmatrix} a & d \\ b & e \\ c & f \end{bmatrix}$; then,

$$\mathbf{AC} = \mathbf{I}_2 \text{ gives } \begin{bmatrix} 1 & -1 & 2 \\ 1 & 0 & 1 \end{bmatrix} \begin{bmatrix} a & d \\ b & e \\ c & f \end{bmatrix} = \begin{bmatrix} 1 & 0 \\ 0 & 1 \end{bmatrix},$$

which gives the equations,

$$a - b + 2c = 1, \quad \textit{and} \quad d - e + 2f = 0, \tag{1.4}$$

$$a + c = 0, \qquad d + f = 1, \tag{1.5}$$

giving the solution matrix,

$$\begin{bmatrix} -\alpha & 1-\beta \\ \alpha - 1 & 1+\beta \\ \alpha & \beta \end{bmatrix},$$

where α and β are arbitrary.

Thus **A** has infinitely many right inverses but not a single left inverse.

The process of finding inverses of rectangular matrices needs more information and is considered later in Section 1.9. The rest of this section deals with the inverse of a square matrix.

Definition 1.7.2 *A square matrix* $\mathbf{B}_{n\times n}$ *is said to be the inverse of* $\mathbf{A}_{n\times n}$ *if and only if* $\mathbf{AB} = \mathbf{BA} = \mathbf{I}_n$.

Then **A** is said to be invertible and **B** is called the inverse of **A** and is denoted by $\mathbf{A}^{-1}$.

Result 1.7.1 $|\mathbf{A}^{-1}| = \frac{1}{|\mathbf{A}|}$.

Proof. From the definition of $\mathbf{A}^{-1}$

$$\mathbf{A}\mathbf{A}^{-1} = I = \mathbf{A}^{-1}\mathbf{A}.$$

Taking determinants and the fact that $|\mathbf{AB}| = |\mathbf{A}||\mathbf{B}|$ yields

$$|\mathbf{A}||\mathbf{A}^{-1}| = 1.$$

Hence,

$$|\mathbf{A}^{-1}| = \frac{1}{|\mathbf{A}|}.$$

Definition 1.7.3 *(i) A nonsingular matrix* $\mathbf{A}$ *is a matrix whose* $|\mathbf{A}| \neq 0$. *(ii) A singular matrix* $\mathbf{A}$ *is a matrix whose* $|\mathbf{A}| = 0$.

Example 1.7.2 $\mathbf{A} = \begin{bmatrix} 1 & 2 & 1 \\ 2 & 1 & 0 \\ 3 & 6 & 3 \end{bmatrix}$ *is singular matrix,*

since $|\mathbf{A}| = 0$.

$\mathbf{B} = \begin{bmatrix} 1 & 0 & 1 \\ 2 & 1 & 0 \\ 0 & 3 & 2 \end{bmatrix}$ *is nonsingular matrix,*

since $|\mathbf{B}| = 8 \neq 0$.

The following matrix is useful for finding the inverse of a matrix.

Definition 1.7.4 *An adjoint matrix of* $\mathbf{A}$, *denoted by adj* $\mathbf{A}$ *is*

$$adj\,\mathbf{A} = \begin{bmatrix} A_{11} & A_{12} & \cdots & A_{1n} \\ A_{21} & A_{22} & \cdots & A_{2n} \\ \vdots & \vdots & \vdots & \vdots \\ A_{m1} & A_{m2} & \cdots & A_{mn} \end{bmatrix}^T_{m\times n},$$

where A_{ij} is the cofactor of the $(i,j)th$ element of $\mathbf{A}$.

Result 1.7.2 For any square matrix of order n,

$$\mathbf{A}\ (adj\,\mathbf{A}) = (adj\,\mathbf{A})\ \mathbf{A} = |\mathbf{A}|\mathbf{I}_n.$$

This yields $\mathbf{A}^{-1} = \frac{1}{|\mathbf{A}|} adj\,\mathbf{A}$.

Thus $\mathbf{A}^{-1} = \frac{1}{|\mathbf{A}|} adj\,\mathbf{A}$ is a working rule.

Theorem 1.7.1 *A square matrix* $\mathbf{A}_{n\times n}$ *possesses an inverse if and only if it is nonsingular.*

Proof. Suppose $\mathbf{A}$ possesses an inverse, i.e., $\mathbf{A}^{-1} = \mathbf{B}$ exists and

$$\mathbf{AB} = \mathbf{BA} = \mathbf{I}_n.$$

Thus $|\mathbf{A}||\mathbf{B}| = |\mathbf{I}_n|$, which means $|\mathbf{A}||\mathbf{B}| = 1$; this gives $|\mathbf{A}| \neq 0$, and hence $\mathbf{A}$ is nonsingular.
Suppose $\mathbf{A}$ is nonsingular, then $|\mathbf{A}| \neq 0$.

Consider $(adj\mathbf{A})(\mathbf{A}) = (\mathbf{A})adj\mathbf{A} = |\mathbf{A}|\mathbf{I}_n$

Since $|\mathbf{A}| \neq 0$,

$$\frac{(\mathbf{A})adj\mathbf{A}}{|\mathbf{A}|} = \frac{(adj\mathbf{A})\mathbf{A}}{|\mathbf{A}|} = \mathbf{I}_n$$

which means

$$\mathbf{A}^{-1} = \frac{adj\mathbf{A}}{|\mathbf{A}|},$$

thus completing the proof.

Result 1.7.3 If $\mathbf{A}_{n\times n}$ and $\mathbf{B}_{n\times n}$ are compatable and invertible then

$$(\mathbf{AB})^{-1} = \mathbf{B}^{-1}\mathbf{A}^{-1}$$

Consider $(\mathbf{AB})(\mathbf{B}^{-1}\mathbf{A}^{-1}) = \mathbf{AI}_n\mathbf{A}^{-1} = \mathbf{I}_n$.

Similarly, $(\mathbf{B}^{-1}\mathbf{A}^{-1})(\mathbf{AB}) = \mathbf{B}^{-1}\mathbf{I}_n\mathbf{B} = \mathbf{I}_n$. Hence

$$(\mathbf{AB})^{-1} = \mathbf{B}^{-1}\mathbf{A}^{-1}.$$

Result 1.7.4 If $\mathbf{A}$ is nonsingular matrix then the following properties hold.

(i) $(\mathbf{A}^{-1})^{-1} = \mathbf{A}$.

(ii) $(\mathbf{A}^T)^{-1} = (\mathbf{A}^{-1})^T$.

(iii) $(\ \overline{\mathbf{A}}\ )^{-1} = \overline{(\mathbf{A}^{-1})}$.

(iv) $(\ \overline{\mathbf{A}^T}\ )^{-1} = \overline{(\mathbf{A}^{-1})}^T$.

Proof.

(i) Since $\mathbf{A}^{-1}\mathbf{A} = \mathbf{I}$ consider

$$\begin{aligned}(\mathbf{A}^{-1})^{-1} &= (\mathbf{A}^{-1})^{-1}\mathbf{I}\\ &= (\mathbf{A}^{-1})^{-1}\mathbf{A}^{-1}\mathbf{A}\\ &= \mathbf{A}.\end{aligned}$$

(ii) Consider $\mathbf{A}\mathbf{A}^{-1} = \mathbf{I} = \mathbf{A}^{-1}\mathbf{A}$. Taking transpose and applying the product rule for transpose gives

$$\begin{aligned}(\mathbf{A}\mathbf{A}^{-1})^T &= \mathbf{I}^T = (\mathbf{A}^{-1}\mathbf{A})^T\\ \Longrightarrow (\mathbf{A}^{-1})^T\mathbf{A}^T &= \mathbf{I} = \mathbf{A}^T(\mathbf{A}^{-1})^T\end{aligned}$$

which means $(\mathbf{A}^T)^{-1} = (\mathbf{A}^{-1})^T$.

(iii) Consider $\mathbf{A}\mathbf{A}^{-1} = \mathbf{I} = \mathbf{A}^{-1}\mathbf{A}$. Taking conjugate and applying the property for conjugates

$$\overline{\mathbf{A}\mathbf{A}^{-1}} = \overline{\mathbf{I}} = \overline{\mathbf{A}^{-1}\mathbf{A}}$$

gives

$$\overline{\mathbf{A}}\,\overline{\mathbf{A}^{-1}} = \mathbf{I} = \overline{\mathbf{A}^{-1}}\,\overline{\mathbf{A}}$$

which gives

$$\overline{\mathbf{A}^{-1}} = (\,\overline{\mathbf{A}}\,)^{-1}.$$

(iv) Applying relations (ii) and (iii) on $\mathbf{A}\mathbf{A}^{-1} = \mathbf{I}$ gives,

$$\overline{\mathbf{A}\mathbf{A}^{-1}} = \overline{\mathbf{I}} = \overline{\mathbf{A}^{-1}\mathbf{A}} \Longrightarrow \overline{\mathbf{A}}\,\overline{\mathbf{A}^{-1}} = \mathbf{I} = \overline{\mathbf{A}^{-1}}\,\overline{\mathbf{A}}.$$

Taking transpose gives the result.

Result 1.7.5 If $\mathbf{A}$ is an involutory matrix, then $\mathbf{A}$ is its own inverse. That is, the inverse of an involutory matrix $\mathbf{A}$ is itself, since $\mathbf{A}^2 = \mathbf{I}$ implies $\mathbf{A}\mathbf{A} = \mathbf{I}$ or $\mathbf{A} = \mathbf{A}^{-1}$.

The following are some methods of finding the inverse of a matrix. The inverse of a matrix $\mathbf{A}_{n\times n}$ can be calculated using the adjoint matrix.

Example 1.7.3 *Find the inverse of* $\mathbf{A} = \begin{bmatrix} 2 & -6 \\ -7 & 3 \end{bmatrix}$ *using* $adj\,\mathbf{A}$.

Clearly $|\mathbf{A}| = -36 \neq 0$ is nonsingular and $\mathbf{A}^{-1}$ exists. Further,

$$A_{11} = 3, \qquad A_{12} = 6$$

$$A_{21} = 7, \qquad A_{22} = 2.$$

Hence by definition, using $adj\,\mathbf{A}$,

$$\mathbf{A}^{-1} = -\frac{1}{36}\begin{bmatrix} 3 & 6 \\ 7 & 2 \end{bmatrix} = \begin{bmatrix} -\frac{1}{12} & -\frac{1}{6} \\ -\frac{7}{36} & -\frac{1}{18} \end{bmatrix}.$$

Remark 1.7.1 *For any square matrix of order two, one can obtain the inverse by the following rule.*

Rule Interchange the position of the main diagonal elements, change the signs of second diagonal elements and divide by $|\mathbf{A}|$.
Verify this rule for the above example.

Example 1.7.4 *Find the inverse of the matrix* $\mathbf{A} = \begin{bmatrix} 0 & 1 & 2 \\ 1 & 2 & 3 \\ 3 & 1 & 1 \end{bmatrix}$.

Here $|\mathbf{A}| = -2$ and hence $\mathbf{A}^{-1}$ exists. Further verify that

$$A_{11} = -1, \quad A_{12} = 8, \quad A_{13} = -5,$$

$$A_{21} = 1, \quad A_{22} = -6, \quad A_{23} = 3,$$

$$A_{31} = -1, \quad A_{32} = 2, \quad A_{33} = -1,$$

$$\mathbf{A}^{-1} = -1/2\begin{bmatrix} -1 & 8 & 5 \\ 1 & -6 & 3 \\ -1 & 2 & -1 \end{bmatrix}^T \quad \text{then,}$$

$$\mathbf{A}^{-1} = 1/2\begin{bmatrix} 1 & -1 & 1 \\ -8 & 6 & -2 \\ -5 & -3 & 1 \end{bmatrix}.$$

Gauss-Jordan method

Inverse of a given square matrix of order n can also be found by applying elementary row (columns) operations. This method is called as Gauss-Jordan method. The method consists of writing a given matrix as $\mathbf{A} = \mathbf{I}_n\mathbf{A}$ and applying row operations $\begin{bmatrix}\mathbf{A} & \mathbf{I}_n\end{bmatrix}$. The matrix can also be written as $\mathbf{A} = \mathbf{A}\mathbf{I}_n$, and column operations are applied $\begin{bmatrix}\mathbf{A} & \mathbf{I}_n\end{bmatrix}$. The following examples illustrate the method.

Example 1.7.5 *Find the inverse of the matrix* $\mathbf{A} = \begin{bmatrix} -2 & 4 \\ 3 & -5 \end{bmatrix}$ *using Gauss-Jordan method.*

Here, $|\mathbf{A}| = -2 \neq 0$, $\mathbf{A}$ is nonsingular and $\mathbf{A}^{-1}$ exists.
Now perform elementary row operations as follows

$$\begin{aligned}
\begin{bmatrix} -2 & 4 \\ 3 & -5 \end{bmatrix} &= \begin{bmatrix} 1 & 0 \\ 0 & 1 \end{bmatrix} \mathbf{A} \\
\approx \begin{bmatrix} 1 & -2 \\ 3 & -5 \end{bmatrix} &= \begin{bmatrix} -\frac{1}{2} & 0 \\ 0 & 1 \end{bmatrix} \mathbf{A}, \quad R_1 \to -\frac{1}{2} R_1 \\
\approx \begin{bmatrix} 1 & -2 \\ 0 & 1 \end{bmatrix} &= \begin{bmatrix} -\frac{1}{2} & 0 \\ \frac{3}{2} & 1 \end{bmatrix} \mathbf{A}, \quad R_2 \to R_2 - 3R_1 \\
\approx \begin{bmatrix} 1 & 0 \\ 0 & 1 \end{bmatrix} &= \begin{bmatrix} \frac{5}{2} & 2 \\ \frac{3}{2} & 1 \end{bmatrix} \mathbf{A}, \quad R_1 \to R_1 + 2R_2
\end{aligned}$$

Hence $\mathbf{A}^{-1} = \begin{bmatrix} \frac{5}{2} & 2 \\ \frac{3}{2} & 1 \end{bmatrix}$.

Example 1.7.6 *Find the inverse of the matrix* $\mathbf{A} = \begin{bmatrix} 1 & 2 & -1 \\ -1 & 1 & 2 \\ 2 & -1 & 1 \end{bmatrix}$ *using row operations.*

Here, $|\mathbf{A}| \neq 0$, and hence $\mathbf{A}^{-1}$ exists. Now,

$$\begin{aligned}
\begin{bmatrix} 1 & 2 & -1 \\ -1 & 1 & 2 \\ 2 & -1 & 1 \end{bmatrix} &= \begin{bmatrix} 1 & 0 & 0 \\ 0 & 1 & 0 \\ 0 & 0 & 1 \end{bmatrix} \mathbf{A} \\
\approx \begin{bmatrix} 1 & 2 & -1 \\ 0 & 3 & 1 \\ 0 & -5 & 3 \end{bmatrix} &= \begin{bmatrix} 1 & 0 & 0 \\ 1 & 1 & 0 \\ -2 & 0 & 1 \end{bmatrix} \mathbf{A}, \quad R_2 \to R_2 + R_1,\ R_3 \to R_3 - 2R_2 \\
\approx \begin{bmatrix} 1 & 2 & -1 \\ 0 & 1 & \frac{1}{3} \\ 0 & -5 & 3 \end{bmatrix} &= \begin{bmatrix} 1 & 0 & 0 \\ \frac{1}{3} & \frac{1}{3} & 0 \\ -2 & 0 & 1 \end{bmatrix} \mathbf{A}, \quad R_2 \to \frac{1}{3} R_2 \\
\approx \begin{bmatrix} 1 & 0 & -\frac{5}{3} \\ 0 & 1 & \frac{1}{3} \\ 0 & 0 & \frac{14}{3} \end{bmatrix} &= \begin{bmatrix} \frac{1}{3} & -\frac{2}{3} & 0 \\ \frac{1}{3} & \frac{1}{3} & 0 \\ -\frac{1}{3} & \frac{5}{3} & 1 \end{bmatrix} \mathbf{A}, \quad R_1 \to R_1 - 2R_2,\ R_3 \to R_3 + 5R_2
\end{aligned}$$

$$\approx \begin{bmatrix} 1 & 0 & -\frac{5}{3} \\ 0 & 1 & \frac{1}{3} \\ 0 & 0 & 1 \end{bmatrix} = \begin{bmatrix} \frac{1}{3} & -\frac{2}{3} & 0 \\ \frac{1}{3} & \frac{1}{3} & 0 \\ -\frac{1}{14} & \frac{5}{14} & \frac{3}{14} \end{bmatrix} \mathbf{A}, \quad R_3 \to \frac{3}{14} R_3$$

$$\approx \begin{bmatrix} 1 & 0 & 0 \\ 0 & 1 & 0 \\ 0 & 0 & 1 \end{bmatrix} = \begin{bmatrix} \frac{3}{14} & -\frac{1}{14} & \frac{5}{14} \\ \frac{5}{14} & \frac{3}{14} & -\frac{1}{14} \\ -\frac{1}{14} & \frac{5}{14} & \frac{3}{14} \end{bmatrix} \mathbf{A}, \quad R_1 \to R_1 + \frac{5}{3} R_3, \; R_2 \to R_2 - \frac{1}{3} R_3$$

$$\text{Hence,} \quad \mathbf{A}^{-1} = \begin{bmatrix} \frac{3}{14} & -\frac{1}{14} & \frac{5}{14} \\ \frac{5}{14} & \frac{3}{14} & -\frac{1}{14} \\ -\frac{1}{14} & \frac{5}{14} & \frac{3}{14} \end{bmatrix}.$$

Example 1.7.7 *Find the inverse of the matrix* $\mathbf{A} = \begin{bmatrix} 3 & -3 & 4 \\ 2 & -3 & 4 \\ 0 & -1 & 1 \end{bmatrix}$ *using column operations.*

Clearly, $\mathbf{A}$ is nonsingular and $\mathbf{A}^{-1}$ exists. Now,

$$\begin{bmatrix} 3 & -3 & 4 \\ 2 & -3 & 4 \\ 0 & -1 & 1 \end{bmatrix} = \mathbf{A} \begin{bmatrix} 1 & 0 & 0 \\ 0 & 1 & 0 \\ 0 & 0 & 1 \end{bmatrix}$$

$$\approx \begin{bmatrix} 1 & -3 & 4 \\ 2 & -3 & 4 \\ 1 & -1 & 1 \end{bmatrix} = \mathbf{A} \begin{bmatrix} -1 & 0 & 0 \\ 0 & 1 & 0 \\ 1 & 0 & 1 \end{bmatrix}, \quad C_1 \to C_1 - C_3, \; C_1 \to -C_1$$

$$\approx \begin{bmatrix} 1 & 0 & 0 \\ 2 & 3 & -4 \\ 1 & 2 & -3 \end{bmatrix} = \mathbf{A} \begin{bmatrix} -1 & -3 & 4 \\ 0 & 1 & 0 \\ 1 & 3 & -3 \end{bmatrix}, \quad C_2 \to C_2 + 3C_1, \; C_3 \to C_3 - 4C_1$$

$$\approx \begin{bmatrix} 1 & 0 & 0 \\ 2 & 1 & -4 \\ 1 & \frac{2}{3} & -3 \end{bmatrix} = \mathbf{A} \begin{bmatrix} -1 & -1 & 4 \\ 0 & \frac{1}{3} & 0 \\ 1 & 1 & -3 \end{bmatrix}, \quad C_2 \to \frac{1}{3} C_2$$

$$\approx \begin{bmatrix} 1 & 0 & 0 \\ 0 & 1 & 0 \\ \frac{1}{3} & \frac{2}{3} & -\frac{1}{3} \end{bmatrix} = \mathbf{A} \begin{bmatrix} -1 & -1 & 0 \\ -\frac{2}{3} & \frac{1}{3} & \frac{4}{3} \\ -1 & 1 & 1 \end{bmatrix}, \quad C_1 \to C_1 - 2C_2, \quad C_3 \to C_3 + 4C_2$$

$$\begin{bmatrix} 1 & 0 & 0 \\ 0 & 1 & 0 \\ -\frac{1}{3} & \frac{2}{3} & 1 \end{bmatrix} = \mathbf{A} \begin{bmatrix} 1 & -1 & 0 \\ -\frac{2}{3} & \frac{1}{3} & -4 \\ -1 & 1 & -3 \end{bmatrix}, \quad C_3 \to -3C_3$$

$$\begin{bmatrix} 1 & 0 & 0 \\ 0 & 1 & 0 \\ 0 & 0 & 1 \end{bmatrix} = \mathbf{A} \begin{bmatrix} 1 & -1 & 0 \\ -2 & 3 & -4 \\ -2 & 3 & -3 \end{bmatrix}, \quad C_1 \to -C_1 + \frac{1}{3}C_3, \; C_2 \to C_2 - \frac{2}{3}C_3$$

$$\text{Finally } \mathbf{A}^{-1} = \begin{bmatrix} 1 & -1 & 0 \\ -2 & 3 & -4 \\ -2 & 3 & -3 \end{bmatrix}.$$

EXERCISE 1.7

1. If k is scalar, prove that $(k\mathbf{A})^{-1} = \frac{1}{k}\mathbf{A}^{-1}$.

2. If A is nonsingular then prove that $\mathbf{AB} = \mathbf{AC}$ implies $\mathbf{B} = \mathbf{C}$. Is the converse true? If not, give an example.

3. Find the inverse of the following matrices:

   (i) $\begin{bmatrix} 5 & 3 \\ 3 & 2 \end{bmatrix}$ (ii) $\begin{bmatrix} 1 & 0 \\ -1 & 2 \end{bmatrix}$ (iii) $\begin{bmatrix} 1 & 2 \\ 3 & 4 \end{bmatrix}$

   (iv) $\begin{bmatrix} 3 & 1 & 1 \\ 2 & 1 & 0 \\ 6 & 3 & 1 \end{bmatrix}$ (v) $\begin{bmatrix} 13 & 1 & 2 \\ 2 & 1 & 0 \\ 5 & 0 & 1 \end{bmatrix}$.

4. Find the inverse of the following matrices using row transformations

   (i) $\begin{bmatrix} 1 & 0 & 1 & 2 \\ 0 & 1 & 3 & 4 \\ 2 & 1 & 6 & 8 \\ 3 & 2 & 9 & 15 \end{bmatrix}$ (ii) $\begin{bmatrix} 2 & 0 & -1 \\ 5 & 1 & 0 \\ 0 & 1 & 3 \end{bmatrix}^T$.

5. Find the inverse of the following matrices using column operations

   (i) $\begin{bmatrix} -1 & -7 \\ 7 & -1 \end{bmatrix}$ (ii) $\begin{bmatrix} 1 & 2 & 3 \\ 2 & 3 & 0 \\ 0 & 1 & 2 \end{bmatrix}$

   (iii) $\begin{bmatrix} 1 & 0 & 0 \\ 2 & 2 & -1 \\ 1 & -1 & 1 \end{bmatrix}$.

1.8 Partitioning of Matrices

Handling matrices of a very large order will be difficult in terms of algebraic operations. Thus it will be convenient to break these large matrices into submatrices in such a way that they will be compatible for performing the chosen algebraic operation. This division of submatrices is done in different ways depending on the operation that is to be executed. This procedure of breaking up a given matrix into submatrices is called partitioning of matrices.

Definition 1.8.1 *A matrix is said to be partitioned if the matrix is divided into non-overlapping rectangular/square blocks with each block forming a reduced dimension matrix, a sub- matrix.*

Example 1.8.1 $\mathbf{A} = \left[\begin{array}{ccc|ccc|ccc} 1 & 2 & 3 & 0 & 4 & 5 & 6 & 0 & 0 \\ 0 & 1 & 2 & 0 & 4 & 3 & 5 & 6 & 0 \\ 0 & 0 & 1 & 1 & 0 & 2 & 4 & 3 & 2 \\ \hline 1 & 0 & 0 & 4 & 3 & 2 & 1 & 5 & 4 \\ 0 & 1 & 1 & 5 & 2 & 2 & 4 & 3 & 2 \\ 1 & 1 & 0 & 6 & 1 & 2 & 1 & 1 & 1 \end{array}\right]$

$$\mathbf{B} = \left[\begin{array}{cc|c} 2 & 4 & 3 \\ 1 & 2 & 1 \\ \hline 4 & 2 & 8 \\ 1 & 4 & 3 \end{array}\right].$$

$\mathbf{A}$ has been divided into six submatrices of the order 3×3, $\mathbf{B}$ has been divided into four submatrices, two of the order 2×2 and two are of the order 2×1.

This partitioning of matrices is done depending upon the operations to be performed.

1. Addition of matrices using partitions

Let $\mathbf{A}_{m \times n}$ and $\mathbf{B}_{m \times n}$ be two matrices of the same order. Then, in order to define addition, they have to be partitioned in such a way that their submatrices are also of the same order. The partitioning may be as follows:

$$\mathbf{A} = \begin{bmatrix} \mathbf{E}_{r \times k} & \mathbf{F}_{r \times n-k} \\ \mathbf{G}_{m-r \times k} & \mathbf{H}_{m-r \times n-k} \end{bmatrix} \text{ and } \mathbf{B} = \begin{bmatrix} \mathbf{P}_{r \times k} & \mathbf{Q}_{r \times n-k} \\ \mathbf{R}_{m-r \times k} & \mathbf{S}_{m-r \times n-k} \end{bmatrix}.$$

Then,

$$\mathbf{A} + \mathbf{B} = \begin{bmatrix} \mathbf{E} + \mathbf{P} & \mathbf{F} + \mathbf{Q} \\ \mathbf{G} + \mathbf{R} & \mathbf{H} + \mathbf{S} \end{bmatrix},$$

that is, the corresponding submatrices are added.

Example 1.8.2 *Let*

$$\mathbf{A} = \begin{bmatrix} \mathbf{E}_{3\times3} & \mathbf{F}_{3\times2} \\ \mathbf{G}_{1\times3} & \mathbf{H}_{1\times2} \end{bmatrix}_{4\times5} \quad \text{and} \quad \mathbf{B} = \begin{bmatrix} \mathbf{P}_{3\times3} & \mathbf{Q}_{3\times2} \\ \mathbf{R}_{1\times3} & \mathbf{S}_{1\times2} \end{bmatrix}_{4\times5},$$

where, $\mathbf{E} = \begin{bmatrix} 1 & 1 & 3 \\ 0 & 4 & 2 \\ 3 & 2 & 3 \end{bmatrix}$, $\mathbf{P} = \begin{bmatrix} 1 & 1 & 2 \\ 2 & 1 & 4 \\ 2 & 0 & 1 \end{bmatrix}$, $\mathbf{F} = \begin{bmatrix} a & b \\ c & d \\ e & f \end{bmatrix}$, $\mathbf{Q} = \begin{bmatrix} d & m \\ r & s \\ b & v \end{bmatrix}$

$\mathbf{G} = \begin{bmatrix} q & q & c \end{bmatrix}$, $\mathbf{R} = \begin{bmatrix} c & b & c \end{bmatrix}$, $\mathbf{H} = \begin{bmatrix} 2 & 1 \end{bmatrix}$ *and* $\mathbf{S} = \begin{bmatrix} 3 & 3 \end{bmatrix}$.

$$\text{Then,} \quad \mathbf{A}+\mathbf{B} = \left[\begin{array}{cccc|c} 2 & 2 & 5 & a+d & b+m \\ 2 & 5 & 6 & c+r & d+s \\ 5 & 2 & 4 & e+b & f+v \\ \hline g+c & q+b & 2c & 5 & 4 \end{array}\right].$$

2. Product of two matrices

Let $\mathbf{A}_{m\times p}$ and $\mathbf{B}_{p\times n}$ be two matrices such that their product $(\mathbf{AB})_{m\times n}$ exists.
To find the product using the partitioning of matrices, care should be taken while doing the partitioning. The partitioning must be done in such a way that the product of each of the submatrices exists. In other words, the partitioning of p-columns of $\mathbf{A}$ and the p-rows of $\mathbf{B}$ must be done in the same way.

$$\text{Let } \mathbf{A} = \begin{bmatrix} \mathbf{E}_{r\times k} & \mathbf{F}_{r\times p-k} \\ \mathbf{G}_{m-r\times k} & \mathbf{H}_{m-r\times p-k} \end{bmatrix}_{m\times p} \quad \text{and } \mathbf{B} = \begin{bmatrix} \mathbf{P}_{k\times l} & \mathbf{Q}_{k\times n-l} \\ \mathbf{R}_{p-k\times l} & \mathbf{S}_{p-k\times n-l} \end{bmatrix}_{p\times n}$$

$$\mathbf{AB} = \begin{bmatrix} (\mathbf{EP}+\mathbf{FR})_{r\times l} & (\mathbf{EQ}+\mathbf{FS})_{r\times n-l} \\ (\mathbf{GP}+\mathbf{HR})_{m-r\times l} & (\mathbf{GQ}+\mathbf{HS})_{m-r\times n-l} \end{bmatrix}_{m\times n}.$$

Example 1.8.3 *Let* $\mathbf{A} = \left[\begin{array}{cc|c} 2 & 1 & 0 \\ -2 & 3 & 0 \\ \hline -1 & 0 & 1 \end{array}\right]$ $\mathbf{B} = \left[\begin{array}{cc|cc} -1 & 1 & -1 & 0 \\ 2 & 1 & 1 & 0 \\ \hline 3 & 2 & 2 & 1 \end{array}\right]$

Find the product of matrices using the partition method.

$$\mathbf{A} = \begin{bmatrix} \mathbf{E} & \mathbf{F} \\ \mathbf{G} & \mathbf{H} \end{bmatrix} \quad \textit{and} \quad \mathbf{B} = \begin{bmatrix} \mathbf{P} & \mathbf{Q} \\ \mathbf{R} & \mathbf{S} \end{bmatrix},$$

hence

$$\mathbf{AB} = \begin{bmatrix} \mathbf{EP+FR} & \mathbf{EQ+FS} \\ \mathbf{GP+HR} & \mathbf{GQ+HS} \end{bmatrix} = \begin{bmatrix} 0 & 3 & -1 & 0 \\ 8 & 1 & 5 & 0 \\ 4 & 1 & 3 & 1 \end{bmatrix}.$$

3. Inverse of a matrix using partitions

Consider a matrix $\mathbf{A}_{n\times n}$ which can be partitioned as

$$\mathbf{A} = \begin{bmatrix} \mathbf{P} & \mathbf{Q} \\ \mathbf{R} & \mathbf{S} \end{bmatrix},$$

where $\mathbf{P}$ is an $r \times r$ matrix, $\mathbf{Q}$ is an $r \times s$ matrix, $\mathbf{R}_{s\times r}$ matrix and $\mathbf{S}_{s\times s}$ matrix such that $r + s = n$.

Assume that $\mathbf{A}^{-1}$ exists and can be partitioned in a similar way as

$$\mathbf{A}^{-1} = \begin{bmatrix} \mathbf{X} & \mathbf{Y} \\ \mathbf{Z} & \mathbf{W} \end{bmatrix},$$

where $\mathbf{X}_{r\times r}$, $\mathbf{Y}_{r\times s}$, $\mathbf{Z}_{s\times r}$ and $\mathbf{W}_{s\times s}$ matrices are to be found.

Observe that $\mathbf{X}$, $\mathbf{Y}$, $\mathbf{Z}$ and $\mathbf{W}$ are of the same order as $\mathbf{P}$, $\mathbf{Q}$, $\mathbf{R}$ and $\mathbf{S}$ respectively. Since $\mathbf{AA}^{-1} = \mathbf{I}$,

$$\mathbf{A} = \left[\begin{array}{c|c} \mathbf{P} & \mathbf{Q} \\ \hline \mathbf{R} & \mathbf{S} \end{array}\right] \left[\begin{array}{c|c} \mathbf{X} & \mathbf{Y} \\ \hline \mathbf{Z} & \mathbf{W} \end{array}\right] = \left[\begin{array}{c|c} \mathbf{I}_1 & \mathbf{O} \\ \hline \mathbf{O} & \mathbf{I}_2 \end{array}\right]$$

where $\mathbf{I}_1$ and $\mathbf{I}_2$ are identity matrices of the order r and s, respectively.

Using the definition of the product of matrices for partitioned matrices gives the following equations.

$$\begin{aligned} \mathbf{PX} + \mathbf{QZ} &= \mathbf{I}_1 \\ \mathbf{PY} + \mathbf{QW} &= \mathbf{O} \\ \mathbf{RX} + \mathbf{SZ} &= \mathbf{O} \\ \mathbf{RY} + \mathbf{SW} &= \mathbf{I}_2. \end{aligned}$$

Solving for $\mathbf{X}$, $\mathbf{Y}$, $\mathbf{Z}$ and $\mathbf{W}$ gives

$$\begin{aligned} \mathbf{W} &= [\mathbf{S} - \mathbf{RP}^{-1}\mathbf{Q}]^{-1} \\ \mathbf{Y} &= -\mathbf{P}^{-1}\mathbf{QW} \\ \mathbf{Z} &= -\mathbf{WRP}^{-1} \\ \mathbf{X} &= \mathbf{P}^{-1} - \mathbf{P}^{-1}\mathbf{QZ}. \end{aligned}$$

Example 1.8.4 *Find the inverse using partitioning method for the matrix*

$$\mathbf{A} = \left[\begin{array}{cc|c} 1 & 1 & 1 \\ 2 & 3 & -1 \\ \hline 3 & 5 & 2 \end{array}\right].$$

Partitioning A as

$$\mathbf{A} = \left[\begin{array}{cc|c} 1 & 1 & 1 \\ 2 & 3 & -1 \\ \hline 3 & 5 & 2 \end{array}\right]$$

$$\mathbf{P} = \begin{bmatrix} 1 & 1 \\ 2 & 3 \end{bmatrix}, \ \mathbf{Q} = \begin{bmatrix} 1 \\ -1 \end{bmatrix}, \ \mathbf{R} = \begin{bmatrix} 3 & 5 \end{bmatrix}, \ \mathbf{S} = \begin{bmatrix} 2 \end{bmatrix}$$

$$\mathbf{P}^{-1} = \begin{bmatrix} 1 & 1 \\ 2 & 3 \end{bmatrix}^{-1} = \begin{bmatrix} 3 & -1 \\ -2 & 1 \end{bmatrix}$$

$$\mathbf{S} - \mathbf{RP}^{-1}\mathbf{Q} = [2] - \begin{bmatrix} 3 & 5 \end{bmatrix} \begin{bmatrix} 3 & -1 \\ -2 & 1 \end{bmatrix} \begin{bmatrix} 1 \\ -1 \end{bmatrix}$$

$$= 2 - [-1 \ 2] \begin{bmatrix} 1 \\ -1 \end{bmatrix} = 2 - [-3] = 5$$

$$\mathbf{W} = [\mathbf{S} - \mathbf{RP}^{-1}\mathbf{Q}]^{-1} = \frac{1}{5}$$

$$\mathbf{Y} = -\mathbf{P}^{-1}\mathbf{QW} = -\frac{1}{5} \begin{bmatrix} 3 & -1 \\ -2 & 1 \end{bmatrix} \begin{bmatrix} 1 \\ -1 \end{bmatrix} = -\frac{1}{5} \begin{bmatrix} 4 \\ -3 \end{bmatrix}$$

$$\mathbf{Z} = -\mathbf{WRP}^{-1} = -\frac{1}{5} \begin{bmatrix} 3 & 5 \end{bmatrix} \begin{bmatrix} 3 & -1 \\ -2 & 1 \end{bmatrix} = -\frac{1}{5} \begin{bmatrix} -1 & 2 \end{bmatrix}$$

$$\begin{aligned} \mathbf{X} = \mathbf{P}^{-1} - \mathbf{P}^{-1}\mathbf{QZ} &= \begin{bmatrix} 3 & -1 \\ -2 & 1 \end{bmatrix} + \frac{1}{5} \begin{bmatrix} 3 & -1 \\ -2 & 1 \end{bmatrix} \begin{bmatrix} 1 \\ -1 \end{bmatrix} \begin{bmatrix} -1 & 2 \end{bmatrix} \\ &= \begin{bmatrix} 3 & -1 \\ -2 & 1 \end{bmatrix} + \frac{1}{5} \begin{bmatrix} 3 & -1 \\ -2 & 1 \end{bmatrix} \begin{bmatrix} -1 & 2 \\ 1 & -2 \end{bmatrix} \\ &= \begin{bmatrix} 3 & -1 \\ -2 & 1 \end{bmatrix} + \frac{1}{5} \begin{bmatrix} 4 & -8 \\ -3 & 6 \end{bmatrix} \\ &= \begin{bmatrix} 2.2 & 0.6 \\ -1.4 & -0.2 \end{bmatrix}. \end{aligned}$$

Thus

$$\mathbf{A}^{-1} = \begin{bmatrix} \mathbf{X} & \mathbf{Y} \\ \mathbf{Z} & \mathbf{W} \end{bmatrix} = \begin{bmatrix} 3.8 & -2.6 & 0.8 \\ -2.6 & 2.2 & -0.6 \\ 0.2 & -0.4 & 0.2 \end{bmatrix} = \begin{bmatrix} 2.2 & 0.6 & -0.8 \\ -1.4 & -0.2 & 0.6 \\ 0.2 & -0.4 & 0.2 \end{bmatrix}.$$

EXERCISE 1.8

1. Find the sum and difference of the following matrices using the partitioning method.

$$\mathbf{C} = \begin{bmatrix} 2 & 0 & 3 \\ 1 & 1 & 1 \\ -6 & 12 & 8 \\ 1 & 1 & 3 \end{bmatrix} \quad \text{and} \quad \mathbf{D} = \begin{bmatrix} 1 & 0 & 5 \\ a & b & c \\ 0 & 2 & 5 \\ 0 & 8 & 9 \end{bmatrix}.$$

2. Find the product of the matrices using the partition method

$$\mathbf{P} = \begin{bmatrix} 3 & 1 & 4 & 2 \\ 2 & 3 & 3 & 2 \\ 1 & 2 & 2 & 3 \\ 5 & 3 & 2 & 3 \\ 4 & 1 & 2 & 1 \end{bmatrix} \quad \text{and} \quad \mathbf{Q} = \begin{bmatrix} 2 & 4 & 1 \\ 1 & 3 & 2 \\ 1 & 2 & 1 \\ 2 & 3 & 5 \end{bmatrix}.$$

3. Find the inverse of the following matrix using the partitioning method

$$\mathbf{A} = \begin{bmatrix} 4 & 0 & 2 & 1 \\ 3 & 2 & 2 & 0 \\ 1 & 3 & 1 & -2 \\ 2 & -2 & 1 & 1 \end{bmatrix}.$$

4. Find the inverse of the following matrix $\mathbf{A}$ using the partitioning method and solve the system $\mathbf{Ax} = \mathbf{b}$, where

$$\mathbf{A} = \begin{bmatrix} 2 & 1 & 1 & -2 \\ 3 & 0 & 2 & 1 \\ 9 & 4 & -1 & 1 \\ 1 & 1 & 2 & 1 \end{bmatrix} \quad \text{and} \quad \mathbf{b} = \begin{bmatrix} 4 \\ 8 \\ 16 \\ 8 \end{bmatrix}.$$

1.9 Advanced Topics: Pseudo Inverse and Congruent Inverse

In Section 1.7, the notion of a left inverse and a right inverse was introduced. Depending on the size, an $m \times n$ matrix may possess an infinite number of left inverses or right inverses. There are many types of inverses defined for a rectangular matrix, which are called a pseudo inverse or a generalized inverse. In this section, the Moore-Penrose inverse for rectangular matrices is defined. The Moore-Penrose inverse, if it exists, is unique.

Next, using the concept of a congruence relation, modular inverse of a matrix is defined. This matrix is used to encrypt messages in cryptography.

Generalized inverse of a matrix

Definition 1.9.1 *Moore-Penrose Inverse. Let $\mathbf{A}_{m\times n}$ be a matrix of rank = r. The Moore-Penrose inverse of A is a matrix of order $n \times m$, denoted by A^+, satisfying the following criteria.*

1. $\mathbf{AA^+A} = \mathbf{A}$.
2. $\mathbf{A^+AA^+} = \mathbf{A^+}$.
3. $\mathbf{AA^+} = (\mathbf{AA^+})^T$.
4. $(\mathbf{A^+A}) = (\mathbf{A^+A})^T$.

Some of the important properties of the Moore-Penrose inverse are

(i) The Moore Penrose inverse is unique, if it exists.

(ii) $(\mathbf{A}^+)^T = (\mathbf{A}^T)^+$

(iii) $(\mathbf{A}^+)^+ = \mathbf{A}$

(iv) $Rank(\mathbf{A}) = Rank(\mathbf{A}^+)$

(v) If $\mathbf{A}$ and $\mathbf{B}$ are two nonsquare matrices and if $\mathbf{AB} = 0$, then $\mathbf{B^+A^+} = 0$

(vi) $\mathbf{A}^T = \mathbf{A}^+$ if and only if $\mathbf{A}^T\mathbf{A}$ is idempotent.

Result 1.9.1 Let $\mathbf{A}$ be a real matrix

(i) if $\mathbf{A}$ has linearly independent columns, then $\mathbf{A}^+ = (\mathbf{A}^T\mathbf{A})^{(-1)}\mathbf{A}^T$ is the left inverse of $\mathbf{A}$ (since $\mathbf{A^+A} = \mathbf{I}$

(ii) if $\mathbf{A}$ has linearly independent rows, then $\mathbf{A}^+ = \mathbf{A}^T(\mathbf{AA}^T)^{(-1)}$ is the right inverse of $\mathbf{A}$ (since $\mathbf{AA^+} = \mathbf{I}$

Example 1.9.1 *If* $\mathbf{a} = (a_1, \cdots a_n)$ *is a nonzero vector then*

$$\mathbf{a}^+ = \frac{1}{\mathbf{aa}^T}\mathbf{a}^T.$$

Verification: *(i)* $\mathbf{aa^+a} = (a_1, \quad a_n)\frac{1}{[aa^T]}\begin{bmatrix} a_1 \\ a_2 \\ \vdots \\ a_n \end{bmatrix}\begin{bmatrix} a_1 & a_2 & \dots & a_n \end{bmatrix} = \mathbf{a}^+.$

Similarly, $\mathbf{a^+aa^+} = \mathbf{a}^+$.
Further, $\mathbf{aa}^+$ *and* $\mathbf{a^+a}$ *are symmetric.*

A systematic procedure to find $\mathbf{A}^+$ is given in later chapters after the singular value decomposition of a matrix is introduced.

Congruent inverse of a matrix

In order to introduce the congruent inverse or the modular inverse of a matrix, the following preliminary definitions are needed.

Definition 1.9.2 $p, r, s \in \mathbb{N}$. *Then r is said to be congruent to s mod p (denoted by r mod $p = s$), if and only if $p/r - s$, that is, p divides $r - s$.*

Definition 1.9.3 ***Modular inverse of*** $\mathbf{k} \in \mathbb{N}$: *The modular inverse of k is l if and only if kl mod $p = 1$.*

Example 1.9.2 $(11)^{-1}$ *mod* $19 = 7$ *since* $(11)7 = 77$ *and* 77 *mod* $19 = 1$.

Result 1.9.2 If congruent inverse or modular inverse of k is l (i.e.)

$$k^{-1} \ mod \ p \ = \ l, \quad 0 \leq l \leq p-1,$$

then $(-k)^{-1}$ *mod* $p = p - l$.

Example 1.9.3 $(-11)^{-1}$ *mod* $19 = (19 - 7) = 12$, *from the above example and result.*

Procedure to find congruent or modular inverse

The following is the procedure to find the congruent or modular inverse of a 2×2 matrix

$$\mathbf{A} = \begin{bmatrix} a & b \\ c & d \end{bmatrix}.$$

Step 1. Find the conventional inverse of **A**. Then

$$\mathbf{A}^{-1} = \frac{1}{k}\begin{bmatrix} d & -b \\ -c & a \end{bmatrix},$$

where $k = ad - bc$.

Step 2. Find l such that kl *mod* $p = 1$ then k^{-1} *mod* $p = l$.

Step 3. Replace $\frac{1}{k}$ in $\mathbf{A}^{-1}$ with l and multiply the matrix by l (i.e.)

$$\mathbf{A}^{-1} = \begin{bmatrix} ld & -bl \\ -cl & al \end{bmatrix}.$$

Step 4. Find *mod* p value for each entry: $\begin{bmatrix} s & t \\ u & v \end{bmatrix}$, where

$$\begin{aligned} ld \ mod \ p = s, &\quad -bl \ mod \ p \ = t, \\ -cl \ mod \ p = u, &\quad al \ mod \ p = v. \end{aligned}$$

This is the modular or congruent inverse of **A**.

Example 1.9.4 *Let* $\mathbf{M} = \begin{bmatrix} 8 & 1 \\ 13 & 3 \end{bmatrix}$; *find modular inverse of* $\mathbf{A}$ *with* $p = 19$.

Step 1. Then $det(M) = 11$ and

$$\mathbf{M}^{-1} = \frac{1}{11}\begin{bmatrix} 3 & -1 \\ -13 & 8 \end{bmatrix}.$$

Step 2. $11^{-1}\ mod\ 19 = 7$.

Step 3. Now

$$\mathbf{M}^{-1} = \begin{bmatrix} 21 & -7 \\ -91 & 56 \end{bmatrix}.$$

Step 4.

(i) $21\ mod\ 19 = 2$ (ii) $-7\ mod\ 19 = 12$ (using result)
(iii) $-91\ mod\ 19 = 4$ (iv) $56\ mod\ 19 = 18$.

Hence, congruent inverse or modular inverse of M is

$$\mathbf{M}_c^{-1} = \begin{bmatrix} 2 & 12 \\ 4 & 18 \end{bmatrix}.$$

Now one can verify that $\mathbf{M}\mathbf{M}_c^{-1} = \begin{bmatrix} 20 & 114 \\ 38 & 210 \end{bmatrix} = \begin{bmatrix} 1 & 0 \\ 0 & 1 \end{bmatrix}$.

EXERCISE 1.9

1. Verify that the **MP** inverse of a column vector **b**, is

$$\mathbf{b}^+ = \frac{1}{(\mathbf{b}^T\mathbf{b})}\mathbf{b}^T.$$

2. Verify if $\mathbf{A} = (1, 2),\ \mathbf{A}^+ = \frac{1}{5}(1, 2)$.

3. Verify that if **a** is a nonzero $1 \times n$ row vector, then

$$(\mathbf{a^T a})^+ = \frac{1}{(\mathbf{a a^T})^2}(\mathbf{a^T a}).$$

4. Find the congruent inverse of

$$\mathbf{N} = \begin{bmatrix} 8 & -1 \\ 13 & -3 \end{bmatrix} \quad \text{with } p = 19.$$

5. Find the congruent inverse of

$$\mathbf{N} = \begin{bmatrix} 8 & -1 \\ 12 & -3 \end{bmatrix} \quad \text{with } p = 11.$$

1.10 Conclusion

This chapter contains the fundamental blocks of the contents of the book. Care has been taken to introduce all the necessary notions that are needed for the smooth flow of results in the following chapters. Determinants and congruent inverse are defined in this chapter and their applications to conics and cryptography respectively are showcased in Chapter 5. The chapter is self-contained and the treatment of the basics relative to vectors and matrices is complete.
The contents of this chapter, which are the basics of linear algebra, are presented in a concise manner. For additional information, the reader can refer to [19, 14]. For information on pseudo inverses refer to [21].

Chapter 2

Linear System of Equations

2.1 Introduction

Advances in technology and sophisticated computers opened up a galaxy of problems that can be studied numerically. Any challenging problem in space exploration, missile technology, genetic coding or the age-old transportation problem involve a whole bunch of equations (obtained through prior knowledge) involving a large number of variables. These equations, if expressed adroitly as a linear system, can be easily studied using matrices. In this chapter, a linear system of equations is introduced in Section 2.2 and the concept of rank of a matrix which gives a very good understanding of the system of equations is discussed in Section 2.3. The row echelon form and the normal form that help finding the rank of a matrix are described in Section 2.4. The theory of linear system of homogeneous and nonhomogeneous equations forms the content of Section 2.5. Cayley–Hamilton theorem is stated and its use in finding the inverse of a matrix is dealt in Section 2.6. The eigen-values and eigen-vectors, along with diagonalization of a matrix are studied in Section 2.7. Section 2.8 deals with singular values and singular vectors and the quadratic forms are dealt in Section 2.9.

2.2 Linear System of Equations

In this section, a linear system of equations is defined and the problems that can arise in solving the system are given by considering equations involving two unknowns. This section provides the necessary motivation for studying the succeeding sections.

Definition 2.2.1 *A linear equation in two unknowns x and y is an equation of the form*

$$ax + by = 0, \text{ where } a \text{ and } b \text{ are constants}$$

If the number of equations involving x and y is more than one, then they can

DOI: 10.1201/9781351014953-2

be considered as a system of equations. In general, a system of m equations in n unknowns is as follows.

Definition 2.2.2 *A linear system of equations or a system of linear equations is a bunch of m linear equations in n unknowns given by*

$$\begin{aligned} a_{11}x_1 + a_{12}x_2 + \cdots + a_{1n}x_n &= b_1 \\ a_{21}x_1 + a_{22}x_2 + \cdots + a_{2n}x_n &= b_2 \\ &\vdots \\ a_{m1}x_1 + a_{m2}x_2 + \cdots + a_{mn}x_n &= b_m \end{aligned} \tag{2.1}$$

It can be observed that in system (2.1) the unknowns $x_1, x_2, \ldots, x_n$ are repeated in every equation. Now their coefficients without changing the order can be written as row vectors $\mathbf{R_1} = (a_{11} \ a_{12} \ \ldots \ a_{1n})$, $\mathbf{R_2} = (a_{21} \ a_{22} \ \ldots \ a_{2n})$... and $\mathbf{R_m} = (a_{m1} \ a_{m2} \ \ldots \ a_{mn})$.

Forming a matrix with the m-rows gives

$$\mathbf{A} = \begin{bmatrix} a_{11} & a_{12} & \cdots & a_{1n} \\ a_{21} & a_{22} & \cdots & a_{2n} \\ & & \vdots & \\ a_{m1} & a_{m2} \ldots & & a_{mn} \end{bmatrix}.$$

By writing $\mathbf{x} = \begin{bmatrix} x_1 \\ x_2 \\ \vdots \\ x_n \end{bmatrix}$ as the unknown $n \times 1$ column vector and using matrix multiplication, system (2.1) can be written as

$$\mathbf{Ax} = \mathbf{b}, \tag{2.2}$$

where $\mathbf{b} = \begin{bmatrix} b_1 \\ b_2 \\ \vdots \\ b_m \end{bmatrix}$ is an $m \times 1$ column vector.

Equation (2.2) describes system (2.1) in an efficient, compact and an elegant manner.

Definition 2.2.3 *A solution of a linear system* (2.2) *is an* $n \times 1$ *column vector* $\mathbf{x} = \begin{bmatrix} s_1 \\ s_2 \\ \vdots \\ s_n \end{bmatrix}$ *satisfying the equation* (2.2).

Remark 2.2.1 *The ordered set of n numbers* $s_1, s_2, \ldots, s_n$ *satisfy all the equations in the system* (2.1). *This can be verified by putting* $x_1 = s_1, x_2 = s_2, \ldots, x_n = s_n$ *in* (2.1).

It is quite obvious that solving a large system of equations could be complicated and various situations may crop up. Most of the situations that need to be understood are illustrated in the following examples.

Example 2.2.1 *Solve the system of equations* $2x + 3y = 5$ *and* $x + y = 2$.

Solution. From the second equation, $y = 2 - x$ and substituting in the first equation gives $-x + 6 = 5$ or $x = 1$ and $y = 1$.
Here the number of unknowns and the number of equations are equal and there is **exactly one solution**.

Example 2.2.2 *Solve* $x + y = 1,\ x - y = 0$ *and* $2x + y = 3$.

Solution. It can be noted that no values of x and y will satisfy all the 3 equations. Hence there is no solution. These equations are said to be **inconsistent**. Observe that there are two unknowns and three equations.

Example 2.2.3 *Solve* $x + y = 1,\ x - y = 0,$ *and* $2x + 2y = 2$.

Solution. The 3 equations have a solution $x = 1/2$ and $y = 1/2$. Here there are 3 equations but still, there is a solution. Discover the reason (Hint. **Redundant** equation).

Example 2.2.4 *Solve* $2x + 3y = 8$.

Solution. There is only one equation and there are two unknowns. Writing $x = \dfrac{(8 - 3y)}{2}$, it can be seen that for every choice of y, there is a solution. In other words, there are **infinite number of solutions**.

All the fore mentioned examples replicate the scenario for the linear system (2.2) however large the values of m and n are!

In order to work with large and very large systems of linear equations, new concepts and tools are introduced in the next two sections.

EXERCISE 2.2

1. Show that the equations

$$\begin{aligned} x + y + z &= 0 \\ -3y + 3z &= 0 \\ y - z &= 0 \end{aligned}$$

have infinitely many solutions.

2. Show that the equations

$$\begin{aligned} x + y + z &= 0 \\ 2x + 5y + z &= 0 \\ x - y + z &= 0 \end{aligned}$$

have only the zero solution (this is called as a trivial solution).

2.3 Rank of a Matrix

In this section, the concept of rank of a matrix is introduced and its properties are given. The rank of a matrix helps to determine the solution structure of the system (2.2).

Definition 2.3.1 *The rank of a matrix* $\mathbf{A}_{m\times n}$ *is denoted by* $\rho(\mathbf{A}) = r$ *and is defined as the number of independent row vectors (column vectors) of the matrix* $\mathbf{A}$.

Alternatively, the following definition of rank of a matrix could be easily verified for matrices of small order.

Definition 2.3.2 *The rank of a matrix* $\mathbf{A}_{m\times n}$ *is said to be* r *if and only if*
(i) Every minor of order $(r+1)$ *and above is zero;*
(ii) There exists at least one minor of order r *which is non zero.*

Example 2.3.1 *Find the rank of the matrix*

$$\mathbf{A} = \begin{bmatrix} -2 & 1 & 2 \\ 1 & 3 & 0 \\ 2 & 0 & 5 \end{bmatrix}.$$

Solution. $|\mathbf{A}| = (-2)(15) - 1(5) + 2(-6) = -30 - 5 - 12 = -47$.
Thus $|\mathbf{A}| \neq 0$ which gives $\rho(\mathbf{A}) = 3$.

Example 2.3.2 *Find the rank of the matrix*

$$\mathbf{A} = \begin{bmatrix} 1 & 4 & 2 \\ 3 & 3 & 4 \\ 2 & -1 & 2 \end{bmatrix}.$$

Solution. Here $|A| = 1[6+4] - 4[6-8] + 2[-3-6] = 10 + 8 - 18 = 0$. Next consider,

$$\begin{vmatrix} 1 & 4 \\ 3 & 3 \end{vmatrix} = 3 - 12 = -9 \neq 0. \text{ Hence, } \rho(\mathbf{A}) = 2.$$

Some of the important properties of the rank of a matrix are stated below. The proofs of these properties are avoided at this stage. Using the material in the next section all the properties mentioned below can be proved easily.

Properties of the rank of a matrix

(i) Rank of a matrix is unique.

(ii) Rank of zero matrix $\mathbf{O} = [0]_{m\times n}$ is defined to be zero.

(iii) For a $\mathbf{A}_{m\times n}$ matrix, the rank, $\rho(\mathbf{A}) \leq min\{m, n\}$.

(iv) Rank of the identity matrix, $\mathbf{I}_n$ is n.

(v) If $\mathbf{A}_{n\times n}$ is nonsingular then, $\rho(\mathbf{A}) = n$.

(vi) $\rho(\mathbf{A}) = r$ implies that there are 'r' linearly independent rows in the matrix.

(vii) $\rho(\mathbf{A}) = r$ implies that every minor of order $(r+1)$ and above is zero.

(viii) If $\rho(\mathbf{A_{m\times n}}) = r$ and $\mathbf{B_{n\times n}}$ and $\mathbf{C_{m\times m}}$ are nonsingular matrices then $\rho(\mathbf{A}) = \rho(\mathbf{CA})$ and $\rho(\mathbf{A}) = \rho(\mathbf{AB})$. This means that rank of a matrix remains unchanged by multiplication by nonsingular matrices.

(ix) if $\rho(\mathbf{A}) = r_1$, and $\rho(\mathbf{B}) = r_2$ Then $\rho(\mathbf{AB}) \leq min\{r_1, r_2\}$.

Example 2.3.3 *Find the rank of the matrix* $\mathbf{A} = \begin{bmatrix} 1 & 1 & 1 \\ x & y & z \\ x^3 & y^3 & z^3 \end{bmatrix}$,

where x, y and z are real numbers.

Solution. By performing elementary column operations it follows that

$$
\begin{aligned}
det(\mathbf{A}) &= \begin{vmatrix} 1 & 0 & 0 \\ x & y-x & z-x \\ x^3 & y^3-x^3 & z^3-x^3 \end{vmatrix} \\
&= 1[(y-x)(z^3-x^3) - (z-x)(y^3-x^3)] \\
&= (y-x)(z-x)[(z^2+xz+x^2) - (y^2+xy+x^2)] \\
&= (y-x)(z-x)[z^2-y^2+x(z-y)] \\
&= (y-x)(z-x)(z-y)[x+y+z] \\
&= (x-y)(y-z)(z-x)(x+y+z).
\end{aligned}
$$

Then the following cases arise:

1. $x+y+z \neq 0$ and $x \neq y \neq z$. Then $|\mathbf{A}| \neq 0$ and $\rho(\mathbf{A}) = 3$ (by definition).

2. $x + y + z = 0$ and $x \neq y \neq z$. Then $|\mathbf{A}| = 0$ hence $\rho(\mathbf{A}) \neq 3$. Then, consider the 2×2 matrix, $\begin{vmatrix} 1 & 1 \\ x & y \end{vmatrix} = y - x \neq 0$. Hence $\rho(\mathbf{A}) = 2$.

3. If $x \neq y$ and $y = z$ or $x = z$ and $z \neq y$ or $x \neq z$ and $x = y$ (i.e. Two of x, y, z are equal and the third is different). Then $\begin{vmatrix} 1 & 1 \\ x & z \end{vmatrix} \neq 0$. Hence $\rho(\mathbf{A}) = 2$.

4. If $x + y + z = 0$ and $x = y = z \neq 0$. Then $|\mathbf{A}| = 0$ and all 2×2 minors have their determinant zero. Hence $\rho(\mathbf{A}) = 1$.

Example 2.3.4 *Prove that the points (x_1, y_1), (x_2, y_2), and (x_3, y_3) are collinear if the rank of the matrix* $\mathbf{A} = \begin{bmatrix} x_1 & y_1 & 1 \\ x_2 & y_2 & 1 \\ x_3 & y_3 & 1 \end{bmatrix}$ *is less than 3.*

Solution. $\rho(\mathbf{A}) < 3$ implies $|\mathbf{A}| = 0$, which implies that the area of the triangle formed by the given points is zero. This means that the points are collinear.

EXERCISE 2.3

1. If $\mathbf{A}$ is a nonzero column matrix of order $m \times 1$ and $\mathbf{B}$ is a nonzero row matrix of order $1 \times n$. Show that $\rho(\mathbf{AB}) = 1$.

2. Determine the rank of the following matrices.

(i) $\begin{bmatrix} 1 & 2 & 3 & 2 \\ 2 & 3 & 5 & 1 \\ 1 & 3 & 4 & 5 \end{bmatrix}$ (ii) $\begin{bmatrix} 1 & 2 & 3 & 0 \\ 2 & 4 & 3 & 2 \\ 3 & 2 & 1 & 3 \\ 6 & 8 & 7 & 5 \end{bmatrix}$

(iii) $\begin{bmatrix} 1^2 & 2^2 & 3^2 & 4^2 \\ 2^2 & 3^2 & 4^2 & 5^2 \\ 3^2 & 4^2 & 5^2 & 6^2 \\ 4^2 & 5^2 & 6^2 & 7^2 \end{bmatrix}$ (iv) $\begin{bmatrix} 1 & 1+i & -i \\ 0 & i & 1+2i \\ 1 & 1+2i & 1+i \end{bmatrix}$

3. Find the rank of the matrix

$$\begin{bmatrix} 0 & c & -b & a' \\ -c & 0 & a & b' \\ b & -a & 0 & c' \\ -a' & -b' & -c' & 0 \end{bmatrix},$$

where $aa' + bb' + cc' = 0$ and a, b, c, a', b' and c' are all positive numbers.

4. Prove that the rank of a matrix A is the number of linearly independent columns (rows) of a matrix.

5. Show that the rank of a matrix of order $n \times n$ having every element as 1 has rank 1.

6. Let $\mathbf{A} = [a_{ij}]$ be a matrix of order n such that $a_{i,i+1} = 1,\ a_{ij} = 0,$ $j - i \neq 1$. Show that the rank of $\mathbf{A}^k$ is $n - k$, $k < n$.

2.4 Echelon Form and Normal Form

In the previous section, it has been observed that the rank of a matrix gives a decent picture about the solutions of the linear system (2.2).

A couple of examples were given to find the rank of a matrix. It is quite clear that such an approach would be extremely unwieldy if the order of the matrix is very large. In this situation, it would be more practical to use the first definition of a rank of a matrix. But the glitch is that just by looking at a matrix it will not be possible to state the rank of a matrix.

Thus in this section, two forms of a matrix are given which provide the rank of a matrix on observation. The first one is called row echelon form and is defined as follows.

Definition 2.4.1 *A matrix is said to be in a row echelon form if and only if*

(i) All zero rows (rows having all zero entries) are below any row with a non zero entry.

(ii) In every nonzero row the first nonzero entry (called leading entry) is one.

(iii) In any two successive nonzero rows, the leading entry (one) in the succeeding row is to the right of the leading entry (one) in the preceding row.

Definition 2.4.2 *A matrix is said to be in a reduced row echelon form if and only if it is in row echelon form and*

(iv) The leading entry (one) in each nonzero row is the only nonzero entry in that column.

Example 2.4.1 *The following matrices are in row echelon form*

$$(i)\ \mathbf{A} = \begin{bmatrix} 0 & 1 & 2 & 1 & 3 \\ 0 & 0 & 0 & 1 & 2 \\ 0 & 0 & 0 & 0 & 1 \\ 0 & 0 & 0 & 0 & 0 \end{bmatrix} \qquad (ii)\ \mathbf{B} = \begin{bmatrix} 1 & 0 & 0 & 2 & 1 & 5 \\ 0 & 1 & 0 & 3 & 2 & 1 \\ 0 & 0 & 1 & 0 & 4 & 2 \\ 0 & 0 & 0 & 1 & 0 & 0 \\ 0 & 0 & 0 & 0 & 0 & 0 \end{bmatrix}$$

Example 2.4.2 *The following matrices are in reduced row echelon form*

$$(i)\ \mathbf{A} = \begin{bmatrix} 0 & 1 & 0 & 0 \\ 0 & 0 & 1 & 0 \\ 0 & 0 & 0 & 1 \\ 0 & 0 & 0 & 0 \end{bmatrix}$$

$$(ii)\ \mathbf{B} = \begin{bmatrix} 1 & 0 & 0 \\ 0 & 1 & 0 \\ 0 & 0 & 1 \\ 0 & 0 & 0 \end{bmatrix}$$

Result 2.4.1 Given any $\mathbf{A}_{m\times n}$ matrix it can be changed into a row echelon form or a reduced row echelon form by a series of elementary row operations.

Result 2.4.2 The number of nonzero rows in an echelon form is the rank of the matrix.

Example 2.4.3 *Reduce the given matrix into echelon form and hence find its rank.*

$$\mathbf{A} = \begin{bmatrix} 2 & 1 & 5 & 3 \\ 8 & 4 & 13 & 7 \\ 4 & 2 & 3 & 1 \\ -8 & -4 & 1 & 3 \end{bmatrix}$$

Solution. Consider the following row operations:
$R_2 \to R_2 - 4R_1$, $R_3 \to R_3 - 2R_1$, $R_4 \to R_4 + 4R_1$

It follows that

$$\mathbf{A} \sim \begin{bmatrix} 2 & 1 & 5 & 3 \\ 0 & 0 & -7 & -5 \\ 0 & 0 & -7 & -5 \\ 0 & 0 & 21 & 15 \end{bmatrix}$$

now, considering the row operations
$R_4 \to R_4 + 3R_3$, $\quad R_3 \to R_3 + (-1)R_2$, then,

$$\mathbf{A} \sim \begin{bmatrix} 2 & 1 & 5 & 3 \\ 0 & 0 & -7 & -5 \\ 0 & 0 & 0 & 0 \\ 0 & 0 & 0 & 0 \end{bmatrix}$$

$R_1 \to 1/2R_1, R_2 \to -1/7R_2$ then $\mathbf{A}$ $\begin{bmatrix} 1 & 1/2 & 5/2 & 3/2//0 & 0 & 1 & 5/7 \\ 0 & 0 & 0 & 0 \\ 0 & 0 & 0 & 0 \end{bmatrix}$

Clearly, **A** has been reduced to a matrix in echelon form and $\rho(\mathbf{A}) = 2$.

Example 2.4.4 *Find the value of q, such that the rank of matrix*

$$\mathbf{A} = \begin{bmatrix} 0 & 1 & 1 & -1 \\ 2 & 0 & 2 & 2 \\ 1 & 2 & q & -1 \\ 2 & -1 & -3 & -1 \end{bmatrix} \text{ is 2.}$$

Solution. Consider $R_1 \leftrightarrow R_2$, then, $\mathbf{A} \sim \begin{bmatrix} 2 & 0 & 2 & 2 \\ 0 & 1 & 1 & -1 \\ 1 & 2 & q & -1 \\ 2 & -1 & -3 & -1 \end{bmatrix}$

$R_1 \to R_1/2$

$$\sim \begin{bmatrix} 1 & 0 & 1 & 1 \\ 0 & 1 & 1 & -1 \\ 1 & 2 & q & -1 \\ 2 & -1 & -3 & -1 \end{bmatrix}$$

$R_3 \to R_3 - R_1; \quad R_4 \to R_4 - 2R_1$

$$\sim \begin{bmatrix} 1 & 0 & 1 & 1 \\ 0 & 1 & 1 & -1 \\ 0 & 2 & q-1 & -2 \\ 0 & -1 & -5 & -3 \end{bmatrix}$$

Since rank of $\mathbf{A} = 2$, determinant of every 3 by 3 matrix $= 0$.

Thus, $\begin{vmatrix} 1 & 1 & -1 \\ 2 & q-1 & -2 \\ -1 & -5 & -3 \end{vmatrix} = 0,$

$1[(q-1)(-3) - 10] + (-1)[-6-2] - 1[-10 + q - 1] = 0$
$-3q - 7 + 8 + 11 - q = 0$
$-4q + 12 = 0 \Rightarrow q = 3.$

Definition 2.4.3 *The* normal form of a matrix *is any of the following matrices:* $\mathbf{I_r}$, $\begin{bmatrix} \mathbf{I_r} & \mathbf{O} \end{bmatrix}$, $\begin{bmatrix} \mathbf{I_r} \\ \mathbf{O} \end{bmatrix}$ *or* $\begin{bmatrix} \mathbf{I_r} & \mathbf{O} \\ \mathbf{O} & \mathbf{O} \end{bmatrix}$, *where* $\mathbf{I_r}$ *is an identity matrix of rank* r *and* **O** *is any* $p \times q$ *zero matrix.*

Example 2.4.5 *The following matrices are in normal form*

(i) $$\begin{bmatrix} 1 & 0 & 0 & 0 & 0 & 0 \\ 0 & 1 & 0 & 0 & 0 & 0 \\ 0 & 0 & 1 & 0 & 0 & 0 \\ 0 & 0 & 0 & 0 & 0 & 0 \\ 0 & 0 & 0 & 0 & 0 & 0 \end{bmatrix} = \begin{bmatrix} \mathbf{I_3} & \mathbf{O_{3\times 3}} \\ \mathbf{O_{2\times 3}} & \mathbf{O_{2\times 3}} \end{bmatrix}$$

(ii) $$\begin{bmatrix} 1 & 0 & 0 & 0 \\ 0 & 1 & 0 & 0 \\ 0 & 0 & 1 & 0 \\ 0 & 0 & 0 & 1 \end{bmatrix} = \mathbf{I_4}$$

(iii) $$\begin{bmatrix} 1 & 0 \\ 0 & 1 \\ 0 & 0 \\ 0 & 0 \end{bmatrix} = \begin{bmatrix} \mathbf{I_2} \\ \mathbf{O_{2\times 2}} \end{bmatrix}$$

(iv) $$\begin{bmatrix} 1 & 0 & 0 & 0 & 0 \\ 0 & 1 & 0 & 0 & 0 \end{bmatrix} = \begin{bmatrix} \mathbf{I_2} & \mathbf{O_{2\times 3}} \end{bmatrix}$$

Result 2.4.3 Every matrix $\mathbf{A_{m\times n}}$ of rank r can be reduced to one of the fore-mentioned normal forms by a sequence of elementary row or column operations.

Example 2.4.6 *Find the rank of the following matrix by reducing it to the normal form*

$$\mathbf{A} = \begin{bmatrix} 2 & 1 & 5 & 3 \\ 8 & 4 & 13 & 7 \\ 4 & 2 & 3 & 1 \\ -8 & -4 & 1 & 3 \end{bmatrix}$$

Solution. Consider the column operation $C_1 \leftrightarrow C_2$,

$$\mathbf{A} \sim \begin{bmatrix} 1 & 2 & 5 & 3 \\ 4 & 8 & 13 & 7 \\ 2 & 4 & 3 & 1 \\ -4 & -8 & 1 & 3 \end{bmatrix}.$$

Next perform the following sequence of operations: $R_2 \to -R_2 + 4R_1$, $R_3 \to R_3 - 2R_1, R_4 \to R_4 + 4R_1$

$$\sim \begin{bmatrix} 1 & 2 & 5 & 3 \\ 0 & 0 & 7 & 5 \\ 0 & 0 & -7 & -5 \\ 0 & 0 & 21 & 15 \end{bmatrix}$$

$R_3 \to R_3 + R_2; \quad R_4 \to R_4 - 3R_2$

$$\sim \begin{bmatrix} 1 & 2 & 5 & 3 \\ 0 & 0 & 7 & 5 \\ 0 & 0 & 0 & 0 \\ 0 & 0 & 0 & 0 \end{bmatrix}$$

$C_2 \to C_2 - 2C_1, C_3 \to C_3 - 5C_1, C_4 \to C_4 - 3C_1$

$$\sim \begin{bmatrix} 1 & 0 & 0 & 0 \\ 0 & 0 & 7 & 5 \\ 0 & 0 & 0 & 0 \\ 0 & 0 & 0 & 0 \end{bmatrix}$$

$C_2 \leftrightarrow C_3$

$$\sim \begin{bmatrix} 1 & 0 & 0 & 0 \\ 0 & 7 & 0 & 5 \\ 0 & 0 & 0 & 0 \\ 0 & 0 & 0 & 0 \end{bmatrix}$$

$C_2 \to (1/7)C_2$

$$\sim \begin{bmatrix} 1 & 0 & 0 & 0 \\ 0 & 1 & 0 & 5 \\ 0 & 0 & 0 & 0 \\ 0 & 0 & 0 & 0 \end{bmatrix}$$

$C_4 \to C_4 - 5C_2$

$$\sim \begin{bmatrix} 1 & 0 & 0 & 0 \\ 0 & 1 & 0 & 0 \\ 0 & 0 & 0 & 0 \\ 0 & 0 & 0 & 0 \end{bmatrix}$$

Thus the above matrix is in the normal form $\begin{bmatrix} \mathbf{I_2} & \mathbf{O_{2\times 2}} \\ \mathbf{O_{2\times 2}} & \mathbf{O_{2\times 2}} \end{bmatrix}$

The rank of the matrix is $\rho(\mathbf{A}) = 2$.

Result 2.4.4 Existence of nonsingular matrices $\mathbf{P}$ and $\mathbf{Q}$.

To reduce a given matrix to its normal form a sequence of row operations $\mathbf{R_1}, \mathbf{R_2}, \ldots, \mathbf{R_i}$ and a sequence of column operations $\mathbf{C_1}, \mathbf{C_2}, \ldots, \mathbf{C_j}$ are performed. This is equivalent to pre- multiplying $\mathbf{A}$ with a sequence of elementary matrices $\mathbf{E_i}, \mathbf{E_{i-1}}, \ldots, \mathbf{E_1}$ and post multiplying $\mathbf{A}$ with a sequence of elementary matrices $\mathbf{E_j}, \mathbf{E_{j-1}}, \ldots, \mathbf{E_1}$. Set $\mathbf{P} = \mathbf{E_i}\mathbf{E_{i-1}}\ldots\mathbf{E_1}$ and $\mathbf{Q} = \mathbf{E_j}\mathbf{E_{j-1}}\ldots\mathbf{E_1}$. Then the matrix $\mathbf{PAQ}$ will be in normal form with $\mathbf{P}$ and $\mathbf{Q}$ being nonsingular matrices. This is illustrated in the following example.

Example 2.4.7

Find non-singular matrices $\mathbf{P}$ and $\mathbf{Q}$ such that $\mathbf{PAQ}$ is in the normal form of the matrix and hence deduce the rank of

$$\mathbf{A} = \begin{bmatrix} 1 & 3 & 6 & -1 \\ 2 & 8 & 10 & 2 \\ 1 & 5 & 4 & 3 \end{bmatrix}.$$

Set $\mathbf{A}=\mathbf{I_3}\ \mathbf{A}\mathbf{I_4}$.

Any row operations done on LHS matrix $\mathbf{A}$ will be done on the matrix $\mathbf{I_3}$ and column operations executed on LHS matrix $\mathbf{A}$ will be operated on the matrix $\mathbf{I_4}$.

$$\begin{bmatrix} 1 & 3 & 6 & -1 \\ 2 & 8 & 10 & 2 \\ 1 & 5 & 4 & 3 \end{bmatrix} = \begin{bmatrix} 1 & 0 & 0 \\ 0 & 1 & 0 \\ 0 & 0 & 1 \end{bmatrix} \mathbf{A} \begin{bmatrix} 1 & 0 & 0 & 0 \\ 0 & 1 & 0 & 0 \\ 0 & 0 & 1 & 0 \\ 0 & 0 & 0 & 1 \end{bmatrix}.$$

$R_2 \rightarrow R_2 - 2R_1, \qquad R_3 \rightarrow R_3 - R_1$

$$\begin{bmatrix} 1 & 3 & 6 & -1 \\ 0 & 2 & -2 & 4 \\ 0 & 2 & -2 & 4 \end{bmatrix} = \begin{bmatrix} 1 & 0 & 0 \\ -2 & 1 & 0 \\ -1 & 0 & 1 \end{bmatrix} \mathbf{A} \begin{bmatrix} 1 & 0 & 0 & 0 \\ 0 & 1 & 0 & 0 \\ 0 & 0 & 1 & 0 \\ 0 & 0 & 0 & 1 \end{bmatrix}$$

$R_3 \rightarrow R_3 - R_2$

$$\begin{bmatrix} 1 & 3 & 6 & -1 \\ 0 & 2 & -2 & 4 \\ 0 & 0 & 0 & 0 \end{bmatrix} = \begin{bmatrix} 1 & 0 & 0 \\ 2 & 1 & 0 \\ 1 & -1 & 1 \end{bmatrix} \mathbf{A} \begin{bmatrix} 1 & 0 & 0 & 0 \\ 0 & 1 & 0 & 0 \\ 0 & 0 & 1 & 0 \\ 0 & 0 & 0 & 1 \end{bmatrix}$$

$R_2 \rightarrow 1/2R_2$

$$\begin{bmatrix} 1 & 3 & 6 & -1 \\ 0 & 1 & -1 & 2 \\ 0 & 0 & 0 & 0 \end{bmatrix} = \begin{bmatrix} 1 & 0 & 0 \\ -1 & \frac{1}{2} & 0 \\ 1 & -1 & 1 \end{bmatrix} \mathbf{A} \begin{bmatrix} 1 & 0 & 0 & 1 \\ 0 & 1 & 0 & 0 \\ 0 & 0 & 1 & 0 \\ 0 & 0 & 0 & 1 \end{bmatrix}$$

$C_2 \rightarrow C_2 - 3C_1,\ C_3 \rightarrow C_3 - 6C_1,\ C_4 \rightarrow C_4 + C_1$

$$\begin{bmatrix} 1 & 3 & 6 & -1 \\ 0 & 0 & -2 & 0 \\ 0 & 2 & 0 & 0 \end{bmatrix} = \begin{bmatrix} 1 & 0 & 0 \\ -1 & 1 & -1 \\ -1 & 0 & 1 \end{bmatrix} \mathbf{A} \begin{bmatrix} 1 & -3 & -6 & 7 \\ 0 & 1 & 0 & -2 \\ 0 & 0 & 1 & 0 \\ 0 & 0 & 0 & 1 \end{bmatrix}$$

$C_3 \rightarrow C_3 + C_2, C_4 \rightarrow C_4 - 2C_2$

$$\begin{bmatrix} 1 & 0 & 0 & 0 \\ 0 & 1 & 0 & 0 \\ 0 & 0 & 0 & 0 \end{bmatrix} = \begin{bmatrix} 1 & 0 & 0 \\ -1 & \frac{1}{2} & 0 \\ 1 & -1 & 1 \end{bmatrix} \mathbf{A} \begin{bmatrix} 1 & -3 & -9 & 7 \\ 0 & 1 & 1 & -2 \\ 0 & 0 & 1 & 0 \\ 0 & 0 & 0 & 1 \end{bmatrix}$$

Thus the nonsingular matrices $\mathbf{P}$ and $\mathbf{Q}$ are

$$\mathbf{P}= \begin{bmatrix} 1 & 0 & 0 \\ -1 & \frac{1}{2} & 0 \\ 1 & -1 & 1 \end{bmatrix} \text{ and } \mathbf{Q}= \begin{bmatrix} 1 & -3 & -9 & 7 \\ 0 & 1 & 1 & -2 \\ 0 & 0 & 1 & 0 \\ 0 & 0 & 0 & 1 \end{bmatrix},$$

and the rank of the matrix, $\rho(\mathbf{A}) = 2$.

A fundamental question that crops up is that whether the row echelon form and the normal form of the matrix depend on the sequence of elementary row operations/column operations performed on the matrix $\mathbf{A}$. The following theorem deals with this question.

Theorem 2.4.1 *The number of nonzero rows in an echelon form and the* $\mathbf{I_r}$ *obtained in a normal form are independent of the sequence of row/column operations performed on a given matrix* $\mathbf{A_{m\times n}}$.
In other words, the row echelon form and normal form of the given matrix $\mathbf{A}$ *are independent of the nonsingular matrices* $\mathbf{P}$ *and* $\mathbf{Q}$ *obtained.*

EXERCISE 2.4

1. Reduce the following matrices to echelon form and hence find their rank

$$\text{(a)} \begin{bmatrix} 1 & 2 & 3 & -2 \\ 2 & 4 & 6 & -1 \\ -1 & -3 & -2 & 2 \\ 2 & 5 & 6 & -4 \end{bmatrix} \qquad \text{(b)} \begin{bmatrix} 1 & -1 & 2 & -1 \\ 4 & 1 & 1 & 2 \\ 1 & 3 & 0 & 4 \\ 1 & 1 & 0 & 2 \end{bmatrix}.$$

2. For what value of p will the matrix

$$\begin{bmatrix} 3 & 1 & p & 8 \\ 2 & 2 & 6 & -1 \\ 8 & 4 & 12 & 15 \\ 5 & 3 & 9 & 7 \end{bmatrix} \text{ have rank} = 2.$$

3. Determine the nonsingular matrices $\mathbf{P}$ and $\mathbf{Q}$ such that $\mathbf{PAQ}$ is in normal form for

$$\text{(a)} \begin{bmatrix} 1 & 1 & 2 & 4 \\ 1 & 2 & 3 & -4 \\ 0 & -1 & -1 & 2 \\ 2 & 1 & 3 & 1 \end{bmatrix} \qquad \text{(b)} \begin{bmatrix} 1 & -1 & -2 \\ 2 & 3 & 1 \\ 1 & -1 & 1 \end{bmatrix}.$$

2.5 Solutions of a Linear System of Equations

A linear system of equations was introduced in Section 2.2 and various possibilities that occur while solving these types of systems have been discussed as special cases. In this section an arbitrary linear system of 'm' equations in

'n' unknowns is again considered. The famous Gauss-elimination method and Gauss-Jordan method will be mentioned in brief (and again will be discussed in detail in Chapter 4) and then theory pertaining to the solutions of the system will be given.

Consider a linear system of 'm' equations in 'n' unknowns given by

$$\mathbf{Ax} = \mathbf{b}, \tag{2.3}$$

where $\mathbf{A} = \begin{bmatrix} a_{11} & a_{12} & \cdots & a_{1n} \\ a_{21} & a_{22} & \cdots & a_{2n} \\ \vdots & \vdots & \vdots & \vdots \\ a_{m1} & a_{m2} & \cdots & a_{mn} \end{bmatrix}_{m \times n}$ is the coefficient matrix,

$\mathbf{x} = [x_1, x_2, \ldots, x_n]^T$ is the unknown column vector(matrix)

and $\mathbf{b} = [b_1, b_2, \ldots, b_m]^T$ is a column vector(matrix).

Definition 2.5.1 *The system* (2.3) *is said to be a nonhomogeneous linear system.*

If $\mathbf{b} = [0, \ldots, 0]^T$ *in the system* (2.3) *then*

$$\mathbf{Ax} = \mathbf{0} \tag{2.4}$$

is said to be a homogeneous linear system.

Definition 2.5.2 *Augmented matrix of the system* (2.3) *is the matrix*

$$\mathbf{K} = [\mathbf{A} : \mathbf{b}].$$

In order to solve the system (2.3), Gauss elimination method is a very natural and a systematic procedure and it is the basis of all other elimination procedures. In this method, the augmented matrix **K** is reduced to row echelon form or an upper triangular matrix and the vector **x** is determined by back substitution. This method is illustrated through the following example.

Example 2.5.1 *Solve the system* $\mathbf{Ax} = \mathbf{b}$, *where*

$$\mathbf{A} = \begin{bmatrix} 2 & 1 & 5 \\ 6 & 2 & 7 \\ 4 & 2 & 3 \end{bmatrix} \quad \textit{and} \quad \mathbf{b} = \begin{bmatrix} 3 \\ 1 \\ 4 \end{bmatrix}.$$

Solution. *Consider the augmented matrix,* $\mathbf{K} = [\mathbf{A} : \mathbf{b}]$

$$\mathbf{K} = \begin{bmatrix} 2 & 1 & 5 & : & 3 \\ 6 & 2 & 7 & : & 1 \\ 4 & 2 & 3 & : & 4 \end{bmatrix}$$

$R_2 \to R_2 - 3R_1$, $R_3 \to R_3 - 2R_1$,

It follows that

$$\mathbf{K} \sim \begin{bmatrix} 2 & 1 & 5 & : 3 \\ 0 & -1 & -8 & : -8 \\ 0 & 0 & -7 & : -2 \end{bmatrix}.$$

The matrix is in an upper triangular form. On doing back substitution, the value of $\mathbf{x} = \begin{bmatrix} \frac{-29}{14} \\ \frac{40}{7} \\ \frac{2}{7} \end{bmatrix}$.

The Gauss-Jordan method also aims at simplifying the given matrix so as to make it easier to find the solutions of a given linear system. The first stage of the Gauss-Jordan method is the same as that of the Gauss elimination method. Instead of doing back substitution once the matrix is available in an upper triangular form, a second set of elementary operations are employed so as to reduce it to an identity matrix.

The following theorem relates the solutions of original system and the simplified system.

Theorem 2.5.1 *Suppose the system* (2.3) *or the augmented matrix* $\mathbf{K}$ *is transformed by a sequence of elementary row operations into the system*

$$\mathbf{A}'\mathbf{x} = \mathbf{b}' \tag{2.5}$$

or the augmented matrix $\mathbf{K}' = [\mathbf{A}' : \mathbf{b}']$. *Then* $\mathbf{x}$ *is a solution of the system* (2.5) *if and only if* $\mathbf{x}$ *is a solution of the system* $\mathbf{A}\mathbf{x} = \mathbf{b}$.

Proof. It is known, from Section 1.5, that every elementary row operation on $\mathbf{K}$ is equivalent to pre multiplication of $\mathbf{K}$ by an elementary matrix. Let p be the number of row operations done on $\mathbf{K}$. Then there exist p elementary nonsingular matrices $\mathbf{E_p}, \mathbf{E_{p-1}}, ...\mathbf{E_1}$ successively pre multiplied on $\mathbf{K}$. Let $\mathbf{H} = \mathbf{E_p}\mathbf{E_{p-1}}...\mathbf{E_1}$. Note that $\mathbf{H}$ is non-singular. Thus

$$\begin{aligned} [\mathbf{A}' : \mathbf{b}'] &= \mathbf{H}[\mathbf{A} : \mathbf{b}] \\ &= [\mathbf{HA} : \mathbf{Hb}] \end{aligned}$$

by the rules of multiplication of partitioned matrices.

This implies that

$$\mathbf{A}' = \mathbf{HA} \quad \textit{and} \quad \mathbf{b}' = \mathbf{Hb}.$$

Let $\mathbf{x}$ be a solution of $\mathbf{A}\mathbf{x} = \mathbf{b}$. Pre-multiplying with $\mathbf{H}$ gives

$\mathbf{HAx} = \mathbf{Hb}$ that is, $\mathbf{A'x} = \mathbf{b'}$. Thus $\mathbf{x}$ is a solution of $\mathbf{A'x} = \mathbf{b'}$.
Next, suppose $\mathbf{x}$ is a solution of $\mathbf{A'x} = \mathbf{b'}$ then $\mathbf{HAx} = \mathbf{Hb}$. Pre-multiplying with $\mathbf{H}^{-1}$ gives $\mathbf{Ax} = \mathbf{b}$ which means $\mathbf{x}$ is a solution of the system (2.3). The proof is complete.

Result 2.5.1 If the linear system (2.3) has two distinct solutions $\mathbf{x}_1$ and $\mathbf{x}_2$ then it has an infinite number of solutions.

Proof. Let $\mathbf{x}_1$ and $\mathbf{x}_2$ be two distinct solutions of the linear system (2.3).
Then $\mathbf{Ax}_1 = \mathbf{b}$ and $\mathbf{Ax}_2 = \mathbf{b}$
Further $(\lambda\mathbf{A})(\mathbf{x}_1) = \mathbf{A}(\lambda\mathbf{x}_1) = \lambda\mathbf{b}$
and $(1-\lambda)(\mathbf{Ax}_2) = \mathbf{A}(1-\lambda)\mathbf{x}_2 = (1-\lambda)\mathbf{b}$
which gives $\mathbf{A}[\lambda\mathbf{x}_1 + (1-\lambda)\mathbf{x}_2)] = \mathbf{b}$.
Thus $\lambda\mathbf{x}_1 + (1-\lambda)\mathbf{x}_2$ is a solution of the linear system (2.3) for any $\lambda \in \mathbb{R}$. Thus if the system (2.3) has two distinct solutions it will have infinitely many solutions.

Theorem 2.5.2 ***Rouche's Theorem***

The nonhomogeneous linear system (2.3) *is consistent if and only if the coefficient matrix* $\mathbf{A}_{m\times n}$ *and the augmented matrix* $\mathbf{K}$ *have the same rank. Otherwise, the system is inconsistent.*

Theorem 2.5.3 *Let the rank of the coefficient matrix* $\mathbf{A}$, $\rho(\mathbf{A}) = p$ *and the rank of the augmented matrix* $\mathbf{K}$, $\rho(\mathbf{K}) = q$. *Then exactly one and only one of the following possibilities hold:*

(i) $p \neq q$, *in which case the system* (2.3) *is inconsistent and there are no solutions.*

(ii) $p = q = n$ *(number of unknowns), then there exists a unique solution for the system* (2.3).

(iii) $p = q \neq n$*(number of unknowns), then there exist infinitely many solutions for the system* (2.3).

Proof. Case (i) $p \neq q$ since the coefficient matrix $\mathbf{A}$ is a part of the augmented matrix $\mathbf{K}$, the rank $\mathbf{A} = p \neq q = rank\ \mathbf{K}$.

The Gauss elimination method or the echelon form of $\mathbf{K}$ implies that the augmented matrix has an extra nonzero row more than the coefficient matrix. In other words there is a row having the equation $0.x_1 + 0.x_2 + ... + 0.x_n = b_{p+1} \neq 0$. Since this can never happen, there is no solution for the system (2.3).

Case (ii) $p = q = n$. This means that the coefficient matrix has n- linearly independent rows and that the inverse $\mathbf{A}^{-1}$ exists. Hence $\mathbf{x} = \mathbf{A}^{-1}\mathbf{b}$ is the only solution of the system (2.3).

Case (iii) $p = q < n$. This means that there are p independent rows and the remaining rows are zero rows in the echelon form. Since there are p non zero rows there are p equations and, hence p-unknowns can be found. The remaining $n-p$ variables have to be chosen arbitrarily and each choice gives a different solution for the system. Since the choices can be infinite, there exist an infinite number of solutions.

If the coefficient matrix is a square matrix $\mathbf{A}_{n\times n}$ then the following theorem holds.

Theorem 2.5.4 *Let $A_{n\times n}$ be the coefficient matrix of the system (2.3). Then the following conditions are equivalent.*

(i) The corresponding homogeneous system (2.4) *has only the zero solution.*

(ii) rank $\mathbf{A} = n$.

(iii) For each $\mathbf{b}_{n\times 1}$ *column vector, the system* $\mathbf{Ax} = \mathbf{b}$ *has exactly one solution.*

(iv) $\mathbf{A}$ *is nonsingular.*

Proof. Assume (i), that is the system $\mathbf{Ax} = \mathbf{0}$ has only one solution. Then by the previous theorem (taking $\mathbf{b} = \mathbf{0}$) implies that rank $\mathbf{A} = n$, that is condition (ii) holds. Next assume that (ii) holds, then rank $\mathbf{A} = n$. Consider the system $\mathbf{Ax} = \mathbf{b}$ and form the augmented matrix $\mathbf{K} = [\mathbf{A} : \mathbf{b}]$.Then clearly rank $\mathbf{A}$=rank $\mathbf{K}$, which implies from the previous theorem that there exists a unique solution for the system $\mathbf{Ax} = \mathbf{b}$.

Assume (iii), that is, there exists a unique solution for the system $\mathbf{Ax} = \mathbf{b}$.

If possible, suppose $\mathbf{A}$ is singular, then clearly $|\mathbf{A}| = 0$ which implies that $rank A < n$ and that there are infinitely many solutions for the system $\mathbf{Ax} = \mathbf{b}$, a contradiction. Hence $\mathbf{A}$ is nonsingular.

Suppose $\mathbf{A}$ is nonsingular, then clearly $\mathbf{A}^{-1}$ exists and $\mathbf{Ax} = \mathbf{0} \implies \mathbf{x} = \mathbf{0}$, hence $\mathbf{x} = \mathbf{0}$ is the only solution of the system $\mathbf{Ax} = \mathbf{0}$, completing the proof.

Result 2.5.2 If $\mathbf{x_1}$ and $\mathbf{x_2}$ are two distinct solutions of the nonhomogeneous linear system (2.3), then $\mathbf{x_1} - \mathbf{x_2}$ is a solution of the homogeneous linear system (2.4)

Proof. Suppose $\mathbf{x_1}$ and $\mathbf{x_2}$ are solutions of the nonhomogeneous linear system (2.3). Then $\mathbf{Ax_1} = \mathbf{b}$ and $\mathbf{Ax_2} = \mathbf{b}$ hold,
Consider $\mathbf{A}(\mathbf{x_1} - \mathbf{x_2}) = \mathbf{Ax_1} - \mathbf{Ax_2} = \mathbf{b} - \mathbf{b} = \mathbf{0}$. Hence the conclusion.

Result 2.5.3 If $\mathbf{x}$ is any solution of the homogeneous linear system (2.4) and $\mathbf{x_0}$ is a solution of the nonhomogeneous linear system(2.3), then every solution of the nonhomogeneous linear system(2.3) is of the form $\mathbf{y} = \mathbf{x} + \mathbf{x_0}$.

Proof. Given $\mathbf{Ax_0} = \mathbf{b}$ and $\mathbf{Ax} = \mathbf{0}$. Thus $\mathbf{Ay} = \mathbf{A}(\mathbf{x_0} + \mathbf{x}) = \mathbf{Ax_0} + \mathbf{Ax} = \mathbf{b}$. This means any solution of system (2.3) is of the form $\mathbf{y} = \mathbf{x_0} + \mathbf{x} = \mathbf{x} + \mathbf{x_0}$.

Cramer's Rule. Let A be a nonsingular matrix then the unique solution $\mathbf{x} = (x_1, x_2, \ldots, x_n)$ of the nonhomogeneous linear system (2.3) is given by $\mathbf{x}_i = \frac{det[\mathbf{A}_i]}{det(\mathbf{A})}, i = 1, 2, \ldots, n$ where $\mathbf{A}_i$ is the matrix obtained by replacing ith column of $\mathbf{A}$ by the column vector $\mathbf{b}$

Example 2.5.2

Solve the system

$$\begin{aligned} x + y + 2z &= 1 \\ 2x - 3y - z &= -3 \\ 2x + 2y + 2z &= 4. \end{aligned}$$

Solution. $A = \begin{bmatrix} 1 & 1 & 2 \\ 2 & -3 & -1 \\ 2 & 2 & 2 \end{bmatrix}$, $\mathbf{x} = \begin{bmatrix} x \\ y \\ z \end{bmatrix}$ *and* $\mathbf{b} = \begin{bmatrix} 1 \\ -3 \\ 4 \end{bmatrix}$.

Here $|\mathbf{A}| = 1(-6+2) - 1(4+2) + 2(4+6) = 10$.

$$|\mathbf{A}_1| = \begin{vmatrix} 1 & 1 & 2 \\ -3 & -3 & -1 \\ 4 & 2 & 2 \end{vmatrix}$$

$$= 1(-6+2) - 1(-6+4) + 2(-6+12) = -4 + 2 + 12 = 10.$$

Hence by Cramer's Rule,

$$x_1 = \frac{|\mathbf{A}_1|}{|\mathbf{A}|} = 1.$$

Next,

$$|\mathbf{A}_2| = \begin{vmatrix} 1 & 1 & 2 \\ 2 & -3 & -1 \\ 2 & 4 & 2 \end{vmatrix}$$

$$= 1(-6+4) - 1(4+2) + 2(8+6) = -2 - 6 + 28 = 20.$$

It follows that

$$y = \frac{|\mathbf{A}_2|}{|\mathbf{A}|} = \frac{20}{10} = 2.$$

Also,

$$|\mathbf{A}_3| = \begin{vmatrix} 1 & 1 & 1 \\ 2 & -3 & -3 \\ 2 & 2 & 4 \end{vmatrix}$$

$= 1(-12+6) - 1(8+6) + 1(4+6) = -6 - 14 + 10 = -10.$

$z = \frac{|\mathbf{A}_3|}{|\mathbf{A}|} = \frac{-10}{10} = -1.$

Hence $\mathbf{x} = \begin{bmatrix} 1 \\ 2 \\ -1 \end{bmatrix}.$

EXERCISE 2.5

1. Find the values of k for which the equations

$$\begin{aligned} x + 2y + z &= k \\ 3x + y - z &= 1 \\ 7x + 4y - z &= k^2 \text{ have a solution.} \end{aligned}$$

   Solve the system for each k.

2. For what values of a and b the equations

$$\begin{aligned} 2x + ay + 2z &= 6 \\ 2x + 2y + z &= b \\ 3x + 5y + z &= 9 \text{ have a solution.} \end{aligned}$$

3. Check the consistency of the following systems and solve them.

$$\text{(a)} \qquad \begin{aligned} 3x + y - 3z &= 13 \\ 26x + 20y - 6z &= 3 \\ 6y - 18z + 1 &= 0. \end{aligned}$$

$$\text{(b)} \qquad \begin{aligned} 2x + 6y + 11 &= 0 \\ x + 2y + z &= 3 \\ 3x + 9y - z &= 4. \end{aligned}$$

4. Discuss for what values of a and b in the equations

$$\begin{aligned} x + 3y + 2z &= 5 \\ 3x + 2y + az &= b \\ x + y + z &= 8 \end{aligned}$$

   have a unique solution, infinitely many solutions and no solution.

5. Solve the homogeneous system

$$\begin{aligned} x + y - 2z + w &= 0 \\ x + 2y - z - w &= 0 \\ 2x + 3y - z + 2w &= 0 \\ 3x + y + 4z + 3w &= 0 \end{aligned}$$

6. Solve the following system of linear equations using Cramers rule.

$$\begin{aligned} &(a)\ x + y + z = 7, 2x + 3y + z = 16, 3x + 4y + z = 22 \\ &(b)\ x + y + z = -1, 3x + 2y + z = -4, 4x + 3y + z = -6 \\ &(c)\ 3x + y + 2z = 6, x + 2y + 4z = 7, 9x + 3y + 2z = 14 \end{aligned}$$

2.6 Cayley–Hamilton Theorem

In this section, the Cayley–Hamilton theorem is given along with proof. Further, application of the theorem in finding the powers of a matrix is presented.

Theorem 2.6.1 *Let $\boldsymbol{A}_{n\times n}$ be a square matrix and $p(\lambda) = det[\lambda \boldsymbol{I} - \boldsymbol{A}] = det[\mathbf{A} - \lambda \mathbf{I}]$ be the characteristic polynomial of $\boldsymbol{A}$. Then $p(\boldsymbol{A}) = 0$.*
(or) Every square matrix satisfies its own characteristic equation.

Proof. Let the characteristic polynomial of $\mathbf{A}$ be

$$p(\lambda) = \lambda^n + c_{n-1}\lambda^{n-1} + c_{n-2}\lambda^{n-2} + ... + c_2\lambda^2 + c_1\lambda + c_0.$$

Set $\mathbf{E} = \lambda\mathbf{I} - \mathbf{A}$. The maximum degree of the element of $adj(\mathbf{E})$ is $n-1$. Hence we can find unique matrices $\mathbf{E}_i$ such that polynomial of

$$(adj(\mathbf{E})) = \mathbf{E}_0 + \mathbf{E}_1\lambda + \mathbf{E}_2\lambda^2 + ... + \mathbf{E}_{n-1}\lambda^{n-1}.$$

Since $\left|\mathbf{E}\right| \mathbf{I} = (adj(\mathbf{E}))\mathbf{E}$ and $p(\lambda) = \left|\mathbf{E}\right|$,

$$(adj(\mathbf{E}))\mathbf{E} = p(\lambda)\mathbf{I} = c_0\mathbf{I} + c_1\mathbf{I}\lambda + c_2\mathbf{I}\lambda^2 + ... + c_{n-1}\mathbf{I}\lambda^{n-1} + \mathbf{I}\lambda^n \tag{2.6}$$

On the other hand

$$\begin{aligned} (adj\mathbf{E})(\mathbf{E}) &= adj\mathbf{E}(\lambda\mathbf{I} - \mathbf{A}) = [\mathbf{E}_0 + \mathbf{E}_1\lambda + ... + \mathbf{E}_{n-1}\lambda^{n-1}][\lambda\mathbf{I} - \mathbf{A}] \\ &= \mathbf{E}_0\lambda + \mathbf{E}_1\lambda^2 + ... + \mathbf{E}_{n-1}\lambda^n - \mathbf{E}_0\mathbf{A} - \mathbf{E}_1 A\lambda - \mathbf{E}_2\mathbf{A}\lambda^2 \\ &\quad - \mathbf{E}_3\mathbf{A}\lambda^3 + ... + \mathbf{E}_{n-1}\mathbf{A}\lambda^{n-1} \\ &= -\mathbf{E}_0\mathbf{A} + [\mathbf{E}_0 - \mathbf{E}_1\mathbf{A}]\lambda + (\mathbf{E}_1 - \mathbf{E}_2\mathbf{A})\lambda^2 + ... + [\mathbf{E}_{n-2} \\ &\quad - \mathbf{E}_{n-1}\mathbf{A}]\lambda^{n-1} + \mathbf{E}_n\lambda^n \end{aligned} \tag{2.7}$$

Since the representations are unique, the equations (2.6) and (2.7) give

$$
\begin{aligned}
-\mathbf{E}_0\mathbf{A} = c_0\mathbf{I} &\implies -\mathbf{E}_0\mathbf{A} = c_0\mathbf{I}\\
\mathbf{E}_0 - \mathbf{E}_1\mathbf{A} = c_1\mathbf{I} &\implies \mathbf{E}_0\mathbf{A} - \mathbf{E}_1\mathbf{A}^2 = c_1\mathbf{A}\\
\mathbf{E}_1 - \mathbf{E}_2\mathbf{A} = c_2\mathbf{I} &\implies \mathbf{E}_1\mathbf{A}^2 - \mathbf{E}_2\mathbf{A}^3 = c_2\mathbf{A}^2\\
\mathbf{E}_2 - \mathbf{E}_3\mathbf{A} = c_3\mathbf{I} &\implies \mathbf{E}_2\mathbf{A}^3 - \mathbf{E}_3\mathbf{A}^4 = c_3\mathbf{A}^3\\
\cdots \qquad &\cdots \qquad \cdots\\
\mathbf{E}_{n-2} - \mathbf{E}_{n-1}\mathbf{A} &\implies \mathbf{E}_{n-2}\mathbf{A}^{n-1} - \mathbf{E}_{n-1}\mathbf{A}^n = c_{n-1}\mathbf{A}^{n-1}\\
\mathbf{E}_{n-1} = \mathbf{I} &\implies \mathbf{E}_{n-1}\mathbf{A}^n = \mathbf{A}^n
\end{aligned}
$$

Now adding the equations on both sides, it follows that

$$\mathbf{0} = p(\mathbf{A}) = c_0\mathbf{I} + c_1\mathbf{A} + \cdots + c_{n-1}\mathbf{A}^{n-1} + \mathbf{A}^n.$$

The proof is complete.

Example 2.6.1 *Verify Cayley–Hamilton theorem for* $\mathbf{A} = \begin{bmatrix} 1 & 3 & 0 & 0\\ 2 & 0 & 0 & 0\\ 0 & 0 & 2 & 3\\ 0 & 0 & 3 & 2 \end{bmatrix}$ *and find its inverse.*

Solution. *Given matrix* $\mathbf{A} = \begin{bmatrix} 1 & 3 & 0 & 0\\ 2 & 0 & 0 & 0\\ 0 & 0 & 2 & 3\\ 0 & 0 & 3 & 2 \end{bmatrix}$

The characteristic matrix is $\boldsymbol{A} - \lambda\boldsymbol{I} = \begin{bmatrix} 1-\lambda & 3 & 0 & 0\\ 2 & -\lambda & 0 & 0\\ 0 & 0 & 2-\lambda & 3\\ 0 & 0 & 3 & 2-\lambda \end{bmatrix}$

Characteristic equation is $|\boldsymbol{A} - \lambda\boldsymbol{I}| = 0 = \begin{vmatrix} 1-\lambda & 3 & 0 & 0\\ 2 & -\lambda & 0 & 0\\ 0 & 0 & 2-\lambda & 3\\ 0 & 0 & 3 & 2-\lambda \end{vmatrix} = 0$

$$\implies (1-\lambda)\begin{vmatrix} -\lambda & 0 & 0\\ 0 & 2-\lambda & 3\\ 0 & 3 & 2-\lambda \end{vmatrix} - 3\begin{vmatrix} 2 & 0 & 0\\ 0 & 2-\lambda & 3\\ 0 & 3 & 2-\lambda \end{vmatrix} = 0$$

$$\implies (1-\lambda)[(-\lambda)[(2-\lambda)^2 - 9]] - 3[2[(2-\lambda)^2 - 9]] = 0$$

$$\lambda^4 - 5\lambda^3 - 7\lambda^2 + 29\lambda + 30 = 0.$$

By Cayley–Hamilton theorem, $\mathbf{A}$ satisfies its characteristic equation.

Therefore $\mathbf{A^4} - \mathbf{5A^3} - 7\mathbf{A^2} + 29\mathbf{A} + 30\mathbf{I} = \mathbf{0}$. To verify this consider

$$\mathbf{A^2} = \mathbf{AA} = \begin{bmatrix} 1 & 3 & 0 & 0 \\ 2 & 0 & 0 & 0 \\ 0 & 0 & 2 & 3 \\ 0 & 0 & 3 & 2 \end{bmatrix} \begin{bmatrix} 1 & 3 & 0 & 0 \\ 2 & 0 & 0 & 0 \\ 0 & 0 & 2 & 3 \\ 0 & 0 & 3 & 2 \end{bmatrix} = \begin{bmatrix} 7 & 3 & 0 & 0 \\ 2 & 6 & 0 & 0 \\ 0 & 0 & 13 & 12 \\ 0 & 0 & 12 & 13 \end{bmatrix},$$

$$\mathbf{A^3} = \mathbf{A^2A} = \begin{bmatrix} 7 & 3 & 0 & 0 \\ 2 & 6 & 0 & 0 \\ 0 & 0 & 13 & 12 \\ 0 & 0 & 12 & 13 \end{bmatrix} \begin{bmatrix} 1 & 3 & 0 & 0 \\ 2 & 0 & 0 & 0 \\ 0 & 0 & 2 & 3 \\ 0 & 0 & 3 & 2 \end{bmatrix} = \begin{bmatrix} 13 & 21 & 0 & 0 \\ 14 & 6 & 0 & 0 \\ 0 & 0 & 62 & 63 \\ 0 & 0 & 63 & 62 \end{bmatrix},$$

$$\mathbf{A^4} = \begin{bmatrix} 7 & 3 & 0 & 0 \\ 2 & 6 & 0 & 0 \\ 0 & 0 & 13 & 12 \\ 0 & 0 & 12 & 13 \end{bmatrix} \begin{bmatrix} 7 & 3 & 0 & 0 \\ 2 & 6 & 0 & 0 \\ 0 & 0 & 13 & 12 \\ 0 & 0 & 12 & 13 \end{bmatrix} = \begin{bmatrix} 55 & 39 & 0 & 0 \\ 26 & 42 & 0 & 0 \\ 0 & 0 & 313 & 312 \\ 0 & 0 & 312 & 313 \end{bmatrix}$$

Now, $\mathbf{A^4} - 5\mathbf{A^3} - 7\mathbf{A^2} + 29\mathbf{A} + 30\mathbf{I}$

$$= \begin{bmatrix} 55 & 39 & 0 & 0 \\ 26 & 42 & 0 & 0 \\ 0 & 0 & 313 & 312 \\ 0 & 0 & 312 & 313 \end{bmatrix} - 5 \begin{bmatrix} 13 & 21 & 0 & 0 \\ 14 & 6 & 0 & 0 \\ 0 & 0 & 62 & 63 \\ 0 & 0 & 63 & 62 \end{bmatrix} - 7 \begin{bmatrix} 7 & 3 & 0 & 0 \\ 2 & 6 & 0 & 0 \\ 0 & 0 & 13 & 12 \\ 0 & 0 & 12 & 13 \end{bmatrix}$$

$$+29 \begin{bmatrix} 1 & 3 & 0 & 0 \\ 2 & 0 & 0 & 0 \\ 0 & 0 & 2 & 3 \\ 0 & 0 & 3 & 2 \end{bmatrix} + 30 \begin{bmatrix} 1 & 0 & 0 & 0 \\ 0 & 1 & 0 & 0 \\ 0 & 0 & 1 & 0 \\ 0 & 0 & 0 & 1 \end{bmatrix} = \begin{bmatrix} 0 & 0 & 0 & 0 \\ 0 & 0 & 0 & 0 \\ 0 & 0 & 0 & 0 \\ 0 & 0 & 0 & 0 \end{bmatrix} = \mathbf{0}.$$

Hence Cayley–Hamilton theorem is verified. To find the inverse of $\mathbf{A}$, consider
$\mathbf{A^4} - 5\mathbf{A^3} - 7\mathbf{A^2} + 29\mathbf{A} + 30\mathbf{I} = \mathbf{0}$.
Now,

$$\mathbf{A^{-1}} = -\frac{1}{30}\{\mathbf{A^3} - 5\mathbf{A^2} - 7\mathbf{A} + 29\mathbf{I}\}$$

$$= \begin{bmatrix} 0 & \frac{1}{2} & 0 & 0 \\ \frac{1}{3} & -\frac{1}{6} & 0 & 0 \\ 0 & 0 & -\frac{2}{5} & \frac{3}{5} \\ 0 & 0 & \frac{3}{5} & -\frac{2}{5} \end{bmatrix}.$$

EXERCISE 2.6

Verify Cayley–Hamilton theorem for each of the following matrices and hence find its inverse

(i) $\mathbf{A} = \begin{bmatrix} 1 & 4 \\ 2 & 3 \end{bmatrix}$ (iii) $\mathbf{A} = \begin{bmatrix} 2 & 1 & 1 \\ 0 & 1 & 0 \\ 1 & 1 & 2 \end{bmatrix}$

(ii) $\mathbf{A} = \begin{bmatrix} 7 & 2 & -2 \\ -6 & -1 & 2 \\ 6 & 2 & -1 \end{bmatrix}$ (iv) $\mathbf{A} = \begin{bmatrix} 1 & 2 & -1 \\ 0 & 1 & -1 \\ 3 & -1 & 1 \end{bmatrix}$.

2.7 Eigen-values and Eigen-vectors

This section deals with only square matrices, say of, order n. Let $\mathbf{A}_{n\times n}$ be any square matrix and $\mathbf{x} \in \mathbb{R}^n$ be an $n \times 1$ vector. Then $\mathbf{Ax} = \mathbf{y} \in \mathbb{R}^n$ is a vector with a direction and magnitude different from $\mathbf{x}$. Taking cue from functions like $f(\mathbf{x}) = x^2, x^3, x^4$ and observing that 0 and 1 remain as fixed points the following question arises. Are there any vectors $\mathbf{x} \in \mathbb{R}^n$ such that $\mathbf{Ax}$ remains in the direction of $\mathbf{x}$?

Further, in modeling of many physical problems matrices arise naturally. These matrices normally depend upon an unknown parameter which is very important. These parameters must be chosen so as to make the matrix singular. Such type of situations can be found in systems exhibiting oscillatory behavior. Some of the applications include vibration of a string, stretching of an elastic membrane, etc.

Another important area is understanding the behavior of the solutions of linear homogeneous differential equations. All the a forementioned topics are studied using the concepts introduced in this section.

Definition 2.7.1 *An eigen-value and eigen-vector of a square matrix* $\boldsymbol{A}$ *are a scalar value* λ *and a nonzero vector* $\boldsymbol{x}$ *satisfying the relation*

$$\mathbf{Ax} = \lambda\mathbf{x}, \qquad \mathbf{x} \neq \mathbf{0}. \tag{2.8}$$

Note. Characteristic values, latent roots, proper values and spectral values are some of the synonyms to eigen-values. The eigen-vectors are also called as characteristic vectors.

From the relation (2.8) it is clear that finding the eigen-values and eigen-vectors means solving the homogeneous linear system

$$(\mathbf{A} - \lambda\mathbf{I})\mathbf{x} = \mathbf{0}, \quad \mathbf{x} \neq \mathbf{0}.$$

Since $\mathbf{x} \neq \mathbf{0}$, this yields that det $(\mathbf{A} - \lambda\mathbf{I}) = 0$, from which the eigen-values λ can be determined.

The above discussion yields the following theorem.

Theorem 2.7.1 *λ is an eigen-value of* $\mathbf{A}$ *if and only if det* $(\boldsymbol{A} - \lambda\boldsymbol{I})$*=0.*

Proof. By definition, λ is an eigen-value of $\mathbf{A}$ if and only if

$$\mathbf{A}\mathbf{x} = \lambda\mathbf{x}, \quad \mathbf{x} \neq \mathbf{0}$$

$$\Longleftrightarrow (\mathbf{A} - \lambda\mathbf{I})\mathbf{x} = \mathbf{0}, \quad \mathbf{x} \neq \mathbf{0}$$

$$\Longleftrightarrow det(\mathbf{A} - \lambda\mathbf{I}) = 0$$

$$\Longleftrightarrow \lambda \text{ is a root of the equation } |\mathbf{A} - \lambda\mathbf{I}| = 0.$$

Note. $|\mathbf{A}-\lambda\mathbf{I}|$ is a polynomial of nth degree. Hence the equation $|\mathbf{A}-\lambda\mathbf{I}| = 0$ yields 'n' eigen-values including multiplicities.

Once the eigen-values are found, the eigen-vector corresponding to the eigen-value can be found by solving the homogeneous linear system

$$(\mathbf{A} - \lambda\mathbf{I})\mathbf{x} = \mathbf{0}, \quad \mathbf{x} \neq \mathbf{0}.$$

The eigen-values together with their corresponding eigen-vectors form the eigensystem of $\mathbf{A}$.

Example 2.7.1 *Find the eigen-values and the associated eigen-vectors for the following matrices. (i)* $\boldsymbol{A} = \begin{bmatrix} 5 & 4 \\ 1 & 2 \end{bmatrix}$ *and (ii)* $\boldsymbol{A} = \begin{bmatrix} 1 & 3 & 2 \\ 0 & 2 & 6 \\ 0 & 0 & 3 \end{bmatrix}$.

Solution. *It is observed that choosing* $\boldsymbol{v} = \begin{bmatrix} 1 \\ -1 \end{bmatrix}$ *yields* $\begin{bmatrix} 5 & 4 \\ 1 & 2 \end{bmatrix}\begin{bmatrix} 1 \\ -1 \end{bmatrix} = \begin{bmatrix} 1 \\ -1 \end{bmatrix}$.

Hence $\lambda = 1$ *is an eigen-value and* $\begin{bmatrix} 1 \\ -1 \end{bmatrix}$ *is an eigen-vector corresponding to the eigen-value* $\lambda = 1$.

(ii) Similarly, $\boldsymbol{x} = \begin{bmatrix} 1 \\ 0 \\ 0 \end{bmatrix}$ *is an eigen-vector corresponding to the eigen-value* $\lambda = 1$,

since $\begin{bmatrix} 1 & 3 & 2 \\ 0 & 2 & 6 \\ 0 & 0 & 3 \end{bmatrix}\begin{bmatrix} 1 \\ 0 \\ 0 \end{bmatrix} = \begin{bmatrix} 1 \\ 0 \\ 0 \end{bmatrix}$.

A systematic approach is required to find the eigen-values and their corresponding eigen-vectors. Taking the idea from Theorem 2.7.1 steps for obtaining the eigen-value and the corresponding eigen-vector are given.

A procedure for finding the eigen-values and related eigen-vectors of a given square matrix $\mathbf{A}$ of order 'n' is as follows.

Algorithm

Step 1. Given a square matrix $\mathbf{A}$ of order n, consider identity matrix, $\mathbf{I}_n$ and write down the matrix $\mathbf{A} - \lambda\mathbf{I}$, λ unknown;

Step 2. Write down the equation given by $\det(\mathbf{A} - \lambda\mathbf{I}) = 0$, which is called as the characteristic equation and the corresponding polynomial is called as the characteristic polynomial;

Step 3. Find the roots of the characteristic equation, which are the eigen values of $\mathbf{A}$. There will be 'n' roots including multiplicities if any, when order of A is n;

Step 4. Corresponding to each eigen-value λ of $\mathbf{A}$, consider the homogeneous system $(\mathbf{A} - \lambda\mathbf{I})\mathbf{x} = 0$ and find the nontrivial solution $\mathbf{x}$. This $\mathbf{x}$ is the eigen-vector corresponding to the eigen-value λ.

Note. If the eigen-values λ are repeated, sometimes distinct eigen-vectors cannot be found. This possibility results in the following definition.

Definition 2.7.2 *The algebraic multiplicity of an eigen-value λ is the number of times λ is a repeated root. The geometric multiplicity of λ is the number of linearly independent eigen-vectors associated with λ.*

Clearly, geometric multiplicity of $\lambda \leq$ algebraic multiplicity of λ.

Example 2.7.2 *Find the eigen-values and the corresponding eigen-vectors of the matrix* $\begin{bmatrix} 1 & 2 & 0 \\ 2 & 1 & 0 \\ 0 & 0 & 2 \end{bmatrix}$.

Solution *Set* $A = \begin{bmatrix} 1 & 2 & 0 \\ 2 & 1 & 0 \\ 0 & 0 & 2 \end{bmatrix}$.

The characteristic equation is given by

$$|\mathbf{A} - \lambda\mathbf{I}| = 0 \text{ i.e. } \begin{vmatrix} 1-\lambda & 2 & 0 \\ 2 & 1-\lambda & 0 \\ 0 & 0 & 2-\lambda \end{vmatrix} = 0.$$

Finding the determinant along the 3rd column yields,

$$(2-\lambda)[(1-\lambda)^2 - 4] = 0.$$

$$(2-\lambda)(\lambda+1)(\lambda-3) = 0.$$

Thus the eigen-values are $\lambda = 2$, $\lambda = -1$, and $\lambda = 3$ with both geometric and algebraic multiplicity of each eigen-value being 1.
To find the eigen-vector corresponding to eigen-value $\lambda = -1$.

Set $\lambda = -1$ in the matrix $(\mathbf{A} - \lambda\mathbf{I})\mathbf{x} = \mathbf{0}$ which gives $\begin{bmatrix} 2 & 2 & 0 \\ 2 & 2 & 0 \\ 0 & 0 & 3 \end{bmatrix} \begin{bmatrix} x_1 \\ x_2 \\ x_3 \end{bmatrix} = \begin{bmatrix} 0 \\ 0 \\ 0 \end{bmatrix}$.

Hence $x_1 + x_2 = 0,\ 3x_3 = 0$ *gives* $x_2 = -x_1, \quad x_3 = 0$. *Choose* $x_1 = k$, *then corresponding to the eigen-value* $\lambda = -1$, *the eigen-vector is* $\begin{bmatrix} k \\ -k \\ 0 \end{bmatrix}$ *or* $\begin{bmatrix} 1 \\ -1 \\ 0 \end{bmatrix}$, *choosing* $k = 1$.

Now consider the eigen-value $\lambda = 2$, *then the matrix* $(\mathbf{A} - 2\boldsymbol{I})\boldsymbol{x} = \boldsymbol{0}$ *yields,* $-x_1 + 2x_2 = 0,\ 2x_1 - x_2 = 0 \iff x_1 = 2x_2, \quad x_2 = 2x_1 \iff x_1 = x_2 = 0$ *and* $x_3 = k$. *Hence the eigen-vector corresponding to the eigen-value* $\lambda = 2$, *choosing k=1 is* $\begin{bmatrix} 0 \\ 0 \\ 1 \end{bmatrix}$.

Corresponding to the eigen-value $\lambda = 3$, *the matrix* $(\mathbf{A} - 3\mathbf{I})\mathbf{x} = \mathbf{0}$ *gives,*

$$-2x_1 + 2x_2 = 0,\ 2x_1 - 2x_2 = 0,\ -x_3 = 0 \iff x_1 = x_2,$$

Hence the eigen-vector corresponding to the eigen-value $\lambda = 3$ *is* $\begin{bmatrix} 1 \\ 1 \\ 0 \end{bmatrix}$.

Thus the eigen-values are $\lambda = -1, 2$ *and* 3, *and the corresponding eigen-vectors are* $\begin{bmatrix} 1 \\ -1 \\ 0 \end{bmatrix}$, $\begin{bmatrix} 0 \\ 0 \\ 1 \end{bmatrix}$ *and* $\begin{bmatrix} 1 \\ 1 \\ 0 \end{bmatrix}$, *respectively.*

Example 2.7.3 *Find the eigen-values and the corresponding eigen-vectors of the matrix* $\begin{bmatrix} 1 & 1 & 0 \\ 1 & 1 & 0 \\ 0 & 0 & 2 \end{bmatrix}$.

Solution. *The characteristic equation is given by*

$$|\mathbf{A} - \lambda\mathbf{I}| = 0 \quad i.e., \quad \begin{vmatrix} 1-\lambda & 1 & 0 \\ 1 & 1-\lambda & 0 \\ 0 & 0 & 2-\lambda \end{vmatrix} = 0.$$

$\implies \lambda(2-\lambda)^2 = 0.$

Hence the eigen-values are $\lambda = 2,\ 2,$ *and* 0.

To find the eigen-vector corresponding to eigen-value $\lambda = 2$.

Set $\lambda = 2$ *in the matrix* $(\mathbf{A} - \lambda\mathbf{I})\mathbf{x} = \mathbf{0}$ *which gives*

$$\begin{bmatrix} -1 & 1 & 0 \\ 1 & -1 & 0 \\ 0 & 0 & 0 \end{bmatrix} \begin{bmatrix} x_1 \\ x_2 \\ x_3 \end{bmatrix} = \begin{bmatrix} 0 \\ 0 \\ 0 \end{bmatrix}.$$

$x_1 - x_2 = 0$, *let* $x_1 = x_2 = k$ *and* $x_3 = l$,

then $\begin{bmatrix} x_1 \\ x_2 \\ x_3 \end{bmatrix} = \begin{bmatrix} k \\ k \\ l \end{bmatrix} = k \begin{bmatrix} 1 \\ 1 \\ 0 \end{bmatrix} + l \begin{bmatrix} 0 \\ 0 \\ 1 \end{bmatrix}$.

Hence the eigen-vectors corresponding to $\lambda = 2$ *are* $\begin{bmatrix}1\\1\\0\end{bmatrix}, \begin{bmatrix}0\\0\\1\end{bmatrix}$.

Now the eigen-vector corresponding to $\lambda = 0$ *is given by,*

$$\begin{bmatrix}1 & 1 & 0\\1 & 1 & 0\\0 & 0 & 2\end{bmatrix}\begin{bmatrix}x_1\\x_2\\x_3\end{bmatrix} = \begin{bmatrix}0\\0\\0\end{bmatrix}.$$

$x_1 + x_2 = 0, x_3 = 0$

$\Longrightarrow x_1 = -x_2,\ x_3 = 0$. *Set* $x_1 = k$, *then* $x_2 = -k$.

Choosing $k = 1$, *the eigen-vector corresponding to the eigen-value* $\lambda = 0$ *is* $\begin{bmatrix}1\\-1\\0\end{bmatrix}$. *Thus* $\begin{bmatrix}1\\1\\0\end{bmatrix}, \begin{bmatrix}0\\0\\1\end{bmatrix}$ *and* $\begin{bmatrix}1\\-1\\0\end{bmatrix}$ *are the eigen-vectors of the considered system.*

Properties of the Characteristic Polynomial

The eigen-values of a square matrix $\mathbf{A}$ of order n depend on the characteristic polynomial $|\mathbf{A}-\lambda\mathbf{I}|$ in λ. The structure of the characteristic polynomial $f(\lambda) = |\mathbf{A} - \lambda\mathbf{I}|$ is as follows.

(i) $f(\lambda)$ is a polynomial of degree n, with the nth degree term given by $(-\lambda)^n$. Thus the coefficient of $(\lambda)^n$ is $(-1)^n$;

(ii) The coefficient of the $(n-1)$th degree term, $(\lambda)^{n-1}$ is the trace of $\mathbf{A}$, tr($\mathbf{A}$)=sum of diagonal elements of $\mathbf{A}$;

(iii) The constant term in the polynomial $f(\lambda)$ is given by $|\mathbf{A}|$.

The next question that arises is that if the eigen-values of a given square matrix $\mathbf{A}$ of order 'n' are known can one determine the eigen-values of its related matrices like $\mathbf{A}^T$, $\mathbf{A}^{-1}$, etc.? The answer is in affirmative.

I Properties of Eigen-values

Let $\mathbf{A}$ be a square matrix of order n and α be and an eigen-value of $\mathbf{A}$ and $\mathbf{x}$ be its corresponding eigen-vector. Then, the following statements hold.

(1) α is also an eigen-value of $\mathbf{A}^T$;

Proof. Consider $|\mathbf{A}^T - \alpha\mathbf{I}| = |(\mathbf{A} - \alpha\mathbf{I})^T| = |\mathbf{A} - \alpha\mathbf{I}| = 0$ since α is an eigen-value of $\mathbf{A}$. Thus $|\mathbf{A}^T - \alpha\mathbf{I}| = 0$ and this implies that α satisfies the characteristic equation of $\mathbf{A}^T$. Hence α is an eigen-value of $\mathbf{A}^T$.

(2) $k\alpha,\ k \pm \alpha$ are the eigen-values of $k\mathbf{A}, k \pm \mathbf{A}$, respectively.

Proof. Since α is an eigen-value of $\mathbf{A}$, $\mathbf{A}\mathbf{x} = \alpha\mathbf{x},\ \mathbf{x} \neq \mathbf{0}$. Consider $(k\mathbf{A})\mathbf{x} =$

$k(\mathbf{Ax}) = k(\alpha\mathbf{x}) = (k\alpha)\mathbf{x},\ \mathbf{x} \neq 0$ and $(k \pm \mathbf{A})\mathbf{x} = k\mathbf{x} \pm \mathbf{Ax} = k\mathbf{x} \pm \alpha\mathbf{x} = (k \pm \alpha)\mathbf{x},\ \mathbf{x} \neq 0.$

(3) α^n is an eigen-value of $\mathbf{A}^n$, $n = 1, 2, ...;$

Proof. (3) proof is simple and follows like the earlier proof.
(4) α^{-1} is an eigen-value of $\mathbf{A}^{-1}$.

Proof. (4) $\mathbf{Ax} = \alpha\mathbf{x},\ \mathbf{x} \neq 0$ Now multiply throughout with $\frac{\mathbf{A}^{-1}}{\alpha} \iff \frac{1}{\alpha}\mathbf{x} = \mathbf{A}^{-1}\mathbf{x},\ \mathbf{x} \neq 0,\ \alpha \neq 0$ from which the result can be concluded.

Note.

(i) The eigen-vector corresponding to α^{-1} is same as the eigen-vector corresponding to α.

(ii) $\mathbf{A}^{-1}$ exists implies that zero is not an eigen-value of $\mathbf{A}$;

(5) $\frac{|\mathbf{A}|}{\alpha}$ is an eigen-value of adj$\mathbf{A}$.

Proof. From (4), $\mathbf{A}^{-1}\mathbf{x} = \frac{1}{\alpha}\mathbf{x},\ \mathbf{x} \neq 0 \iff \frac{adj\mathbf{A}}{|\mathbf{A}|}\mathbf{x} = \frac{1}{\alpha}\mathbf{x},\ \mathbf{x} \neq 0 \iff adj\mathbf{Ax} = \frac{|\mathbf{A}|}{\alpha}\mathbf{x},\ \mathbf{x} \neq 0;$

(6) The eigen-value of $\mathbf{B} = c_0\mathbf{A}^n + c_1\mathbf{A}^{n-1} + ... + c_{n-1}\mathbf{A} + c_n\mathbf{I}$ is $c_0\alpha^n + c_1\alpha^{n-1} + ... + c_{n-1}\alpha + c_n$.

Proof. Follows from the application of the properties (2) and (3);

(7) if α is a complex number and $\mathbf{x}$ is its corresponding complex eigen-vector, then $\bar{\alpha}$ is also an eigen-value and $\bar{\mathbf{x}}$ is its corresponding eigen-vector. ($\mathbf{A}$ is a real matrix.)

Proof. Given $\mathbf{Ax} = \alpha\mathbf{x},\ \mathbf{x} \neq \mathbf{0},$
Taking conjugates yields, $\bar{\mathbf{A}}\bar{\mathbf{x}} = \bar{\alpha}\bar{\mathbf{x}},\ \bar{\mathbf{x}} \neq \mathbf{0}, \implies \mathbf{A}\bar{\mathbf{x}} = \bar{\alpha}\bar{\mathbf{x}}.$

II Properties of Eigen-vectors

(1) Let $\alpha_1 \neq \alpha_2$ be any two eigen-values of $\mathbf{A}$ then their corresponding eigen-vectors $\mathbf{x}_1$ and $\mathbf{x}_2$ are distinct.

Proof. If possible let $\mathbf{x}$ be an eigen-vector for both α_1 and α_2. Then, $\mathbf{Ax} = \alpha_1\mathbf{x}$ and $\mathbf{Ax} = \alpha_2\mathbf{x}\ \ \mathbf{x} \neq \mathbf{0}$, this implies $(\alpha_1 - \alpha_2)\mathbf{x} = \mathbf{0},\ \ \mathbf{x} \neq \mathbf{0} \iff \alpha_1 - \alpha_2 = 0 \iff \alpha_1 = \alpha_2$. Thus $\mathbf{x}_1$ and $\mathbf{x}_2$ are distinct.

(2) Let $\alpha_1 \neq \alpha_2$ be any two eigen-values of $\mathbf{A}$ then their corresponding eigen-vectors, $\mathbf{x}_1$ and $\mathbf{x}_2$ are linearly independent.

Proof. To show $\mathbf{x}_1$ and $\mathbf{x}_2$ are linearly independent, consider the linear combination $c_1\mathbf{x}_1 + c_2\mathbf{x}_2 = \mathbf{0}$.

$$\text{Then} \quad c_1\alpha_1\mathbf{x}_1 + c_2\alpha_1\mathbf{x}_2 = \mathbf{0}. \tag{2.9}$$

Consider $\mathbf{A}(c_1\mathbf{x}_1 + c_2\mathbf{x}_2) = c_1\mathbf{A}\mathbf{x}_1 + c_2\mathbf{A}\mathbf{x}_2$

$$= c_1\alpha_1\mathbf{x}_1 + c_2\alpha_2\mathbf{x}_2 = \mathbf{0}. \tag{2.10}$$

Subtracting the relation (2.10) from (2.9) gives $c_2(\alpha_1 - \alpha_2)\mathbf{x}_2 = \mathbf{0}$ implies $c_2 = 0$. Substituting the value of c_2 in (2.9) provides $c_1 = 0$. Hence $\mathbf{x}_1$ and $\mathbf{x}_2$ are linearly independent.

III Eigen-values and eigen-vectors of various matrices

(1) Eigen-values of an upper triangular matrix are its diagonal elements.

Proof. Consider the characteristic equation $|\mathbf{A} - \alpha\mathbf{I}| = 0$. The determinant gives $(a_{11} - \alpha)(a_{22} - \alpha)...(a_{nn} - \alpha) = 0$ and the result follows.

(2) Let $\mathbf{A}$ be a real symmetric matrix, then
(a) The eigen-values of $\mathbf{A}$ are real.
(b) The eigen-vectors of two distinct eigen-values of $\mathbf{A}$ are orthogonal.

Proof. If possible let $\alpha \in \mathbb{C}$ be an eigen-value and $\mathbf{x}$ be its corresponding eigen-vector of a real symmetric matrix $\mathbf{A}$. Then

$$\mathbf{A}\mathbf{x} = \alpha\mathbf{x} \tag{2.11}$$

and $\bar{\mathbf{A}}\bar{\mathbf{x}} = \bar{\alpha}\bar{\mathbf{x}}$ which implies

$$\mathbf{A}\bar{\mathbf{x}} = \bar{\alpha}\bar{\mathbf{x}} \tag{2.12}$$

Taking the transpose of (2.12) yields

$$\bar{\mathbf{x}}^{\mathbf{T}}\mathbf{A} = \bar{\mathbf{x}}^{\mathbf{T}}\bar{\alpha}. \tag{2.13}$$

Here, pre-multiplying (2.11) with $\bar{\mathbf{x}}^{\mathbf{T}}$ and post multiplying (2.13) with $\mathbf{x}$ yields $\bar{\mathbf{x}}^{\mathbf{T}}\mathbf{A}\mathbf{x} = \alpha\bar{\mathbf{x}}^{\mathbf{T}}\mathbf{x}$, $\bar{\mathbf{x}}^{\mathbf{T}}\mathbf{A}\mathbf{x} = \bar{\mathbf{x}}^{\mathbf{T}}\bar{\alpha}\mathbf{x}$.

On subtraction, the above two relations give $(\alpha - \bar{\alpha})\bar{\mathbf{x}}^{\mathbf{T}}\mathbf{x} = 0$ implies $\alpha = \bar{\alpha}$ as $\mathbf{x} \neq 0$. Hence α is real.
(b) Let α_1 and α_2 be two distinct eigen-values of $\mathbf{A}$ and $\mathbf{x_1}$ and $\mathbf{x_2}$ be their corresponding eigen-vectors. Then

$$\mathbf{A}\mathbf{x_1} = \alpha_1\mathbf{x_1} \tag{2.14}$$

$$\mathbf{A}\mathbf{x_2} = \alpha_2\mathbf{x_2} \tag{2.15}$$

Taking transpose for the equation (2.14) gives

$$\mathbf{x_1}^T\mathbf{A} = \mathbf{x_1}^T\alpha_1 \tag{2.16}$$

pre-multiplying (2.15) with ${\mathbf{x_1}}^T$ and post multiplying (2.16) with $\mathbf{x_2}$ and subtracting yields

$${\mathbf{x_1}}^T(\alpha_1 - \alpha_2)\mathbf{x_2} = 0.$$

Thus ${\mathbf{x_1}}^T\mathbf{x_2} = 0$ since $\alpha_1 \neq \alpha_2$. Hence $\mathbf{x_1}$ and $\mathbf{x_2}$ are orthogonal.

(3)(a) A Hermitian matrix has only real eigen-values.
(b) The eigen-vectors of distinct eigen-values of a Hermitian matrix are orthogonal.

Proof. Follows similar pattern as in (2) and is left as an exercise.

(4) Let $\mathbf{A}$ be a skew-Hermitian matrix. Its eigen-values are purely imaginary or zero.

Proof. Follows similar pattern as in (2) and is left as an exercise.

(5) If α is an eigen-value of a Unitary matrix, then $|\alpha| = 1$.

Proof. By definition

$$\mathbf{Ax} = \alpha\mathbf{x}, \quad \mathbf{x} \neq 0.$$

Taking conjugate transpose leads to

$$\begin{aligned}
\mathbf{\bar{x}^T\bar{A}^T} &= \bar{\alpha}\mathbf{\bar{x}^T} \\
\iff \mathbf{\bar{x}^T\bar{A}^TAx} &= \bar{\alpha}\alpha\mathbf{\bar{x}^Tx} \\
\mathbf{\bar{x}^Tx} &= |\alpha|^2\mathbf{\bar{x}^Tx} \\
\implies |\alpha|^2 = 1 &\iff |\alpha| = 1,
\end{aligned}$$

completing the proof.

6) Both α and $\frac{1}{\alpha}$ are eigen-values of an orthogonal matrix.

Proof. $\mathbf{A}$ is an orthogonal matrix implies $\mathbf{A}^T = \mathbf{A}^{-1}$. The proof follows from $I(1)$ and $I(4)$.

7) Let $\mathbf{P}$ be any nonsingular matrix such that $\mathbf{A} = \mathbf{PBP}^{-1}$. Then $\mathbf{A}$ and $\mathbf{B}$ have the same eigen-values but their eigen-vectors are different.

Proof. Let $\mathbf{A} = \mathbf{PBP}^{-1}$, where $\mathbf{P}$ is a nonsingular matrix and let α be an eigen-value of $\mathbf{A}$. Then $\mathbf{Ax} = \alpha\mathbf{x}, \quad \mathbf{x} \neq \mathbf{0}$. Now suppose there is a vector $\mathbf{y}$ such that $\mathbf{x} = \mathbf{Py}$. Then

$$\mathbf{Ax} = \mathbf{APy}$$

and

$$\mathbf{By} = \mathbf{P}^{-1}\mathbf{APy} \iff \mathbf{P}^{-1}\mathbf{Ax} = \mathbf{P}^{-1}\alpha\mathbf{x} = \alpha\mathbf{P}^{-1}\mathbf{x} = \alpha\mathbf{y}.$$

Hence α is an eigen-value for $\mathbf{B}$ but its corresponding eigen-vector is $\mathbf{y}$.

Observation. A similarity transformation preserves the eigen-values. This leads to the eigen-value decomposition theorem–which transforms a given matrix to a diagonal form.

Definition 2.7.3 *Modal matrix is a matrix consisting of all the eigen-vectors of* **A** *which are linearly independent.*

Theorem 2.7.2 *Suppose $\boldsymbol{A}$, a square matrix of order n, has n linearly independent eigen-vectors and $\boldsymbol{P}$ is the matrix having these eigen-vectors as columns. Then $\boldsymbol{D} = \boldsymbol{P}^{-1}\boldsymbol{A}\boldsymbol{P}$ is a diagonal matrix.*

Proof. Let $\alpha_1,\ \alpha_2,\ \alpha_3,\ ...,\alpha_n$ be the eigen-values of $\mathbf{A}$ and $\mathbf{x_1},\ \mathbf{x_2},\ \mathbf{x_3},\ ...,\mathbf{x_n}$ be the corresponding linearly independent eigen-vectors. then $A\mathbf{x}_i = \alpha_i\mathbf{x}_i,\quad i = 1, 2, \ldots, n$. where x_i is the ith column of $\mathbf{P}$. Clearly, by multiplication of partition matrices,

$$\begin{aligned}\mathbf{AP} &= \mathbf{A}[\mathbf{x_1},\ ...,\mathbf{x_n}] \\ &= [\mathbf{Ax_1},\ \mathbf{Ax_2},\ ...,\mathbf{Ax_n}] \\ &= [\alpha_1\mathbf{x}_1,\ \alpha_2\mathbf{x}_2,\ ...,\alpha_n\mathbf{x}_n] \\ &= [\mathbf{x}_1,\ \mathbf{x}_2,\ ...,\mathbf{x}_n]Diag[\alpha_1,\alpha_2,\ ...,\alpha_n] \\ &= \mathbf{PD},\end{aligned}$$

where $\mathbf{D} = diag\ [\alpha_1, \alpha_2,\ ..., \alpha_n]$ is a diagonal matrix consisting of eigen-values of $\mathbf{A}$. Thus $\mathbf{AP}$=$\mathbf{PD}$ gives $\mathbf{A} = \mathbf{PDP}^{-1}$, concluding the proof.

Observation.

1. Diagonal matrix $\mathbf{D}$ exists when all the eigen-values are real and distinct.
2. The modal matrix consisting of the eigen-vectors is not unique.
3. The order of the eigen-values and the eigen-vectors must be maintained.
4. If the eigen-values are repeated then the modal matrix may not exist.
5. If the rank of the matrix is less than the order of the matrix then the modal matrix does not exist.

Definition 2.7.4 *A square matrix* **A** *of order n is said to be diagonalizable if there exists a modal matrix* **P** *such that $\mathbf{D} = \mathbf{P}^{-1}\mathbf{AP}^{-1}$, where* **D** *consists of the eigen-values of* **A** *and is a diagonal matrix.*

Example 2.7.4 *Reduce the matrix $\mathbf{A} = \begin{bmatrix} 5 & 2 \\ 2 & 2 \end{bmatrix}$ into diagonal form by using a modal matrix.*

Solution. The characteristic matrix is $\mathbf{A} - \lambda\mathbf{I} = \begin{bmatrix} 5-\lambda & 2 \\ 2 & 2-\lambda \end{bmatrix}$

The characteristic equation $|\mathbf{A} - \lambda\mathbf{I}| = \begin{vmatrix} 5-\lambda & 2 \\ 2 & 2-\lambda \end{vmatrix} = 0$

$$(5-\lambda)(2-\lambda) - 4 = 0$$

$$\lambda^2 - 7\lambda + 6 = 0$$
$$(\lambda - 6)(\lambda - 1) = 0$$
$$\lambda = 1,\ 6 \text{ are eigen-values.}$$

(i) To find the eigen-vector corresponding to λ=1. Let $\mathbf{x}$ be a nonzero vector such that $(\mathbf{A}\ \text{-}\mathbf{I}\ )\mathbf{x} = 0$

$$\begin{bmatrix} 4 & 2 \\ 2 & 1 \end{bmatrix} \begin{bmatrix} x_1 \\ x_2 \end{bmatrix} = \begin{bmatrix} 0 \\ 0 \end{bmatrix}$$

$R_2 \rightarrow R_2 - \frac{R_1}{2}$

$$\begin{bmatrix} 4 & 2 \\ 0 & 0 \end{bmatrix} \begin{bmatrix} x_1 \\ x_2 \end{bmatrix} = \begin{bmatrix} 0 \\ 0 \end{bmatrix}$$

$4x_1 + 2x_2 = 0$

Take x_2=k, then $4x_1 + 2k = 0 \Rightarrow 4x_1 = -2\text{k} \Rightarrow x_1 = -\frac{k}{2}$

Hence $\begin{bmatrix} x_1 \\ x_2 \end{bmatrix} = \begin{bmatrix} -\frac{k}{2} \\ k \end{bmatrix} = \frac{k}{2} \begin{bmatrix} -1 \\ 2 \end{bmatrix}$

$\mathbf{x} = \begin{bmatrix} -1 \\ 2 \end{bmatrix}$ is an eigen-vector corresponding to λ=1

(ii) eigen-vector corresponding to λ=6. Let $\mathbf{x}$ be a nonzero vector such that $(\mathbf{A} - \mathbf{6I})\mathbf{x} = 0$

$$\begin{bmatrix} -1 & 2 \\ 2 & -4 \end{bmatrix} \begin{bmatrix} x_1 \\ x_2 \end{bmatrix} = \begin{bmatrix} 0 \\ 0 \end{bmatrix}$$

$R_2 \rightarrow R_2 + 2R_1$

$$\begin{bmatrix} -1 & 2 \\ 0 & 0 \end{bmatrix} \begin{bmatrix} x_1 \\ x_2 \end{bmatrix} = \begin{bmatrix} 0 \\ 0 \end{bmatrix}$$

$-x_1 + 2x_2 = 0$

Take x_2=k, then x_1=2k

Hence $\begin{bmatrix} x_1 \\ x_2 \end{bmatrix} = \begin{bmatrix} 2k \\ k \end{bmatrix} = \text{k} \begin{bmatrix} 2 \\ 1 \end{bmatrix}$

$\mathbf{x}_2 = \begin{bmatrix} 2 \\ 1 \end{bmatrix}$ is an eigen-vector corresponding to $\lambda = 6$

Take $\mathbf{P} = \begin{bmatrix} -1 & 2 \\ 2 & 1 \end{bmatrix}$.

Then adj$\mathbf{P} = \begin{bmatrix} 1 & -2 \\ -2 & -1 \end{bmatrix}$, $|\mathbf{P}| = -1 - 4 = -5$, and

$\mathbf{P}^{-1} = \frac{adj\mathbf{P}}{|\mathbf{P}|} = \frac{-1}{5} \begin{bmatrix} 1 & -2 \\ -2 & -1 \end{bmatrix}$

Then $\mathbf{P}$ diagonalizes $\mathbf{A}$ and the diagonal form is given by $\mathbf{P}^{-1}\ \mathbf{A}\ \mathbf{P} = \mathbf{D}$

$$\frac{-1}{5}\begin{bmatrix} 1 & -2 \\ -2 & -1 \end{bmatrix}\begin{bmatrix} 5 & 2 \\ 2 & 2 \end{bmatrix}\begin{bmatrix} -1 & 2 \\ 2 & 1 \end{bmatrix}=\begin{bmatrix} 1 & 0 \\ 0 & 6 \end{bmatrix}$$

Example 2.7.5 *Reduce the matrix* $\boldsymbol{A} = \begin{bmatrix} 7 & 4 & -4 \\ 4 & 1 & 8 \\ -4 & 8 & 1 \end{bmatrix}$ *into diagonal form by using a modal matrix.*

Given matrix is $\boldsymbol{A} = \begin{bmatrix} 7 & 4 & -4 \\ 4 & 1 & 8 \\ -4 & 8 & 1 \end{bmatrix}$

The characteristic equation is $|\boldsymbol{A} - \lambda \boldsymbol{I}| = 0$

$$\Rightarrow \begin{vmatrix} 7-\lambda & 4 & -4 \\ 4 & 1-\lambda & 8 \\ -4 & 8 & 1-\lambda \end{vmatrix} = 0$$

$$\Rightarrow (7-\lambda)[(1-\lambda)^2 - 64] - 4[4(1-\lambda) + 32] - 4[32 + 4[(1-\lambda)] = 0$$

$$\Rightarrow (\lambda - 9)(81 - \lambda^2) = 0$$

$$\Rightarrow \lambda = -9, 9, 9.$$

Therefore the eigen-values of $\boldsymbol{A}$ *are* $9, 9, -9$

To find an eigen-vector corresponding to the eigen-value $\lambda = 9$

Consider the system of homogeneous equation $[\mathbf{A} - \lambda \mathbf{I}]\mathbf{x} = 0$

$$\begin{bmatrix} 7-9 & 4 & -4 \\ 4 & 1-9 & 8 \\ -4 & 8 & 1-9 \end{bmatrix}\begin{bmatrix} x \\ y \\ z \end{bmatrix} = \begin{bmatrix} 0 \\ 0 \\ 0 \end{bmatrix}$$

$$\begin{bmatrix} -2 & 4 & -4 \\ 4 & -8 & 8 \\ -4 & 8 & -8 \end{bmatrix}\begin{bmatrix} x \\ y \\ z \end{bmatrix} = \begin{bmatrix} 0 \\ 0 \\ 0 \end{bmatrix}$$

$R_2 \rightarrow R_2 + 2R_1,\ R_3 \rightarrow R_3 - 2R_1$

$$\begin{bmatrix} -2 & 4 & -4 \\ 0 & 0 & 0 \\ 0 & 0 & 0 \end{bmatrix}\begin{bmatrix} x \\ y \\ z \end{bmatrix} = \begin{bmatrix} 0 \\ 0 \\ 0 \end{bmatrix}$$

Let $z = k, y = l$ *(parameters)*

$-2x + 4y - 4z = 0$

$-x + 2y - 2z = 0$

$x = (2y - 2z) = (2l - 2k)$

$$\begin{bmatrix} x \\ y \\ z \end{bmatrix} \begin{bmatrix} (2l-2k) \\ l \\ k \end{bmatrix} = l\begin{bmatrix} 2 \\ 1 \\ 0 \end{bmatrix} + k\begin{bmatrix} -2 \\ 0 \\ 1 \end{bmatrix}$$

Therefore $\begin{bmatrix} 2 \\ 1 \\ 0 \end{bmatrix}, \begin{bmatrix} -2 \\ 0 \\ 1 \end{bmatrix}$ *are the eigen-vectors corresponding to the eigen-value, i.e.* $\lambda = 9$ *repeated twice.*

To find an eigen-vector corresponding the eigen-value $\lambda = -9$
Consider the system of homogeneous equation $[\mathbf{A} - \lambda \mathbf{I}]\mathbf{x} = 0$
$[\mathbf{A} + 9\mathbf{I}]\mathbf{x} = 0$

$$\begin{bmatrix} 16 & 4 & -4 \\ 4 & 10 & 8 \\ -4 & 8 & 10 \end{bmatrix} \begin{bmatrix} x \\ y \\ z \end{bmatrix} = \begin{bmatrix} 0 \\ 0 \\ 0 \end{bmatrix}$$

$R_1 \to \frac{1}{4}R_1, R_2 \to R_2 - R_1, R_3 \to R_3 + R_1$

$$\begin{bmatrix} 4 & 1 & -1 \\ 0 & 9 & 9 \\ 0 & 9 & 9 \end{bmatrix} \begin{bmatrix} x \\ y \\ z \end{bmatrix} = \begin{bmatrix} 0 \\ 0 \\ 0 \end{bmatrix}$$

$R_3 \to R_3 - R_2, R_2 \to 1/9R_2$

$$\begin{bmatrix} 4 & 1 & -1 \\ 0 & 1 & 1 \\ 0 & 0 & 0 \end{bmatrix} \begin{bmatrix} x \\ y \\ z \end{bmatrix} = \begin{bmatrix} 0 \\ 0 \\ 0 \end{bmatrix}$$

Let $z = k$ *(parameter)*
$y + z = 0$
$y = -z = -k$
$4x = -y + z,\ x = \frac{1}{4}(-y + z) = \frac{1}{4}(k + k) = \frac{1}{2}k$

$$\begin{bmatrix} x \\ y \\ z \end{bmatrix} = \begin{bmatrix} \frac{1}{2}k \\ -k \\ k \end{bmatrix} = \frac{1}{2}k \begin{bmatrix} 1 \\ -2 \\ 2 \end{bmatrix}$$

Therefore $\begin{bmatrix} 1 \\ -2 \\ 2 \end{bmatrix}$ *is an eigen-vector corresponding to* $\lambda = -9$
Writing the three eigen-vectors as the three columns,
the modal matrix is $\boldsymbol{K} = \begin{bmatrix} 2 & -2 & 1 \\ 1 & 0 & -2 \\ 0 & 1 & 2 \end{bmatrix}$.

Now, $\boldsymbol{K}^{-1} = \frac{1}{9} \begin{bmatrix} 2 & 5 & 4 \\ -2 & 4 & 5 \\ 1 & -2 & 2 \end{bmatrix}$ *and*

$$\boldsymbol{D} = \frac{1}{9} \begin{bmatrix} 2 & 5 & 4 \\ -2 & 4 & 5 \\ 1 & -2 & 2 \end{bmatrix} \begin{bmatrix} 7 & 4 & -4 \\ 4 & 1 & 8 \\ -4 & 8 & 1 \end{bmatrix} \begin{bmatrix} 2 & -2 & 1 \\ 1 & 0 & -2 \\ 0 & 1 & 2 \end{bmatrix}$$

$$\boldsymbol{D} = \begin{bmatrix} 9 & 0 & 0 \\ 0 & 9 & 0 \\ 0 & 0 & -9 \end{bmatrix} = diag(9, 9, -9).$$

Note. Theorem 2.7.2 reduces to a special form if **A** is a symmetric matrix and this is given below.

Theorem 2.7.3 *If* $\boldsymbol{A}$ *is a symmetric matrix then there exists an orthogonal matrix* $\boldsymbol{S}$ *such that* $\boldsymbol{A} = \boldsymbol{SDS}^T$, *where* $\boldsymbol{D}$ *is a diagonal matrix consisting of the distinct eigen-values of* $\boldsymbol{A}$ *and* $\boldsymbol{S}$ *is the matrix whose columns are eigen-vectors corresponding to eigen-values of* $\boldsymbol{A}$.

Proof. In property III(2), it has been shown that the eigen-vectors of a symmetric matrix are orthogonal. Since for an orthogonal matrix $\mathbf{P}$, $\mathbf{P}^{-1} = \mathbf{P}^T$, the result follows from Theorem 2.7.2.

When the rank $A < n$ and when there are only s linearly independent eigen-vectors, with $s < n$, then in place of diagonal matrix, there will be s blocks each having one eigen-value.

Each block is called a Jordan block and is as follows.

Definition 2.7.5 *Jordan block is a square matrix having an eigen-value* α *on the main diagonal and* 1 *on the super diagonal. All other elements are zero.*

Example 2.7.6 $\begin{bmatrix} 2 & 1 & 0 \\ 0 & 2 & 0 \\ 0 & 0 & 2 \end{bmatrix}$, $\begin{bmatrix} 2 & 0 & 0 \\ 0 & -4 & 1 \\ 0 & 0 & -4 \end{bmatrix}$ *are Jordan blocks.*

The following theorem describes the Jordan canonical form.

Theorem 2.7.4 *Jordan Canonical form: If a square matrix of order 'n' has a characteristic polynomial*

$$C(x) = det(\mathbf{A} - x\mathbf{I})) = \Pi_{i=1}^{s}(x - \alpha_i)^{r_i}$$

Then $\mathbf{A}$ *is similar to a matrix with* α_i *on the main diagonal and one's on the super diagonal and zeros elsewhere.*

Observation.

1. The α_i's are the eigen-values of $\mathbf{A}$ and r_i are its algebraic multiplicities.

EXERCISE 2.7

1. Find the eigen-values and the corresponding eigen-vectors for the following matrices

(i) $A=\begin{bmatrix} 4 & 1 & -5 \\ -3 & 0 & 5 \\ 3 & 3 & -2 \end{bmatrix}$ (ii) $A=\begin{bmatrix} 0 & 1 \\ -6 & 5 \end{bmatrix}$

(iii) $A=\begin{bmatrix} 2 & 1 & 6 \\ 0 & -5 & 3 \\ 0 & 0 & 4 \end{bmatrix}$ (iv) $A=\begin{bmatrix} 2 & 3 \\ 2 & 1 \end{bmatrix}$

(v) $A=\begin{bmatrix} 7 & 1 \\ 10 & 4 \end{bmatrix}$ (vi) $A=\begin{bmatrix} 1 & 2 & 3 \\ 0 & 4 & 5 \\ 0 & 0 & 6 \end{bmatrix}$

2. Find the modal matrix and hence reduce the following matrices into diagonal from

(i) $A=\begin{bmatrix} 2 & 2 & 2 \\ 2 & 2 & 2 \\ 2 & 2 & 2 \end{bmatrix}$ (ii) $A=\begin{bmatrix} 3 & 2 & -3 \\ -3 & -4 & 9 \\ -1 & -2 & 5 \end{bmatrix}$

(iii) $A=\begin{bmatrix} 7 & 4 & -4 \\ 4 & 1 & 8 \\ -4 & 8 & 1 \end{bmatrix}$ (iv) $A=\begin{bmatrix} 1 & 0 & 1 \\ 0 & -1 & 3 \\ 0 & 0 & 6 \end{bmatrix}$

2.8 Singular Values and Singular Vectors

This section mainly deals with rectangular matrices. A matrix A of order $m \times n$ can be considered as a mapping from $\mathbb{R}^n$ to $\mathbb{R}^m$. Systems of algebraic equations in n unknowns and m equations or algebraic equations with m unknowns and n equations $(m \neq n)$ give rise to matrices of order $m \times n$ or $n \times m$. In this case in place of eigen-values, singular values are introduced. Their corresponding vectors are called as singular vectors. A formal definition is as follows.

Definition 2.8.1 *A singular value and a pair of singular vectors of a square or a rectangular matrix* **A** *are a nonnegative scalar* σ *and two nonzero vectors,* **u** *and* **v** *such that*

$$\mathbf{Av} = \sigma \mathbf{u}$$

and

$$\mathbf{A^T u} = \sigma \mathbf{v},$$

where $\mathbf{A^T}$ *is the transpose of* **A**.

The following is the method to find singular values and singular vectors.

Method to find singular values and singular vectors.

Step 1 Given a matrix $\mathbf{A}$ (square or rectangular), find $\mathbf{A^T A}$.

Step 2 Find the eigen-values α of $\mathbf{A^T A}$ and the corresponding eigen-vectors, $\mathbf{v}$. Then the singular value of $\mathbf{A}$, $\sigma = \sqrt{\alpha}$ or $\sigma^2 = \alpha$.

Step 3 Find the eigen-values of $\mathbf{AA^T}$ and the corresponding eigen-vectors, $\mathbf{u}$.

Step 4 Then $\mathbf{Av} = \sigma\mathbf{u}$.

Step 5 σ is the singular value and $\mathbf{v}$ and $\mathbf{u}$ are the corresponding singular vectors of $\mathbf{A}$.

Example 2.8.1 *Find the singular values and singular vectors for the matrix* $\boldsymbol{A} = \begin{bmatrix} 1 & 1 \\ 2 & 2 \\ 1 & 1 \end{bmatrix}$

Step 1. $\boldsymbol{A}^T\boldsymbol{A} = \begin{bmatrix} 1 & 2 & 1 \\ 1 & 2 & 1 \end{bmatrix} \begin{bmatrix} 1 & 1 \\ 2 & 2 \\ 1 & 1 \end{bmatrix} = \begin{bmatrix} 6 & 6 \\ 6 & 6 \end{bmatrix}$

Step 2. *The characteristic equation of* $\boldsymbol{A}^T\boldsymbol{A}$ *is* $\left|\boldsymbol{A}^T\boldsymbol{A} - \lambda\boldsymbol{I}\right| = 0$

$$\begin{vmatrix} 6-\lambda & 6 \\ 6 & 6-\lambda \end{vmatrix} = 0$$

$$(6-\lambda)^2 - 36 = 0$$

$\lambda = 12, 0$ *are the eigen-values of* $\boldsymbol{A}^T\boldsymbol{A}$.

Hence $\sigma_1 = \sqrt{12} = 2\sqrt{3}, \sigma_2 = 0$ *are the singular values of* $\boldsymbol{A}$.

The eigen-vectors of $\boldsymbol{A}^T\boldsymbol{A}$ *corresponding to* $\lambda = 12$ *and* $\lambda = 0$ *are* $\begin{bmatrix} 1 \\ 1 \end{bmatrix}$ *and* $\begin{bmatrix} -1 \\ 1 \end{bmatrix}$, *respectively and the orthonormal vectors are* $\mathbf{v_1} = \begin{bmatrix} \frac{1}{\sqrt{2}} \\ \frac{1}{\sqrt{2}} \end{bmatrix}$ *and* $\mathbf{v_2} = \begin{bmatrix} \frac{-1}{\sqrt{2}} \\ \frac{1}{\sqrt{2}} \end{bmatrix}$, *respectively.*

Step 3. $\boldsymbol{A}\boldsymbol{A}^T = \begin{bmatrix} 1 & 1 \\ 2 & 2 \\ 1 & 1 \end{bmatrix} \begin{bmatrix} 1 & 2 & 1 \\ 1 & 2 & 1 \end{bmatrix} = \begin{bmatrix} 2 & 4 & 2 \\ 4 & 8 & 4 \\ 2 & 4 & 2 \end{bmatrix}$

has eigen-values 12, 0 and 0.

The eigen-vector associated to the eigen-value $\lambda_1 = 2\sqrt{3}$ *of* $\boldsymbol{A}\boldsymbol{A}^T$ *is* $\begin{bmatrix} 1 \\ 2 \\ 1 \end{bmatrix}$ *and the corresponding normalized vector* $\mathbf{u}_1$ *is* $\begin{bmatrix} \frac{1}{\sqrt{6}} \\ \frac{2}{\sqrt{6}} \\ \frac{1}{\sqrt{6}} \end{bmatrix}$.

Normalized eigen-vectors corresponding to $\lambda = 0$ *are* $\mathbf{u_2} = \begin{bmatrix} \frac{-2}{\sqrt{5}} \\ \frac{1}{\sqrt{5}} \\ 0 \end{bmatrix}$ *and* $\mathbf{u_3} = \begin{bmatrix} \frac{-2}{\sqrt{5}} \\ 0 \\ \frac{1}{\sqrt{5}} \end{bmatrix}$.

It can be easily shown that $\mathbf{Av_1} = \begin{bmatrix} 1 & 1 \\ 2 & 2 \\ 1 & 1 \end{bmatrix} \begin{bmatrix} \frac{1}{\sqrt{2}} \\ \frac{1}{\sqrt{2}} \end{bmatrix} = 2\sqrt{3} \begin{bmatrix} \frac{1}{\sqrt{6}} \\ \frac{2}{\sqrt{6}} \\ \frac{1}{\sqrt{6}} \end{bmatrix}$
$= \sigma_1 \mathbf{u}_1.$

Similarly, $\mathbf{Av_2} = \sigma_2 \mathbf{u}_2$. *Now* $\sigma_1 = 2\sqrt{3}$ *and* $\sigma_2 = 0$ *are the singular values of* $\boldsymbol{A}$ *and the corresponding singular vectors are* $\mathbf{u}_1, \mathbf{v_1}$ *and* $\mathbf{u}_2, \mathbf{v_2}$.

EXERCISE 2.8

Find the singular values and singular vectors for the following matrices

(i) $\mathbf{A} = \begin{bmatrix} 2 & 1 & 2 \\ 2 & 1 & 2 \end{bmatrix}$ (ii) $\mathbf{A} = \begin{bmatrix} 1 & 0 & 1 & 0 \\ 0 & 1 & 0 & 1 \end{bmatrix}$

2.9 Quadratic Forms

In this section, a bilinear form is introduced. A special case of the bilinear form is a quadratic form and is discussed in detail. Quadratic forms arise naturally in problems in physics, mechanics, statistics and in problems involving maxima and minima, that is, optimization problems. Quadratic forms reduced to canonical forms give insights to the conics considered. The nature of quadratic forms is also discussed.

Definition 2.9.1 *A bilinear form is a function* $f : \mathbb{R}^n \times \mathbb{R}^m \to \mathbb{R}$ *defined by* $f(\mathbf{y}, \mathbf{x}) = \mathbf{y}^T \boldsymbol{A}\mathbf{x}$, *where* $\boldsymbol{A}$ *is an* $n \times m$ *matrix.*

Definition 2.9.2 *A quadratic form is a homogeneous expression of second degree in n variables* $x_1,\ x_2, \cdots,\ x_n$ *of the form*

$$\boldsymbol{Q} = \mathbf{x^T A x} = \sum_{j=1}^{n} \sum_{i=1}^{n} a_{ij} x_i x_j, \tag{2.17}$$

where $a_{ij}, i, j = 1,\ 2, \cdots,\ n$ *are constants.*

Let $\mathbf{x} = (x_1,\ x_2, \cdots,\ x_n)^T$. Then $\mathbf{x}^T = (x_1,\ x_2, \cdots,\ x_n)$

and $\mathbf{x^T A x} = \begin{bmatrix} x_1 & x_2 \ldots & x_n \end{bmatrix} \begin{bmatrix} a_{11} & a_{12} \ldots & a_{1n} \\ a_{21} & a_{22} \ldots & a_{2n} \\ & \vdots & \\ a_{n1} & a_{n2} \ldots & a_{nn} \end{bmatrix} \begin{bmatrix} x_1 \\ x_2 \\ \vdots \\ x_n \end{bmatrix}$

$= \sum_{j=1}^{n} \sum_{i=1}^{n} a_{ij} x_i x_j$

$= a_{11}x_1^2 + a_{12}x_1x_2 + \cdots + a_{1n}x_1x_n + a_{21}x_2x_1 + a_{22}x_2^2 + \cdots a_{2n}x_2x_n + \cdots + a_{n1}x_nx_1 + \cdots + a_{nn}x_n^2$

$= a_{11}x_1^2 + (a_{12} + a_{21})x_1x_2 + \cdots + (a_{1n} + a_{n1})x_1x_n + \cdots + a_{nn}x_n^2.$

The matrix $\mathbf{A}$ need not be symmetric but a nonsymmetric matrix can be replaced by a symmetric matrix with identical results.

Example 2.9.1 1. *$x^2 + 6xy - 4y^2$ is a quadratic form in 2 variables and the matrix representation is*

$$\begin{bmatrix} x & y \end{bmatrix} \begin{bmatrix} 1 & 3 \\ 3 & -4 \end{bmatrix} \begin{bmatrix} x \\ y \end{bmatrix}$$

2. *$4x^2 + 2y^2 - 8z^2 + 2xy + 5yz - 4zx$ is a quadratic form in 3 variables. The matrix representation is as follows:*

$$\begin{bmatrix} x & y & z \end{bmatrix} \begin{bmatrix} 4 & 1 & -2 \\ 1 & 2 & 5/2 \\ -2 & 5/2 & -8 \end{bmatrix} \begin{bmatrix} x \\ y \\ z \end{bmatrix}$$

3. *$2x^2 + y^2 + z^2 + 5w^2 + 2xy + 8zw$ is a quadratic expression in 4 variables. The matrix representation is*

$$\begin{bmatrix} x & y & z & w \end{bmatrix} \begin{bmatrix} 2 & 1 & 0 & 0 \\ 1 & 1 & 0 & 0 \\ 0 & 0 & 1 & 4 \\ 0 & 0 & 4 & 5 \end{bmatrix} \begin{bmatrix} x \\ y \\ z \\ w \end{bmatrix}$$

The observation that a real quadratic form can be expressed as a matrix is stated as the following theorem.

Theorem 2.9.1 *Every real quadratic form in n variables $x_1,\ x_2, \cdots,\ x_n$ can be expressed in the form $\mathbf{x^T A x}$, where $\mathbf{x} = (x_1 \quad x_2 \quad \cdots \quad x_n)^T$ is a column matrix and $\boldsymbol{A}$ is a real symmetric square matrix of order n. This matrix $\boldsymbol{A}$ is called as the matrix of the quadratic form $\mathbf{x^T A x}$.*

Definition 2.9.3 Rank of a Quadratic form

Let $\mathbf{x^T A x}$ be a real quadratic form. The rank r of $\boldsymbol{A}$ is called the rank of the quadratic form.

If $\rho(\boldsymbol{A}) = r < n$ or $|\boldsymbol{A}| = 0$ then the quadratic form is said to be singular; otherwise it is nonsingular.

Definition 2.9.4 Canonical Form or Normal Form of a Quadratic form
Let $\mathbf{x^T A x}$ *be a quadratic form in* n *variables. Then there exists a real non-singular orthogonal matrix,* $\boldsymbol{P}$ *(can also be treated as an orthogonal linear transformation) such that* $\mathbf{x} = \mathbf{Py}$ *transforms* $\mathbf{x^T A x}$ *to the form* $\mathbf{y^T B y}$. *The form* $\mathbf{y^T B y} = \lambda_1 y_1^2 + \lambda_2 y_2^2 + \cdots + \lambda_n y_n^2$ *is called the canonical form of* $\mathbf{x^T A x}$, *where* $\boldsymbol{B} = diag[\lambda_1 \;\; \lambda_2 \;\; \cdots \;\; \lambda_n]$.

Procedure to reduce the given quadratic form to the canonical form when all the n eigen-vectors can be found

Step 1. Write the matrix $\mathbf{A}$

Step 2. Find the eigen-values of $\mathbf{A}$

Step 3. Find the corresponding eigen-vectors if all the n (order of the matrix) eigen-vectors are found proceed further. Otherwise canonical form is not possible

Step 4. Normalize them. (i.e.) If $\begin{bmatrix} x_1 \\ x_2 \\ x_3 \end{bmatrix}$ is an eigen-vector then $\mathbf{x} = \frac{1}{\sqrt{x_1^2+x_2^2+x_3^2}} \begin{bmatrix} x_1 \\ x_2 \\ x_3 \end{bmatrix}$, is normalized eigen-vector.

Step 5. Form the matrix $\mathbf{P}$ by the eigen-vectors.

Step 6. Find $\mathbf{P}^{-1}$

Step 7. Calculate $\mathbf{B} = \mathbf{P}^{-1}\mathbf{AP}$

Then $\mathbf{B} = diag[\lambda_1 \cdots \lambda_n]$, where $\lambda_1, \; \lambda_2, \cdots \; \lambda_n$ are the eigen-values of $\mathbf{A}$ and $\mathbf{B}$ is in canonical form.

Example 2.9.2 *Reduce the Quadratic form* $7x^2 + 6y^2 + 5z^2 - 4xy - 4yz$ *to the canonical form.*

Solution: *Its matrix is* $\boldsymbol{A} = \begin{bmatrix} 7 & -2 & 0 \\ -2 & 6 & -2 \\ 0 & -2 & 5 \end{bmatrix}$

$$
\begin{aligned}
&|\boldsymbol{A} - \lambda \boldsymbol{I}| = 0 \\
&\quad\Longrightarrow \begin{vmatrix} 7-\lambda & -2 & 0 \\ -2 & 6-\lambda & -2 \\ 0 & -2 & 5-\lambda \end{vmatrix} = 0 \\
&\qquad (7-\lambda)[(6-\lambda)(5-\lambda) - 4] + 2[2(\lambda - 5)] = 0 \\
&\quad\Longrightarrow (7-\lambda)[30 - 11\lambda + \lambda^2 - 4] + 4(\lambda - 5) = 0 \\
&\qquad (7-\lambda)[\lambda^2 - 11\lambda + 26] + 4\lambda - 20 = 0 \\
&\qquad 7\lambda^2 - 77\lambda + 182 - \lambda^3 + 11\lambda^2 - 26\lambda + 4\lambda - 20 = 0 \\
&\quad\Longrightarrow -\lambda^3 + 18\lambda^2 - 99\lambda + 162 \\
&\quad\Longrightarrow \lambda^3 - 18\lambda^2 + 99\lambda - 162 \\
&\quad\Longrightarrow (\lambda - 3)(\lambda - 6)(\lambda - 9) \\
&\quad\Longrightarrow \lambda = 3,\ 6,\ 9 \quad \textit{are the eigen-values.}
\end{aligned}
$$

The canonical form is $3u^2 + 6v^2 + 9w^2$

Index and signature of a quadratic form

Let the quadratic form $\mathbf{x}^{\mathbf{T}}\mathbf{A}\mathbf{x}$ be transformed into the canonical form $\mathbf{y}^{\mathbf{T}}\mathbf{B}\mathbf{y}$. The canonical form has only r terms, if the rank of $\mathbf{A} = \rho(\mathbf{A}) = r$. The terms can be either positive or negative.

Definition 2.9.5 *The number of positive terms in the normal form of the quadratic form is called the index (denoted by i) of the quadratic form.*

Example 2.9.3
1. *Suppose the canonical form of a quadratic form is given by* $2x^2 + 3y^2 - z^2$ *then the index of the quadratic form is* 2.
2. *Suppose the canonical form of a quadratic form is given by* $4x^2+6y^2+z^2+w^2$ *then the index of the quadratic form is* 4.

Theorem 2.9.2 *The number of positive terms (negative terms) in any two canonical forms are the same.*

Definition 2.9.6 Signature of a Quadratic Form
Let r be the rank of the quadratic form and i be the index of the quadratic form (i.e.) the number of positive terms in the canonical form.
Then the signature of the quadratic form (denoted by s) is defined as the excess number of positive terms over the number of negative terms, that is,

$$s = i - (r - i) = 2i - r.$$

Nature of a quadratic form

The nature of the quadratic form can be determined by the index or rank of the matrix $\mathbf{A}$ or in terms of the eigen-values of the matrix $\mathbf{A}$.
Let $\rho(\mathbf{A}) = r$ and index of the quadratic form be i.
Then, the real quadratic form is said to be

1. Positive Definite if $r = i = n$ or in terms of eigen-values, all the eigen-values of $\mathbf{A}$ are positive.

2. Negative Definite if $i = 0$ and $r = n$ or, all the eigen-values of $\mathbf{A}$ are negative.

3. Positive Semi Definite if $i = r < n$ or if the eigen-values of $\mathbf{A}$ are positive and at least one eigen-value is 0.

4. Negative Semi Definite if $i = 0$ and $r < n$ or the eigen-values of $\mathbf{A}$ are negative and at least one of them is 0.

5. Indefinite in all other cases that is, the eigen-values are both positive and negative.

Theorem 2.9.3 (Sylvester's Law of Inertia) *The signature of a quadratic form is independent of all canonical forms, (i.e.) the signature does not change with different canonical forms.*

Example 2.9.4 *Find the index, signature and the nature of the quadratic form* $x^2 + y^2 + 2z^2 + 4xy$.
Solution. *Let* $\mathbf{x} = \begin{bmatrix} x & y & z \end{bmatrix}^T$

then the matrix associated with the quadratic form is $\mathbf{A} = \begin{bmatrix} 1 & 2 & 0 \\ 2 & 1 & 0 \\ 0 & 0 & 2 \end{bmatrix}$

From Example 2.7.2 the eigen-values are $\lambda = -1,\ 2,\ 3$ *and the corresponding eigen-vectors are* $\begin{bmatrix} 1 \\ -1 \\ 0 \end{bmatrix}, \begin{bmatrix} 0 \\ 0 \\ 1 \end{bmatrix}$ *and* $\begin{bmatrix} 1 \\ 1 \\ 0 \end{bmatrix}$, *respectively.*

Now normalizing the vectors gives $\frac{1}{\sqrt{2}} \begin{bmatrix} 1 \\ -1 \\ 0 \end{bmatrix}, \begin{bmatrix} 0 \\ 0 \\ 1 \end{bmatrix}$ *and* $\frac{1}{\sqrt{2}} \begin{bmatrix} 1 \\ 1 \\ 0 \end{bmatrix}$

Let $\mathbf{P} = \begin{bmatrix} \frac{1}{\sqrt{2}} & 0 & \frac{1}{\sqrt{2}} \\ \frac{-1}{\sqrt{2}} & 0 & \frac{1}{\sqrt{2}} \\ 0 & 1 & 0 \end{bmatrix}$, $\mathbf{P}^{-1} = \mathbf{P}^T$ *and* $\mathbf{P}^{-1}\mathbf{A}\mathbf{P} = \begin{bmatrix} -1 & 0 & 0 \\ 0 & 2 & 0 \\ 0 & 0 & 3 \end{bmatrix}$

Now the canonical form of the quadratic form is $-u^2 + 2v^2 + 3w^2$
Hence the index is 2 and the signature is 1.
The nature of the quadratic form is indefinite.

Example 2.9.5 *Find the nature of the quadratic form* $4x_1^2+x_2^2+4x_3^2+2x_1x_2+4x_1x_3$.

Solution. *The matrix corresponding to the quadratic form is* $\begin{bmatrix} 4 & 1 & 2 \\ 1 & 1 & 0 \\ 2 & 0 & 4 \end{bmatrix}$

The eigen-values are $\lambda_1 \approx 0.56,\ \lambda_2 \approx 2.33,\ \lambda_3 \approx 6.10$
Since all the eigen-values of $\mathbf{A}$ *are positive the quadratic form is positive definite.*

EXERCISE 2.9

1. Find the index, signature and nature of the quadratic form by finding the normal form.

1. $x^2 + y^2 + 2z^2 + 2xy$.
2. $5x^2 + 2y^2 + 4xy$.
3. $7x^2 + y^2 + z^2 + 8xy - 8xz + 16yz$.

2.10 Conclusion

In this chapter, the various setups of linear system of equations are described and related theoretical concepts are given. Various approaches to reducing a given matrix to a diagonal, upper or lower triangular form are given and Jordan blocks are introduced. Eigen-values and eigen-vectors along with their properties are discussed. Singular values and singular vectors are introduced. These concepts are used in studying principal component analysis in Section 5.9. A study of quadratic forms is initiated.

This chapter is a continuation of Chapter 1 and deals with the fundamentals of linear algebra. For further reading, the readers are suggested to refer to [14] and [19].

Chapter 3

Vector Spaces

3.1 Introduction

In this chapter, the theoretical foundation for the concepts introduced in earlier chapters is given. Section 3.2 deals with a vector space and its subspace. In Section 3.3 the linear independence of vectors, the basis and the dimension of a vector space are discussed. In Section 3.4, the change of basis matrix is described. Linear transformation is introduced in Section 3.5. The matrix of linear transformation is described in Section 3.6. In Section 3.7, an inner product space and its important properties are given. Section 3.8 deals with the Gram-Schmidt orthogonalization process and singular value decomposition theorem. Finally, an attempt is made to link linear algebra with differential equations in Section 3.9.

3.2 Vector Space and Subspaces

In this section, the theoretical foundation for the structure of vectors and matrices developed in Chapter 1 is given.

Let V be any nonempty set of elements and let two operations called vector addition, '+' and scalar multiplication, '.' on elements of V be defined such that,
(i) if $\mathbf{x} \in V$ and $\mathbf{y} \in V$ then $\mathbf{x} + \mathbf{y} \in V$ (an operation '+' satisfying this property is called a binary operation); and
(ii) for any real number $\alpha \in \mathbb{R}$, $\alpha\mathbf{x} \in V$.

Definition 3.2.1 *The triad $(V, +, .)$ is a real vector space (or a vector space in short) if the operations of '+' and '.' satisfy the following properties.*
Let $\mathbf{x}, \mathbf{y}, \mathbf{z} \in V$ and $\alpha, \beta \in \mathbb{R}$. Then

(i) $\mathbf{x} + \mathbf{y} \in V$ *(Closure under addition);*

(ii) $\mathbf{x} + \mathbf{y} = \mathbf{y} + \mathbf{x}$ *(Commutative);*

DOI: 10.1201/9781351014953-3

(iii) $(\mathbf{x}+\mathbf{y})+\mathbf{z}=\mathbf{x}+(\mathbf{y}+\mathbf{z})$ *(Associative);*

(iv) *There exists an additive identity* $\mathbf{0} \in V$ *such that* $\mathbf{x}+\mathbf{0}=\mathbf{0}+\mathbf{x}=\mathbf{x}$, *for all* $\mathbf{x} \in V$;

(v) *To each* $\mathbf{x} \in V$, *there exists a* $\mathbf{y} \in V$ *such that* $\mathbf{x}+\mathbf{y}=\mathbf{y}+\mathbf{x}=\mathbf{0}$, *which is called the additive inverse of* $\mathbf{x}$ *and is denoted by* $-\mathbf{x}$;

(vi) $\alpha(\mathbf{x}+\mathbf{y})=\alpha\mathbf{x}+\alpha\mathbf{y}$;

(vii) $(\alpha+\beta)\mathbf{x}=\alpha\mathbf{x}+\beta\mathbf{x}$;

(viii) $(\alpha\beta)\mathbf{x}=\alpha(\beta\mathbf{x})$;

(ix) *There exists* $1 \in \mathbb{R}$ *such that* $1\ \mathbf{x}=\mathbf{x} \in V$.

Example 3.2.1 *1. Let* $V=\{\mathbf{x}=(x_1\ \ x_2\ \ ...\ \ x_n) : x_i \in \mathbb{R}\}$, *be the set of all vectors in* $\mathbb{R}^n$. *Let the vector addition + and scalar multiplication . be defined componentwise. Then* $(V,+,.)$ *is a vector space.*
2. Let $V=\{(a_{ij})_{m\times n} : a_{ij} \in \mathbb{R}\}$ *be the set of all* $m \times n$ *matrices with entries in* $\mathbb{R}$. *Let the binary operations + and . be defined entrywise. Then* $(V,+,.)$ *is a vector space.*
3. Let $V=\{p(x)=a_0+a_1x+a_2x^2+\cdots+a_nx^n : a_0,a_1,\ldots,a_n \in \mathbb{R}\}$, *be set of all polynomials of degree* $\leq n$, *with real constants. Let the operations + and . be defined as follows: for any two polynomials* $p,q \in V$, $(p+q)(x)=p(x)+q(x)$ *and* $(\alpha p)(x)=\alpha p(x)$. *Then* $(V,+,.)$ *is a vector space.*

Sometimes it may not be necessary to work with the whole space. For example, suppose one is working in $\mathbb{R}^3$, but the information given is a subset of $\mathbb{R}^3$. Then the question is whether this subset of $\mathbb{R}^3$ is a vector space or not and what is the criteria for verification. The following definition answers this question.

Definition 3.2.2 ***Subspace***
Let $(V,+,.)$ *be a vector space and* W *be a subset of* V. *Then* $(W,+,.)$ *is a subspace of* V *if and only if* $(W,+,.)$ *is a vector space in its own right.*

Example 3.2.2 $V=\{(x_1\ \ x_2\ \ x_3) : x_i \in \mathbb{R}, i=1,2,3\}=\mathbb{R}^3$. *Then* $V=\mathbb{R}^3$. *Let* $W=\{(x_1\ \ x_2\ \ 0)/x_i \in \mathbb{R}, i=1,2\}$. *Then clearly* $W \subseteq V$ *and* $(W,+,.)$ *is a subspace of* V.

Example 3.2.3 *Consider* $V=\left\{\begin{bmatrix} a & b \\ c & d \end{bmatrix} : a,b,c,d \in \mathbb{R}\right\}$ *with the definitions of matrix addition and scalar multiplication. Then* V *is a vector space.*
Now set $W=\left\{\begin{bmatrix} a & 0 \\ 0 & b \end{bmatrix} : a,b \in \mathbb{R}\right\}$. *Then* W *is a subspace of* V *with the operations of matrix addition and scalar multiplication of* V.

The following theorem gives the necessary and sufficient condition for a subset W of V to be a subspace.

Theorem 3.2.1 *A nonempty subset W of a vector space V over $\mathbb{R}$ is a subspace of V iff the following conditions are satisfied.*

(i) $\mathbf{w}_1, \mathbf{w}_2 \in W \implies \mathbf{w}_1 + \mathbf{w}_2 \in W$

(ii) $\alpha \in \mathbb{R}, \mathbf{w} \in W \implies \alpha\mathbf{w} \in W.$

Proof. Suppose W is a subspace of V. Then by definition of a subspace, W is a vector space with respect to the operations of + and . defined on V. Thus for $\mathbf{w}_1, \mathbf{w}_2 \in W$, $\mathbf{w}_1 + \mathbf{w}_2 \in W$ and $\alpha\mathbf{w} \in W$ for $\alpha \in \mathbb{R}$.
Conversely suppose (i) and (ii) are satisfied then consider $\mathbf{w}_1, \mathbf{w}_2, \mathbf{w}_3 \in W \implies \mathbf{w}_1, \mathbf{w}_2, \mathbf{w}_3 \in V$. Since V is a vector space, $\mathbf{w}_1 + (\mathbf{w}_2 + \mathbf{w}_3) = (\mathbf{w}_1 + \mathbf{w}_2) + \mathbf{w}_3$ and $\mathbf{w}_1 + \mathbf{w}_2 = \mathbf{w}_2 + \mathbf{w}_1$, and $(\alpha\beta)\mathbf{w}_1 = \alpha(\beta\mathbf{w}_1)$
$\alpha(\mathbf{w}_1 + \mathbf{w}_2) = \alpha\mathbf{w}_1 + \alpha\mathbf{w}_2$
$(\alpha + \beta)\mathbf{w}_1 = \alpha\mathbf{w}_1 + \beta\mathbf{w}_1$. Using (i) and (ii) in combination, all the above elements are in W.
To show that additive inverse and additive identity exist,
taking $\alpha = -1$ in (ii), yields $(-1)\mathbf{w} \in W$ by (ii)
and by (i) $\mathbf{w} - \mathbf{w} \in W \implies \mathbf{0} \in W$. Hence the proof.

Observation. The two conditions in the above theorem can be written as a single one as $\alpha\mathbf{w}_1 + \mathbf{w}_2 \in W$.

Theorem 3.2.2 *Suppose W_1 and W_2 are subspaces of V. Then $W_1 \cap W_2$ is also a subspace of V.*

Proof is given as exercise.

Definition 3.2.3 *If W_1 and W_2 are any two subspaces of V then the sum of the two subspaces is denoted by $W_1 + W_2$ and is defined as $W_1 + W_2 = \{\mathbf{w}_1 + \mathbf{w}_2 / \mathbf{w}_1 \in W_1$ and $\mathbf{w}_2 \in W_2\}$.*

Theorem 3.2.3 *$W_1 + W_2$ is a subspace of V.*

Proof is given as exercise.

Definition 3.2.4 *A vector space V is said to be a direct sum of two subspaces W_1 and W_2, denoted by $W_1 \oplus W_2$, if and only if $V = W_1 + W_2$ and $W_1 \cap W_2 = \{0\}$.*

Example 3.2.4 *Let $W_1 = \begin{bmatrix} a & 0 \\ 0 & b \end{bmatrix}$, $W_2 = \begin{bmatrix} 0 & c \\ d & 0 \end{bmatrix}$ then $W_1 \bigoplus W_2 = \begin{bmatrix} a & c \\ d & b \end{bmatrix}$ is the space $\mathbb{R}^{2\times 2}$.*

EXERCISE 3.2

1. Show that the set of all 3 column vectors $S = \{(x \quad \sin x \quad \cos x) : x \in \mathbb{R}\}$ is not closed under vector addition.

2. Show that the line $\{(0 \quad 0 \quad z) : z \in \mathbb{R}\}$ and the plane $\{(0 \quad y \quad z) : y,\ z \in \mathbb{R}\}$ are subspaces of $\mathbb{R}^3$.

3. Show that the solutions of the system $\mathbf{Ax} = 0$ where $\mathbf{A}$ is an $m \times n$ matrix form a vector space under vector addition and scalar multiplication.

4. Does the set of all 2×2 matrices of the form $\begin{bmatrix} a & b \\ c & d \end{bmatrix}$, where $a \geq 0$ with usual operations in $\mathbb{R}^{2\times 2}$ form a vector space? If not explain why?

5. Show that the set of all positive real numbers V with operations $\bigoplus$ and $\bigodot$ defined by $x, y \in V$, $x \bigoplus y = xy$ and $c \bigodot x = x^c$ is a vector space.

6. Show that $\{\begin{bmatrix} a & b \\ c & d \end{bmatrix}$ such that $a + b = c\}$ is a subspace of $\mathbb{R}^{2\times 2}$.

7. Let $W_1 = \left\{\begin{bmatrix} a & 0 \\ 0 & b \end{bmatrix} : \ a,\ b \in \mathbb{R}\right\}$ and $W_2 = \left\{\begin{bmatrix} 0 & c \\ d & 0 \end{bmatrix} : \ c,\ d \in \mathbb{R}\right\}$ then show that $W_1 \bigoplus W_2 = \mathbb{R}^{2\times 2}$.

8. Let $W_1 = \left\{\begin{bmatrix} a & 0 \\ 0 & b \end{bmatrix} : \ a,\ b \in \mathbb{R}\right\}$ and $W_2 = \left\{\begin{bmatrix} 0 & c \\ d & 0 \end{bmatrix} : \ c,\ d \in \mathbb{R}\right\}$ then show that $W_1 \bigoplus W_2 = \begin{bmatrix} a & 0 \\ c & b \end{bmatrix}$ is a subspace of $\mathbb{R}^{2\times 2}$.

9. Show that the space $\mathbb{R}^3$ is a direct sum of two planes xy-plane and z-plane, $W_1 = \{(x \quad y \quad 0) : x, y \ \in \mathbb{R}\}$ and $W_2 = \{(0 \quad 0 \quad z) : z \ \in \mathbb{R}\}$.

10. Show that $M = \{(x \quad 0) : x \in \mathbb{R}\}$ is a subspace of $\mathbb{R}^2$.

11. Find two subspaces H and K of $\mathbb{R}^2$ such that $H \bigoplus K = \mathbb{R}^2$ and find two subspaces S_1 and S_2 of $\mathbb{R}^3$ such that $S_1 \bigoplus S_2 = \mathbb{R}^3$.

12. Show that the following sets are subspaces of 3rd order matrices, $\mathbb{R}^3 \times \mathbb{R}^3$.

    (i) all diagonal matrices of order 3×3

    (ii) all symmetric matrices of order 3×3.

13. Are $H = \{(x \quad y) : x \geq 0, y \geq 0, x, y \in \mathbb{R}\} \bigcup \{\{x \quad y\} : x \leq 0, y \leq 0, x, y \in \mathbb{R}\}$ and $K = \{(x \quad y) : x \geq 0, y \leq 0, x, y \in \mathbb{R}\} \bigcup \{\{x \quad y\} : x \leq 0, y \geq 0, x, y \in \mathbb{R}\}$ are subspaces of $\mathbb{R}^2$. Give reasons.

14. If W_1 and W_2 are subspaces of V then show that

    (i) $W_1 \bigcap W_2$ is a subspace of V.

    (ii) $W_1 + W_2$ is also a subspace of V.

    (iii) $W_1 \bigcup W_2$ need not be a subspace of V by an example.

3.3 Linear Independence, Basis and Dimension

A vector space may have a finite number of elements or an infinite number of elements. In the latter situation, it is quite complex and complicated to verify the properties or study the vector space. Thus it would be convenient if there is a set of finite number of elements that represent the vector space. In this section, this set is discovered. The following definitions are a step in that direction.

Let $(V, +, .)$ be a real vector space.

Definition 3.3.1 *Linear Combination. Let* $C = \{\mathbf{v}_1, \mathbf{v}_2, \ldots, \mathbf{v}_n\}$ *be a subset of the vector space* V*. Then a linear combination of the vectors in set* C *is a vector* $\mathbf{v} = \alpha_1\mathbf{v}_1 + \alpha_2\mathbf{v}_2 + \cdots + \alpha_n\mathbf{v}_n \in V$ *where* $\alpha_1, \alpha_2, \ldots, \alpha_n \in \mathbb{R}$.

Example 3.3.1 *Let* $C = \{(1\ \ 1), (3\ \ 0)\}$*, then* $\mathbf{v} = 3(1\ \ 1) + 4(3\ \ 0) = (15\ \ 3)$ *is a linear combination of the vectors in* C.

Definition 3.3.2 *Linear span is the collection of all possible linear combinations of the set of all vectors in* C *and is denoted by* S_c.

Example 3.3.2 $C=\{(1\ \ 0\ \ 0), (0\ \ 1\ \ 0), (0\ \ 0\ \ 1)\}$. *Then* $S_c = \mathbb{R}^3$*, verify!*

Example 3.3.3 *Let* $C = \{(1\ 1\ 0), (1\ 0\ 1), (0\ 1\ 1)\}$.
Then $S_c = \mathbb{R}^3$*. Verify!*

Definition 3.3.3 *Linearly independent set. A subset* $S = \{\mathbf{v}_1, \mathbf{v}_2, \ldots, \mathbf{v}_n\}$ *of* V *is said to be linearly independent if and only if for* $\alpha_i \in \mathbb{R}$ *for all* $i = 1, 2, \ldots, n$ *and* $\alpha_1\mathbf{v}_1+\alpha_2\mathbf{v}_2+\cdots+\alpha_n\mathbf{v}_n = \mathbf{0}$ *imply* $\alpha_i = 0$ *for all* $i = 1, 2, \ldots, n$.

Example 3.3.4 $S_1 = \{(1\ \ 0\ \ 0), (0\ \ 1\ \ 0), (0\ \ 0\ \ 1)\}$ *and* $S_2 = \{(0\ 1\ 1), (1\ 0\ 1), (1\ 1\ 0)\}$ *are linearly independent sets. Verify!*

Definition 3.3.4 *Linearly dependent set. A set* S *is said to be linearly dependent if it is not linearly independent.*

Aliter definition

A set S is said to be linearly dependent if there exist scalars $\alpha_i, i = 1, 2, \ldots, n$, not all zero, such that $\alpha_1\mathbf{v}_1 + \alpha_2\mathbf{v}_2 + \cdots + \alpha_n\mathbf{v}_n = 0$.

Example 3.3.5 *C={(1 1), (2 0), (5 3)} is a linearly dependent set. Verify!*

Result. Linear span S_c is a subspace of V.

Proof. Let $C = \{\mathbf{v}_1, \mathbf{v}_2, \ldots, \mathbf{v}_n\}$ be any set of vectors and $S_c = \{\mathbf{v}/\mathbf{v} = \alpha_1\mathbf{v}_1 + \alpha_2\mathbf{v}_2 + \cdots + \alpha_n\mathbf{v}_n \textit{ where } \mathbf{v}_i \in V,\ i = 1, 2, \ldots, n\}$
Consider $\mathbf{u}, \mathbf{v} \in S_c$.
Then $\mathbf{u} = \sum_{i=1}^{n} \alpha_i\mathbf{v}_i$ and $\mathbf{v} = \sum_{i=1}^{n} \beta_i\mathbf{v}_i$ where $\alpha_i, \beta_i \in \mathbb{R}$, i=1,2,. . . ,n. Then

$$\mathbf{u} + \mathbf{v} = \sum_{i=1}^{n}(\alpha_i + \beta_i)\mathbf{v}_i = \sum_{i=1}^{n} r_i\mathbf{v}_i,$$

where $r_i = \alpha_i + \beta_i \in \mathbb{R}$ which means that $\mathbf{u} + \mathbf{v} \in S_c$.
Consider $\mathbf{v} \in S_c$ then for any $\alpha \in \mathbb{R}$.

$$\alpha\mathbf{v} = \alpha\sum_{i=1}^{n} \alpha_i\mathbf{v}_i = \sum_{i=1}^{n}(\alpha\alpha_i)\mathbf{v}_i \in S_c.$$

Hence by Theorem 3.2.1, S_c is a subspace of V.

A combination of the above two notions gives rise to one of the most important and fundamental concepts in linear algebra and is defined below.

Definition 3.3.5 *A subset $\mathcal{B} = \{\mathbf{v}_1, \mathbf{v}_2, \ldots, \mathbf{v}_n\}$ of a vector space V is said to be a basis for V if and only if*

(i) $\mathcal{B}$ is a linearly independent set and

(ii) The space V is a linear span of $\mathcal{B}$.

Example 3.3.6 *1. The sets S_1 and S_2 given in Example 3.3.4 are two different bases for $\mathbb{R}^3$.*

2. $\mathcal{B} = \{1, x, x^2, \ldots, x^n\}$ forms a basis for the vector space of all polynomials of degree $\leq$ 'n' with the definition of '+' and '.' defined as in Example 3.2.1 (3).

3. $\mathcal{B} = \left\{\begin{bmatrix}1 & 0\\0 & 0\end{bmatrix}, \begin{bmatrix}0 & 1\\0 & 0\end{bmatrix}, \begin{bmatrix}0 & 0\\1 & 0\end{bmatrix}, \begin{bmatrix}0 & 0\\0 & 1\end{bmatrix}\right\}$ forms a basis for the vector space consisting of all 2×2 matrices with '+' and '.' defining matrix addition and scalar multiplication of a matrix.

4. The eigen-vectors of a given $n \times n$ matrix $\mathbf{A}$ with 'n' distinct eigen-values form a basis for $\mathbb{R}^n$.

5. *Every nonsingular matrix of order* 3 *forms a basis for using its* 3 *column vectors or its* 3 *row vectors* $\mathbb{R}^3$.

Observation. From the above examples it can be observed that the basis of a vector space is not unique. In fact, there can be infinitely many bases, but the number of elements in any basis is unique. This leads to the following definition.

Definition 3.3.6 *The number of elements in any basis of a vector space is called the dimension of V.*

Example 3.3.7 *(i) The dimension of* $\mathbb{R}^n$ *is* n.

(ii) The dimension of the vector space of all 2×2 *matrices is 4.*

(iii) The dimension of the vector space of all polynomials of degree $\leq n$ *is* $n+1$ *from Example 3.3.6 (2).*

Observation.

1. The number of elements in any basis of the vector space of all polynomials, for every degree cannot be finite based on earlier presentation. Hence a basis can have finite, or infinite number of elements.

2. If the basis has a finite number of elements, the basis is said to be finite and the vector space is called as a finite dimensional vector space. Otherwise, the basis is infinite and the vector space is said to be infinite dimensional.

The following properties relative to a basis are stated without proof.

Result. Suppose that $dim V = n$.

1. Let $\mathcal{B} = \{\mathbf{v}_1, \mathbf{v}_2, \ldots, \mathbf{v}_m\}$ be any subset of V having m linearly independent vectors, $m < n$. Then $\mathcal{B}$ can be extended to form a basis, that is there exist vectors $\{\mathbf{v}_{m+1}, \mathbf{v}_{m+2}, \ldots, \mathbf{v}_n\}$ such that $\{\mathbf{v}_1, \mathbf{v}_2, \ldots, \mathbf{v}_m, \mathbf{v}_{m+1}, \mathbf{v}_{m+2}, \ldots, \mathbf{v}_n\}$ is a basis for V.

2. If $S_c = \{\mathbf{v}_1, \mathbf{v}_2, \ldots, \mathbf{v}_p\}$ is a linear span of V, $p > n$ then there exist $\mathcal{B} \subseteq S_c$ such that $\mathcal{B} = \{\mathbf{v}_{i_1}, \mathbf{v}_{i_2}, \ldots, \mathbf{v}_{i_n}\}$ is a basis for V where $i_1, i_2, \ldots i_n$ are distinct elements from $\{1, 2, \ldots, p\}$.

Theorem 3.3.1 *If* W_1 *and* W_2 *are any two finite dimensional subspaces of a vector space V, then* $W_1 + W_2$ *is also a finite dimensional subspace and*

$$dim(W_1 + W_2) = dim W_1 + dim W_2 - dim(W_1 \cap W_2).$$

Proof. Let $dim W_1 = m$ and $dim W_2 = n$. Now Theorem 3.2.2 and Theorem 3.2.3 guarantee that $W_1 \cap W_2$ and $W_1 + W_2$ are subspaces of V. Let

$dim(W_1 \cap W_2) = p$. Then clearly $p \leq m$ and $p \leq n$ (why?). Let $\mathcal{B}_0 = \{\mathbf{v}_1, \mathbf{v}_2, \ldots, \mathbf{v}_p\}$ be a basis for $W_1 \cap W_2$. Since $\mathcal{B}_0$ is a linearly independent subset of W_1 and W_2, it follows from the above observation that $\mathcal{B}_0$ can be extended to $\mathcal{B}_1 = \{\mathbf{v}_1, \mathbf{v}_2, \ldots, , \mathbf{v}_p, \mathbf{w}_{1p+1}, \mathbf{w}_{1p+2}, \ldots, \mathbf{w}_{1m}\}$ and $\mathcal{B}_2 = \{\mathbf{v}_1, \mathbf{v}_2, \ldots, , \mathbf{v}_p, \mathbf{w}_{2\,p+1}, \mathbf{w}_{2\,p+2}, \ldots, \mathbf{w}_{2\,n}\}$ to form a basis for W_1 and W_2, respectively.

To show the vectors in $\mathcal{B}_0$, $\mathcal{B}_1$ and $\mathcal{B}_2$ are linearly independent. Consider

$$\sum_{i=1}^{p} a_i \mathbf{v}_i + \sum_{i=p+1}^{m} b_i \mathbf{w}_{1i} + \sum_{i=p+1}^{n} c_i \mathbf{w}_{2i} = 0 \tag{3.1}$$

and where $a_1, a_1, \ldots, a_p, b_{p+1}, \ldots, b_m, c_{p+1}, c_{p+2}, \ldots, c_n$ are constants. That gives

$$\sum_{i=p+1}^{m} b_i \mathbf{w}_{1i} = -\sum_{i=1}^{p} a_i \mathbf{v}_i - \sum_{i=p+1}^{n} c_i \mathbf{w}_{2i}.$$

Now left-hand side is a vector in W_1 and right-hand side is a vector in W_2. Thus

$$\sum_{i=p+1}^{m} b_i \mathbf{w}_{1i} \in W_1 \cap W_2.$$

Hence there exist constants $\delta_1, \delta_2, \ldots, \delta_p$ such that

$$\sum_{i=p+1}^{m} b_i \mathbf{w}_{1i} = \sum_{i=1}^{p} \delta_i \mathbf{v}_i$$

$$\Longrightarrow \sum_{i=1}^{p} \delta_i \mathbf{v}_i + \sum_{i=p+1}^{m} b_1 \mathbf{w}_{1i} = 0.$$

Since $\mathbf{v}_1, \mathbf{v}_2 \cdots \mathbf{v}_p$ and $\mathbf{w}_{1p+1}, \ldots, \mathbf{w}_{1m}$ is a basis for W_1, it implies that $\delta_i = 0,\ b_i = 0,\ \ i = 1, 2, \ldots, m$.

Substituting in relation (3.1), $b_i = 0, \ \ i = 1, 2, \ldots, m$, yield

$$\sum_{i=1}^{p} a_i \mathbf{v}_i + \sum_{i=p+1}^{n} c_i \mathbf{w}_{2i} = 0.$$

Now $\mathcal{B}_2 = \{\mathbf{v}_1, \mathbf{v}_2, \ldots, \mathbf{v}_p, \mathbf{w}_{2\,p+1}, \mathbf{w}_{2\,p+2}, \ldots, \mathbf{w}_{2\,n}\}$ is a basis for W_2 which gives $a_i = 0$ and $c_i = 0$ for i=1,2,...,p, p+1,...,n. Set $\mathcal{B} = \mathcal{B}_0 \cup \mathcal{B}_1 \cup \mathcal{B}_2$ then clearly every vector in $W_1 + W_2$ can be expressed as a linear combination of elements in $\mathcal{B}$.

Thus,

$$\mathcal{B} = \{\mathbf{v}_1, \mathbf{v}_2, \ldots, \mathbf{v}_p, \mathbf{w}_{1p+1}, \ldots, \mathbf{w}_{1m}, \mathbf{w}_{2\,p+1}, \mathbf{w}_{2\,p+2}, \ldots, \mathbf{w}_{2\,n}\}$$

forms a basis for $W_1 + W_2$. Also,

$$\begin{aligned} dim(W_1 + W_2) &= p + (m - p) + (n - p) \\ &= m + n - p \\ &= dimW_1 + dimW_2 - dim(W_1 \cap W_2). \end{aligned}$$

Hence, the proof is complete.

EXERCISE 3.3

1. Show that $\{(1\ \ -2\ \ 1),\ (2\ \ 3\ \ -4), (7\ \ 6\ \ 5),\ (3\ \ -2\ \ 5)\}$ is a linearly dependent set. Find a basis for $\mathbb{R}^3$ that has the vector $(2\ \ 0\ \ 1)^T$.

2. Show that $\{1,\ t,\ t^2\}$ is a linearly independent set.

3. Verify $\{1, t, t^2, \ldots, t^n\}$ is a basis for a vector space of all polynomials of degree $\leq n$.

4. Add a polynomial to the set $\{1,\ 1 + x^2\}$ so that it forms a basis for $\mathbb{P}_2$, all polynomials of degree ≤ 2.

5. Find a basis for each of these subspaces of 3×3 matrices

   (i) All diagonal matrices
   (ii) All symmetric matrices
   (iii) All skew symmetric matrices.

6. Find the dimensions of the subspaces of $\mathbb{R}^2$ spanned by the vectors

   (a) $(0\ \ 0),\ (3\ \ 4),\ (9\ \ 16)$
   (b) $(2\ \ 4),\ (7\ \ 3)$.

7. Find the dimensions of the subspaces of $\mathbb{R}^4$ spanned by the vectors

   (a) $(1\ \ 0\ \ 0\ \ 1),\ (0\ \ 1\ \ 0\ \ 0),\ (1\ \ 1\ \ 1\ \ 1),\ (0\ \ 1\ \ 1\ \ 1)$
   (b) $(1\ \ -1\ \ 0\ \ 2),\ (3\ \ -1\ \ 2\ \ 1),\ (1\ \ 0\ \ 0\ \ 1)$
   (c) $(-2\ \ 4\ \ 6\ \ 4),\ (0\ \ 1\ \ 2\ \ 0),\ (-1\ \ 2\ \ 3\ \ 2), (-3\ \ 2\ \ 5\ \ 6),$
   $(-2\ \ -1\ \ 0\ \ 4)$.

8. Find the dimension of the solution space W of 4×5 matrix

$$\begin{bmatrix} 1 & 2 & 3 & -1 & 1 \\ 0 & 2 & 4 & 1 & 3 \\ 3 & 1 & -2 & 0 & 5 \\ 4 & 3 & 1 & 2 & 3 \end{bmatrix} \begin{bmatrix} x_1 \\ x_2 \\ x_3 \\ x_4 \\ x_5 \end{bmatrix} = \begin{bmatrix} 0 \\ 0 \\ 0 \\ 0 \\ 0 \end{bmatrix}.$$

3.4 Change of Basis–Matrix

It has been discussed in the previous section that a basis can be considered as a representative of any vector in a vector space. A basis can also be considered as a coordinate system for a vector space by putting an additional condition which is given below.

Definition 3.4.1 *An ordered basis $\mathcal{B}$ is a basis whose vectors are arranged in a definite order.*

Example 3.4.1 $\mathcal{B}_1 = \{(1\ \ 0\ \ 0),\ (0\ \ 1\ \ 0),\ (0\ \ 0\ \ 1)\}$ *is an ordered basis and* $\mathcal{B}_2 = \{(0\ \ 1\ \ 0),\ (1\ \ 0\ \ 0),\ (0\ \ 0\ \ 1)\}$ *is another ordered basis. In fact, there are four more different ordered bases with these elements.*

Let $\mathcal{B} = \{\mathbf{v}_1, \mathbf{v}_2, \cdots, \mathbf{v}_n\}$ be an ordered basis. Then every vector can be written as a unique linear combination of the vectors in $\mathcal{B}$, say $\mathbf{v} = \alpha_1\mathbf{v}_1 + \alpha_2\mathbf{v}_2 + \cdots + \alpha_n\mathbf{v}_n$ (verify!). The coefficients $\alpha_1, \alpha_2, \ldots, \alpha_n$ are written as a column and are called as the coordinates of the vector $\mathbf{v}$ with respect to the basis $\mathcal{B}$ and is written as

$$[\mathbf{v}]_{\mathcal{B}} = \begin{bmatrix} \alpha_1 \\ \alpha_2 \\ \vdots \\ \alpha_n \end{bmatrix}. \tag{3.2}$$

As a vector space can have more than one basis it would be convenient if there exists a tool that would link the coordinates of a vector represented in one basis to the coordinates of the vector obtained by using a different basis.

In other words, the question is the availability of a tool that can give the coordinates of a vector with respect to a basis $\mathcal{B}'$ when its coordinates with respect to a basis $\mathcal{B}$ are known. The answer is affirmative and the procedure is given below.

Let $\mathcal{B} = \{\mathbf{v}_1, \mathbf{v}_2, \ldots, \mathbf{v}_n\}$ and $\mathcal{B}' = \{\mathbf{w}_1, \mathbf{w}_2, \cdots, \mathbf{w}_n\}$ be two different ordered bases of an n-dimensional vector space $\mathbf{V}$.

Then each vector in $\mathcal{B}'$ can be written as a linear combination of vectors in $\mathcal{B}$ as

$$\mathbf{w}_j = \sum_{i=1}^{n} c_{ij}\mathbf{v}_i, \quad j = 1, 2, 3, \ldots, n. \tag{3.3}$$

Let $\mathbf{v} \in V$. Then $\mathbf{v}$ can be expressed in terms of vectors of both bases $\mathcal{B}$ and $\mathcal{B}'$ as follows.

$$\mathbf{v} = \sum_{i=1}^{n} \alpha_i\mathbf{v}_i \quad = \sum_{j=1}^{n} \beta_j\mathbf{w}_j \tag{3.4}$$

using (3.3) in (3.4) gives

$$\mathbf{v} = \sum_{j=1}^{n} \beta_j \mathbf{w}_j = \sum_{j=1}^{n} \beta_j \sum_{i=1}^{n} c_{ij}\mathbf{v}_i = \sum_{i=1}^{n} \mathbf{v}_i \sum_{j=1}^{n} \beta_j c_{ij} = \sum_{i=1}^{n} (\sum_{j=1}^{n} c_{ij}\beta_j)\mathbf{v_i}.$$

Comparing the coefficients yields

$$\alpha_i = \sum_{j=1}^{n} c_{ij}\beta_j$$

or using vector notation for $\mathbf{v} \in V$

$$[\mathbf{v}]_{\mathcal{B}} = \mathbf{C}[\mathbf{v}]_{\mathcal{B}'}, \tag{3.5}$$

where $\mathbf{C} = [c_{ij}]_{n\times n}$.

This procedure results in the following theorem.

Theorem 3.4.1 *Let V be a vector space of finite dimension n. Let $\mathcal{B}$ and $\mathcal{B}'$ be two bases for V. Then there exists a unique nonsingular matrix $\mathbf{C}$ of order n such that, for any vector $\mathbf{v} \in V$,*

$$[\mathbf{v}]_{\mathcal{B}} = \mathbf{C}\,[\mathbf{v}]_{\mathcal{B}'}.$$

Example 3.4.2 *Let $\mathcal{B} = \{(1\ \ 1\ \ 0),\ (1\ \ 0\ \ 1), (0\ \ 1\ \ 1)\}$ be a basis for $\mathbb{R}^3$. If $\mathbf{v} = (2\ \ 4\ \ 8)$, find the coordinates of $\mathbf{v}$ with respect to $\mathcal{B}$.*

Solution.

Let $\mathbf{v} = a(1\ \ 1\ \ 0) + b(1\ \ 0\ \ 1) + c(0\ \ 1\ \ 1)$, then this is equivalent to solving the system $\mathbf{Ax} = \mathbf{b}$ where $\mathbf{x} = (a\ \ b\ \ c), \mathbf{b} = (2\ \ 4\ \ 8)$ and $\mathbf{A} = \begin{bmatrix} 1 & 1 & 0 \\ 1 & 0 & 1 \\ 0 & 1 & 1 \end{bmatrix}$.

Using the augmented matrix

$\begin{bmatrix} 1 & 1 & 0:2 \\ 1 & 0 & 1:4 \\ 0 & 1 & 1:8 \end{bmatrix}$ *and reducing to echelon form yields* $\begin{bmatrix} a \\ b \\ c \end{bmatrix}_{\mathcal{B}} = \begin{bmatrix} -1 \\ 3 \\ 5 \end{bmatrix}$.

Example 3.4.3 *Let $\mathcal{B}'$ be the basis in the above example and $\mathcal{B}$ $=\{(1\ \ 0\ \ 0),\ (0\ \ 1\ \ 0),\ (0\ \ 0\ \ 1)\}$ be the standard basis for $\mathbb{R}^3$. Verify Theorem 3.4.1 by finding the nonsingular matrix $\mathbf{C}$ such that $[\mathbf{v}]_{\mathcal{B}'} = C[\mathbf{v}]_{\mathcal{B}}$ for $\mathbf{v} = [2, 4, 8]$.*

Solution.

Consider the augmented matrix $\mathbf{K} = [\mathcal{B}' : \mathcal{B}]$

$$\begin{bmatrix} 1 & 1 & 0:1 & 0 & 0 \\ 1 & 0 & 1:0 & 1 & 0 \\ 0 & 1 & 1:0 & 0 & 1 \end{bmatrix}$$

$R_2 \to R_2 - R_1$

$$\begin{bmatrix} 1 & 1 & 0 : 1 & 0 & 0 \\ 0 & -1 & 1 : -1 & 1 & 0 \\ 0 & 1 & 1 : 0 & 0 & 1 \end{bmatrix}$$

$R_3 \to R_3 + R_2,\ R_1 \to R_1 + R_2$

$$\begin{bmatrix} 1 & 0 & 1 : 0 & 1 & 0 \\ 0 & -1 & 1 : -1 & 1 & 0 \\ 0 & 0 & 2 : -1 & 1 & 1 \end{bmatrix}$$

$R_2 \to (-1)R_2,\ R_3 \to \frac{1}{2}R_3$

$$\begin{bmatrix} 1 & 0 & 1 : 0 & 1 & 0 \\ 0 & 1 & -1 : 1 & -1 & 0 \\ 0 & 0 & 1 : \frac{-1}{2} & \frac{1}{2} & \frac{1}{2} \end{bmatrix}$$

$R_2 \to R_2 + R_3$

$$\begin{bmatrix} 1 & 0 & 1 : 0 & 1 & 0 \\ 0 & 1 & 0 : \frac{1}{2} & \frac{-1}{2} & \frac{1}{2} \\ 0 & 0 & 1 : \frac{-1}{2} & \frac{1}{2} & \frac{1}{2} \end{bmatrix}$$

$R_1 \to R_1 - R_3$

$$\begin{bmatrix} 1 & 0 & 0 : \frac{1}{2} & \frac{1}{2} & \frac{-1}{2} \\ 0 & 1 & 0 : \frac{1}{2} & \frac{-1}{2} & \frac{1}{2} \\ 0 & 0 & 1 : \frac{-1}{2} & \frac{1}{2} & \frac{1}{2} \end{bmatrix}.$$

$$\textit{Hence} \quad \mathbf{C} = \begin{bmatrix} \frac{1}{2} & \frac{1}{2} & \frac{-1}{2} \\ \frac{1}{2} & \frac{-1}{2} & \frac{1}{2} \\ \frac{-1}{2} & \frac{1}{2} & \frac{1}{2} \end{bmatrix}.$$

$$[\mathbf{v}]_{\mathcal{B}} = \begin{bmatrix} 2 \\ 4 \\ 8 \end{bmatrix}$$

$$[\mathbf{v}]_{\mathcal{B}'} = \begin{bmatrix} \frac{1}{2} & \frac{1}{2} & \frac{-1}{2} \\ \frac{1}{2} & \frac{-1}{2} & \frac{1}{2} \\ \frac{-1}{2} & \frac{1}{2} & \frac{1}{2} \end{bmatrix} \begin{bmatrix} 2 \\ 4 \\ 8 \end{bmatrix}$$

$$= \begin{bmatrix} -1 \\ 3 \\ 5 \end{bmatrix}, \textit{ verifying the theorem.}$$

EXERCISE 3.4

1. Find the change of basis matrix, that is, a nonsingular matrix $\mathbf{C}$ such that $[\mathbf{v}]_{\mathcal{B}'} = \mathbf{C}[\mathbf{v}]_{\mathcal{B}}$ for the bases $\mathcal{B} = \{(1 \ \ 0), \ (1 \ \ 1)\}$ and $\mathcal{B}' = \{(1 \ \ 3), \ (-2 \ \ 6)\}$.

2. Let $\mathcal{B}' = \{(1 \ \ 3), \ (2 \ \ 8)\}$. If $\mathbf{v} = (4 \ \ 2)$, find the coordinates of $\mathbf{v}$ with respect to basis of $\mathcal{B}'$.

3. Let $\mathcal{B} = \{(1 \ \ 1 \ \ 0), \ (2 \ \ 0 \ \ 1), \ (0 \ \ 1 \ \ 1)\}$ and $\mathcal{B}' = \{(0 \ \ 2 \ \ 0), \ (1 \ \ 4 \ \ 3), \ (-1 \ \ 2 \ \ 1)\}$ find the change of basis matrix.

4. $\mathcal{B} = \{(1 \ \ 0 \ \ 0 \ \ 0), \ (0 \ \ 1 \ \ 0 \ \ 0), \ (0 \ \ 0 \ \ 1 \ \ 0), \ (0 \ \ 0 \ \ 0 \ \ 1)\}$ and $\mathcal{B}' = \{(1 \ \ 0 \ \ 0 \ \ 1), \ (0 \ \ 1 \ \ 0 \ \ 0), \ (1 \ \ 1 \ \ 1 \ \ 1), \ (0 \ \ 1 \ \ 1 \ \ 1)\}$, Find change of basis matrix.

5. $\mathcal{B} = \{(1 \ \ 0 \ \ 0 \ \ 1), \ (0 \ \ 1 \ \ 0 \ \ 0), \ (1 \ \ 1 \ \ 1 \ \ 1), \ (0 \ \ 1 \ \ 1 \ \ 1)\}$ and $\mathcal{B}' = \{(2 \ \ 2 \ \ 0 \ \ 0), \ (-1 \ \ 1 \ \ 1 \ \ 2), \ (0 \ \ 0 \ \ 2 \ \ 2), \ (2 \ \ 1 \ \ 2 \ \ 1)\}$, Find the change of basis matrix.

3.5 Linear Transformations

In Section 1.3, matrix multiplication was described and it can be observed that an $n \times 1$ column vector, (i.e.) $\mathbf{x}$, when multiplied by an $m \times n$ matrix $\mathbf{A}$, yields an $m \times 1$-vector, (i.e.) $\mathbf{y}$, that is, $\mathbf{Ax} = \mathbf{y}$.

In other words, a matrix behaves like a mapping taking a vector in $\mathbb{R}^n$ to a vector in $\mathbb{R}^m$. Further, matrix addition and scalar multiplication are distributive in nature. Taking cue from matrices the idea is to introduce a mapping from one vector space into another such that the properties or rules satisfied by a matrix are preserved. This leads to the following definition of a mapping which generalizes the concept of a matrix.

Definition 3.5.1 *A linear transformation T is a mapping from a vector space, V, into another vector space, W, denoted by $T : V \to W$ satisfying*

(i) $T(\mathbf{v}_1 + \mathbf{v}_2) = T\mathbf{v}_1 + T\mathbf{v}_2$ *for all* $\mathbf{v}_1, \mathbf{v}_2 \in V$*;*

(ii) $(\alpha T)(\mathbf{v}) = (\alpha)(T\mathbf{v})$*, for all* $\alpha \in \mathbb{R}$*,* $\mathbf{v} \in V$*.*

Observation. The addition and scalar multiplication on left-hand side are that in V and on right-hand side is that in W and hence may be different.

Example 3.5.1 *Let* $T : \mathbb{R}^2 \to \mathbb{R}^2$ *be such that* *(a)* $T(x\ \ y)=(y\ \ x)$

(b) $T(x\ \ y) = T(x\ \ 0)$.

Both the examples are linear transformations. Verify.

Example 3.5.2 $T : \mathbb{R}^3 \to \mathbb{R}^3$ *defined by*

(a) $T(x\ \ y\ \ z)=(ax\ \ by\ \ cz)$

(b) $T(x\ \ y\ \ z) = (x+y\ \ y+z\ \ z+x)$

are linear transformations. Verify.

Example 3.5.3 $T : \mathbb{R}^3 \to \mathbb{R}^2$ *defined by*
$T(x\ \ y\ \ z) = (x\ \ y)$ *is a linear transformation.*

Example 3.5.4 $T : \mathbb{R}^{2\times 2} \to \mathbb{R}$ *defined by*
(a) $T\begin{bmatrix} a & b \\ c & d \end{bmatrix} = a+b+c+d$ *is a linear transformations.*

Solution. *Consider* $A = \begin{bmatrix} a_1 & a_2 \\ a_3 & a_4 \end{bmatrix}$ *and* $B = \begin{bmatrix} b_1 & b_2 \\ b_3 & b_4 \end{bmatrix}$

Then, $T(A+B) = T\begin{bmatrix} a_1+b_1 & a_2+b_2 \\ a_3+b_3 & a_4+b_4 \end{bmatrix}$

$$\begin{aligned} &= a_1 + b_1 + a_2 + b_2 + a_3 + b_3 + a_4 + b_4 \\ &= (a_1 + a_2 + a_3 + a_4) + (b_1 + b_2 + b_3 + b_4) \\ &= T(A) + T(B). \end{aligned}$$

Also, $(\alpha T)(A) = (\alpha T)\begin{bmatrix} a_1 & a_2 \\ a_3 & a_4 \end{bmatrix} = T\begin{bmatrix} \alpha a_1 & \alpha a_2 \\ \alpha a_3 & \alpha a_4 \end{bmatrix}$

$$\begin{aligned} &= \alpha a_1 + \alpha a_2 + \alpha a_3 + \alpha a_4) \\ &= \alpha(a_1 + a_2 + a_3 + a_4) \\ &= \alpha T(A). \end{aligned}$$

Observations.

(1) A linear transformation $T : V \to V$ is called a linear operator.

(2) The identity mapping $I : V \to V$ defined by $I(\mathbf{v}) = \mathbf{v}$ is a linear operator.

(3) The zero mapping $O : V \to V$ defined by $O(\mathbf{v}) = \mathbf{0}$ is a linear operator.

(4) T($\mathbf{0}$)=$\mathbf{0}$, for any linear transformation T.

Example 3.5.5 $T : \mathbb{R} \to \mathbb{R}$ *defined by*
$T(x) = x+c$ *is not a linear transformation since*
$T(x+y) = x+y+c \neq Tx + Ty = (x+c) + (y+c)$.

The above example leads to a very interesting definition which plays a vital role in applications relating to computer graphics.

Definition 3.5.2 *An affine transformation is a mapping $T : V \to V$ defined by $T\boldsymbol{v} = \boldsymbol{v} + \boldsymbol{u}_0$, where $\boldsymbol{u}_0$ is a fixed vector.*

Observation.

(1) In $\mathbb{R}$, a linear transformation is a straight line passing through the origin ($T(0) = 0$ holds).

(2) In $\mathbb{R}$, an affine transformation is a straight line not passing through the origin ($T(0) \neq 0$ holds).

With the definition of a linear transformation in place, the next question that arises is about the domain, range and null space of a linear transmission T. These are given below.

Let $T : V \to W$ be a linear transformation defined from a vector space V to another vector space W (with underlying elements in $\mathbb{R}$).

Definition 3.5.3 *The null space or kernel of T is defined as the set of all vectors, $\mathbf{v} \in V$ such that $T\mathbf{v} = \mathbf{0}$, and is denoted by $N(T) = \{\mathbf{v} : T\mathbf{v} = \mathbf{0}\}$.*

Definition 3.5.4 *The range of T is the set of all $\mathbf{w} \in W$ such that $T\mathbf{v} = \mathbf{w}$ and is denoted by $R(T) = \{T\mathbf{v} : \mathbf{v} \in V\}$.*

Observation. $N(T) \subseteq V$ and $R(T) \subseteq W$.

Theorem 3.5.1 *If $T : V \to W$ is a linear transformation, then N(T) is a subspace of V and R(T) is a subspace of W.*

Proof. By Theorem 3.2.1, it is sufficient to show that $\alpha\mathbf{v}_1 + \mathbf{v}_2 \in N(T)$ whenever $\mathbf{v}_1, \mathbf{v}_2 \in N(T)$.

Let $\mathbf{v}_1, \mathbf{v}_2 \in N(T)$ then $T(\mathbf{v}_1) = T(\mathbf{v}_2) = \mathbf{0}$. Consider

$$\begin{aligned} T(\alpha\mathbf{v}_1 + \mathbf{v}_2) &= \alpha T(\mathbf{v}_1) + T(\mathbf{v}_2) \\ &= \alpha(\mathbf{0}) + \mathbf{0} \\ &= \mathbf{0} \end{aligned}$$

which yields $\alpha\mathbf{v}_1 + \mathbf{v}_2 \in N(T)$ and thus N(T) is a subspace of V.

Let $\mathbf{w}_1, \mathbf{w}_2 \in R(T)$ then there exist $\mathbf{v}_1$ and $\mathbf{v}_2 \in V$ such that $T\mathbf{v}_1 = \mathbf{w}_1$ and $T\mathbf{v}_2 = \mathbf{w}_2$. Consider

$$\begin{aligned} \alpha\mathbf{w}_1 + \mathbf{w}_2 &= \alpha T\mathbf{v}_1 + T\mathbf{v}_2 \\ &= T(\alpha\mathbf{v}_1) + T(\mathbf{v}_2) \\ &= T(\alpha\mathbf{v}_1 + \mathbf{v}_2) \in R(T). \end{aligned}$$

The above relations follow from the fact that $\alpha\mathbf{v}_1 + \mathbf{v}_2 \in V$. From the observation following Theorem 3.2.1, R(T) is a subspace of W and the proof is complete.

The importance of a basis studied in Section 3.3 can be observed when studying linear transformations. A linear transformation must map every element in V to an element in W. But because of the basis it is enough for a linear transformation to map every element in the basis to an element in W (Why?).

The following theorem characterizes the importance of the basis. It is stated without proof.

Theorem 3.5.2 *Let V be a vector space with dim$V$=n, and $\{\mathbf{v}_1, \mathbf{v}_2, \dots, \mathbf{v}_n\}$ be an ordered basis for V. Further, let W be another vector space and $\mathbf{w}_1, \mathbf{w}_2, \dots, \mathbf{w}_n \in W$ be any given n vectors. Then there exists a unique linear transformation T from V to W such that $T\mathbf{v}_i = \mathbf{w}_i$, $i = 1, 2, 3, \dots, n$.*

The result given below will link up the concepts of a basis and a linear transformation. In order to do so the following definitions are in order.

Definition 3.5.5 *The dimension of the null space of a linear transformation T is called nullity of T and is denoted by dim $N(T)$ or $n(T)$.*

Definition 3.5.6 *The dimension of the range space of a linear transformation T is called the rank of T and is denoted by dim $R(T)$ or $r(T)$.*

Theorem 3.5.3 *Let V and W be vector spaces with dim$V$ $= n < \infty$ and $T : V \to W$ be a linear transformation. Then $r(T) + n(T) = $ dim$V$.*

Proof. From Theorem 3.5.1, N(T) is a subspace of V, hence dim N(T)=m (say) $\leq n$. Let $\mathcal{B} = \{\mathbf{v}_1, \mathbf{v}_2, \dots, \mathbf{v}_m\}$ is a basis for N(T). Extend $\mathcal{B}$ to form a basis for V by adding $n - m$ vectors say $\mathbf{v}_{m+1}, \mathbf{v}_{m+2}, \dots, \mathbf{v}_n$. Then $\mathcal{B}' = \{\mathbf{v}_1, \mathbf{v}_2, \dots, \mathbf{v}_m, \mathbf{v}_{m+1}, \dots, \mathbf{v}_n\}$ is a basis for V. To conclude the result of the theorem it is sufficient to show that $r(T) = n - m$. In other words, it is enough to show that the set $\hat{\mathcal{B}} = \{T\mathbf{v}_{m+1}, T\mathbf{v}_{m+2}, \dots, T\mathbf{v}_n\}$ is a basis for R(T). The proof of linear independence of the set $\hat{\mathcal{B}}$ is as follows.

Let $\alpha_1, \alpha_2, \dots, \alpha_{n-m}$ be $n - m$ scalars such that

$$\alpha_1 T\mathbf{v}_{m+1} + \alpha_2 T\mathbf{v}_{m+2} + \cdots + \alpha_{n-m} T\mathbf{v}_n = \mathbf{0}.$$

Then

$$T(\alpha_1 \mathbf{v}_{m+1} + \alpha_2 \mathbf{v}_{m+2} + \cdots + \alpha_{n-m} \mathbf{v}_n) = \mathbf{0}.$$

This implies that

$$\mathbf{v} = \alpha_1 \mathbf{v}_{m+1} + \alpha_2 \mathbf{v}_{m+2} + \cdots + \alpha_{n-m} \mathbf{v}_n \in N(T),$$

from which $\mathbf{v}$ can be written as a linear combination of elements of the basis $\mathcal{B}$ of $N(T)$.

Hence $\mathbf{v} = \alpha_1\mathbf{v}_{m+1} + \alpha_2\mathbf{v}_{m+2} + \cdots + \alpha_{n-m}\mathbf{v}_n = \beta_1\mathbf{v}_1 + \beta_2\mathbf{v}_2 + \cdots + \beta_m\mathbf{v}_m$ for some scalars $\beta_1, \beta_2, \ldots, \beta_m$, which gives,

$$-\{\beta_1\mathbf{v}_1 + \beta_2\mathbf{v}_2 + \cdots + \beta_m\mathbf{v}_m\} + \alpha_1\mathbf{v}_{m+1} + \alpha_2\mathbf{v}_{m+2} + \cdots + \alpha_{n-m}\mathbf{v}_n = 0$$

which implies $\beta_1 = \beta_2 = ... = \beta_m = \alpha_1 = \alpha_2 = \cdots = \alpha_{n-m} = 0$ as $\{\mathbf{v}_1, \mathbf{v}_2, \ldots, \mathbf{v}_n\}$ is a basis for V.

Next, to show that $\{T\mathbf{v}_{m+1}, T\mathbf{v}_{m+2}, \ldots, T\mathbf{v}_n\}$ spans R(T). Consider $\mathbf{w} \in R(T)$ then $\mathbf{w} = T\mathbf{v}$, for some $\mathbf{v} \in V$.

Let $\mathbf{v} = \gamma_1\mathbf{v}_1 + \gamma_2\mathbf{v}_2 + \cdots + \gamma_n\mathbf{v}_n$. Then

$$\begin{aligned}\mathbf{w} &= T(\gamma_1\mathbf{v}_1 + \gamma_2\mathbf{v}_2 + \cdots + \gamma_n\mathbf{v}_n)\\ &= \gamma_1 T(\mathbf{v}_1) + \gamma_2 T(\mathbf{v}_2) + \cdots + \gamma_m T(\mathbf{v}_m) + \gamma_{m+1} T(\mathbf{v}_{m+1}) + \cdots + \gamma_n T(\mathbf{v}_n)\\ &= \gamma_{m+1} T(\mathbf{v}_{m+1}) + \cdots + \gamma_n T(\mathbf{v}_n),\end{aligned}$$

since $T(\mathbf{v}_1) = T(\mathbf{v}_2) = \cdots = T(\mathbf{v}_m) = 0$, and the proof is complete.

Result.

1. Let T be a linear transformation from a vector space V to a vector space W whose zero elements are given by $\mathbf{0_v}$ and $\mathbf{0_w}$, respectively. Then $T(\mathbf{0_v}) = \mathbf{0_w}$.

   Proof. $T(\mathbf{0_v}) + \mathbf{0_w} \;=\; T(\mathbf{0_v}) = T(\mathbf{0_v} + \mathbf{0_v}) \;=\; T(\mathbf{0_v}) + T(\mathbf{0_v})$

   Thus, $T(\mathbf{0_v}) = \mathbf{0_w}$, since $T(\mathbf{0_v}) \in W$.
   In other words,

2. If $T(\mathbf{0_v}) \neq \mathbf{0_w}$, then T is not a linear transformation.

Minimal polynomial

Consider a linear operator T (which is a linear transformation from V to V) defined on a finite dimensional vector space V with dim $V = n$.

Corresponding to the linear operator T there exists a matrix $\mathbf{A}$ of order $n \times n$.

From Cayley–Hamilton theorem, every square matrix $\mathbf{A}$ satisfies its own characteristic equation. Thus if $p(\lambda)$ is the characteristic polynomial of $\mathbf{A}$ then $p(\mathbf{A}) = 0$, (also $p(T) = 0$).

The question that arises is whether there is a polynomial of lesser degree $q(\mathbf{A})$ that divides the characteristic polynomial and is such that $q(\mathbf{A}) = 0$. The answer is yes and leads to the following definition.

Definition 3.5.7 *Minimal polynomial for a linear operator T(and to the corresponding matrix $\mathbf{A}$) is a polynomial $q(\lambda)$ of smallest degree such that*

1. *$q(T) = 0$ and $q(\mathbf{A}) = 0$;*

2. *The coefficient of its highest degree term is '1';*

3. *q divides the characteristic polynomial (i.e.) $q(\lambda)$ divides $p(\lambda)$.*

Observation.

1. If $p(\lambda)$ is the characteristic polynomial of $\mathbf{A}$ and all the eigen-values of $\mathbf{A}$ are distinct then the minimal polynomial is the characteristic polynomial.

2. If $p(\lambda)$ is the characteristic polynomial of $\mathbf{A}$ and some or all of the eigen-values of $\mathbf{A}$ are repeated then there exists a minimal polynomial $q(\lambda)$ such that $q(\mathbf{A}) = 0$ and q divides p. This result leads to the Jordan blocks described in Chapter 2.

Example 3.5.6 *Find the minimal polynomial of*

$$A = \begin{bmatrix} 1 & 0 & 0 \\ 2 & 3 & 0 \\ 3 & 4 & 5 \end{bmatrix}.$$

Solution.
The characteristic equation of $\mathbf{A}$ is

$$|A - \lambda I| = 0$$

$$\left|\begin{bmatrix} 1-\lambda & 0 & 0 \\ 2 & 3-\lambda & 0 \\ 3 & 4 & 5-\lambda \end{bmatrix}\right| = 0$$

$$(1-\lambda)(3-\lambda)(5-\lambda) = 0$$

The characteristic polynomial is $p(\lambda) = (1-\lambda)(3-\lambda)(5-\lambda)$.

Here all the eigen-values different. So the minimal polynomial is the characteristic polynomial.

Thus the minimal polynomial is $q(\lambda) = (1-\lambda)(3-\lambda)(5-\lambda)$.

Example 3.5.7 *Find the minimal polynomial of*

$$A = \begin{bmatrix} 1 & 0 & 0 \\ 1 & -2 & 0 \\ 1 & 0 & -2 \end{bmatrix}.$$

Solution.
The characteristic equation of $\mathbf{A}$ is $|A - \lambda I|=0$

$$\left|\begin{bmatrix} 1-\lambda & 0 & 0 \\ 1 & -2-\lambda & 0 \\ 1 & 0 & -2-\lambda \end{bmatrix}\right| = 0$$

$$(\lambda - 1)(\lambda + 2)(\lambda + 2) = 0$$

The eigen-values are $1, -2, -2$.
The characteristic polynomial is $p(\lambda) = (\lambda - 1)(\lambda + 2)(\lambda + 2)$ and the minimal polynomial is $q(\lambda) = (\lambda - 1)(\lambda + 2)$.

EXERCISE 3.5

1. Let $T : \mathbb{R}^2 \to \mathbb{R}^3$ be a linear transformation given by

$$T(x_1 \;\; x_2) = (x_1 + x_2 \;\; x_1 - x_2 \;\; x_2)$$

   Then find Rank T.

2. If $T : \mathbb{R}^2 \to \mathbb{R}^2$ is defined by $T(x_1 \;\; x_2) = (x_1^2 \;\; x_2)$. Is T is a linear transformation? if not why?

3. If $T : \mathbb{R}^2 \to \mathbb{R}^2$ is defined by $T(x_1 \;\; x_2) = (x_1 - x_2 \;\; 0)$. Show that T is a linear transformation.

4. If $T : \mathbb{R}^3 \to \mathbb{R}^2$ is defined by $T(x \;\; y \;\; z) = (x + 1 \;\; y + z)$. Show that T is not a linear transformation.

5. Let $T : \mathbb{R}^3 \to \mathbb{R}^3$ be a linear transformation defined by $T(x \;\; y \;\; z) = (x + 2y - z \;\; y + z \;\; x - 2z)$ then find the rank of T.

6. Which of the following are linear transformations?

   (a) $T : \mathbb{R}^2 \to \mathbb{R}^2$ such that $T(x \;\; y) = (x \;\; x)$;

   (b) $T : \mathbb{R}^2 \to \mathbb{R}^2$ such that $T(x \;\; y) = (x + y \;\; x - y)$;

   (c) $T : \mathbb{R}^2 \to \mathbb{R}^2$ such that $T(x \;\; y) = (x + y \;\; siny)$.

3.6 Matrices of Linear Transformations

In this section corresponding to every linear transformation, a matrix associated with it is described. Specific examples of linear transformations like projection, rotation, dilation, etc. are discussed. As observed in previous sections bases of both the domain and the codomain vector spaces play a vital role in obtaining the correspondence between a linear transformation and a matrix. Procedures to obtain the matrix corresponding to a linear transformation are given.

In Section 1.2, vectors of various types have been defined. These specific examples are considered to obtain the desired matrix.

Example 3.6.1 *The projection map* $T : \mathbb{R}^3 \to \mathbb{R}^2$ *is defined by*

$$T(x \;\; y \;\; z) = (x \;\; y).$$

The matrix corresponding to this transformation is given by

$$A = \begin{bmatrix} 1 & 0 & 0 \\ 0 & 1 & 0 \\ 0 & 0 & 0 \end{bmatrix}.$$

Example 3.6.2 *The dilation map* $T : \mathbb{R}^2 \to \mathbb{R}^2$ *is defined by*

$$T(x \;\; y) = (\alpha x \;\; \alpha y) \;\; \alpha > 1$$

and the associated matrix is

$$\begin{bmatrix} \alpha & 0 \\ 0 & \alpha \end{bmatrix}.$$

Example 3.6.3 *The rotation map* $T : \mathbb{R}^2 \to \mathbb{R}^2$ *is defined by*

$$T(x \;\; y) = (xcos\theta + ysin\theta \quad -xsin\theta + ycos\theta),$$

where θ *is the angle of rotation of the vector* $(x \;\; y)$ *from the origin. The corresponding matrix is given by*

$$\begin{bmatrix} cos\theta & sin\theta \\ -sin\theta & cos\theta \end{bmatrix}.$$

Example 3.6.4 *The reflection map with respect to x-axis* $T : \mathbb{R}^2 \to \mathbb{R}^2$ *is defined by*

$$T(x \;\; y) = (x \;\; -y)$$

and the corresponding matrix is

$$A = \begin{bmatrix} 1 & 0 \\ 0 & -1 \end{bmatrix}$$

Example 3.6.5 *The shear in x-direction* $L : \mathbb{R}^2 \to \mathbb{R}^2$ *is defined as*

$$L(x \;\; y) = [x + ky \;\; y]$$

and the associated linear transformation is

$$A = \begin{bmatrix} 1 & k \\ 0 & 1 \end{bmatrix},$$

where k *is any scalar.*

Observation. All the above examples are considered with respect to the standard basis in $\mathbb{R}^2$ or $\mathbb{R}^3$. The question that arises naturally is the role of a basis in finding the matrix corresponding to a linear transformation. The following theorem is a useful result in this direction.

Theorem 3.6.1 *Let V and W be finite dimensional vector spaces with $dimV = n$ and $\mathcal{B} = \{\mathbf{v}_1, \mathbf{v}_2, \ldots, \mathbf{v}_n\}$ be a basis of V. Further, let $T : V \to W$ be a linear transformation. Then, for any $\mathbf{v} \in V$,*

$$T\mathbf{v} = c_1 T(\mathbf{v}_1) + c_2 T(\mathbf{v}_2) + \cdots + c_n T(\mathbf{v}_n),$$

where $c_1, c_2, c_3, \cdots, c_n \in \mathbb{R}$ are unique scalars.

Observation. The theorem states that the linear transformation is completely determined for every vector $\mathbf{v} \in V$ once the values of $T(\mathbf{v}_1), T(\mathbf{v}_2), \ldots, T(\mathbf{v}_n)$ corresponding to the basis $\mathcal{B}$ are known. The proof is simple and is left as an exercise.

The unique matrix corresponding to a linear transformation from a vector space V to a vector space W is given below.

Theorem 3.6.2 *Let V and W be finite dimensional vector spaces with dim $V = n$ and dim $W = m$. Further let $\mathcal{B} = \{\mathbf{v}_1, \mathbf{v}_2, \ldots, \mathbf{v}_n\}$ and $\mathcal{B}' = \{\mathbf{w}_1, \mathbf{w}_2, \ldots, \mathbf{w}_m\}$ bases for V and W, respectively. Let $T : V \to W$ be a linear transformation. Then, there exists a unique matrix $A_{m\times n}$ whose jth column is the coordinate vector of $T(\mathbf{v}_j)$ with respect to the basis $\mathcal{B}'$ on W and the matrix $\mathbf{A}$ is such that for $\mathbf{v} \in V$*

$$[T(\mathbf{v})]_{\mathcal{B}'} = \mathbf{A}[\mathbf{v}]_{\mathcal{B}},$$

where $[\mathbf{v}]_{\mathbb{B}}$ and $[T\mathbf{v}]_{\mathbb{B}'}$ are the coordinate vectors of $\mathbf{v}$ and $T\mathbf{v}$ with respect to the basis $\mathcal{B}$ and $\mathcal{B}'$, respectively. This matrix is unique and jth column $\mathbf{A}_j$ is given by

$$\begin{bmatrix} c_{1j} \\ c_{2j} \\ \cdot \\ \cdot \\ \cdot \\ c_{mj} \end{bmatrix}_{\mathcal{B}'},$$

where $c_{1j}, c_{2j}, c_{3j}, \cdots c_{mj}$ are constants obtained by the relation

$$T(\mathbf{v}_j) = c_{1j}\mathbf{w}_1 + c_{2j}\mathbf{w}_2 + \cdots + c_{mj}\mathbf{w}_m.$$

Definition 3.6.1 *The matrix of linear transformation $T : V \to W$ is given* *by*

$$\begin{bmatrix} c_{11} & c_{12} & . & . & . & c_{1n} \\ c_{21} & c_{22} & . & . & . & c_{2n} \\ . & . & . & . & . & . \\ . & . & . & . & . & . \\ c_{m1} & c_{m2} & . & . & . & c_{mn} \end{bmatrix},$$

where

$$\mathbf{A}_j = \begin{bmatrix} c_{1j} \\ c_{2j} \\ . \\ . \\ . \\ c_{mj} \end{bmatrix}_{\mathcal{B}'} = [T\mathbf{v}_j]_{\mathcal{B}'}$$

The following is a procedure for finding the matrix of linear transformation.

Step 1 Given $T : V \to W$, basis $\mathcal{B} = \{\mathbf{v}_1, \mathbf{v}_2, \ldots, \mathbf{v}_n\} \subseteq V$, $\mathcal{B}' = \{\mathbf{w}_1, \mathbf{w}_2, \ldots, \mathbf{w}_m\} \subseteq W$.

Step 2 Compute $T(\mathbf{v}_j), \quad j = 1, 2, 3, \ldots, n$.

Step 3 Find coordinate vector for $[T\mathbf{v}_j)]_{\mathcal{B}'}$. That is

$$T(\mathbf{v}_j) = c_{1j}\mathbf{w}_1 + c_{2j}\mathbf{w}_2 + \cdots + c_{mj}\mathbf{w}_m$$

and find $c_{1j}, c_{2j}, \cdots c_{mj}$.

Step 4 The matrix of linear transformation is $\mathbf{A} = [\mathbf{A}_1\ \mathbf{A}_2\ \mathbf{A}_3 \cdots \mathbf{A}_j \cdots\ \mathbf{A}_n]$ with $[T(\mathbf{v}_j)]_{\mathcal{B}'} = \mathbf{A}_j = [c_{1j}\ \ c_{2j}\ \ \cdots c_{mj}]^T$.

Below is a simple procedure to find the matrix $\mathbf{A}$.
If $T : \mathbb{R}^n \to \mathbb{R}^m$ to find $\mathbf{A}$ that is to find
To find $c_{1j}, c_{2j}, \ldots, c_{mj}, \quad j = 1, 2, 3, \ldots, n$ write the augmented matrix

$$\left[\begin{array}{cccccc:cccc} \mathbf{w}_1 & \mathbf{w}_2 & . & . & . & \mathbf{w}_n & T(\mathbf{v}_1) & T(\mathbf{v}_2) & . & .T(\mathbf{v}_n) \\ & & & & & & & & & \\ & & & & & & & & & \\ & & & & & & & & & \end{array}\right]$$

Next transform the augmented matrix to reduced row echelon form $[\mathbf{I}_n : \mathbf{A}]$. Then $\mathbf{A}$ is the required matrix.

Observation. Finding the coordinate vector of $T(\mathbf{v}_j)$ means solving the equation.

$$[\mathbf{w}_1\ \mathbf{w}_2\ \mathbf{w}_3\ \cdots\ \mathbf{w}_m]\mathbf{x}_j = T(\mathbf{v}_j), \quad j = 1, 2, \ldots, n.$$

This can be done at one go by considering the augmented matrix as above, where $\mathbf{x}_j$ is the jth column of $\mathbf{A}$ (to be found).

Example 3.6.6 *Let* $L : \mathbb{R}^3 \to \mathbb{R}^2$ *be denoted by* $L\begin{bmatrix} x \\ y \\ z \end{bmatrix} = \begin{bmatrix} y+z \\ x-y \end{bmatrix}$.

Let $S = \{\mathbf{v}_1,\ \mathbf{v}_2,\ \mathbf{v}_3\}$ *and* $T = \{\mathbf{w}_1,\ \mathbf{w}_2\}$ *be the bases for* $\mathbb{R}^3$ *and* $\mathbb{R}^2$*, respectively,*

where $\mathbf{v}_1 = \begin{bmatrix} 1 \\ 0 \\ 0 \end{bmatrix},\ \mathbf{v}_2 = \begin{bmatrix} 0 \\ 1 \\ 0 \end{bmatrix},\ \mathbf{v}_3 = \begin{bmatrix} 0 \\ 0 \\ 1 \end{bmatrix}$ *and* $\mathbf{w}_1 = \begin{bmatrix} 1 \\ 0 \end{bmatrix},\ \mathbf{w}_2 = \begin{bmatrix} 0 \\ 1 \end{bmatrix}$.

Find the matrix of L *with respect to* S *and* T.

Solution.

Calculating

$$L(\mathbf{v}_1) = \begin{bmatrix} 0+0 \\ 1-0 \end{bmatrix} = \begin{bmatrix} 0 \\ 1 \end{bmatrix},\ L(\mathbf{v}_2) = \begin{bmatrix} 1+0 \\ 0-1 \end{bmatrix} = \begin{bmatrix} 1 \\ -1 \end{bmatrix},\ L(\mathbf{v}_3) = \begin{bmatrix} 0+1 \\ 0-0 \end{bmatrix} = \begin{bmatrix} 1 \\ 0 \end{bmatrix},$$

$$[L(\mathbf{v}_1)]_T = \begin{bmatrix} 0 \\ 1 \end{bmatrix},\ [L(\mathbf{v}_2)]_T = \begin{bmatrix} 1 \\ -1 \end{bmatrix},\ [L(\mathbf{v}_3)]_T = \begin{bmatrix} 1 \\ 0 \end{bmatrix},\ \textit{Hence, } \mathbf{A} = \begin{bmatrix} 0 & 1 & 1 \\ 1 & -1 & 0 \end{bmatrix}$$

Example 3.6.7 *Let* $L : \mathbb{R}^3 \to \mathbb{R}^2$ *denoted as* $L\begin{bmatrix} x \\ y \\ z \end{bmatrix} = \begin{bmatrix} x+y \\ y-z \end{bmatrix}$. *Let* $S = \{\mathbf{v}_1,\ \mathbf{v}_2,\ \mathbf{v}_3\}$ *and* $T = \{\mathbf{w}_1,\ \mathbf{w}_2\}$*, respectively,*

where $\mathbf{v}_1 = \begin{bmatrix} 1 \\ 0 \\ 1 \end{bmatrix},\ \mathbf{v}_2 = \begin{bmatrix} 0 \\ 1 \\ 1 \end{bmatrix},\ \mathbf{v}_3 = \begin{bmatrix} 1 \\ 1 \\ 1 \end{bmatrix}$ *and* $\mathbf{w}_1 = \begin{bmatrix} 1 \\ 2 \end{bmatrix},\ \mathbf{w}_2 = \begin{bmatrix} -1 \\ 1 \end{bmatrix}$.

Find the matrix of L *with respect to* S *and* T.

Solution.

Calculating

$$L(\mathbf{v}_1) = \begin{bmatrix} 1 \\ -1 \end{bmatrix}, L(\mathbf{v}_2) = \begin{bmatrix} 1 \\ 0 \end{bmatrix}, L(\mathbf{v}_3) = \begin{bmatrix} 2 \\ 0 \end{bmatrix}.$$

To find the coordinate vectors, we write

$$L(\mathbf{v}_1) = \begin{bmatrix} 1 \\ -1 \end{bmatrix} = a_1\mathbf{w}_1 + a_2\mathbf{w}_2 = a_1\begin{bmatrix} 1 \\ 2 \end{bmatrix} + a_2\begin{bmatrix} -1 \\ 1 \end{bmatrix},$$

$$L(\mathbf{v}_2) = \begin{bmatrix} 1 \\ 0 \end{bmatrix} = b_1\mathbf{w}_1 + b_2\mathbf{w}_2 = b_1\begin{bmatrix} 1 \\ 2 \end{bmatrix} + b_2\begin{bmatrix} -1 \\ 1 \end{bmatrix},$$

$$L(\mathbf{v}_3) = \begin{bmatrix} 2 \\ 0 \end{bmatrix} = c_1\mathbf{w}_1 + c_2\mathbf{w}_2 = c_1\begin{bmatrix} 1 \\ 2 \end{bmatrix} + c_2\begin{bmatrix} -1 \\ 1 \end{bmatrix}.$$

This requires solving three linear systems. Thus, form the matrix

$$\begin{bmatrix} 1 & -1 & : & 1 & : & 1 & : & 2 \\ 2 & 1 & : & -1 & : & 0 & : & 0 \end{bmatrix}$$

$R_2 \to R_2 - 2R_1$

$$\begin{bmatrix} 1 & -1 & : & 1 & : & 1 & : & 2 \\ 0 & 3 & : & -3 & : & -2 & : & -4 \end{bmatrix}$$

$R_1 \to 3R_1 + R_2$

$$\begin{bmatrix} 3 & 0 & : & 0 & : & 1 & : & 2 \\ 0 & 3 & : & -3 & : & -2 & : & -4 \end{bmatrix}$$

$R_1 \to R_1/3,\ R_2 \to R_2/3$

$$\begin{bmatrix} 1 & 0 & : & 0 & : & \frac{1}{3} & : & \frac{2}{3} \\ 0 & 1 & : & -1 & : & \frac{-2}{3} & : & \frac{-4}{3} \end{bmatrix}$$

$$[L(x)]_T = \begin{bmatrix} 0 & \frac{1}{3} & \frac{2}{3} \\ -1 & \frac{-2}{3} & \frac{-4}{3} \end{bmatrix} [x]_S$$

Therefore, the matrix $\mathbf{A} = \begin{bmatrix} 0 & \frac{1}{3} & \frac{2}{3} \\ -1 & \frac{-2}{3} & \frac{-4}{3} \end{bmatrix}$ *of L w.r.t S and T.*

Example 3.6.8 *Let* $\mathbf{P}$ *be any vector in* $\mathbb{R}^3$ *and* $\mathcal{B} = \{\mathbf{v}_1, \mathbf{v}_2, \mathbf{v}_3\}$ $\mathcal{B}' = \{\mathbf{w}_1, \mathbf{w}_2, \mathbf{w}_3\}$ *be any two orthonormal bases of* $\mathbb{R}^3$ *then find the co-ordinates of* $\mathbf{p}$ *with respect to* $\mathcal{B}$ *and* $\mathcal{B}'$ *and find the change of basis matrix.*

Solution.

To find the coordinates of $\mathbf{p}$ *with respect to the basis* $\mathcal{B}$, *write* $\mathbf{p}$ *as*

$\mathbf{p} = a\mathbf{v}_1 + b\mathbf{v}_2 + c\mathbf{v}_3$

$\mathbf{p}.\mathbf{v}_1 = a,\ \mathbf{p}.\mathbf{v}_2 = b,\ \mathbf{p}.\mathbf{v}_3 = c.$

Thus co-ordinate vector of $\mathbf{p}$ *with respect to the basis* $\mathcal{B}$ *is*

$$[\mathbf{p}]_{\mathcal{B}} = (\mathbf{p}.\mathbf{v}_1\ \mathbf{p}.\mathbf{v}_2\ \mathbf{p}.\mathbf{v}_3)^T$$

Similarly the co-ordinate vector of $\mathbf{p}$ *with respect to basis* $\mathcal{B}'$ *is*

$$[\mathbf{p}]_{\mathcal{B}'} = (\mathbf{p}.\mathbf{w}_1\ \mathbf{p}.\mathbf{w}_2\ \mathbf{p}.\mathbf{w}_3)^T.$$

Working in a similar manner, the change of basis matrix from $\mathcal{B} \to \mathcal{B}'$ *is*

$$\begin{bmatrix} \mathbf{v}_1.\mathbf{w}_1 & \mathbf{v}_1.\mathbf{w}_2 & \mathbf{v}_1.\mathbf{w}_3 \\ \mathbf{v}_2.\mathbf{w}_1 & \mathbf{v}_2.\mathbf{w}_2 & \mathbf{v}_2.\mathbf{w}_3 \\ \mathbf{v}_3.\mathbf{w}_1 & \mathbf{v}_3.\mathbf{w}_2 & \mathbf{v}_3.\mathbf{w}_3 \end{bmatrix}$$

EXERCISE 3.6

1. Find the matrix corresponding to the linear transformation $T : \mathbb{R}^2 \to \mathbb{R}^2$ of the reflection map with respect to y-axis.

2. Find the matrix corresponding to the shear in y-direction.

3. Prove Theorem 3.6.1.

4. The matrix linear transmission T on $\mathbb{R}^3$ defined as $T(x \ \ y \ \ z) = (2y + z \ \ x - 4y \ \ 3x)$ with respect to the basis $\mathcal{B} = \{(1 \ \ 1 \ \ 1), \ (1 \ \ 1 \ \ 0), \ (1 \ \ 0 \ \ 0)\}$ then find $[T]_{\mathcal{B}}$.

5. Let T be a linear transformation from $\mathbb{R}^3 \to \mathbb{R}^2$ defined by $T(x \ \ y \ \ z) = (x + y \ \ y - z)$. Then find the matrix of T with respect to the ordered bases $\{(1 \ \ 1 \ \ 1), \ (1 \ \ -1 \ \ 0), \ (0 \ \ 1 \ \ 0)\}$ and $\{(1 \ \ 1), \ (1 \ \ 0)\}$.

6. Let T be a linear transformation on $\mathbb{R}^3$ given by $T(x \ \ y \ \ z) = (2x \ \ 4x - y \ \ 2x + 3y - z)$ then show that the matrix of the linear transformation is invertible.

7. The matrix linear transformation $T : \mathbb{R}^3 \to \mathbb{R}^2$ with $T \begin{bmatrix} 1 \\ 0 \\ 0 \end{bmatrix} = \begin{bmatrix} 7 \\ 11 \end{bmatrix}$, $T \begin{bmatrix} 0 \\ 1 \\ 0 \end{bmatrix} = \begin{bmatrix} 6 \\ 9 \end{bmatrix}$, and $T \begin{bmatrix} 0 \\ 0 \\ 1 \end{bmatrix} = \begin{bmatrix} -13 \\ 17 \end{bmatrix}$. Find the matrix $\mathbf{A}$ of T with respect to the basis $\{(1 \ \ 0\}, (1 \ \ 1)\}$.

8. $T : \mathbb{R}^2 \to \mathbb{R}^2$ such that $T(x \ \ y) = (x - y \ \ 2x + y)$. Find the matrix which represents T.
   (a) Relative to the basis $\{(1 \ \ 0), \ (0 \ \ 1)\}$
   (b) Relative to the basis $\{(1 \ \ 2), \ (1 \ \ 1)\}$

3.7 Inner Product Space

In Section 1.2, the dot product of two vectors is given and it can be observed that in $\mathbb{R}^2$, every orthogonal basis $\mathcal{B} = \{\mathbf{v}_1, \mathbf{v}_2\}$ satisfies the property $\mathbf{v}_1.\mathbf{v}_2 = 0$. In other words, these basis vectors are perpendicular (orthogonal) to each other in $\mathbb{R}^2$. This property holds in $\mathbb{R}^3$ and also in $\mathbb{R}^n$, for any $n < \infty$. The main idea of this section is to describe a notion that characterizes this concept of orthogonality in a vector space. The following definition of an inner product suits the bill and is useful for defining a basis in a vector space.

Definition 3.7.1 *An inner product on a vector space V is a mapping from $V \times V$ into $\mathbb{R}$, denoted by $\langle \mathbf{x}, \mathbf{y} \rangle$, for every pair of vectors $\mathbf{x}$ and $\mathbf{y}$, satisfying the following properties:*

(i) $\langle \mathbf{x}, \mathbf{x} \rangle \geq 0$ and $\langle \mathbf{x}, \mathbf{x} \rangle = 0$ if and only if $\mathbf{x} = \mathbf{0}, \quad \mathbf{x} \in V$;

(ii) $\langle \alpha\mathbf{x}, \mathbf{y}\rangle = \alpha\langle \mathbf{x}, \mathbf{y}\rangle$ for all $\mathbf{x}, \mathbf{y} \in V$ and

(iii) $\langle \mathbf{x}+\mathbf{y}, \mathbf{z}\rangle = \langle \mathbf{x}, \mathbf{y}\rangle + \langle \mathbf{x}, \mathbf{z}\rangle$ for all $\mathbf{x}, \mathbf{y}, \mathbf{z} \in V$.

Observation. Note that as the underlying space is $\mathbb{R}$, the definition of the inner product had been restricted to $\mathbb{R}$. In general, an inner product definition caters to complex numbers with the following extra condition.

(iv) $\langle \mathbf{x}, \mathbf{y}\rangle = \langle \overline{\mathbf{y}, \mathbf{x}}\rangle$ for all $\mathbf{x}, \mathbf{y} \in V$. In this situation, the inner product is a mapping from $V \times V \to \mathbb{C}$.

Example 3.7.1 *Let* $V=\mathbb{R}^n$ *then for* $\mathbf{x} = (x_1 \quad x_2 \quad \dots \quad x_n)$, $\mathbf{y} = (y_1 \quad y_2 \quad \dots \quad y_n)$, $\langle \mathbf{x}, \mathbf{y}\rangle = \mathbf{x}.\mathbf{y} = \sum_{i=1}^{n} x_i y_i$.

Definition 3.7.2 *A Vector space V with an inner product defined on it is called an inner product space.*

Example 3.7.2 *(i)* $\mathbb{R}^n$, $0 < n < \infty$ *is an inner product space with inner product defined as above.*
(ii) Let $\mathcal{K} = \{f : [a, b] \to \mathbb{R}$ *such that f is continuous*$\}$. *For* $f, g \in \mathcal{K}$, *define,*

$$\langle f, g\rangle = \int_a^b f(x)g(x)dx.$$

Then $\mathcal{K}$ is an inner product space.

Observation.

1. Note that the inner product induces a distance called norm (see Section 1.2) as $\langle \mathbf{x}, \mathbf{x}\rangle = ||\mathbf{x}||^2$.

2. In Example 3.7.2 (i), it can be seen that the inner product in $\mathbb{R}^n$ is nothing but the dot product in $\mathbb{R}^n$ and hence the concept of angle comes into picture and hence the idea of perpendicular vectors can be considered. This notion is extended to a vector space as follows.

Let the vector space (V,+,.) with an associated inner product $\langle ., .\rangle$ be an inner product space.

Definition 3.7.3 *Two vectors* $\mathbf{x}, \mathbf{y} \in V$ *are said to be orthogonal or perpendicular if and only if* $\langle \mathbf{x}, \mathbf{y}\rangle = \mathbf{0}$ *and is denoted by* $\mathbf{x} \perp \mathbf{y}$.

Definition 3.7.4 *Let* $S \subseteq V$. *Then S is said to be an orthogonal set if every two distinct vectors* $\mathbf{x}, \mathbf{y} \in S$ *are orthogonal (i.e.)*

$$\langle \mathbf{x}, \mathbf{y}\rangle = 0 \quad for \quad \mathbf{x}, \mathbf{y} \neq 0, \quad \mathbf{x}, \mathbf{y} \in S.$$

Definition 3.7.5 *An orthogonal complement of* S *is a set (denoted by* $S^{\perp}$*) of vectors that are perpendicular to every vector in* S*. (i.e.)*

$$S^{\perp} = \{\mathbf{x} \in V / \langle \mathbf{x}, \mathbf{y} \rangle = 0, \quad \textit{for all } \mathbf{y} \in S\}.$$

Example 3.7.3 $\{\mathbf{0}\}^{\perp} = V, \quad V^{\perp} = \{\mathbf{0}\}$.

Definition 3.7.6 *Two sets* S_1*,* S_2 *are said to be mutually orthogonal, denoted by* $S_1 \perp S_2$*, if for all* $\mathbf{x} \in S_1$ *and* $\mathbf{y} \in S_2$*,* $\langle \mathbf{x}, \mathbf{y} \rangle = 0$.

Example 3.7.4 *Let* $S_1 = xy - plane$*,* $S_2 = z - axis$*. Then* $S_1^{\perp} = S_2$*, and* $S_2^{\perp} = S_1$*. Also* $\mathbb{R}^3 = S_1 \bigoplus S_2$*.*

(See Section 3.2 for the definition of direct sum.)

Definition 3.7.7 *An orthogonal set* $S \subseteq V$ *is said to be orthonormal if every vector in* S *is a unit vector.*

Example 3.7.5 $S = \{(1 \ \ 0 \ \ 0), \ (0 \ \ 1 \ \ 0), (0 \ \ 0 \ \ 1)\}$ *is an orthonormal set in* $\mathbb{R}^3$*.*

Definition 3.7.8 *A subset* S *(*$\subseteq V$*) is said to be an orthonormal basis for* V *if it is an ordered basis and is also an orthonormal set.*

Example 3.7.6 *The preceding example forms an orthonormal basis.*

Diagonalizing a rectangular matrix

In Section 1.3, a matrix has been introduced as an arrangement of vectors in columns. In Section 2.7, it was shown that a matrix consisting of linearly independent eigen-vectors diagonalizes the matrix. These eigen-vectors form a basis. If $\mathbf{A}$ is symmetric, the eigen-vectors can be made orthonormal and the matrix consisting of orthonormal eigen-vectors is an orthogonal matrix, $\mathbf{Q}$, and $\mathbf{A}_{n\times n}$ can be diagonalized by $\mathbf{\Lambda} = \mathbf{Q}^T\mathbf{A}\mathbf{Q}$.

In general, if one wants an orthonormal basis for diagonalizing a rectangular matrix $\mathbf{A}_{m\times n}$, it is essential to consider singular values and singular vectors.

Consider $\{\mathbf{v}_1, \mathbf{v}_2, \ldots, \mathbf{v}_n\}$ eigen-vectors of $\mathbf{A}^T\mathbf{A}$ and $\{\mathbf{u}_1, \mathbf{u}_2, \mathbf{u}_3, \ldots, \mathbf{u}_n\}$ eigen-vectors of $\mathbf{A}\mathbf{A}^T$ with the eigen-values $\lambda_i = \sigma_i^2$. Then the $\mathbf{v}_i$'s and $\mathbf{u}_i$'s are linked by the relation,

$$\mathbf{A}\mathbf{v}_1 = \sigma_1\mathbf{u}_1, \quad \mathbf{A}\mathbf{v}_2 = \sigma_2\mathbf{u}_2, \ldots, \mathbf{A}\mathbf{v}_n = \sigma_n\mathbf{u}_n$$

which can be written in the matrix notation $\mathbf{A}\mathbf{V} = \mathbf{U}\Sigma$ *or* $\mathbf{A} = \mathbf{U}\Sigma\mathbf{V}^T$ as $\mathbf{U}$ and $\mathbf{V}$ are orthogonal matrices.

Any matrix $\mathbf{A}_{m\times n}$ represents a linear transformation from a vector space V of dimension n to a vector space W of dimension m.

To diagonalize $\mathbf{A}$ the following theorem is essential.

Theorem 3.7.1 ***Singular Value Decomposition.*** *Let* $\{\mathbf{v}_1, \mathbf{v}_2, \ldots, \mathbf{v}_n\}$ *be the eigen-vectors of* $(\mathbf{A}^T\mathbf{A})_{n\times n}$ *and* $\{\mathbf{u}_1, \mathbf{u}_2, \ldots, \mathbf{u}_m\}$ *be the eigen-vectors corresponding to* $(\mathbf{A}\mathbf{A}^T)_{m\times m}$. *Let* $\lambda_i = \sigma_i^2$ *be the eigen-values of* $\mathbf{A^T A}$. *Then* σ_i*'s are called as the singular values of* A *and* $\{\mathbf{v}_1, \mathbf{v}_2, \ldots, \mathbf{v}_n\}$ *and* $\{\mathbf{u}_1, \mathbf{u}_2, \ldots, \mathbf{u}_m\}$ *are the singular vectors of* $\mathbf{A}$. *Further,*

$$A = U\Sigma V^T = [\mathbf{u}_1, \mathbf{u}_2, \ldots, \mathbf{u}_m]_{m\times m} \begin{bmatrix} \sigma_1 & 0 & . & . & 0 & 0 & 0 \\ 0 & \sigma_2 & . & . & 0 & . & . \\ . & & & & & . & \\ 0 & . & . & . & \sigma_m & 0 & 0 \end{bmatrix}_{m\times n} [\mathbf{v}_1, \mathbf{v}_2, \ldots, \mathbf{v}_n]^T_{n\times n}$$

.

Example 3.7.7 *Let* $\mathbf{A} = \begin{bmatrix} 1 & 0 & 0 \\ 0 & 2 & 1 \end{bmatrix}_{2\times 3}$

Then $\mathbf{A}^T = \begin{bmatrix} 1 & 0 \\ 0 & 2 \\ 0 & 1 \end{bmatrix}_{3\times 2}$, $\mathbf{A}\mathbf{A}^T = \begin{bmatrix} 1 & 0 \\ 0 & 5 \end{bmatrix}$

The eigen-values $\mathbf{A}\mathbf{A}^T$ *are 5 and 1.*

The eigen-vector corresponding to $\lambda = 5$ *is* $\mathbf{u}_1 = \begin{bmatrix} 0 \\ 1 \end{bmatrix}$ *and*

the eigen-vector corresponding to $\lambda = 1$ *is* $\mathbf{u}_2 = \begin{bmatrix} 1 \\ 0 \end{bmatrix}$.

The normalized vector matrices are $\mathbf{u}_1 = \begin{bmatrix} 0 \\ 1 \end{bmatrix}$ *and* $\mathbf{u}_2 = \begin{bmatrix} 1 \\ 0 \end{bmatrix}$, *hence set*

$$\mathbf{U} = \begin{bmatrix} 0 & 1 \\ 1 & 0 \end{bmatrix}.$$

Next consider $\mathbf{A}^T\mathbf{A} = \begin{bmatrix} 1 & 0 & 0 \\ 0 & 4 & 2 \\ 0 & 2 & 1 \end{bmatrix}$.

The eigen-values are 5, 1 and 0.

Corresponding to the eigen-values for $\lambda = 5$, $\mathbf{v}_1 = \begin{bmatrix} 0 \\ 2 \\ 1 \end{bmatrix}$; *for* $\lambda = 1$, $\mathbf{v}_2 = \begin{bmatrix} 1 \\ 0 \\ 0 \end{bmatrix}$;

and for $\lambda = 0$, $\mathbf{v}_3 = \begin{bmatrix} 0 \\ -0.5 \\ 1 \end{bmatrix}$

Normalizing the vectors gives $\mathbf{v}_1 = \begin{bmatrix} 0 \\ 0.8944 \\ 0.4472 \end{bmatrix}$, $\mathbf{v}_2 = \begin{bmatrix} 1 \\ 0 \\ 0 \end{bmatrix}$, *and* $\mathbf{v}_3 = \begin{bmatrix} 0 \\ -0.4472 \\ 0.8944 \end{bmatrix}$,

Then the matrix is $\mathbf{V} = \begin{bmatrix} 0 & 1 & 0 \\ 0.8944 & 0 & -0.4472 \\ 0.4472 & 0 & 0.8944 \end{bmatrix}$

Further, $\sum = \begin{bmatrix} \sqrt{5} & 0 & 0 \\ 0 & \sqrt{1} & 0 \end{bmatrix} = \begin{bmatrix} 2.2361 & 0 & 0 \\ 0 & 1 & 0 \end{bmatrix}.$

Verification

$\mathbf{U}\sum\mathbf{V}^T = \begin{bmatrix} 0 & 1 \\ 1 & 0 \end{bmatrix} \begin{bmatrix} 2.2361 & 0 & 0 \\ 0 & 1 & 0 \end{bmatrix} \begin{bmatrix} 0 & 0.8944 & 0.4472 \\ 1 & 0 & 0 \\ 0 & -0.4472 & 0.8944 \end{bmatrix} = \begin{bmatrix} 1 & 0 & 0 \\ 0 & 2 & 1 \end{bmatrix} =$ **A**.

Observation. The matrices **U** and **V** are orthogonal matrices.

EXERCISE 3.7

1. Show that K in Example 3.7.2(ii) is an inner product space with the given definition. (Verify the definition in Example 3.7.2(ii) is an inner product.)
2. Find the value of the inner product
   (i) $\langle 3\mathbf{x} + 4\mathbf{y}, 2\mathbf{x} + 9\mathbf{y} \rangle$ in $\mathbb{R}$.
   (ii) $\langle (6 - 8i)\mathbf{z}, 5i + 2\mathbf{x} \rangle$ in $\mathbb{C}$.
3. Show that $\mathbf{x} \perp \mathbf{y}$ implies $\mathbf{y} \perp \mathbf{x}$.
4. Show that $\mathbf{x} \perp \mathbf{y}$ for all $\mathbf{x} \in V$, then $\mathbf{x} = \mathbf{0}$.
5. If $S = \{\mathbf{v}_1, \mathbf{v}_2, \ldots, \mathbf{v}_n\}$ is a set of mutually orthogonal vectors in $\mathbb{R}^n$, show that S is a basis for $\mathbb{R}^n$.
6. Find the singular value decomposition for $A = \begin{bmatrix} 3 & 3 \\ 2 & -2 \end{bmatrix}$.

3.8 Gram–Schmidt Orthogonalization

In this section, a process to obtain an orthonormal basis is described. This procedure is known as Gram–Schmidt orthogonalization process and is useful to represent functions in terms of a basis.

Let V be an inner product space and $\{\mathbf{x}_1, \mathbf{x}_2, \ldots, \mathbf{x}_n\}$ be a linearly independent subset of V. Then, there exists an orthonormal subset of nonzero vectors $\{\mathbf{e}_1, \mathbf{e}_2, \ldots, \mathbf{e}_n\}$ such that span $\{\mathbf{e}_1, \mathbf{e}_2, \ldots, \mathbf{e}_n\}$ = span $\{\mathbf{x}_1, \mathbf{x}_2, \ldots, \mathbf{x}_n\}$. The procedure is as follows.

Step 1: Set $\mathbf{e}_1 = \frac{1}{\|\mathbf{x}_1\|}\mathbf{x}_1$.

Step 2: Write $\mathbf{v}_2 = x_2 - \langle \mathbf{x}_2, \mathbf{e}_1 \rangle \mathbf{e}_1$.

Consider $\langle \mathbf{v}_2, \mathbf{e}_1 \rangle = \langle \mathbf{x}_2, \mathbf{e}_1 \rangle - \langle \mathbf{x}_2, \mathbf{e}_1 \rangle \langle \mathbf{e}_1, \mathbf{e}_1 \rangle = 0$.

Hence $\mathbf{v}_2$ perpendicular to $\mathbf{e}_1$ and $\mathbf{v}_2 \neq 0$. Set

$$\mathbf{e}_2 = \frac{1}{\|\mathbf{v}_2\|}\mathbf{v}_2.$$

Step 3. Write $\mathbf{v}_3 = \mathbf{x}_3 - \langle \mathbf{x}_3, \mathbf{e}_1 \rangle \mathbf{e}_1 + \langle \mathbf{x}_3, \mathbf{e}_2 \rangle \mathbf{e}_2$
then $\langle \mathbf{v}_3, \mathbf{e}_1 \rangle = \langle \mathbf{v}_3, \mathbf{e}_2 \rangle = 0$ and $\mathbf{v}_3 \neq 0$. Thus $\mathbf{v}_3 \perp \mathbf{e}_1$ and $\mathbf{v}_3 \perp \mathbf{e}_2$.

Now set

$$\mathbf{e}_3 = \frac{1}{\|\mathbf{v}_3\|}\mathbf{v}_3.$$

Proceeding in a similar fashion,

Step *n*:

$$\mathbf{v}_n = \mathbf{x}_n - \sum_{k=1}^{n-1} \langle \mathbf{x}_n, \mathbf{e}_k \rangle \mathbf{e}_k$$

with $\mathbf{v}_n \neq 0$ and $\mathbf{v}_n$ is orthogonal to $\mathbf{e}_1, \mathbf{e}_2, \mathbf{e}_3, \ldots, \mathbf{e}_{n-1}$
and

$$\mathbf{e}_n = \frac{1}{\|\mathbf{v}_n\|}\mathbf{v}_n.$$

Thus $\{\mathbf{e}_1, \mathbf{e}_2, \mathbf{e}_3, \ldots, \mathbf{e}_n\}$ is an orthonormal set in V.

Observation.

1. The proof of the equivalence of the linear span is given as exercise. It follows by induction.

2. If V is of finite dimension n, then $\{\mathbf{e}_1, \mathbf{e}_2, \ldots, \mathbf{e}_n\}$ is an orthonormal basis for V.

3. The procedure carries on for any n, hence it holds for a countably infinite set $\{\mathbf{x}_1, \mathbf{x}_2, \ldots, \mathbf{x}_n, \ldots\}$ (this set is called as a sequence).

4. Looking at the vectors $\{\mathbf{x}_1, \mathbf{x}_2, \mathbf{x}_3, \ldots, \mathbf{x}_n\}$ as representing n directions, it can be stated that at each step i, the components in the previous $i-1$ orthogonal directions are being subtracted from the vector $\mathbf{x}_i$.

5. At any point of the procedure, $\mathbf{v}_i \neq 0$.

Result. Let V be an inner product space and $\{\mathbf{e}_1, \mathbf{e}_2, \mathbf{e}_3, \ldots, \mathbf{e}_n\}$ be an orthonormal subset of non-zero vectors. If $x \in span\{\mathbf{e}_1, \mathbf{e}_2, \ldots, \mathbf{e}_n\}$, then

$$\mathbf{x} = \sum_{i=1}^{n} \langle \mathbf{x}, \mathbf{e}_i \rangle \mathbf{e}_i.$$

The coefficients $\langle \mathbf{x}, \mathbf{e}_i \rangle$ in the above linear combination are called as Fourier coefficients.

EXERCISE 3.8

1. In Gram–Schmidt procedure prove that

$$span\{\mathbf{x}_1, \mathbf{x}_2, \ldots, \mathbf{x}_n\} = span\{\mathbf{e}_1, \mathbf{e}_2, \ldots, \mathbf{e}_n\}.$$

2. Consider any 3×3 nonsingular matrix and construct the orthonormal set of vectors.

3. Let $S = \{(1 \quad -2), (2 \quad 3)\}$ and W = span S. Construct the orthonormal set of vectors. Find the Fourier coefficients for the vectors $x = (10 \quad -6)$.

3.9 Linking Linear Algebra to Differential Equations

This section is meant for readers having knowledge of differential equations up to undergraduate level. In this section, a brief note is given relating a couple of topics in linear algebra to differential equations.

In Chapter 8, a differential system $\mathbf{x}' = \mathbf{A}\mathbf{x}$ where $\mathbf{A}$ is an $n \times n$ constant matrix and $\mathbf{x}$ is an $n \times 1$ column vector is considered. It will be shown that the properties of this system are studied using the eigen-values of $\mathbf{A}$. The same can be observed for systems dealt in other chapters also.

Differential equations involving variable coefficients do not fit into the above setup. Their study involves the notion of orthogonal sets. Some results dealing with them are given below.

Definition 3.9.1 *Let $L_2[a, b] = \{x : [a, b] \to \mathbb{R} / \quad x \text{ is continuous}\}$ with an inner product defined by*

$$\langle x, y \rangle = \int_a^b x(t)y(t)dt.$$

Then $L_2[a, b]$ is an inner product space.

It is called as Hilbert space on addition of a new concept called completeness, which is beyond the scope of this book.

Definition 3.9.2 *Power series is a linear combination of a countably infinite set of functions of the form,* $S = \{1, x, x^2, \ldots, x^n, \ldots\}$.

Thus a power series is given by

$$\sum_{n=0}^{\infty} a_n x^n.$$

Many linear differential equations of variable coefficients like $y'' + xy' + xy = 0$ are assumed to have a power series solution and the coefficients of the power series are found.

There are many differential equations whose solutions exhibit specific behavior. A couple of them are discussed below.

Definition 3.9.3 *Legendre differential equation is given by* $(1-x^2)y'' - 2xy' + n(n+1)y = 0$, *where* n *is a whole number.*

Then an orthogonal sequence of functions denoted by $P_n(x)$ are solutions of Legendre differential equation. The $P_n(x)$ are defined by

$$P_n(x) = \sum_{k=0}^{N} (-1)^k \frac{(2n-2k)!}{2^n k!(n-k)!(n-2k)!} x^{n-2k},$$

where

$$N = \begin{cases} \dfrac{n}{2}, & n \text{ is even}, \\ \dfrac{(n-1)}{2}, & n \text{ is odd}. \end{cases} \tag{3.6}$$

Then $P_0(x) = 1$, $P_1(x) = x$, $P_2(x) = \frac{1}{2}(3x^2 - 1)$, $P_3(x) = \frac{1}{2}(5x^2 - 3x)$, and so on. $P_n(x)$ is called a Legendre polynomial of order n. The space in which $P_n(x)$ is defined for each n is $L_2[-1, 1]$ and the inner product is defined by

$$\langle x, y \rangle = \int_{-1}^{1} x(t)y(t)dt.$$

Definition 3.9.4 *A Bessel's differential equation is of the form*

$$x^2 \frac{d^2y}{dx^2} + x\frac{dy}{dx} + (x^2 - n^2)y = 0, \quad n \text{ is a real number}.$$

Then there is an orthogonal sequence of functions denoted by $J_n(x)$ which are solutions of this equation and are defined by

$$J_n(x) = \sum_{k=0}^{\infty}(-1)^k(\frac{x}{2})^{n+2k}\frac{1}{k!\Gamma(n+k+1)}.$$

These functions are in $L_2[0,1]$ and the inner product is defined by

$$\int_0^1 xJ_n(\alpha x)J_n(\beta x)dx.$$

Each $J_n(x)$ is called as a Bessel's function of first kind of order n.

Similarly there exist orthogonal sequences called Hermite polynomials, Lagrange polynomials which are solutions of Hermite differential equations and Lagrange differential equations, respectively.

The readers are suggested to read books on special functions and also differential equations for more information.

3.10 Conclusion

This chapter begins with fundamental concepts like a vector space, subspace, linear independence, basis and dimension. The concept of change of basis of matrix, widely used in application is given in detail. Various linear transformations used in applications, for example, in topics relating to computer graphics and robotics are discussed.

The concept of a basis is extended from a finite dimensional space to an infinite dimensional space and orthonormal sets are introduced. Again these sets are linked to special functions and differential equations. Singular value decomposition theorem, an important result, used in principal component analysis, is given. For further reading the readers are suggested to refer to [3], [7], [10], [14] and [19].

Chapter 4

Numerical Methods in Linear Algebra

4.1 Introduction

Numerical methods are essential for handling large and very large systems of equations. This chapter concentrates on the well-known and useful techniques that are needed to solve these linear systems. Numerical computation induces error and the elements of these concepts form the content of Section 4.2. Section 4.3 deals with the standard techniques that reduce a given system to a diagonal or an upper or a lower triangular system. Section 4.4 introduces the iterative methods used to solve a system of equations. In Section 4.5, Householder transformation, which is an orthogonal transformation, is given. This technique mitigates the propagation of errors. In Section 4.6, another orthogonal transformation, plane rotation is utilized to reduce a symmetric matrix to a tridiagonal form. Section 4.7 deals with QR method. The bounds of eigen-values and a method of finding the largest eigen-value form the content of Section 4.8. Krylov subspace methods, which are solvers, are briefly touched in Section 4.9.

4.2 Elements of Computation and Errors

As this chapter concentrates on numerical techniques, it is essential to understand how digital computers handle computation and its impact on the problems considered.

Real numbers representation

The representation of a real number on a digital computer has a specific format called floating-point number. Here the number is expressed as a fraction or an integer or an exponent.

DOI: 10.1201/9781351014953-4

Definition 4.2.1 *A floating- point number is written as*

$$r = \pm 0.a_1a_2...a_m \times 10^n \tag{4.1}$$

where $a_1, a_2, ...a_m$ *are integers such that* $0 < a_1 \leq 9$ *and* $0 \leq a_i \leq 9$, $i = 2, 3, ..., m$, *and* n *is an integer between* $-99 \leq n \leq 99$.
The equation (4.1) can be rewritten as

$$r = a \times 10^n,$$

where a *is called mantissa and* n *is called the characteristic or exponent. The number of digits in* a, *that is,* m, *is called the precision of the floating point number.*

Definition 4.2.2 *m-digit arithmetic is the arithmetic involving m-digit floating point numbers.*

Observation. High-speed scientific computation uses 15 digit precision arithmetic.

Example 4.2.1 *If* r=*0.29345675* $\times 10^{-6}$ *then, mantissa* a=*0.29345675,* n=$-$*6 is the exponent* m=*8 is the precision.*

Example 4.2.2 $\frac{1}{374} = 0.0026737968 = 0.26737968 \times 10^{-2}$ *in floating point arithmetic. Here, mantissa is* a=*0.26737968, exponent is* n=$-$*2 and* m=*8 is the precision.*

Every digital computer has a certain range of numbers that can be handled. Suppose a system has the capacity to work with numbers, n ranging from 10^{-7} to 10^8 *i.e.* $10^{-7} < n < 10^8$. Then any number $m > 10^8$ is beyond the capacity of the system and is termed as an overflow while all numbers less than 10^{-7} are termed as underflow. These numbers cannot be handled by the considered system.

When arithmetic operations are performed the machine restricts the numbers to be within its prescribed range by either chopping off the digits or by rounding off the number. Both these operations lead to errors and together are termed as round off errors.

Thus solving problems numerically generates errors automatically and they pile up at every stage. It becomes essential to understand the errors and hence the following definitions.

Definition 4.2.3 *Absolute error is the difference between the actual quantity* Q *and the calculated quantity* $\bar{Q}$, *that is* $\epsilon = Q - \bar{Q}$.

Definition 4.2.4 *Relative error is the ratio of the absolute error and the actual quantity,*

$$RE = \frac{Q - \bar{Q}}{Q}$$

Example 4.2.3 *Suppose a company has a turnover of Rs 20 crores and during calculations Rs 10 lakhs is missing. Find the absolute error and the relative error.*

Solution. *Here the absolute error is Rs 10 lakhs which appears like a significant amount. But the relative error* $RE = \frac{1000000}{200000000} = \frac{1}{200} = 0.005$.

In order to understand the effect of the errors the following notions are defined.

Definition 4.2.5 *An ill-conditioned problem is a numerical problem wherein small changes in the data result in a large relative change in the computed solution.*

Definition 4.2.6 *A well-conditioned problem is a numerical problem whose computed solution is insensitive to small changes in the data of the problem.*

To understand the sensitivity of the problem certain results related to matrices and norms are required. These are stated without proof and are given as problems in the exercise.

Result 4.2.1 If $\mathbf{A}$ is a square matrix of order n and $||\mathbf{A}|| < 1$, where $||\mathbf{A}||$ is a matrix norm then the infinite series $\mathbf{I} + \mathbf{A} + ... + \mathbf{A}^n + ...$ converges to the sum $(\mathbf{I} - \mathbf{A})^{-1}$.

Result 4.2.2 If $||\mathbf{A}|| < 1$ and the matrix $(\mathbf{I} - \mathbf{A})$ is nonsingular, then

(i) $||(\mathbf{I} - \mathbf{A})^{-1}|| \leq \frac{1}{||\mathbf{I}-\mathbf{A}||}$, whenever $||\mathbf{I}|| = 1$.

(ii) $||\mathbf{I} - (\mathbf{I} - \mathbf{A})^{-1}|| \leq \frac{||\mathbf{A}||}{||\mathbf{I}-\mathbf{A}||}$, whenever $||\mathbf{I}|| = 1$.

Observation. $||\mathbf{I}|| = 1$ in 1-,2- and $\infty-$ norms.

The conditioning number of a matrix in case of perturbations involved when solving a linear system of equations is discussed below.

Linear system of equations

Consider a linear system of equations given by

$$\mathbf{Ax} = \mathbf{b}, \tag{4.2}$$

where $\mathbf{A}_{n\times n}$ matrix and $\mathbf{x}_{n\times 1}$ and $\mathbf{b}_{n\times 1}$ are column vectors. While solving the system (4.2), errors can perturb the matrix $\mathbf{A}$ or the vector $\mathbf{b}$.

Case (i) Suppose $\mathbf{A}$ is perturbed and the new system is

$$(\mathbf{A} + \mathbf{A}')\mathbf{x} = \mathbf{b}. \tag{4.3}$$

Let $\mathbf{x} + \mathbf{x}'$ be the corresponding solution.

Then,

$$(\mathbf{A} + \mathbf{A}')(\mathbf{x} + \mathbf{x}') = \mathbf{b} \tag{4.4}$$

is satisfied. Now subtracting (4.2) from (4.4) and simplifying gives

$$\mathbf{A}\mathbf{x}' + \mathbf{A}'(\mathbf{x} + \mathbf{x}') = 0$$

or

$$\mathbf{x}' = (I + \mathbf{A}^{-1}\mathbf{A}')^{-1}\mathbf{A}^{-1}\mathbf{A}'\mathbf{x}.$$

This gives

$$\begin{aligned}\tfrac{||\mathbf{x}'||}{||\mathbf{x}||} &= ||\mathbf{A}^{-1}\mathbf{A}'||\ ||(I + \mathbf{A}^{-1}\mathbf{A}')^{-1}|| \\ &\leq \tfrac{||\mathbf{A}^{-1}\mathbf{A}'||}{||(I+\mathbf{A}^{-1}\mathbf{A}')||}.\end{aligned}$$

Now set $\mu = \frac{||\mathbf{A}'||}{||\mathbf{A}||}, \epsilon = \frac{||\mathbf{x}'||}{||\mathbf{x}||}$ and $k(\mathbf{A}) = ||\mathbf{A}||\ ||\mathbf{A}^{-1}||$. Multiplying and dividing the rhs by $||\mathbf{A}||$ gives an estimate of relative error, $\epsilon \leq k(\mathbf{A})\mu(1 - k(\mathbf{A})\mu)^{-1}$.

Observation. $\mu = \frac{||\mathbf{A}'||}{||\mathbf{A}||}$ gives the relative change in $\mathbf{A}$.

Definition 4.2.7 $k(\mathbf{A}) = ||\mathbf{A}||\ ||\mathbf{A}^{-1}||$ *is called the condition number of* $\mathbf{A}$.

It is the deciding factor in determining whether the relative error 'ϵ' is large or small.

Case (ii) Suppose the perturbation occurs in $\mathbf{b}$ and the new system is

$$\mathbf{A}\mathbf{x} = \mathbf{b} + \mathbf{b}' \tag{4.5}$$

and the solution for the system (4.5) is given by $\mathbf{x} + \mathbf{x}'$. Then,

$$\mathbf{A}(\mathbf{x} + \mathbf{x}') = \mathbf{b} + \mathbf{b}'. \tag{4.6}$$

Subtracting (4.2) from (4.6) gives

$$\mathbf{A}\mathbf{x}' = \mathbf{b}'$$

or

$$\mathbf{x}' = \mathbf{A}^{-1}\mathbf{b}'. \tag{4.7}$$

Taking norm on both sides of (4.7),

$$||\mathbf{x}'|| \leq ||\mathbf{A}^{-1}||\ ||\mathbf{b}'||,$$

also,

$$||\mathbf{b}|| \leq ||\mathbf{A}||\ ||\mathbf{x}||.$$

Now the relative error

$$\epsilon = \frac{||\mathbf{x}'||}{||\mathbf{x}||} \leq ||\mathbf{A}^{-1}||\frac{||\mathbf{A}||}{||\mathbf{b}||}\ ||\mathbf{b}'||,$$

$$\gamma = \frac{||\mathbf{b}'||}{||\mathbf{b}||} : \text{ relative change in } \mathbf{b}$$

and

$$k(\mathbf{A}) = ||\mathbf{A}||\ ||\mathbf{A}^{-1}||.$$

Then, the relative error $\epsilon \leq k(\mathbf{A})\gamma$.

Example 4.2.4 *Suppose* $\mathbf{A} = \begin{bmatrix} 1 & 1 \\ 2 & 1 \end{bmatrix}$ *and* $\mathbf{b} = \begin{bmatrix} 3.5 \\ 4 \end{bmatrix}$.

(i) Solve the system $\mathbf{Ax} = \mathbf{b}$.

(ii) Suppose $\mathbf{A}$ *is perturbed by* $\mathbf{A}' = \begin{bmatrix} 0.001 & -0.002 \\ 0.003 & -0.003 \end{bmatrix}$ *solve the system* $(\mathbf{A} + \mathbf{A}')\mathbf{x} = \mathbf{b}$ *and state whether the system* $\mathbf{Ax} = \mathbf{b}$ *is well-conditional or not.*

Solution.

(i) Consider $\mathbf{Ax} = \mathbf{b}$. *Then* $\mathbf{x} = \begin{bmatrix} 0.5 & 3 \end{bmatrix}^T$.

(ii) The solution of the system $(\mathbf{A} + \mathbf{A}')\tilde{\mathbf{x}} = \mathbf{b}$, *that is,*

$$\begin{bmatrix} 1.001 & 0.998 \\ 2.003 & 0.997 \end{bmatrix} \tilde{\mathbf{x}} = \begin{bmatrix} 3.5 \\ 4 \end{bmatrix}.$$

is $\tilde{\mathbf{x}} = \begin{bmatrix} 0.502 \\ 3.0035 \end{bmatrix}$.

Since the difference between $\mathbf{x}$ *and* $\tilde{\mathbf{x}}$ *is a very small vector the problem is well conditioned.*

Example 4.2.5 *Let* $\mathbf{A} = \begin{bmatrix} 2 & 2 \\ 2 & 2.001 \end{bmatrix}$, $\mathbf{b} = \begin{bmatrix} 3.5 \\ 4 \end{bmatrix}$ *(i) Solve the system* $\mathbf{Ax} = \mathbf{b}$.

(ii) Suppose $\mathbf{A}'$ *is as in above example. Then solve the system* $(\mathbf{A} + \mathbf{A}')\tilde{\mathbf{x}} = \mathbf{b}$.

(iii) State the conditionality of the problem.

Solution. *The system (i)* $\mathbf{Ax} = \mathbf{b}$ *has the solution* $\mathbf{x} = \begin{bmatrix} -498.25 \\ 500 \end{bmatrix}$.

(ii) The system $(\mathbf{A} + \mathbf{A}')\tilde{\mathbf{x}} = \mathbf{b}$ *has the solution* $\tilde{\mathbf{x}} = \begin{bmatrix} 250 \\ -248.6236 \end{bmatrix}$.

(iii) Since the difference between $\mathbf{x}$ *and* $\tilde{\mathbf{x}}$ *is large the problem is ill-conditioned.*

EXERCISE 4.2

1. Verify Result 1.

2. Verify Result 2.

3. Find the condition number for the matrix (i) $\mathbf{A} = \begin{bmatrix} 3 & 1 \\ 2 & 2 \end{bmatrix}$ and (ii) $\begin{bmatrix} 2 & 1 & 3 \\ 2 & 2 & 1 \\ 1 & 2 & 3 \end{bmatrix}$ in 2-norm.

4. Consider $\mathbf{A} = \begin{bmatrix} 1 & 3 \\ 2 & 4 \end{bmatrix}$ and $\mathbf{b} = \begin{bmatrix} 5 \\ 8 \end{bmatrix}$. Solve the system $\mathbf{Ax} = \mathbf{b}$. State the conditioning of the system $\mathbf{Ax} = \mathbf{b}$ when perturbed with $\mathbf{A}' = \begin{bmatrix} 0.002 & -0.002 \\ 0.003 & -0.004 \end{bmatrix}$.

5. Suppose the matrix $\mathbf{A} = \begin{bmatrix} 3 & 3 \\ 3 & 3.003 \end{bmatrix}$ and $\mathbf{b} = \begin{bmatrix} 5 \\ 8 \end{bmatrix}$. Taking $\mathbf{A}'$ as in problem 4 find the conditionality of the problem.

6. In problem 4 change $\mathbf{b}$ to $\mathbf{b}' = \begin{bmatrix} 5.02 \\ 8.01 \end{bmatrix}$ find the conditionality of the problem.

4.3 Direct Methods for Solving a Linear System of Equations

In this section, a few numerical procedures are given to solve a linear system of equations,

$$\mathbf{Ax} = \mathbf{b}, \tag{4.8}$$

where $\mathbf{A}$ is an $m \times n$ matrix, $\mathbf{x}$ is an $n \times 1$ vector and $\mathbf{b}$ is an $m \times 1$ vector.

Gauss elimination method is a method imitating the systematic approach taught at elementary level to solve a system of equations. As mentioned in Section 4.2, arithmetic operations propagate errors and to reduce its impact following notions are defined below. Consider the matrix $\mathbf{A} = (a_{ij})_{m \times n}$.

Definition 4.3.1 *Pivot is that element denoted by p, which is the largest in magnitude among all elements of* (a_{ij}) *of the* j*th column considered for* $i \geq j$. *That is,* $|p| = max\{|a_{ij}| : i \geq j\}$.

Observation. After fixing the p, elementary row operations are performed to shift the element p to (j,j) position.

Let

$$\mathbf{A} = \begin{bmatrix} 1 & 1 & 0 \\ -1 & 2 & 3 \\ -4 & 3 & 1 \\ 1 & 5 & -2 \end{bmatrix}$$

to find the pivot p of 1st column, consider $|p| = max\{1, |-1|, |-4|, 1\}$=4

pivot of 1st column is p = -4

pivot of 2nd column, consider $|p| = max\{2, 3, 5\}$=5

pivot of 2nd column, p=5

Similarly, pivot of 3rd column is p = -2.

Definition 4.3.2 *Partial pivoting is the process wherein the search for the pivot is restricted to each column.*

Definition 4.3.3 *Complete pivoting is the approach where the pivot is chosen as the largest element in magnitude among all the mn elements of the matrix.*

Observation. The pivot p is such that $|p|$=$max\{|a_{ij}|, \ i, j = 1, 2, \ldots, n\}$= a_{kl} (say).

1. The pivot p is shifted to the (1,1) position by applying both elementary row operation and elementary column operation of interchanging R_1 and R_k rows and C_1 and C_l columns.

2. At every stage of the method, complete pivoting is done by reducing the order of the matrix by 1, every time.

Gauss Elimination Method

The idea in Gauss elimination method is to consider the augmented matrix $\mathbf{K} = [\mathbf{A} : \mathbf{b}]$ and perform elementary row operations to transform $\mathbf{K}$ to $\mathbf{K}' = [\mathbf{U} : \mathbf{c}]$ where $\mathbf{U}$ is an upper triangular matrix.

Algorithm 1

The algorithm of the Gauss elimination method is as follows.

Step 1. Consider the augmented matrix $\mathbf{K} = [\mathbf{A} : \mathbf{b}]$.

Step 2. Find the pivot p for 1st column and perform elementary row transformation so that p is in (1,1) position.

Step 3. Perform row operations $R_i \to R_i + \frac{-a_{i1}}{|\mathbf{p}|} R_1$, i=2,3,...,$n$ giving the matrix

$$\mathbf{K}^{(1)} = \left[\begin{array}{cccc|c} a_{11}^{(1)} & a_{12}^{(1)} & \cdots & a_{1n}^{(1)} & b_1^{(1)} \\ 0 & a_{22}^{(1)} & \cdots & a_{2n}^{(1)} & b_2^{(1)} \\ \vdots & \vdots & \vdots & \vdots & \vdots \\ 0 & a_{m2}^{(1)} & \cdots & a_{mn}^{(1)} & b_m^{(1)} \end{array}\right]$$

Step 4. Repeat steps 2 and 3 for column 2 excluding row1.

This gives

$$\mathbf{K}^{(2)} = \begin{bmatrix} a_{11}^{(2)} & a_{12}^{(2)} & \cdots & a_{1n}^{(2)} & \vdots & b_1^{(2)} \\ 0 & a_0^{(2)} & \cdots & a_{2n}^{(2)} & \vdots & b_2^{(2)} \\ \vdots & \vdots & \vdots & \vdots & \vdots & \vdots \\ 0 & 0 & \cdots & a_{mn}^{(2)} & \vdots & b_m^{(2)} \end{bmatrix}$$

Step 5. Repeat steps 2 and 3 for column 3 excluding rows 1 and 2.

Step 6. Repeat step 5 excluding one more row with each step, until $\mathbf{K}' = [\mathbf{U} : \mathbf{c}]$.

Now the system $\mathbf{Ux} = \mathbf{c}$ is solved by back substitution and $\mathbf{x}$ can be found, thus solving the system $\mathbf{Ax} = \mathbf{b}$.

Observation.

(1) No pivoting is required if the matrix is diagonally dominant (largest number in magnitude is on the main or if the order of the matrix is small).

(2) If the pivot is very small, it leads to errors and may be ill-conditioned and complete pivoting may have to be used.

(3) Only elementary row operations must be performed as each column represents coefficients of one variable.

Example 4.3.1 *Solve the system* $\mathbf{Ax} = \mathbf{B}$ *by Gauss elimination method, where*

$$\mathbf{A} = \begin{bmatrix} 1 & 0 & 2 & 1 \\ 2 & 1 & 2 & 2 \\ 3 & 2 & 1 & 3 \\ 1 & 1 & 2 & 2 \end{bmatrix}, \ \mathbf{b} = \begin{bmatrix} 3 & 7 & 11 & 6 \end{bmatrix}^T, \ \mathbf{x} = \begin{bmatrix} x_1 & x_2 & x_3 & x_4 \end{bmatrix}^T.$$

Solution. *Let*

$$\mathbf{K} = \begin{bmatrix} 1 & 0 & 2 & 1 & \vdots & 3 \\ 2 & 1 & 2 & 2 & \vdots & 7 \\ 3 & 2 & 1 & 3 & \vdots & 11 \\ 1 & 1 & 2 & 2 & \vdots & 6 \end{bmatrix}$$

$$R_2 \to R_2 - 2R_1, \quad R_3 \to R_3 - 3R_1, \quad R_4 \to R_4 - R_1$$

$$\begin{bmatrix} 1 & 0 & 2 & 1 & \vdots & 3 \\ 0 & 1 & -2 & 0 & \vdots & 1 \\ 0 & 2 & -5 & 0 & \vdots & 2 \\ 0 & 1 & 0 & 1 & \vdots & 3 \end{bmatrix}$$

$$R_3 \to R_3 - 2R_2, \quad R_4 \to R_4 - R_2$$

$$\left[\begin{array}{cccc|c} 1 & 0 & 2 & 1 & 3 \\ 0 & 1 & -2 & 0 & 1 \\ 0 & 0 & -1 & 0 & 0 \\ 0 & 0 & 2 & 1 & 2 \end{array}\right]$$

$$R_3 \to -R_3$$

$$\left[\begin{array}{cccc|c} 1 & 0 & 2 & 1 & 3 \\ 0 & 1 & -2 & 0 & 1 \\ 0 & 0 & 1 & 0 & 0 \\ 0 & 0 & 2 & 1 & 2 \end{array}\right]$$

$$R_4 \to R_4 - 2R_3$$

$$\left[\begin{array}{cccc|c} 1 & 0 & 2 & 1 & 3 \\ 0 & 1 & -2 & 0 & 1 \\ 0 & 0 & 1 & 0 & 0 \\ 0 & 0 & 0 & 1 & 2 \end{array}\right]$$

by back substitution,

$$x_4 = 2, \quad x_3 = 0, \quad x_2 = 1, \quad x_1 = 1.$$

Example 4.3.2 *Solve by Gauss elimination method the linear system of equations*

$$\begin{aligned} 2x_1 + 4x_2 + x_3 &= 4 \\ x_1 + 3x_2 + 2x_3 &= 4 \\ 3x_1 + 5x_2 + 4x_3 &= 8 \end{aligned}$$

Solution. *Let*

$$\mathbf{K} = \left[\begin{array}{ccc|c} 2 & 4 & 1 & 4 \\ 1 & 3 & 2 & 4 \\ 3 & 5 & 4 & 8 \end{array}\right]$$

$$R_2 \leftrightarrow R_1$$

$$\left[\begin{array}{ccc|c} 1 & 3 & 2 & 4 \\ 2 & 4 & 1 & 4 \\ 3 & 5 & 4 & 8 \end{array}\right]$$

$$R_2 \to R_2 - 2R_1, \quad R_3 \to R_3 - 3R_1$$

$$\left[\begin{array}{ccc|c} 1 & 3 & 2 & 4 \\ 0 & -2 & -3 & -4 \\ 0 & -4 & -2 & -4 \end{array}\right]$$

$$R_3 \to R_3 - 2R_2$$

$$\begin{bmatrix} 1 & 3 & 2 & \mid & 4 \\ 0 & -2 & -3 & \mid & -4 \\ 0 & 0 & 4 & \mid & 4 \end{bmatrix}$$

$$R_2 \to \tfrac{-1}{2}R_2, \quad R_3 \to \tfrac{1}{4}R_3$$

$$\begin{bmatrix} 1 & 3 & 2 & \mid & 4 \\ 0 & 1 & \frac{3}{2} & \mid & 2 \\ 0 & 0 & 1 & \mid & 1 \end{bmatrix}$$

by back substitution,

$$x_3 = 1, \;\; x_2 = \frac{1}{2}, \;\; x_1 = \frac{1}{2}.$$

Gauss Jordan method is an extension of the Gauss elimination method. In place of reducing the augmented matrix $\mathbf{K} = [\mathbf{A} : \mathbf{b}]$ to an upper triangular matrix $\mathbf{K}' = [\mathbf{U} : \mathbf{c}]$ additional elementary row operations are performed to transfer $\mathbf{K} = [\mathbf{A} : \mathbf{b}]$ to a diagonal matrix $\widetilde{\mathbf{K}} = [\mathbf{D} : \mathbf{d}]$. The Algorithm 1 is extended by repeating steps 2 and 3 without excluding any row.

Result 4.3.1 The Gauss-Jordan method is well utilized to find the inverse of a matrix.

Procedure. Replace the column vector $\mathbf{b}$ in the augmented matrix $\mathbf{K}$ by an identity matrix of same order as $A_{n\times n}$, that is, set $\mathbf{K} = [\mathbf{A} : \mathbf{I_n}]$.

Perform steps 2 and 3 in Algorithm 1 for all rows $i = 1, 2, \ldots, n$ so that

$$\mathbf{K} = [\mathbf{A} : \mathbf{I_n}] \to \widetilde{\mathbf{K}} = [\mathbf{I_n} : \mathbf{B}],$$

where $\mathbf{AB} = \mathbf{I_n} = \mathbf{BA}$.

Example 4.3.3 *Solve the system* $\mathbf{Ax} = \mathbf{b}$ *by Gauss-Jordan method, where,*

$$\mathbf{A} = \begin{bmatrix} 2 & 2 & 4 \\ 11 & 12 & 2 \\ 9 & 8 & 3 \end{bmatrix}, \quad \mathbf{b} = \begin{bmatrix} 18 \\ 40 \\ 35 \end{bmatrix}.$$

Solution. *Let* $\mathbf{x} = \begin{bmatrix} x \\ y \\ z \end{bmatrix}$ *and*

$$\mathbf{K} = \begin{bmatrix} 2 & 2 & 4 & \mid & 18 \\ 11 & 12 & 2 & \mid & 40 \\ 9 & 8 & 3 & \mid & 35 \end{bmatrix}$$

$$R_1 \to \tfrac{1}{2}R_1$$

$$\begin{bmatrix} 1 & 1 & 2 & | & 9 \\ 11 & 12 & 2 & | & 40 \\ 9 & 8 & 3 & | & 35 \end{bmatrix}$$

$$R_2 \to R_2 - 11R_1, \quad R_3 \to R_3 - 9R_1$$

$$\begin{bmatrix} 1 & 1 & 2 & | & 9 \\ 0 & 1 & -20 & | & -59 \\ 0 & -1 & -15 & | & -46 \end{bmatrix}$$

$$R_1 \to R_1 - R_2, \quad R_3 \to R_3 + R_2$$

$$\begin{bmatrix} 1 & 0 & 22 & | & 68 \\ 0 & 1 & -20 & | & -59 \\ 0 & 0 & -35 & | & -105 \end{bmatrix}$$

$$R_3 \to \tfrac{-1}{35} R_3$$

$$\begin{bmatrix} 1 & 0 & 22 & | & 68 \\ 0 & 1 & -20 & | & -59 \\ 0 & 0 & 1 & | & 3 \end{bmatrix}$$

$$R_1 \to R_1 - 22R_3, \quad R_2 \to R_2 + 20R_3$$

$$\begin{bmatrix} 1 & 0 & 0 & | & 2 \\ 0 & 1 & 0 & | & 1 \\ 0 & 0 & 1 & | & 3 \end{bmatrix}.$$

Hence, $x = 2,\ y = 1,\ z = 3.$

Example 4.3.4 *Solve by Gauss-Jordan method*

$$\begin{aligned} x + y + 2z + 2w &= 13, \\ y + z + 3w &= 12, \\ 2x + 2z + w &= 11, \\ x + 2y + z + w &= 9. \end{aligned}$$

Solution. *The augmented matrix is given by*

$$\mathbf{K} = \begin{bmatrix} 1 & 1 & 2 & 2 & | & 13 \\ 0 & 1 & 1 & 3 & | & 12 \\ 2 & 0 & 2 & 1 & | & 11 \\ 1 & 2 & 1 & 1 & | & 9 \end{bmatrix}$$

$$R_3 \to R_3 - 2R_1, \quad R_4 \to R_4 - R_1$$

$$\begin{bmatrix} 1 & 1 & 2 & 2 & | & 13 \\ 0 & 1 & 1 & 3 & | & 12 \\ 0 & -2 & -2 & -3 & | & -15 \\ 0 & 1 & -1 & -1 & | & -4 \end{bmatrix}$$

$$R_3 \to R_3 + 2R_2, \quad R_4 \to R_4 - R_2$$

$$\left[\begin{array}{cccc|c} 1 & 1 & 2 & 2 & 13 \\ 0 & 1 & 1 & 3 & 12 \\ 0 & 0 & 0 & 3 & 9 \\ 0 & 0 & -2 & -4 & -16 \end{array}\right]$$

$$R_4 \to \tfrac{-1}{2} R_4$$

$$\left[\begin{array}{cccc|c} 1 & 1 & 2 & 2 & 13 \\ 0 & 1 & 1 & 3 & 12 \\ 0 & 0 & 0 & 3 & 9 \\ 0 & 0 & 1 & 2 & 8 \end{array}\right]$$

$$R_4 \leftrightarrow R_3$$

$$\left[\begin{array}{cccc|c} 1 & 1 & 2 & 2 & 13 \\ 0 & 1 & 1 & 3 & 12 \\ 0 & 0 & 1 & 2 & 8 \\ 0 & 0 & 0 & 3 & 9 \end{array}\right]$$

$$R_4 \to \tfrac{1}{3} R_4$$

$$\left[\begin{array}{cccc|c} 1 & 1 & 2 & 2 & 13 \\ 0 & 1 & 1 & 3 & 12 \\ 0 & 0 & 1 & 2 & 8 \\ 0 & 0 & 0 & 1 & 3 \end{array}\right].$$

Hence, by back substitution, $w = 3, \quad z = 2, \quad y = 1, \quad x = 2.$

From the Gauss elimination method, it can be easily concluded that a triangular system can be solved with much less effort. This led to the notion of splitting a given matrix **A** into a product of a lower and an upper triangular matrix. This technique is called as the decomposition method or the factorization method.

Factorization method-LU decomposition method

The matrix **A** in (4.8) is to be decomposed as

$$\mathbf{A} = \mathbf{LU}$$

where $\mathbf{A} = [a_{ij}]_{n\times n}$, $\mathbf{L} = (l_{ij}),\ l_{ij} = 0,\ i < j$ and $\mathbf{U} = (u_{ij}),\ u_{ij} = 0,\ i > j$. This yields

$$\begin{bmatrix} a_{11} & a_{12} & \cdots & a_{1n} \\ a_{21} & a_{22} & \cdots & a_{2n} \\ \vdots & \vdots & \vdots & \vdots \\ a_{n2} & a_{n2} & \cdots & a_{nn} \end{bmatrix} = \begin{bmatrix} l_{11} & 0 & \cdots & 0 \\ l_{21} & l_{22} & \cdots & 0 \\ \vdots & \vdots & \vdots & \vdots \\ l_{n1} & l_{n2} & \cdots & l_{nn} \end{bmatrix} \begin{bmatrix} u_{11} & u_{12} & \cdots & u_{1n} \\ 0 & u_{22} & \cdots & u_{2n} \\ \vdots & \vdots & \vdots & \vdots \\ 0 & 0 & \cdots & u_{nn} \end{bmatrix}$$

This gives a system of $n^2 + n$ equations,

$$l_{i1}u_{1j} + l_{i2}u_{2j} + \ldots\ldots + l_{in}u_{nj} = a_{ij},\ j = 1, 2, \ldots, n \tag{4.9}$$

where $l_{ij} = 0$ for $i < j$ and $u_{ij} = 0$ for $i > j$

To obtain a unique solution the 'n' diagonal elements may be fixed. Thus set $u_{ii} = 1, i = 1, 2, 3, \ldots, n$.

Then the system (4.9) reduces to the form

$$u_{ii} = 1,\ \ i = 1, 2, \ldots, n$$

After computing the matrices $\mathbf{L}$ and $\mathbf{U}$, using (4.9) $\mathbf{Ax} = \mathbf{b}$ reduces to $\mathbf{LUx} = \mathbf{b}$, set

$$\mathbf{Ux} = \mathbf{y}. \tag{4.10}$$

Then

$$\mathbf{Ly} = \mathbf{b}, \tag{4.11}$$

is a lower triangular system. Equation (4.11) can be easily solved by forward substitution and $\mathbf{y}$ can be found, say $\mathbf{y} = \begin{bmatrix} c_1 \\ c_2 \\ \vdots \\ c_n \end{bmatrix}$

Now substituting $\mathbf{y}$ in equation (4.10) gives

$$\mathbf{Ux} = \begin{bmatrix} c_1 \\ c_2 \\ \vdots \\ c_n \end{bmatrix} \tag{4.12}$$

The upper triangular system (4.12) can be solved by back substitution and x, the solution can be easily found.

Before working out some examples, the following points must be noted.

Observation.

(1) LU-factorization is guaranteed if $\mathbf{A}$ is a positive definite matrix. (This is only a sufficient condition but is not necessary).

(2) LU decomposition is possible if all the principal minors of $\mathbf{A}$ are different from zero. That is, $a_{11} \neq 0$, $\begin{vmatrix} a_{11} & a_{12} \\ a_{21} & a_{22} \end{vmatrix} \neq 0, \ldots, |\mathbf{A}| \neq 0$.

(3) If $u_{ii} = 1, i = 1, 2, 3, \ldots, n$, the method is called Doolittle's method.

(4) If $l_{ii} = 1, i = 1, 2, 3, \ldots, n$, this method is called Crout's method.

(5) If $u_{ii} = 0$ for some i, Doolittle's method fails and if $l_{ii} = 0$, for some i, Crout's method fails.

Example 4.3.5 *Solve the following system of equations using LU decomposition.*

$$\begin{aligned} 2x + 6y + 8w &= 10, \\ x + y + 3z + w &= 5, \\ 4x + 2y + 8z + 11w &= 13, \\ 3x + y + 2z + 6w &= 5. \end{aligned}$$

Solution. *Let* $\mathbf{A} = \mathbf{LU}$ *that is,*

$$\begin{bmatrix} 2 & 6 & 0 & 8 \\ 1 & 1 & 3 & 1 \\ 4 & 2 & 8 & 11 \\ 3 & 1 & 2 & 6 \end{bmatrix} = \begin{bmatrix} 1 & 0 & 0 & 0 \\ l_{21} & 1 & 0 & 0 \\ l_{31} & l_{32} & 1 & 0 \\ l_{41} & l_{42} & l_{43} & 1 \end{bmatrix} \begin{bmatrix} u_{11} & u_{12} & u_{13} & u_{14} \\ 0 & u_{22} & u_{23} & u_{24} \\ 0 & 0 & u_{33} & u_{34} \\ 0 & 0 & 0 & u_{44} \end{bmatrix}.$$

Then simplifying,

$$\begin{aligned} u_{11} &= 2 & l_{21}u_{11} = 1 \quad & l_{21} = \frac{1}{2} \\ u_{12} &= 6 & l_{31}u_{11} = 4 \quad & l_{31} = 2 \\ u_{13} &= 0 & & \\ u_{14} &= 8 & l_{41}u_{11} = 3 \quad & l_{41} = \frac{3}{2}, \end{aligned}$$

$$\begin{aligned} l_{21}u_{12} + u_{22} = 1 &\implies \tfrac{1}{2} \cdot 6 + u_{22} = 1 \implies u_{22} = -2 \\ l_{21}u_{13} + u_{23} = 3 &\implies \tfrac{1}{2} \cdot 0 + u_{23} = 3 \implies u_{23} = 3 \\ l_{21}u_{14} + u_{24} = 1 &\implies \tfrac{1}{2} \cdot 8 + u_{24} = 1 \implies u_{24} = -3. \end{aligned}$$

Thus,

$$\mathbf{L} = \begin{bmatrix} 1 & 0 & 0 & 0 \\ 0.5 & 1 & 0 & 0 \\ 2 & 5 & 1 & 0 \\ \frac{3}{2} & 4 & \frac{10}{7} & 1 \end{bmatrix} \qquad \mathbf{U} = \begin{bmatrix} 2 & 6 & 0 & 8 \\ 0 & -2 & 3 & -3 \\ 0 & 0 & -7 & 10 \\ 0 & 0 & 0 & \frac{-58}{7} \end{bmatrix},$$

that is solving $\mathbf{Ly} = \mathbf{b}$,

$$\mathbf{Ly} = \begin{bmatrix} 1 & 0 & 0 & 0 \\ 0.5 & 1 & 0 & 0 \\ 2 & 5 & 1 & 0 \\ \frac{3}{2} & 4 & \frac{10}{7} & 1 \end{bmatrix} \begin{bmatrix} y_1 \\ y_2 \\ y_3 \\ y_4 \end{bmatrix} = \begin{bmatrix} 10 \\ 5 \\ 13 \\ 5 \end{bmatrix}$$

by forward substitution gives $\mathbf{y} = \begin{bmatrix} 10 \\ 0 \\ -7 \\ 0 \end{bmatrix}$. *Next substituting for* $\mathbf{Ux} = \mathbf{y}$

$$\begin{bmatrix} 2 & 6 & 0 & 8 \\ 0 & -2 & 3 & -3 \\ 0 & 0 & -7 & 10 \\ 0 & 0 & 0 & \frac{-58}{7} \end{bmatrix} \begin{bmatrix} x \\ y \\ z \\ w \end{bmatrix} = \begin{bmatrix} 10 \\ 0 \\ -7 \\ 0 \end{bmatrix}$$

gives by backward substitution,

$$\mathbf{x} = \begin{bmatrix}0.5 & 1.5 & 1 & 0\end{bmatrix}^T,$$

the solution of the given system of equations.

Algorithm 2.

The LU decomposition can also be obtained by using Gauss elimination method.

Procedure. Given a matrix $\mathbf{A}_{n\times n}$ reduce it to an upper triangular matrix $\mathbf{U}$.

In the process, the lower triangular matrix is obtained as follows.
Set $l_{ii} = 1$, i=1,2,...,n.
$l_{ij} = 0$, for $i > j$
and $l_{ij} = c_{ij}$ for $i < j$.
where c_{ij} is obtained in the process of making $u_{ij} = 0$ for $i < j$,
i.e. $R_i \to R_i - c_{ij}R_j$ where c_{ij} is chosen in such a way that $u_{ij} = 0$.

Example 4.3.6 *Find the LU decomposition of the matrix* $\mathbf{A} = \begin{bmatrix}1 & 2 & 3\\ 3 & 2 & 1\\ 1 & 1 & 1\end{bmatrix}$ *using the approach of Gauss elimination method (without pivoting).*

Solution. *Consider set* $l_{11} = 1$, $\mathbf{A} = \begin{bmatrix}1 & 2 & 3\\ 3 & 2 & 1\\ 1 & 1 & 1\end{bmatrix}$

Using Gaussian elimination method

$$R_2 \to R_2 - 3R_1 \ \textit{which gives} \ l_{21} = 3$$

$$\begin{bmatrix}1 & 2 & 3\\ 0 & -4 & -8\\ 1 & 1 & 1\end{bmatrix}$$

next

$$R_3 \to R_3 - R_1 \ \textit{thus} \ l_{31} = 1$$

$$\begin{bmatrix}1 & 2 & 3\\ 0 & -4 & -8\\ 0 & -1 & -2\end{bmatrix}$$

next

$$R_3 \to R_3 - \frac{1}{4}R_1 \ \textit{hence} \ l_{32} = \frac{1}{4}$$

$$\begin{bmatrix}1 & 2 & 3\\ 0 & -4 & -8\\ 0 & 0 & 0\end{bmatrix}.$$

$\mathbf{L} = \begin{bmatrix} 1 & 0 & 0 \\ 3 & 1 & 0 \\ 1 & \frac{1}{4} & 1 \end{bmatrix}$ *and* $\mathbf{U} = \begin{bmatrix} 1 & 2 & 3 \\ 0 & -4 & -8 \\ 0 & 0 & 0 \end{bmatrix}$ *which is the LU decomposition of* $\mathbf{A}$.

Example 4.3.7 *Solve the system*

$$\begin{aligned} x + 2y + z + w &= 4, \\ x - 2y + z - w &= 0, \\ 3x + 2y + z - 4w &= -4, \\ x + 2y - z - 2w &= -4. \end{aligned}$$

Solution. *In matrix notation*

$$\begin{bmatrix} 1 & 2 & 1 & 1 \\ 1 & -2 & 1 & -1 \\ 3 & 2 & 1 & -4 \\ 1 & 2 & -1 & -2 \end{bmatrix} \begin{bmatrix} x \\ y \\ z \\ w \end{bmatrix} = \begin{bmatrix} 4 \\ 0 \\ -4 \\ -4 \end{bmatrix}$$

To find LU decomposition of $\mathbf{A}$

Consider $\begin{bmatrix} 1 & 2 & 1 & 1 \\ 1 & -2 & 1 & -1 \\ 3 & 2 & 1 & -4 \\ 1 & 2 & -1 & -2 \end{bmatrix}$ *and apply Gauss elimination method without pivoting.*

Initially set $l_{11} = 1$ *and* $l_{21} = 1$

$$R_2 \to R_2 - R_1$$

$$\begin{bmatrix} 1 & 2 & 1 & 1 \\ 0 & -4 & 0 & -2 \\ 3 & 2 & 1 & -4 \\ 1 & 2 & -1 & -2 \end{bmatrix}$$

next

$$R_3 \to R_3 - 3R_1 \ \text{that gives} \ l_{31} = 3$$

$$\begin{bmatrix} 1 & 2 & 1 & 1 \\ 0 & -4 & 0 & -2 \\ 0 & -4 & -2 & -7 \\ 1 & 2 & -1 & -2 \end{bmatrix}$$

setting

$$R_4 \to R_4 - R_1 \ \text{that gives} \ l_{41} = 1$$

$$\begin{bmatrix} 1 & 2 & 1 & 1 \\ 0 & -4 & 0 & -2 \\ 0 & -4 & -2 & -7 \\ 0 & 0 & -2 & -3 \end{bmatrix}$$

next

$$R_3 \to R_3 - R_2 \ \textit{gives} \ l_{32} = 1$$

$$\begin{bmatrix} 1 & 2 & 1 & 1 \\ 0 & -4 & 0 & -2 \\ 0 & 0 & -2 & -5 \\ 0 & 0 & -2 & -3 \end{bmatrix}$$

Again

$$R_4 \to R_4 - R_1 \ \textit{that gives} \ l_{43} = 1$$

$$\begin{bmatrix} 1 & 2 & 1 & 1 \\ 0 & -4 & 0 & -2 \\ 0 & 0 & -2 & -5 \\ 0 & 0 & 0 & 2 \end{bmatrix}$$

$$\textit{Thus } \mathbf{U} = \begin{bmatrix} 1 & 2 & 1 & 1 \\ 0 & -4 & 0 & -2 \\ 0 & 0 & -2 & -5 \\ 0 & 0 & 0 & 2 \end{bmatrix} \ \textit{and } \mathbf{L} = \begin{bmatrix} 1 & 0 & 0 & 0 \\ 1 & 1 & 0 & 0 \\ 3 & 1 & 1 & 0 \\ 1 & 0 & 1 & 1 \end{bmatrix}$$

Now solving for $\mathbf{Ax} = \mathbf{B}$ *reduces to* $\mathbf{LUx} = \mathbf{b}$, *set* $\mathbf{y} = \mathbf{Ux}$
Now solving for $\mathbf{Ly} = \mathbf{b}$

$$\begin{bmatrix} 1 & 0 & 0 & 0 \\ 1 & 1 & 0 & 0 \\ 3 & 1 & 1 & 0 \\ 1 & 0 & 1 & 1 \end{bmatrix} \begin{bmatrix} y_1 \\ y_2 \\ y_3 \\ y_4 \end{bmatrix} = \begin{bmatrix} 4 \\ 0 \\ -4 \\ -4 \end{bmatrix}$$

$$\textit{gives } \mathbf{y} = \begin{bmatrix} 4 \\ -4 \\ -12 \\ 4 \end{bmatrix}.$$

Solving for $\mathbf{Ux} = \mathbf{b}$

$$\begin{bmatrix} 1 & 2 & 1 & 1 \\ 0 & -4 & 0 & -2 \\ 0 & 0 & -2 & -5 \\ 0 & 0 & 0 & 2 \end{bmatrix} \begin{bmatrix} x \\ y \\ z \\ w \end{bmatrix} = \begin{bmatrix} 4 \\ -4 \\ -12 \\ 4 \end{bmatrix}$$

gives $\begin{bmatrix} x & y & z & w \end{bmatrix}^T = \begin{bmatrix} 1 & 0 & 1 & 2 \end{bmatrix}$.

In case **A** is a symmetric and a positive definite matrix, the LU decomposition becomes simplified and is described in the following method.

Cholesky Method. If **A** is a symmetric and a positive definite matrix, then

$$\mathbf{A} = \mathbf{LL^T} \tag{4.13}$$

where **L** is a lower triangular matrix.

The same procedure as LU decomposition method is used to solve the system $\mathbf{Ax} = \mathbf{b}$.

Observation.

(1) It is possible to write $\mathbf{A} = \mathbf{UU^T}$ and solve for $\mathbf{U}$.

(2) This method is also called as square root method.

(3) Since $\mathbf{A} = \mathbf{LL^T}$ one can find $\mathbf{A}^{-1}$ after finding $\mathbf{L}$. Thus $\mathbf{A}^{-1} = (\mathbf{L^T})^{-1}\mathbf{L}^{-1}$.

Computational complexity

The number of operations required to obtain the solution in the fore-mentioned method is given here without detail.

In Gauss elimination method, the number of operations required is $(n/3)[(n^2 + 3n - 1)]$.
Thus for large 'n ', the computational complexity is given by $\approx n^3$.
For large 'n ', the computational complexity for

(1) Gauss-Jordan method is $n^3/2$

(2) LU decomposition method is $n^3/3$

(3) Cholesky method is $n^3/6$

EXERCISE 4.3

1. Use Gauss elimination method to solve the system

$$\begin{aligned} 3x + 2y + 2z &= 3 \\ x + y + 2z &= 1 \\ 2x + 3y + 4z &= 5 \end{aligned}$$

2. Find the inverse of A using Gauss Jordan method where

$$A = \begin{bmatrix} 1 & 2 & 3 \\ 4 & 5 & 4 \\ 7 & 8 & 7 \end{bmatrix}$$

3. Find a decomposition for A using Cholesky method where

$$A = \begin{bmatrix} 1 & 4 & 5 \\ 4 & 25 & 26 \\ 5 & 26 & 45 \end{bmatrix}$$

4. Find LU decomposition for A using Gauss elimination method for A, where

$$A = \begin{bmatrix} 1 & 4 & 3 \\ 3 & 5 & 1 \\ -1 & 1 & 3 \end{bmatrix}$$

4.4 Iterative Methods

In this section, an attempt is made to solve a linear system of equations using iterative techniques. All iterative techniques start with an initial approximation. Successive approximations are found using the previous approximation. The convergence of the sequence of approximations is important and criteria are given for their convergence.

Of the many iterative methods available, the Gauss-Jacobi and Gauss-Siedel are well utilized and are discussed first and criteria of when these sequence of approximations converge is stated. Successive over relaxation (SoR) technique, which is developed on Gauss-Siedel method is given.

Gauss-Jacobi method

In Gauss-Jacobi method, the system

$\mathbf{Ax} = \mathbf{b}$ is rewritten as

$$\mathbf{Ax} = (\mathbf{D} + \mathbf{L} + \mathbf{U})\mathbf{x} = \mathbf{b}, \tag{4.14}$$

where $\mathbf{D}$, $\mathbf{L}$ and $\mathbf{U}$ are diagonal, lower triangular and upper triangular matrices respectively.

The system (4.14) is grouped as

$$\mathbf{Dx} = -(\mathbf{L} + \mathbf{U})\mathbf{x} + \mathbf{b}$$

Then

$$\mathbf{x} = -\mathbf{D}^{-1}(\mathbf{L} + \mathbf{U})\mathbf{x} + \mathbf{D}^{-1}\mathbf{b}$$

Now replace $\mathbf{x}$ on rhs with an initial approximation $\mathbf{x}^{(0)}$ to give

$$\mathbf{x}^{(1)} = -\mathbf{D}^{-1}(\mathbf{L} + \mathbf{U})\mathbf{x}^{(0)} + \mathbf{D}^{-1}\mathbf{b},$$

replacing $\mathbf{x}^{(0)}$ with the new estimate $\mathbf{x}^{(1)}$, the next iteration is

$$\mathbf{x}^{(2)} = -\mathbf{D}^{-1}(\mathbf{L} + \mathbf{U})\mathbf{x}^{(1)} + \mathbf{D}^{-1}\mathbf{b}$$

proceeding in a similar fashion gives

$$\mathbf{x}^{(\mathbf{k}+1)} = -\mathbf{D}^{-1}(\mathbf{L} + \mathbf{U})\mathbf{x}^{(\mathbf{k})} + \mathbf{D}^{-1}\mathbf{b}. \tag{4.15}$$

Observation.

(i) The sequence of iterates converge if the matrix $\mathbf{A}$ is strictly diagonally dominant.

(ii) This method is called as the method of simultaneous displacement, as the approximation $\mathbf{x^{(k)}}$ is completely used to find the new approximation $\mathbf{x^{(k+1)}}$.

Example 4.4.1 *Solve the following equations by using the Gauss Jacobi method*

$$\begin{aligned} 2x - y &= 1 \\ -x + 6y - 2z &= 3 \\ 4x - 3y + 8z &= 9. \end{aligned}$$

Solution. *Let* $\mathbf{Ax} = \mathbf{b}$

$$\begin{bmatrix} 2 & -1 & 0 \\ -1 & 6 & -2 \\ 4 & -3 & 8 \end{bmatrix} \begin{bmatrix} x \\ y \\ z \end{bmatrix} = \begin{bmatrix} 1 \\ 3 \\ 9 \end{bmatrix}$$

$(\mathbf{D} + \mathbf{L} + \mathbf{U})\mathbf{x} = \mathbf{b}$

$$\left(\begin{bmatrix} 2 & 0 & 0 \\ 0 & 6 & 0 \\ 0 & 0 & 8 \end{bmatrix} + \begin{bmatrix} 0 & 0 & 0 \\ -1 & 0 & 0 \\ 4 & -3 & 0 \end{bmatrix} + \begin{bmatrix} 0 & -1 & 0 \\ 0 & 0 & -2 \\ 0 & 0 & 0 \end{bmatrix} \right) \mathbf{x} = \begin{bmatrix} 1 \\ 3 \\ 9 \end{bmatrix}$$

$$\begin{bmatrix} 2 & 0 & 0 \\ 0 & 6 & 0 \\ 0 & 0 & 8 \end{bmatrix} \mathbf{x} = -\left(\begin{bmatrix} 0 & 0 & 0 \\ -1 & 0 & 0 \\ 4 & -3 & 0 \end{bmatrix} + \begin{bmatrix} 0 & -1 & 0 \\ 0 & 0 & -2 \\ 0 & 0 & 0 \end{bmatrix} \right) \mathbf{x} + \begin{bmatrix} 1 \\ 3 \\ 9 \end{bmatrix}$$

$\mathbf{x} = -\mathbf{D}^{-1}(\mathbf{L} + \mathbf{U})\mathbf{x} + \mathbf{D}^{-1}\mathbf{b}$

$$\mathbf{x} = -\begin{bmatrix} \frac{1}{2} & 0 & 0 \\ 0 & \frac{1}{6} & 0 \\ 0 & 0 & \frac{1}{8} \end{bmatrix} \begin{bmatrix} 0 & -1 & 0 \\ -1 & 0 & -2 \\ 4 & -3 & 0 \end{bmatrix} \cdot \mathbf{x} + \begin{bmatrix} \frac{1}{2} & 0 & 0 \\ 0 & \frac{1}{6} & 0 \\ 0 & 0 & \frac{1}{8} \end{bmatrix} \begin{bmatrix} 1 \\ 3 \\ 9 \end{bmatrix}.$$

$\mathbf{x}^{(0)} = \begin{bmatrix} 0 & 0 & 0 \end{bmatrix}^T$

$$\mathbf{x}^{(1)} = \begin{bmatrix} 0.5 \\ 0.5 \\ 1.125 \end{bmatrix}$$

$$\mathbf{x}^{(2)} = -\begin{bmatrix} \frac{1}{2} & 0 & 0 \\ 0 & \frac{1}{6} & 0 \\ 0 & 0 & \frac{1}{8} \end{bmatrix} \begin{bmatrix} 0 & -1 & 0 \\ -1 & 0 & -2 \\ 4 & -3 & 0 \end{bmatrix} \begin{bmatrix} 0.5 \\ 0.5 \\ 1.125 \end{bmatrix} + \begin{bmatrix} \frac{1}{2} & 0 & 0 \\ 0 & \frac{1}{6} & 0 \\ 0 & 0 & \frac{1}{8} \end{bmatrix} \begin{bmatrix} 1 \\ 3 \\ 9 \end{bmatrix} =$$

$$\begin{bmatrix} 0.75 \\ 0.9583 \\ 1.0625 \end{bmatrix}$$

$$\mathbf{x}^{(3)} = -\begin{bmatrix} \frac{1}{2} & 0 & 0 \\ 0 & \frac{1}{6} & 0 \\ 0 & 0 & \frac{1}{8} \end{bmatrix} \begin{bmatrix} 0 & -1 & 0 \\ -1 & 0 & -2 \\ 4 & -3 & 0 \end{bmatrix} \begin{bmatrix} 0.75 \\ 0.9583 \\ 1.0625 \end{bmatrix} + \begin{bmatrix} \frac{1}{2} & 0 & 0 \\ 0 & \frac{1}{6} & 0 \\ 0 & 0 & \frac{1}{8} \end{bmatrix} \begin{bmatrix} 1 \\ 3 \\ 9 \end{bmatrix} = \begin{bmatrix} 0.9791 \\ 0.9792 \\ 1.1093 \end{bmatrix}$$

$$\mathbf{x}^{(4)} = -\begin{bmatrix} \frac{1}{2} & 0 & 0 \\ 0 & \frac{1}{6} & 0 \\ 0 & 0 & \frac{1}{8} \end{bmatrix} \begin{bmatrix} 0 & -1 & 0 \\ -1 & 0 & -2 \\ 4 & -3 & 0 \end{bmatrix} \begin{bmatrix} 0.9791 \\ 0.9792 \\ 1.1093 \end{bmatrix} + \begin{bmatrix} \frac{1}{2} & 0 & 0 \\ 0 & \frac{1}{6} & 0 \\ 0 & 0 & \frac{1}{8} \end{bmatrix} \begin{bmatrix} 1 \\ 3 \\ 9 \end{bmatrix} = \begin{bmatrix} 0.9791 \\ 1.0329 \\ 1.0025 \end{bmatrix}$$

$$\mathbf{x}^{(5)} = -\begin{bmatrix} \frac{1}{2} & 0 & 0 \\ 0 & \frac{1}{6} & 0 \\ 0 & 0 & \frac{1}{8} \end{bmatrix} \begin{bmatrix} 0 & -1 & 0 \\ -1 & 0 & -2 \\ 4 & -3 & 0 \end{bmatrix} \begin{bmatrix} 0.9791 \\ 1.0329 \\ 1.0025 \end{bmatrix} + \begin{bmatrix} \frac{1}{2} & 0 & 0 \\ 0 & \frac{1}{6} & 0 \\ 0 & 0 & \frac{1}{8} \end{bmatrix} \begin{bmatrix} 1 \\ 3 \\ 9 \end{bmatrix} = \begin{bmatrix} 0.9896 \\ 1.033 \\ 1.0027 \end{bmatrix}$$

$$\mathbf{x}^{(6)} = -\begin{bmatrix} \frac{1}{2} & 0 & 0 \\ 0 & \frac{1}{6} & 0 \\ 0 & 0 & \frac{1}{8} \end{bmatrix} \begin{bmatrix} 0 & -1 & 0 \\ -1 & 0 & -2 \\ 4 & -3 & 0 \end{bmatrix} \begin{bmatrix} 0.9896 \\ 1.033 \\ 1.0027 \end{bmatrix} + \begin{bmatrix} \frac{1}{2} & 0 & 0 \\ 0 & \frac{1}{6} & 0 \\ 0 & 0 & \frac{1}{8} \end{bmatrix} \begin{bmatrix} 1 \\ 3 \\ 9 \end{bmatrix} = \begin{bmatrix} 1.0165 \\ 0.99916 \\ 1.0176 \end{bmatrix}.$$

Hence the solution is,

$$\mathbf{x} = \begin{bmatrix} 1 \\ 1 \\ 1 \end{bmatrix}.$$

Gauss-Siedel Method. The difference between Gauss-Jacobi method and Gauss-Siedel method is that the new information obtained is immediately used in the Gauss-Siedel method.

It can be observed that in solving the system (4.8) using the relation (4.15), the first component of $\mathbf{x}^{(\mathbf{k+1})}$ is found in the first step and 2nd component of $\mathbf{x}^{(\mathbf{k+1})}$ is found in second step and so on.

This observation is fruitfully utilized by considering the setup

$$\mathbf{Ax} = (\mathbf{D} + \mathbf{L} + \mathbf{U})\mathbf{x} = \mathbf{b},$$

and rearranging it as

$$(\mathbf{D} + \mathbf{L})\mathbf{x} = -\mathbf{Ux} + \mathbf{b},$$

giving the iterative scheme as

$$(\mathbf{D} + \mathbf{L})\mathbf{x}^{(\mathbf{k+1})} = -\mathbf{Ux}^{(\mathbf{k})} + \mathbf{b},$$

or

$$\mathbf{x^{(k+1)}} = -(\mathbf{D}+\mathbf{L})^{-1}\mathbf{U}\mathbf{x^{(k)}} + (\mathbf{D}+\mathbf{L})^{-1}\mathbf{b}. \tag{4.16}$$

The relation (4.16) is the iterative scheme for Gauss-Siedel Method.

Observation.

(1) The Gauss-Siedel method is known as the method of successive displacement.

(2) This method converges if $\mathbf{A}$ is strictly diagonally dominant.

Example 4.4.2 *Solve by Gauss-Siedel method*

$$\begin{aligned} 5x_1 + 3x_2 + 2x_3 + x_4 &= 5 \\ x_1 + 5x_2 + 2x_3 + 2x_4 &= 6 \\ x_1 + 2x_2 + 6x_3 + x_4 &= 7 \\ 2x_1 + x_2 + x_3 + 5x_4 &= 8. \end{aligned}$$

Solution.
Here

$$\mathbf{D}+\mathbf{L} = \begin{bmatrix} 5 & 0 & 0 & 0 \\ 1 & 5 & 0 & 0 \\ 1 & 2 & 6 & 0 \\ 2 & 1 & 1 & 5 \end{bmatrix},$$

$$(\mathbf{D}+\mathbf{L})^{-1} = \begin{bmatrix} 0.2 & 0 & 0 & 0 \\ -0.04 & 0.2 & 0 & 0 \\ -0.02 & -0.0667 & 0.1667 & 0 \\ -0.068 & -0.0267 & -0.0333 & 0.2 \end{bmatrix},$$

$$\mathbf{U} = \begin{bmatrix} 0 & 3 & 2 & 1 \\ 0 & 0 & 2 & 2 \\ 0 & 0 & 0 & 1 \\ 0 & 0 & 0 & 0 \end{bmatrix},$$

$$-(\mathbf{D}+\mathbf{L})^{-1}\mathbf{U} = \begin{bmatrix} 0 & -0.6 & -0.4 & -0.2 \\ 0 & 0.12 & -0.32 & -0.36 \\ 0 & 0.06 & 0.1733 & -0.0133 \\ 0 & 0.204 & 0.1893 & 0.1547 \end{bmatrix},$$

$$(\mathbf{D}+\mathbf{L})^{-1}\mathbf{b} = (\mathbf{D}+\mathbf{L})^{-1}\begin{bmatrix} 5 & 6 & 7 & 8 \end{bmatrix}^T = \begin{bmatrix} 1 \\ 1 \\ 0.667 \\ 0.8667 \end{bmatrix},$$

substituting these values in formula (4.16) with initial approximation $x^{(0)} = (\mathbf{D}+\mathbf{L})^{-1}\mathbf{b}$.

Suppose the initial approximation is

$$\mathbf{x}^{(0)} = \begin{bmatrix} 1 \\ 1 \\ 0.667 \\ 0.8667 \end{bmatrix}$$

$$\mathbf{x}^{(1)} = \begin{bmatrix} 0 & -0.6 & -0.4 & -0.2 \\ 0 & 0.12 & -0.32 & -0.36 \\ 0 & 0.06 & 0.1733 & -0.0133 \\ 0 & 0.204 & 0.1893 & 0.1547 \end{bmatrix} \begin{bmatrix} 1 \\ 1 \\ 0.667 \\ 0.8667 \end{bmatrix} + \begin{bmatrix} 1 \\ 1 \\ 0.667 \\ 0.8667 \end{bmatrix} = \begin{bmatrix} -0.04 \\ 0.5946 \\ 0.8307 \\ 1.331 \end{bmatrix},$$

$$\mathbf{x}^{(2)} = \begin{bmatrix} 0 & -0.6 & -0.4 & -0.2 \\ 0 & 0.12 & -0.32 & -0.36 \\ 0 & 0.06 & 0.1733 & -0.0133 \\ 0 & 0.204 & 0.1893 & 0.1547 \end{bmatrix} \begin{bmatrix} -0.04 \\ 0.5946 \\ 0.8307 \\ 1.331 \end{bmatrix} + \begin{bmatrix} 1 \\ 1 \\ 0.667 \\ 0.8667 \end{bmatrix} = \begin{bmatrix} 0.0448 \\ 0.3264 \\ 0.8286 \\ 1.3512 \end{bmatrix},$$

$$\mathbf{x}^{(3)} = \begin{bmatrix} 0 & -0.6 & -0.4 & -0.2 \\ 0 & 0.12 & -0.32 & -0.36 \\ 0 & 0.06 & 0.1733 & -0.0133 \\ 0 & 0.204 & 0.1893 & 0.1547 \end{bmatrix} \begin{bmatrix} 0.0448 \\ 0.3264 \\ 0.8286 \\ 1.3512 \end{bmatrix} + \begin{bmatrix} 1 \\ 1 \\ 0.667 \\ 0.8667 \end{bmatrix} = \begin{bmatrix} 0.2025 \\ 0.2876 \\ 0.8119 \\ 1.2992 \end{bmatrix},$$

$$\mathbf{x}^{(4)} = \begin{bmatrix} 0 & -0.6 & -0.4 & -0.2 \\ 0 & 0.12 & -0.32 & -0.36 \\ 0 & 0.06 & 0.1733 & -0.0133 \\ 0 & 0.204 & 0.1893 & 0.1547 \end{bmatrix} \begin{bmatrix} 0.2025 \\ 0.2876 \\ 0.8119 \\ 1.2992 \end{bmatrix} + \begin{bmatrix} 1 \\ 1 \\ 0.667 \\ 0.8667 \end{bmatrix} = \begin{bmatrix} 0.2428 \\ 0.307 \\ 0.8074 \\ 1.28 \end{bmatrix},$$

$$\mathbf{x}^{(5)} = \begin{bmatrix} 0 & -0.6 & -0.4 & -0.2 \\ 0 & 0.12 & -0.32 & -0.36 \\ 0 & 0.06 & 0.1733 & -0.0133 \\ 0 & 0.204 & 0.1893 & 0.1547 \end{bmatrix} \begin{bmatrix} 0.2428 \\ 0.307 \\ 0.8074 \\ 1.28 \end{bmatrix} + \begin{bmatrix} 1 \\ 1 \\ 0.667 \\ 0.8667 \end{bmatrix} = \begin{bmatrix} 0.2368 \\ 0.3177 \\ 0.808 \\ 1.2802 \end{bmatrix},$$

$$\mathbf{x}^{(6)} = \begin{bmatrix} 0 & -0.6 & -0.4 & -0.2 \\ 0 & 0.12 & -0.32 & -0.36 \\ 0 & 0.06 & 0.1733 & -0.0133 \\ 0 & 0.204 & 0.1893 & 0.1547 \end{bmatrix} \begin{bmatrix} 0.2368 \\ 0.3177 \\ 0.808 \\ 1.2802 \end{bmatrix} + \begin{bmatrix} 1 \\ 1 \\ 0.667 \\ 0.8667 \end{bmatrix} = \begin{bmatrix} 0.2301 \\ 0.3187 \\ 0.8088 \\ 1.2825 \end{bmatrix},$$

$$\mathbf{x}^{(7)} = \begin{bmatrix} 0 & -0.6 & -0.4 & -0.2 \\ 0 & 0.12 & -0.32 & -0.36 \\ 0 & 0.06 & 0.1733 & -0.0133 \\ 0 & 0.204 & 0.1893 & 0.1547 \end{bmatrix} \begin{bmatrix} 0.2301 \\ 0.3187 \\ 0.8088 \\ 1.2825 \end{bmatrix} + \begin{bmatrix} 1 \\ 1 \\ 0.667 \\ 0.8667 \end{bmatrix} = \begin{bmatrix} 0.228 \\ 0.3277 \\ 0.8089 \\ 1.2832 \end{bmatrix},$$

$$\mathbf{x}^{(8)} = \begin{bmatrix} 0 & -0.6 & -0.4 & -0.2 \\ 0 & 0.12 & -0.32 & -0.36 \\ 0 & 0.06 & 0.1733 & -0.0133 \\ 0 & 0.204 & 0.1893 & 0.1547 \end{bmatrix} \begin{bmatrix} 0.228 \\ 0.3277 \\ 0.8089 \\ 1.2832 \end{bmatrix} + \begin{bmatrix} 1 \\ 1 \\ 0.667 \\ 0.8667 \end{bmatrix} = \begin{bmatrix} 0.2292 \\ 0.3173 \\ 0.8089 \\ 1.2831 \end{bmatrix},$$

$$\mathbf{x}^{(9)} = \begin{bmatrix} 0 & -0.6 & -0.4 & -0.2 \\ 0 & 0.12 & -0.32 & -0.36 \\ 0 & 0.06 & 0.1733 & -0.0133 \\ 0 & 0.204 & 0.1893 & 0.1547 \end{bmatrix} \begin{bmatrix} 0.2292 \\ 0.3173 \\ 0.8089 \\ 1.2831 \end{bmatrix} + \begin{bmatrix} 1 \\ 1 \\ 0.667 \\ 0.8667 \end{bmatrix} = \begin{bmatrix} 0.2294 \\ 0.3173 \\ 0.8089 \\ 1.283 \end{bmatrix},$$

$$\mathbf{x}^{(10)} = \begin{bmatrix} 0 & -0.6 & -0.4 & -0.2 \\ 0 & 0.12 & -0.32 & -0.36 \\ 0 & 0.06 & 0.1733 & -0.0133 \\ 0 & 0.204 & 0.1893 & 0.1547 \end{bmatrix} \begin{bmatrix} 0.2294 \\ 0.3173 \\ 0.8089 \\ 1.283 \end{bmatrix} + \begin{bmatrix} 1 \\ 1 \\ 0.667 \\ 0.8667 \end{bmatrix} = \begin{bmatrix} 0.2295 \\ 0.3173 \\ 0.8089 \\ 1.283 \end{bmatrix},$$

$$\mathbf{x}^{(11)} = \begin{bmatrix} 0 & -0.6 & -0.4 & -0.2 \\ 0 & 0.12 & -0.32 & -0.36 \\ 0 & 0.06 & 0.1733 & -0.0133 \\ 0 & 0.204 & 0.1893 & 0.1547 \end{bmatrix} \begin{bmatrix} 0.2295 \\ 0.3173 \\ 0.8089 \\ 1.283 \end{bmatrix} + \begin{bmatrix} 1 \\ 1 \\ 0.667 \\ 0.8667 \end{bmatrix} = \begin{bmatrix} 0.2295 \\ 0.3173 \\ 0.8089 \\ 1.283 \end{bmatrix},$$

Thus,
$x_1 = 0.2295, \quad x_2 = 0.3173, \quad x_3 = 0.8089, \quad x_4 = 1.283.$

Verification

$$\mathbf{x} = \mathbf{A}^{-1}\mathbf{b} = \begin{bmatrix} 0.2295 \\ 0.3173 \\ 0.8089 \\ 1.283 \end{bmatrix}.$$

Example 4.4.3 *Solve the system Gauss-Siedel method*

$$\begin{aligned} 2x_1 + x_2 + x_3 + x_4 &= 5 \\ x_1 + 3x_2 + x_3 + x_4 &= 6 \\ x_1 + x_2 + 4x_3 + x_4 &= 7 \\ x_1 + x_2 + x_3 + 5x_4 &= 8. \end{aligned}$$

Solution. *Here*

$$\mathbf{D}+\mathbf{L} = \begin{bmatrix} 2 & 0 & 0 & 0 \\ 1 & 3 & 0 & 0 \\ 1 & 1 & 4 & 0 \\ 1 & 1 & 1 & 5 \end{bmatrix},$$

$$\text{and } (\mathbf{D}+\mathbf{L})^{-1} = \begin{bmatrix} 0.5 & 0 & 0 & 0 \\ -0.1667 & 0.3333 & 0 & 0 \\ 0.0833 & -0.0833 & 0.25 & 0 \\ -0.05 & -0.05 & -0.05 & 0.2 \end{bmatrix},$$

$$\mathbf{U} = \begin{bmatrix} 0 & 1 & 1 & 1 \\ 0 & 0 & 1 & 1 \\ 0 & 0 & 0 & 1 \\ 0 & 0 & 0 & 0 \end{bmatrix},$$

$$-(\mathbf{D}+\mathbf{L})^{-1}\mathbf{U} = \begin{bmatrix} 0 & -0.5 & -0.5 & -0.5 \\ 0 & 0.1667 & -0.1667 & -0.1667 \\ 0 & 0.0833 & 0.1667 & -0.0833 \\ 0 & 0.05 & 0.1 & 0.15 \end{bmatrix},$$

$$(\mathbf{D}+\mathbf{L})^{-1}\mathbf{b} = \begin{bmatrix} 2.5 \\ 1.1667 \\ 0.8333 \\ 0.7 \end{bmatrix}.$$

Now substituting these values in formula (4.16) with initial approximation $\mathbf{x}^{(0)} = (\mathbf{D}+\mathbf{L})^{-1}\mathbf{b}$.

Suppose the initial approximation is

$$\mathbf{x}^{(0)} = \begin{bmatrix} 2.5 \\ 1.1667 \\ 0.8333 \\ 0.7 \end{bmatrix},$$

$$\mathbf{x}^{(1)} = \begin{bmatrix} 0 & -0.5 & -0.5 & -0.5 \\ 0 & 0.1667 & -0.1667 & -0.1667 \\ 0 & 0.0833 & 0.1667 & -0.0833 \\ 0 & 0.05 & 0.1 & 0.15 \end{bmatrix} \begin{bmatrix} 2.5 \\ 1.1667 \\ 0.8333 \\ 0.7 \end{bmatrix} + \begin{bmatrix} 2.5 \\ 1.1667 \\ 0.8333 \\ 0.7 \end{bmatrix} = \begin{bmatrix} 1.15 \\ 1.1056 \\ 1.011 \\ 0.9467 \end{bmatrix},$$

$$\mathbf{x}^{(2)} = \begin{bmatrix} 0 & -0.5 & -0.5 & -0.5 \\ 0 & 0.1667 & -0.1667 & -0.1667 \\ 0 & 0.0833 & 0.1667 & -0.0833 \\ 0 & 0.05 & 0.1 & 0.15 \end{bmatrix} \begin{bmatrix} 1.15 \\ 1.1056 \\ 1.011 \\ 0.9467 \end{bmatrix} + \begin{bmatrix} 2.5 \\ 1.1667 \\ 0.8333 \\ 0.7 \end{bmatrix} = \begin{bmatrix} 0.9683 \\ 1.0246 \\ 1.0151 \\ 0.9984 \end{bmatrix},$$

$$\mathbf{x}^{(3)} = \begin{bmatrix} 0 & -0.5 & -0.5 & -0.5 \\ 0 & 0.1667 & -0.1667 & -0.1667 \\ 0 & 0.0833 & 0.1667 & -0.0833 \\ 0 & 0.05 & 0.1 & 0.15 \end{bmatrix} \begin{bmatrix} 0.9683 \\ 1.0246 \\ 1.0151 \\ 0.9984 \end{bmatrix} + \begin{bmatrix} 2.5 \\ 1.1667 \\ 0.8333 \\ 0.7 \end{bmatrix} = \begin{bmatrix} 0.981 \\ 1.0019 \\ 1.0047 \\ 1.0025 \end{bmatrix},$$

$$\mathbf{x}^{(4)} = \begin{bmatrix} 0 & -0.5 & -0.5 & -0.5 \\ 0 & 0.1667 & -0.1667 & -0.1667 \\ 0 & 0.0833 & 0.1667 & -0.0833 \\ 0 & 0.05 & 0.1 & 0.15 \end{bmatrix} \begin{bmatrix} 0.981 \\ 1.0019 \\ 1.0047 \\ 1.0025 \end{bmatrix} + \begin{bmatrix} 2.5 \\ 1.1667 \\ 0.8333 \\ 0.7 \end{bmatrix} = \begin{bmatrix} 0.9955 \\ 0.99991 \\ 1.0007 \\ 1.0009 \end{bmatrix},$$

$$\mathbf{x}^{(5)} = \begin{bmatrix} 0 & -0.5 & -0.5 & -0.5 \\ 0 & 0.1667 & -0.1667 & -0.1667 \\ 0 & 0.0833 & 0.1667 & -0.0833 \\ 0 & 0.05 & 0.1 & 0.15 \end{bmatrix} \begin{bmatrix} 0.9955 \\ 0.99991 \\ 1.0007 \\ 1.0009 \end{bmatrix} + \begin{bmatrix} 2.5 \\ 1.1667 \\ 0.8333 \\ 0.7 \end{bmatrix} = \begin{bmatrix} 0.9997 \\ 0.9996 \\ 1 \\ 1.0002 \end{bmatrix},$$

$$\mathbf{x}^{(6)} = \begin{bmatrix} 0 & -0.5 & -0.5 & -0.5 \\ 0 & 0.1667 & -0.1667 & -0.1667 \\ 0 & 0.0833 & 0.1667 & -0.0833 \\ 0 & 0.05 & 0.1 & 0.15 \end{bmatrix} \begin{bmatrix} 0.9997 \\ 0.9996 \\ 1 \\ 1.0002 \end{bmatrix} + \begin{bmatrix} 2.5 \\ 1.1667 \\ 0.8333 \\ 0.7 \end{bmatrix} = \begin{bmatrix} 1.0001 \\ 0.9999 \\ 1 \\ 1 \end{bmatrix},$$

$$\mathbf{x}^{(7)} = \begin{bmatrix} 0 & -0.5 & -0.5 & -0.5 \\ 0 & 0.1667 & -0.1667 & -0.1667 \\ 0 & 0.0833 & 0.1667 & -0.0833 \\ 0 & 0.05 & 0.1 & 0.15 \end{bmatrix} \begin{bmatrix} 1.0001 \\ 0.9999 \\ 1 \\ 1 \end{bmatrix} + \begin{bmatrix} 2.5 \\ 1.1667 \\ 0.8333 \\ 0.7 \end{bmatrix} = \begin{bmatrix} 1.0001 \\ 1 \\ 1 \\ 1 \end{bmatrix},$$

$$\mathbf{x}^{(8)} = \begin{bmatrix} 0 & -0.5 & -0.5 & -0.5 \\ 0 & 0.1667 & -0.1667 & -0.1667 \\ 0 & 0.0833 & 0.1667 & -0.0833 \\ 0 & 0.05 & 0.1 & 0.15 \end{bmatrix} \begin{bmatrix} 1.0001 \\ 1 \\ 1 \\ 1 \end{bmatrix} + \begin{bmatrix} 2.5 \\ 1.1667 \\ 0.8333 \\ 0.7 \end{bmatrix} = \begin{bmatrix} 1 \\ 1 \\ 1 \\ 1 \end{bmatrix},$$

$$\mathbf{x}^{(9)} = \begin{bmatrix} 0 & -0.5 & -0.5 & -0.5 \\ 0 & 0.1667 & -0.1667 & -0.1667 \\ 0 & 0.0833 & 0.1667 & -0.0833 \\ 0 & 0.05 & 0.1 & 0.15 \end{bmatrix} \begin{bmatrix} 1 \\ 1 \\ 1 \\ 1 \end{bmatrix} + \begin{bmatrix} 2.5 \\ 1.1667 \\ 0.8333 \\ 0.7 \end{bmatrix} = \begin{bmatrix} 1 \\ 1 \\ 1 \\ 1 \end{bmatrix}.$$

Therefore $x_1 = 1, \;\; x_2 = 1, \;\; x_3 = 1, \;\; x_4 = 1.$

Verification

$$\mathbf{x} = \mathbf{A}^{-1}\mathbf{b} = \begin{bmatrix} 1 \\ 1 \\ 1 \\ 1 \end{bmatrix}.$$

The next method described in this section needs the following definitions.

Definition 4.4.1 *Permutation matrix* $\mathbf{U}$ *is a matrix having exactly 1 in each row and column and all other entries are zero.*

Definition 4.4.2 *Property A. A real matrix* $\mathbf{B}$ *is said to have property* $\mathbf{A}$ *iff there exists a permutation matrix* $\mathbf{U}$ *such that* $\mathbf{UBU^T} = \begin{bmatrix} \mathbf{A}_{11} & \mathbf{A}_{12} \\ \mathbf{A}_{21} & \mathbf{A}_{22} \end{bmatrix}$ *where* $\mathbf{A}_{11}$ *and* $\mathbf{A}_{22}$ *are diagonal matrices.*

Note. This property $\mathbf{A}$ helps in solving the parameters 'w', required in the following successive over relaxation method.

Successive over relaxation technique

Successive over relaxation (SOR) technique is a generalization of the Gauss-Siedel method and can be successfully applied if the given matrix $\mathbf{A}$ is symmetric and has the property 'A'.

Given the system $\mathbf{Ax} = \mathbf{b}$,
the Jacobi method is first developed and the iteration matrix is $\mathbf{J} = \mathbf{D}^{-1}(\mathbf{L} + \mathbf{U})$

Consider $|\mathbf{J} - \lambda\mathbf{I}| = 0$ and let $\lambda_1, \lambda_2, ..., \lambda_n$ be the eigen-values of the iteration matrix J. Let $\eta = \rho(\mathbf{J}) = \max\{|\lambda_1|, |\lambda_2|, ..., |\lambda_n|\}$ be the spectral radius of $\mathbf{J}$.
The successive over relaxation method is developed as follows.

Introduce a new vector at each iteration

$$\widetilde{\mathbf{x}}^{(k+1)} = -\mathbf{D}^{-1}\mathbf{L}\mathbf{x}^{(k+1)} - \mathbf{D}^{-1}\mathbf{U}\mathbf{x}^{(k)} + \mathbf{D}^{-1}\mathbf{b}, \tag{4.17}$$

(observe right hand side is similar to Jacobi method iterate) and the $(k+1)$th approximation is defined as

$$\mathbf{x}^{(\mathbf{k+1})} = (1-w)\mathbf{x}^{(k)} + w\widetilde{\mathbf{x}}^{(k+1)},$$

where w is called as the relaxation operator. Here the $(k+1)$th iterate is the weighted mean of the **k**th iterate and the constructed iterate $\widetilde{\mathbf{x}}^{(k+1)}$. Substituting the value of $\widetilde{\mathbf{x}}^{(k+1)}$ in (4.17),

$$\mathbf{x}^{(k+1)} = (1-w)\mathbf{x}^{(k)} + w(-\mathbf{D}^{-1}\mathbf{L}\mathbf{x}^{(k+1)} - \mathbf{D}^{-1}\mathbf{U}\mathbf{x}^{(k)} + \mathbf{D}^{-1}\mathbf{b})$$

$$(\mathbf{I} + w\mathbf{D}^{-1}\mathbf{L})\mathbf{x}^{(k+1)} = [(1-w)\mathbf{I} - w\mathbf{D}^{-1}\mathbf{U}]\mathbf{x}^{(k)} + w\mathbf{D}^{-1}\mathbf{b} \tag{4.18}$$

$$\mathbf{D}^{-1}(\mathbf{D} + w\mathbf{L})\mathbf{x}^{(k+1)} = \mathbf{D}^{-1}[(1-w)\mathbf{D} - w\mathbf{U}]\mathbf{x}^{(k)} + w\mathbf{D}^{-1}\mathbf{b}$$

Thus

$$\mathbf{x}^{(k+1)} = (\mathbf{D}+w\mathbf{L})^{-1}[(1-w)\mathbf{D} - w\mathbf{U}]\mathbf{x}^{(k)} + w(\mathbf{D} + w\mathbf{L})^{-1}\mathbf{b}$$

Set

$$\mathbf{J} = (\mathbf{D} + \mathbf{wL})^{-\mathbf{1}}[(\mathbf{1} - \mathbf{w})\mathbf{D} - \mathbf{wU}]$$

and

$$\mathbf{c} = \mathbf{w}(\mathbf{D} + \mathbf{wL})^{-\mathbf{1}}\mathbf{b}$$

Then the iteration scheme is

$$\mathbf{x}^{(\mathbf{k+1})} = \mathbf{J}\mathbf{x}^{(\mathbf{k})} + \mathbf{c}, \quad \mathbf{k} = \mathbf{0}, \mathbf{1}, \mathbf{2}, \ldots. \tag{4.19}$$

Observation.

1. If $w = 1$, the iteration scheme (4.18) reduces to Gauss-Seidel method.

2. The weights w are non-negative for $0 \leq w \leq 1$, and if $w < 1$, the method is called a successive under relaxation method.

3. If $w > 1$, the method is called a successive over relaxation method.

Note. To find '**w**', let $\eta = \rho(J)$ spectral radius of the matrix **J** in Jacobi method. Then $w = 2/\eta^2(1 \pm \sqrt{1-\eta^2})$.

Also, observe that $\eta < 1$, else w becomes a complex number and SOR cannot be used.

Example 4.4.4 *Solve the system* $\mathbf{Ax} = \mathbf{b}$ *using successive over relaxation method.* $\mathbf{A} = \begin{bmatrix} 3 & -1 & 0 \\ -1 & 3 & -1 \\ 0 & -1 & 3 \end{bmatrix}$, $\mathbf{x} = \begin{bmatrix} x \\ y \\ z \end{bmatrix}$ *and* $\mathbf{b} = \begin{bmatrix} 5 \\ 4 \\ 11 \end{bmatrix}$

Solution. To apply the iterative formula (4.19) first w must be calculated for matrix $\mathbf{J} = \mathbf{D}^{-1}(\mathbf{L}+\mathbf{U})$ in Jacobi method. Here

$$\mathbf{D} = \begin{bmatrix} 3 & 0 & 0 \\ 0 & 3 & 0 \\ 0 & 0 & 3 \end{bmatrix} \mathbf{L} = \begin{bmatrix} 0 & 0 & 0 \\ -1 & 0 & 0 \\ 0 & -1 & 0 \end{bmatrix} \mathbf{U} = \begin{bmatrix} 0 & -1 & 0 \\ 0 & 0 & -1 \\ 0 & 0 & 0 \end{bmatrix}$$

$$\mathbf{D}^{-1}(\mathbf{L}+\mathbf{U}) = \begin{bmatrix} 1/3 & 0 & 0 \\ 0 & 1/3 & 0 \\ 0 & 0 & 1/3 \end{bmatrix} \begin{bmatrix} 0 & -1 & 0 \\ -1 & 0 & -1 \\ 0 & -1 & 0 \end{bmatrix}$$

$$= \begin{bmatrix} 0 & -1/3 & 0 \\ -1/3 & 0 & -1/3 \\ 0 & -1/3 & 0 \end{bmatrix}$$

The largest eigen-value is $\eta = \sqrt{2}/3$ $w_{opt} = \dfrac{2}{\eta^2}\left(1-\sqrt{1-\eta^2}\right) = \dfrac{2}{(\sqrt{2}/3)^2}\left(1-\sqrt{1-\dfrac{2}{9}}\right) = 1.062746$

The recurrence formula is given by $\mathbf{x}^{(k+1)} = \mathbf{Jx}^{(k)} + \mathbf{c}$, where

$$\mathbf{J} = (\mathbf{D} + w\mathbf{L})^{-1}\left[(1-w)\mathbf{D} - w\mathbf{U}\right]$$

$$(\mathbf{D} + \mathbf{wL})^{-1} = \begin{bmatrix} 3 & 0 & 0 \\ -w & 3 & 0 \\ 0 & -w & 3 \end{bmatrix}^{-1} = \begin{bmatrix} 1/3 & 0 & 0 \\ w/9 & 1/3 & 0 \\ \dfrac{w^2}{27} & w/9 & 1/3 \end{bmatrix}$$

$$((1-w)\mathbf{D} - w\mathbf{U}] = \begin{bmatrix} 3(1-w) & w & 0 \\ 0 & 3(1-w) & w \\ 0 & 0 & 3(1-w) \end{bmatrix}$$

$$\mathbf{J} = (\mathbf{D} + w\mathbf{L})^{-1}\left((1-w)\mathbf{D} - w\mathbf{U}\right]$$

$$= \begin{bmatrix} 1-w & \dfrac{w}{3} & 0 \\ \dfrac{w(1-w)}{3} & \dfrac{w^2-9w+9)}{9} & \dfrac{w}{3} \\ \dfrac{w^2(1-w)}{9} & \dfrac{w^3+9w(1-w)}{27} & \dfrac{w^2-9w+9}{9} \end{bmatrix}$$

$$\mathbf{c} = w\begin{pmatrix} 1/3 & 0 & 0 \\ w/9 & 1/3 & 0 \\ \dfrac{w^2}{27} & \frac{w}{9} & \dfrac{1}{3} \end{pmatrix}\begin{pmatrix} 5 \\ 4 \\ 11 \end{pmatrix} = \begin{pmatrix} \dfrac{5w}{3} \\ \dfrac{w(5w+12)}{9} \\ \dfrac{w(5w^2+12w+99)}{27} \end{pmatrix}$$

substituting the value of $w_{opt} = w = 1.062746$

$$\mathbf{J} = \begin{bmatrix} -0.0627 & 0.35423 & 0 \\ -0.02221 & 0.06278 & 0.35423 \\ -0.007867 & 0.02224 & 0.06278 \end{bmatrix}$$

and

$$\mathbf{c} = \begin{bmatrix} 1.77117 \\ 2.04434 \\ 4.62074 \end{bmatrix}$$

Substituting for $\mathbf{J}$ and $\mathbf{c}$ in

$$\mathbf{x}^{(k+1)} = \mathbf{J}\mathbf{x}^{(k)} + c$$

gives, by taking $\mathbf{x}^{(0)} = [0 \quad 0 \quad 0]^T$

$$\mathbf{x}^{(1)} = \mathbf{c} = \begin{bmatrix} 1.77117 \\ 2.04434 \\ 4.62074 \end{bmatrix},$$

$$\mathbf{x}^{(2)} = \begin{pmatrix} 2.38428 \\ 0.77011 \\ 4.94236 \end{pmatrix},$$

substituting $\mathbf{x}^{(2)}$ in the recursive formula gives

$$\mathbf{x}^{(3)} = \begin{pmatrix} 2.95716 \\ 3.97875 \\ 4.99611 \end{pmatrix}$$

Similarly

$$\mathbf{x}^{(4)} = \begin{pmatrix} 2.99514 \\ 3.99816 \\ 4.99961 \end{pmatrix}$$

which is approximately equal to the real value of

$$\mathbf{x} = \begin{pmatrix} x \\ y \\ z \end{pmatrix} = \begin{pmatrix} 3 \\ 4 \\ 5 \end{pmatrix}.$$

EXERCISE 4.4

1. Solve the following systems using
   (i) Gauss-Jacobi method (ii) Gauss-Siedel method

   (a) $3x + y + z = 4$
   $x + 4y + 2z = 4$
   $2x + y + 5z = 5$

   (b) $2x + y + 2z = 15$
   $2x + 9y + 3z = 25$
   $2x + 4y + 8z = 30$

2. Find the solution for the following system using (i) Gauss-elimination method (ii) Gauss-Jacobi method (iii) Gauss-Siedel method.

   What is your inference.

$$\begin{aligned} 2x + y + 2z &= 4 \\ x + 3y + 4z &= 6 \\ 4x + 2y + 2z &= 7. \end{aligned}$$

3. Using SOR technique solve the system $Ax = b$ where

$$A = \begin{bmatrix} 3 & -2 & 1 \\ -2 & 3 & -2 \\ 1 & -2 & 3 \end{bmatrix}, x = \begin{bmatrix} x \\ y \\ z \end{bmatrix}, b = \begin{bmatrix} -1 \\ 7 \\ -7 \end{bmatrix}$$

.

4.5 Householder Transformation

In Chapter 1, it has been shown that every elementary operation can be represented by an elementary matrix and the product of these elementary matrices is a nonsingular matrix '**P**' that when operated on the matrix reduces to an upper triangular matrix or a diagonal matrix.

As mathematical operations lead to round off errors, an orthogonal matrix would be an appropriate transformation matrix to mitigate propagation of errors. This is because for an orthogonal matrix $\mathbf{H}$, $\|\mathbf{x}\| = \|\mathbf{Hx}\|$, that is, norms are preserved for an orthogonal matrix.

In this section, of the many approaches known, the Householders transformation method is discussed.

Definition 4.5.1 *Householder transformation is a matrix* $\mathbf{H} \in \mathbb{R}^{n\times n}$ *defined by* $\mathbf{H} = \mathbf{I} - \mathbf{2ww^T}$ *where* $\mathbf{w} \in \mathbb{R}^{\mathbf{n}}$ *such that* $\mathbf{w^T w} = \mathbf{1}$, *and* $\mathbf{I}$ *is an* $n \times n$ *identity matrix of order 'n'.*

Observation. The Householder transformation $\mathbf{H}$ satisfies the following properties (1) $\mathbf{H}^T\mathbf{H} = \mathbf{I}$, (2) $\mathbf{H}^T = \mathbf{H^{-1}}$.

Theorem 4.5.1 *For any non zero vector* $\mathbf{x} \in \mathbb{R}^{\mathbf{n}}$, *there exists a Householder transformation* $\mathbf{H}$ *such that* $\mathbf{Hx} = a\mathbf{e_1}$, *where* $\mathbf{e_1} = (1,\ 0,\ 0,\ 0)^T \in \mathbb{R}^{\mathbf{n}}$, *for some constant* a.

Proof. Given $\mathbf{x} \in \mathbb{R}^n$, finding the Householder transformation $\mathbf{H}$ means finding a vector $\mathbf{w}$ such that

$$\mathbf{Hx} = (\mathbf{I} - \mathbf{2ww^T})\mathbf{x} = a\mathbf{e_1},$$

where $\mathbf{I}$ is an nxn identity matrix, and $\mathbf{e_1} = (1,\ 0,\ 0,\ 0)^T \in \mathbb{R}^{\mathbf{n}}$.

Let $\mathbf{x} = (x_1\ x_2\ x_n)$. Then

$$\|\mathbf{x}\|_{\mathbf{2}}^{\mathbf{2}} = x_1^2 + x_2^2 + \cdots + x_n^2 = k^2(say).$$

Also,

$$k^2 = \mathbf{x}^T\mathbf{x} = \mathbf{x}^T\mathbf{H}^T\mathbf{Hx} = a^2\mathbf{e}_1^T\mathbf{e}_1 = a^2.$$

Thus

$$a = \pm k = \pm\|\mathbf{x}\|_2.$$

To find $\mathbf{w}$, consider

$$\mathbf{Hx} = a\mathbf{e_1}, \text{ that is, } (\mathbf{I} - \mathbf{2ww^T})\mathbf{x} = a\mathbf{e_1},$$

where $\mathbf{w} = (w_1,\ w_2,\ w_3,\ w_n)^T$

$$\begin{bmatrix} 1-2w_1^2 & -2w_1w_2 & \cdots & -2w_1w_n \\ -2w_2w_1 & 1-2w_2^2 & \cdots & -2w_2w_n \\ \vdots & \vdots & \vdots & \vdots \\ -2w_nw_1 & -2w_nw_2 & \cdots & 1-2w_n^2 \end{bmatrix} \begin{bmatrix} x_1 \\ x_2 \\ \vdots \\ x_n \end{bmatrix} = a \begin{bmatrix} 1 \\ 0 \\ \vdots \\ 0 \end{bmatrix}$$

which is,

$$\begin{bmatrix} x_1 + (-2w_1)w_1x_1 + (-2w_1)(w_2x_2) & \cdots & (-2w_1)(w_nx_n) \\ (-2w_2)w_1x_1 + x_2 + (-2w_2)(w_2x_2) & \cdots & (-2w_2)(w_nx_n) \\ \vdots & \vdots & \vdots \\ (-2w_n)(w_1x_1) + (-2w_n)(w_2x_2) + & \cdots & x_n - (2w_n)(w_nx_n) \end{bmatrix} = \begin{bmatrix} a \\ 0 \\ 0 \\ 0 \end{bmatrix}$$

Thus,

$$x_1 - 2w_1(\mathbf{w}^T\mathbf{x}) = a = \pm k,$$
$$x_i - 2w_i(\mathbf{w}^T\mathbf{x}) = 0, \quad i = 2, \ldots, n \tag{4.20}$$

where,

$$\mathbf{w^T x} = w_1x_1 + w_2x_2 + \cdots + w_nx_n.$$

To solve $\mathbf{w}$ from the 'n' equations in (4.20),
set

$$\mathbf{w}^T\mathbf{x} = l,$$

Then,

$$2w_1 l = x_1 \mp k \tag{4.21}$$

and

$$2w_i l = x_i, \quad i = 2, \ldots, k. \tag{4.22}$$

Squaring and adding the equations in (4.21) and (4.22) give,

$$4(w_1^2 + \cdots + w_n^2)l^2 = x_1^2 + \cdots + x_n^2 + k^2 \pm 2x_1k = 2k^2 \mp 2x_1k$$

or

$$\begin{aligned} 2l^2 &= k(k \mp x_1) \\ &= k(k - (sgnx_1)x_1]. \end{aligned}$$

where $sgnx_1 = 1, -1$, according as $x_1 \geq 0,\ x_1 < 0$.
Set $\mathbf{w} = \frac{\mathbf{u}}{2l}$, where $\mathbf{u} = (x_1 - k,\ x_2,\ x_n)^T$ and $a = \pm k = -(\mp k)$ $= -(sgnx_1)\|\mathbf{x}\|_2$. Thus,

$$\begin{aligned} \mathbf{Hx} &= (\mathbf{I} - 2\mathbf{ww}^T)\mathbf{x} = (\mathbf{x} - 2(\mathbf{w}^T\mathbf{x})\mathbf{w}) \\ &= \mathbf{x} - 2\frac{(u^T\mathbf{x})\mathbf{u}}{||\mathbf{u}||^2} \\ &= (x_1\ \ x_n)^T - (x_1 \mp k\ \ x_2\ \ x_n)^T \\ &= -(\pm k\ \ 0\ \ 0) = -(sgnx_1)\|\mathbf{x}\|_2\mathbf{e}_1. \end{aligned}$$

Therefore, $\mathbf{Hx} = a\mathbf{e}_1$, where $a = -(sgnx_1)\|\|\mathbf{x}\|_2$

Algorithm for Householder method

Step 1. Given $\mathbf{x} \in \mathbb{R}^\mathbf{n}$ find $\|\mathbf{x}\|_2^2 = x_1^2 + \cdots + x_n^2 = k^2$ (say).

Step 2. Write $2l^2 = k(k + (sgnx_1)x_1]$.

Step 3. Write $\mathbf{u} = (x_1 + k\ \ x_2\ \ x_n)^T$ and $\mathbf{w} = \frac{1}{2l}\mathbf{u}$.

Step 4. $\mathbf{H} = \mathbf{I} - 2\frac{\mathbf{uu}^T}{||\mathbf{u}||^2}$.

Observation. The above theorem can be utilized for any mxn matrix by considering each column of the matrix or a column of its submatrix.

Householder Method to reduce a given matrix to an upper triangular form

Consider a matrix $\mathbf{A}_{n\times n} = (a_{ij}), \quad i = j = 1, 2, \ldots, n.$
Goal is to find a transformation matrix $\mathbf{H}$ such that

$$\mathbf{HA} = \mathbf{C} = (c_{ij}), \quad i = j = 1, 2, ...n$$

such that $c_{ij} = 0$ for $i > j$
Step 1. Set $\mathbf{x} = (a_{11}\, a_{21}\, a_{n1})^T$ (1st column of $\mathbf{A}$)

To find a transformation $\mathbf{H_n}$ such that $\mathbf{H_n x} = c_{11}\mathbf{e_1}, \mathbf{e_1} \in \mathbb{R}^n$
Using the algorithm for Householder transformation, with
$c_{11} = -(sgn\ a_{11})\|x\|_2$.

Applying the algorithm gives

$$\mathbf{H}_n\mathbf{A} = \begin{bmatrix} c_{11} & c_{12} & . & . & . & c_{1n} \\ 0 & a_{22}^{(1)} & . & . & . & a_{2n}^{(1)} \\ . & . & . & . & . & . \\ . & . & . & . & . & . \\ 0 & a_{n1}^{(1)} & . & . & . & a_{nn}^{(1)} \end{bmatrix},$$

now consider the submatrix

$$\mathbf{A}_{n-1,n-1} = \begin{bmatrix} a_{22}^{(1)} & . & . & . & a_{2n}^{(1)} & \\ a_{32}^{(1)} & . & . & . & a_{3n}^{(1)} & \\ . & . & . & . & . & . \\ . & . & . & . & . & . \\ a_{n2}^{(1)} & . & . & . & a_{nn}^{(1)} & \end{bmatrix},$$

Again set

$$\mathbf{x} = \begin{bmatrix} a_{22}^{(1)} & . & . & .a_{n1}^{(1)} \end{bmatrix}^T \in \mathbb{R}^{n-1}$$

Applying Householder Algorithm

$$\mathbf{H_{n-1}x} = c_{22}\mathbf{e_1},\ \ \mathbf{e_1} \in \mathbb{R}^{n-1}$$

and

$$c_{22} = (-sgn a_{22}^{(1)})\|\mathbf{x}\|_\mathbf{2}$$

This gives, after applying the algorithm

$$\mathbf{H}_{n-1}\mathbf{A}_{n-1,n-1} = \begin{bmatrix} c_{22} & c_{23} & . & . & . & c_{2n} \\ 0 & a_{33}^{(2)} & . & . & . & a_{3n} \\ . & . & . & . & . & . \\ . & . & . & . & . & . \\ 0 & a_{n3}^{(2)} & . & . & . & a_{nn}^{(2)} \end{bmatrix},$$

This step is equivalent to considering the matrix

$$\begin{bmatrix} 1 & 0 \\ 0 & \mathbf{H}_{n-1} \end{bmatrix} \mathbf{H}_n \mathbf{A} = \begin{bmatrix} c_{11} & c_{12} & . & . & . & c_{1n} \\ 0 & c_{22} & . & . & . & c_{2n} \\ . & . & . & . & . & . \\ . & . & . & . & . & . \\ 0 & a_{n2}^{(2)} & . & . & . & a_{nn}^{(2)} \end{bmatrix},$$

Now set

$$\mathbf{x} = [a_{33}^{(2)} ......a_{n3}^{(2)}]^T \in \mathbb{R}^{n-2}$$

the goal is to find the matrix $\mathbf{H}_{n-2} \in \mathbb{R}^{n-2 \times n-2}$ such that

$$\mathbf{H}_{n-2}\mathbf{A}_{n-2,n-2} = c_{33}\mathbf{e}_1, \quad \mathbf{e}_1 \in \mathbb{R}^{n-2}$$

with $c_{33} = -sgn a_{33}^{(2)} \|\mathbf{x}\|_2$.

Using Householder transformation gives

$$\mathbf{H}_{n-2}\mathbf{A}_{n-2,n-2} = \begin{bmatrix} c_{33} & c_{34} & . & . & . & c_{3n} \\ 0 & a_{44}^{(3)} & . & . & . & a_{4n}^{(4)} \\ . & . & . & . & . & . \\ . & . & . & . & . & . \\ 0 & a_{n4}^{(3)} & . & . & . & a_{nn}^{(3)} \end{bmatrix},$$

The matrix $\mathbf{A}$ is transformed as

$$\begin{bmatrix} \mathbf{I}_2 & 0 \\ 0 & \mathbf{H}_{n-2} \end{bmatrix} \begin{bmatrix} 1 & 0 \\ 0 & \mathbf{H}_{n-1} \end{bmatrix} \mathbf{H}_n \mathbf{A} = \begin{bmatrix} c_{11} & c_{12} & c_{13} & . & . & c_{1n} \\ 0 & c_{22} & c_{23} & . & . & c_{2n} \\ . & . & c_{33} & . & . & . \\ . & . & . & . & . & . \\ 0 & 0 & 0 & a_{n4}^{(3)} & . & a_{nn}^{(3)} \end{bmatrix}$$

For simplicity sake set $\mathbf{Q}_{k+1} =$

$$\begin{bmatrix} \mathbf{I}_k & 0 \\ 0 & \mathbf{H}_{n-k} \end{bmatrix},$$

where $\mathbf{H_{n-k}}$ is a Householder transformation of order $\mathbf{n-k}$.
Then $\mathbf{H} = \mathbf{Q_{n-1}Q_{n-2}......Q_1}$ is the orthogonal transformation reducing $\mathbf{A}$ to a triangular form.

Example 4.5.1 *Reduce the matrix* $\mathbf{A}$ *to an upper triangular matrix using the Householder method*

$$\mathbf{A} = \begin{bmatrix} 2 & 2 & 3 \\ 2 & 3 & 0 \\ 1 & -2 & 1 \end{bmatrix}.$$

Solution.

Write $\mathbf{A}_1 = \begin{bmatrix} 2 & 2 & 3 \\ 2 & 3 & 0 \\ 1 & -2 & 1 \end{bmatrix}$ and consider $\mathbf{x} = \begin{bmatrix} 2 \\ 2 \\ 1 \end{bmatrix}$,

Then $||\mathbf{x}|| = \sqrt{9} = 3.$

$$\mathbf{u} = \mathbf{x} + (sgn\mathbf{x}_1)||\mathbf{x}||e_1 = \begin{bmatrix} 2 \\ 2 \\ 1 \end{bmatrix} + 3 \begin{bmatrix} 1 \\ 0 \\ 0 \end{bmatrix} = \begin{bmatrix} 5 \\ 2 \\ 1 \end{bmatrix}$$

$$\begin{aligned} \mathbf{H}_3 &= \mathbf{I} - 2\frac{\mathbf{u} \cdot \mathbf{u}^T}{||\mathbf{u}||^2} \\ &= \begin{bmatrix} 1 & 0 & 0 \\ 0 & 1 & 0 \\ 0 & 0 & 1 \end{bmatrix} - \frac{2}{30} \begin{bmatrix} 5 \\ 2 \\ 1 \end{bmatrix} \cdot \begin{bmatrix} 5 & 2 & 1 \end{bmatrix} = \begin{bmatrix} \frac{-2}{3} & \frac{-2}{3} & \frac{-1}{3} \\ \frac{-2}{3} & \frac{11}{15} & \frac{-2}{15} \\ \frac{-1}{3} & \frac{-2}{15} & \frac{14}{15} \end{bmatrix} \end{aligned}$$

$$\mathbf{H}_3\mathbf{A}_1 = \begin{bmatrix} \frac{-2}{3} & \frac{-2}{3} & \frac{-1}{3} \\ \frac{-2}{3} & \frac{11}{15} & \frac{-2}{15} \\ \frac{-1}{3} & \frac{-2}{15} & \frac{14}{15} \end{bmatrix} \begin{bmatrix} 2 & 2 & 3 \\ 2 & 3 & 0 \\ 1 & -2 & 1 \end{bmatrix} = \begin{bmatrix} -3 & \frac{-8}{3} & \frac{-7}{3} \\ 0 & \frac{17}{15} & \frac{-32}{15} \\ 0 & \frac{-44}{15} & \frac{-1}{15} \end{bmatrix}$$

Now removing 1st *row and* 1st *column, holds*

$$\mathbf{A}_2 = \begin{bmatrix} \frac{17}{15} & \frac{-32}{15} \\ \frac{-44}{15} & \frac{-1}{15} \end{bmatrix}$$

$$\mathbf{x} = \begin{bmatrix} \frac{17}{15} \\ \frac{-44}{15} \end{bmatrix}$$

$||\mathbf{x}|| = \sqrt{9.8889} = 3.1447$

$$\mathbf{u} = \mathbf{x} + (sgn\mathbf{x}_1)||\mathbf{x}||e_1 = \begin{bmatrix} \frac{17}{15} \\ \frac{-44}{15} \end{bmatrix} + (3.1447) \begin{bmatrix} 1 \\ 0 \end{bmatrix} = \begin{bmatrix} 4.278 \\ \frac{-44}{15} \end{bmatrix}$$

$$\begin{aligned} \mathbf{H}_2 &= \mathbf{I} - 2\frac{\mathbf{u} \cdot \mathbf{u}^T}{||\mathbf{u}||^2} \\ &= \begin{bmatrix} 1 & 0 \\ 0 & 1 \end{bmatrix} - \frac{2}{26.9057} \begin{bmatrix} 4.278 \\ \frac{44}{15} \end{bmatrix} \cdot \begin{bmatrix} 4.278 & \frac{44}{15} \end{bmatrix} = \begin{bmatrix} -0.3604 & 0.9328 \\ 0.9328 & 0.3604 \end{bmatrix} \end{aligned}$$

$$\mathbf{H}_2\mathbf{A}_2 = \begin{bmatrix} -0.3604 & 0.9328 \\ 0.9328 & 0.3604 \end{bmatrix} \begin{bmatrix} \frac{17}{15} & \frac{-32}{15} \\ \frac{-44}{15} & \frac{-1}{15} \end{bmatrix} = \begin{bmatrix} -3.1447 & 0.7067 \\ 0 & -2.014 \end{bmatrix}$$

Now removing 1st *row and* 1st *column from* $\mathbf{A}_2$ *gives*

$$\mathbf{A}_1 = \begin{bmatrix} -2.014 \end{bmatrix}$$

$$\mathbf{x} = \begin{bmatrix} -2.014 \end{bmatrix}$$

$||\mathbf{x}|| = 2.014$
$\mathbf{u} = \mathbf{x} + (sgn\mathbf{x}_1)||\mathbf{x}||e_1 = \begin{bmatrix}-4.028\end{bmatrix}$

$$\begin{aligned}\mathbf{H}_1 &= \mathbf{I} - 2\frac{\mathbf{u}\cdot\mathbf{u}^T}{||\mathbf{u}||^2} \\ &= \begin{bmatrix}1\end{bmatrix} - \frac{2}{1}\begin{bmatrix}1\end{bmatrix} = \begin{bmatrix}-1\end{bmatrix}\end{aligned}$$

Now

$$\mathbf{H}_1\mathbf{H}_2\mathbf{H}_3\mathbf{A} =$$

$$\begin{bmatrix}1 & 0 & 0\\ 0 & 1 & 0\\ 0 & 0 & -1\end{bmatrix}\begin{bmatrix}1 & 0 & 0\\ 0 & 0.53 & -0.848\\ 0 & -0.848 & -0.53\end{bmatrix}\begin{bmatrix}\frac{2}{3} & \frac{2}{3} & \frac{1}{3}\\ \frac{2}{3} & \frac{-1}{3} & \frac{-2}{3}\\ \frac{1}{3} & \frac{-2}{3} & \frac{2}{3}\end{bmatrix}\begin{bmatrix}2 & 2 & 3\\ 2 & 3 & 0\\ 1 & -2 & 1\end{bmatrix} =$$

$$\begin{bmatrix}-3 & \frac{-8}{3} & \frac{-7}{3}\\ 0 & -0.31447 & 0.7067\\ 0 & 0 & 2.014\end{bmatrix} = \mathbf{R}$$

which is an upper triangular matrix.

Now,

$$\mathbf{Q} = \mathbf{H}_1\mathbf{H}_2\mathbf{H}_3 = \begin{bmatrix}1 & 0 & 0\\ 0 & 1 & 0\\ 0 & 0 & -1\end{bmatrix}\begin{bmatrix}\frac{2}{3} & \frac{2}{3} & \frac{1}{3}\\ \frac{2}{3} & \frac{-1}{3} & \frac{-2}{3}\\ \frac{1}{3} & \frac{-2}{3} & \frac{2}{3}\end{bmatrix}\begin{bmatrix}1 & 0 & 0\\ 0 & 0.53 & -0.848\\ 0 & -0.848 & -0.53\end{bmatrix}$$

$$= \begin{bmatrix}\frac{-2}{3} & -0.0707 & 0.742\\ \frac{-2}{3} & -0.3887 & -0.636\\ \frac{-1}{3} & 0.9187 & -0.212\end{bmatrix} \text{ is an orthogonal matrix.}$$

Verify $\mathbf{QR} = \mathbf{A}$.

Observation.

1. For any square matrix of order **n**, the total number of operations of multiplications and square roots required to reduce **A** to a triangular matrix is $2/n^3 + 3/2n^2 + 11/6n - 4 = 2/3n^3$ if n is very large.

2. This approach can be applied for any matrix of the form $\mathbf{A_{mxn}}$.

EXERCISE 4.5

1. Transform the matrix $A = \begin{bmatrix}1 & 2 & 1\\ 2 & 2 & 3\\ 1 & 3 & 3\end{bmatrix}$ to an upper triangular matrix using Householder method. Find the orthogonal transformation.

2. Solve the system $\begin{bmatrix} 3 & 2 & 2 \\ 1 & 1 & 2 \\ 2 & 3 & 4 \end{bmatrix} \begin{bmatrix} x \\ y \\ z \end{bmatrix} = \begin{bmatrix} 10 \\ 5 \\ 11 \end{bmatrix}$ by reducing the matrix to an upper triangular form using Householder method. Also find the orthogonal transformation.

4.6 Tridiagonalization of a Symmetric Matrix by Plane Rotation

Definition 4.6.1 *An orthogonal matrix* $\mathbf{R_{ij}}$ *of order 'n' is said to be a plane rotation or a plane transformation in the* (i, j)th *plane if*

(i) $\mathbf{R}_{ij}$ *is an identity matrix except in columns* i *and* j

(ii) $r_{ii} = cos\theta$, $r_{ij} = -sin\theta$, $r_{ji} = sin\theta$, $r_{jj} = cos\theta$

Example 4.6.1

$$\mathbf{P}_{13} = \begin{bmatrix} cos\theta & 0 & -sin\theta \\ 0 & 1 & 0 \\ sin\theta & 0 & cos\theta \end{bmatrix}$$

is a plane transformation in the $(1, 3)$ *plane.*

Observation. 1. Plane rotation matrix is an orthogonal matrix.

Algorithm for tridiagonalization of a symmetric matrix of order n

Step 1. Given $\mathbf{A} = (a_{ij})_{n \times n}$. To reduce $\mathbf{A}$ to a matrix $\mathbf{C} = (c_{ij})_{n \times n}$, where $c_{ij} = 0$ for i doesnot belong to $\{j-1, j, j+1\}$.

Step 2. Consider $\mathbf{R}_{23} = (r_{ij})_{n \times n}$ with $r_{22} = cos\theta$, $r_{23} = -sin\theta$, $r_{32} = -sin\theta$, $r_{33} = cos\theta$ and rest is an identity matrix.

Then $\mathbf{R}_{23}^T \mathbf{A} \mathbf{R}_{23} =$

$$\begin{bmatrix} b_{11} & b_{12} & . & . & . & b_{1n} \\ b_{12} & b_{22} & . & . & . & b_{2n} \\ f(\theta) & b_{32} & . & . & . & b_{3n} \\ . & . & . & . & . & . \\ . & . & . & . & . & . \\ . & . & . & . & . & . \\ b_{n1} & b_{n2} & . & . & . & b_{nn} \end{bmatrix}$$

Step 3. Find θ by setting (3,1) element $f(\theta) = 0$.

Step 4. Repeat step 2 with matrices $\mathbf{R}_{24}, .., \mathbf{R}_{2n}$ and their transposes to get $\mathbf{R}_{2n}^T\mathbf{R}_{2n-1}^T...(\mathbf{R}_{24}^T\mathbf{R}_{23}^T\mathbf{A}\mathbf{R}_{23})\mathbf{R}_{24}...\mathbf{R}_{2n-1}, ..., \mathbf{R}_{2n}$

$$= \begin{bmatrix} c_1 & \alpha_1 & . & . & . & 0 \\ \alpha_1 & \alpha_{22} & \alpha_{23} & . & . & \alpha_{2n} \\ 0 & \alpha_{32} & . & . & . & \alpha_{3n} \\ \vdots & \vdots & . & . & . & \vdots \\ 0 & \alpha_{n2} & . & . & . & \alpha_{nn} \end{bmatrix} = \mathbf{B}\ (say).$$

Step 5. Repeat the steps 3 & 4 to eliminate elements $\alpha_{42}, ..., \alpha_{n2}$ and $\alpha_{24}, ..., \alpha_{2n}$ by premultiplying $\mathbf{B}$ with $\mathbf{R}_{34}, \mathbf{R}_{35}, \ldots, \mathbf{R}_{3n}$ successively and post multiply with $\mathbf{R}_{34}^T, R_{35}^T, \ldots, \mathbf{R}_{3n}^T$ successively.

Step 6. Repeat the procedure to obtain $\mathbf{C}$, in the form

$$\begin{bmatrix} c_1 & \alpha_1 & 0 & . & . & 0 \\ \alpha_1 & c_2 & \alpha_1 & . & . & . \\ 0 & \alpha_2 & \alpha_3 & . & . & \alpha_{n-1} \\ . & . & . & . & . & \alpha_n c_n \\ . & 0 & 0 & . & . & \alpha_{n-1} \end{bmatrix}$$

The following example will illustrate the procedure.

Example 4.6.2 *Reduce the following matrix to a tridiagonal matrix using plane rotation,*

$$\mathbf{A} = \begin{bmatrix} 1 & 1 & 1 & 1 \\ 1 & 2 & 3 & 4 \\ 1 & 3 & 6 & 10 \\ 1 & 4 & 30 & 20 \end{bmatrix}$$

Solution.

Consider

$$R_{23} = \begin{bmatrix} 1 & 0 & 0 & 0 \\ 0 & \cos\theta & -\sin\theta & 0 \\ 0 & \sin\theta & \cos\theta & 0 \\ 0 & 0 & 0 & 1 \end{bmatrix}$$

Then

$$\mathbf{A}_1 = \mathbf{R}_{23}^T\mathbf{A}\mathbf{R}_{23} = \begin{bmatrix} 1 & 0 & 0 & 0 \\ 0 & cos\theta & sin\theta & 0 \\ 0 & -sin\theta & cos\theta & 0 \\ 0 & 0 & 0 & 1 \end{bmatrix}^T \begin{bmatrix} 1 & 1 & 1 & 1 \\ 1 & 2 & 3 & 4 \\ 1 & 3 & 6 & 10 \\ 1 & 4 & 10 & 20 \end{bmatrix} \begin{bmatrix} 1 & 0 & 0 & 0 \\ 0 & cos\theta & -sin\theta & 0 \\ 0 & sin\theta & cos\theta & 0 \\ 0 & 0 & 0 & 1 \end{bmatrix}$$

$$= \begin{bmatrix} 1 & cos\theta + sin\theta & -sin\theta + cos\theta & 1 \\ cos\theta + sin\theta & 2cos^2\theta + 6sin^2\theta + 3sin2\theta & 3cos^2\theta + 4cos\theta sin\theta - 3sin^2\theta & 4cos\theta + 10sin\theta \\ -sin\theta + cos\theta & 4cos\theta sin\theta + 2cos^2\theta 3sin^2\theta & 2sin^2\theta + 6cos^2\theta - 6cos\theta sin\theta & -4sin\theta + 10cos\theta \\ 1 & 4cos\theta + 10sin\theta & -4sin\theta + 10cos\theta & 20 \end{bmatrix}$$

Equating the element (1,3) position to zero, gives $\tan\theta = 1$, that is $\theta = \frac{\pi}{4}$.

$$\mathbf{A}_1 = \mathbf{R}_{23}^T\mathbf{A}\mathbf{R}_{23} = \begin{bmatrix} 1 & \sqrt{2} & 0 & 1 \\ \sqrt{2} & 7 & 2 & 7\sqrt{2} \\ 0 & 2 & 1 & 3\sqrt{2} \\ 1 & 7\sqrt{2} & 3\sqrt{2} & 20 \end{bmatrix}$$

As the matrix is symmetric so the (3,1) element is also reduced to zero. Now to reduce the element in the (1,4) position to zero, we take $\mathbf{R}_{24}$

$$\mathbf{R}_{24} = \begin{bmatrix} 1 & 0 & 0 & 0 \\ 0 & cos\theta & 0 & -sin\theta \\ 0 & 0 & 1 & 0 \\ 0 & sin\theta & 0 & cos\theta \end{bmatrix}$$

$\mathbf{A}_2 = \mathbf{R}_{24}^{-1}\mathbf{A}_1\mathbf{R}_{24} = \mathbf{R}_{24}^T\mathbf{A}_1\mathbf{R}_{24}$ and equating

$$\begin{bmatrix} 1 & 0 & 0 & 0 \\ 0 & cos\theta & 0 & sin\theta \\ 0 & 0 & 1 & 0 \\ 0 & -sin\theta & 0 & cos\theta \end{bmatrix} \begin{bmatrix} 1 & 1.414 & 0 & 1 \\ 1.414 & 7 & 2 & 9.8995 \\ 0 & 2 & 1 & 4.2426 \\ 1 & 9.8995 & 4.2426 & 20 \end{bmatrix} \begin{bmatrix} 1 & 0 & 0 & 0 \\ 0 & cos\theta & 0 & -sin\theta \\ 0 & 0 & 1 & 0 \\ 0 & sin\theta & 0 & cos\theta \end{bmatrix}$$

Computing $\mathbf{A}_2 = \mathbf{R}_{34}^T\mathbf{A}_1\mathbf{R}_{34}$ and equating the entry in $(1,4)$ position to zero gives,

$$\sin\theta = 0.5774 \quad \text{and} \quad \cos\theta = 0.8165$$

substituting these values in $\mathbf{A}_2$ yields

$$\mathbf{A}_2 = \begin{bmatrix} 1 & 1.7321 & 0 & 0 \\ 1.7321 & 20.6681 & 4.0827 & 9.4281 \\ 0 & 4.0821 & 1 & 2.3093 \\ 0 & 9.4281 & 2.3093 & 6.333 \end{bmatrix}$$

To annihilate the element in the $(2,4)$ position of $\mathbf{A}_2$

$$\text{choose } \mathbf{R}_{34} = \begin{bmatrix} 1 & 0 & 0 & 0 \\ 0 & 1 & 0 & 0 \\ 0 & 0 & \cos\theta & -\sin\theta \\ 0 & 0 & \sin\theta & \cos\theta \end{bmatrix}$$

Computing $\mathbf{A}_3 = \mathbf{R}_{34}^T\mathbf{A}_2\mathbf{R}_{34}$ and equating the entry in $(2,4)$ position to zero gives $\sin\theta = 0.9177$ and $\cos\theta = 0.3974$ substituting in $\mathbf{A}_3$ gives the required matrix in tridiagonal form

$$\mathbf{A}_3 = \mathbf{R}_{34}^T\mathbf{R}_{24}^T\mathbf{R}_{23}^T\mathbf{A}\mathbf{R}_{23}\mathbf{R}_{24}\mathbf{R}_{34}$$
$$= \begin{bmatrix} 1 & 1.7321 & 0 & 0 \\ 1.7321 & 20.6667 & 10.2740 & 0 \\ 0 & 10.2740 & 7.1754 & 0.3646 \\ 0 & 0 & 0.3646 & 0.1580 \end{bmatrix}.$$

Observation.

1. If $\mathbf{A}$ is a symmetric matrix then Householder transformation also can be applied in a similar manner to obtain a tridiagonal matrix.

2. If $\mathbf{A}$ is any square matrix, then $\mathbf{A}$ can be reduced to a triangular form using plane rotations.

EXERCISE 4.6

1. Show that the plane rotation matrix P_{24}-4×4 matrix is an orthogonal matrix.

2. Using plane rotation reduce the matrix

$$\mathbf{A} = \begin{bmatrix} 1 & \sqrt{2} & \sqrt{2} & 2 \\ \sqrt{2} & -\sqrt{2} & -1 & \sqrt{2} \\ \sqrt{2} & -1 & \sqrt{2} & \sqrt{2} \\ 2 & \sqrt{2} & \sqrt{2} & -3 \end{bmatrix}$$

to a tridiagonal form.

4.7 QR Decomposition

In this section, the QR decomposition is described and using the Householder method the decomposition is obtained for an example illustrating the technique.

Definition 4.7.1 *The QR decomposition of a given matrix* $\mathbf{A}$ *is the product of an orthogonal matrix* $\mathbf{Q}$ *and an upper triangular matrix* R*, that is* $\mathbf{A} = \mathbf{QR}$.

Observation. The orthogonal matrix $\mathbf{Q}$ in this section is obtained using Householder method.

An upper triangular matrix **R** is in any of the following forms for an $m \times n$ matrix **A**.

1. If **A** is an $m \times n$ matrix and $m > n$ then the triangular form holds for an $n \times n$ matrix and all rows $n+1$, $n+2$, ..., m become zero rows.

$$\begin{bmatrix} c_{11} & c_{12} & . & . & . & c_{1n} \\ 0 & c_{22} & . & . & . & c_{2n} \\ . & . & . & . & . & . \\ . & . & . & . & . & . \\ 0 & 0 & . & . & . & c_{nn} \\ 0 & 0 & . & . & . & 0 \\ \vdots & \vdots & . & . & . & \vdots \\ 0 & 0 & . & . & . & 0 \end{bmatrix}$$

2. If $m < n$, then there exists an $m \times m$ triangular matrix and rest of the columns exist.

$$\begin{bmatrix} c_{11} & c_{12} & . & . & . & c_{1m} & . & . & . & c_{1n} \\ 0 & c_{22} & . & . & . & c_{2m} & . & . & . & c_{2n} \\ . & . & . & . & . & . & . & . & . & . \\ . & . & . & . & . & . & . & . & . & . \\ . & . & . & . & . & . & . & . & . & . \\ 0 & 0 & . & . & . & c_{mm} & . & . & . & c_{nn} \end{bmatrix}$$

Example 4.7.1 *Obtain QR decomposition for*

$$\mathbf{A} = \begin{bmatrix} 3 & 5 \\ 4 & 3 \end{bmatrix}$$

using Householder's Algorithm

Solution.

$$||\mathbf{x}|| = \sqrt{x_1^2 + x_2^2} = \sqrt{9+16} = 5$$

$$||\mathbf{x}||(sgn x_1)x_1 = 25 + 5.3 = 40$$

$$\mathbf{u} = \mathbf{x} - (sgn x_1)||\mathbf{x}||e_1 = \begin{bmatrix} 3 \\ 4 \end{bmatrix} - 5\begin{bmatrix} 1 \\ 0 \end{bmatrix} = \begin{bmatrix} -2 \\ 4 \end{bmatrix}$$

$$\mathbf{H_2} = I_2 - 2\frac{\mathbf{u}\cdot\mathbf{u}^T}{||\mathbf{u}||^2} = \begin{bmatrix} 1 & 0 \\ 0 & 1 \end{bmatrix} - \frac{2}{10}\begin{bmatrix} -2 \\ 4 \end{bmatrix}\begin{bmatrix} -2 & 4 \end{bmatrix} = \frac{1}{5}\begin{bmatrix} 3 & 4 \\ 4 & -3 \end{bmatrix}.$$

$$\textit{Now}\ \ \mathbf{H_2 A} = \frac{1}{5}\begin{bmatrix} 3 & 4 \\ 4 & -3 \end{bmatrix}\begin{bmatrix} 3 & 5 \\ 4 & 3 \end{bmatrix} = \begin{bmatrix} 5 & \frac{27}{5} \\ 0 & \frac{11}{5} \end{bmatrix}$$

Now set $\mathbf{Q} = \mathbf{H}_2$ *and then* $\mathbf{QA} = \mathbf{R}$
Further $\mathbf{Q}^T\mathbf{R}=\mathbf{A}$, *verify.*

EXERCISE 4.7

1. Obtain the QR decomposition for the matrix $A = \begin{bmatrix} 2 & 1 & 2 \\ 2 & 3 & 1 \\ 1 & 4 & 3 \end{bmatrix}$ using Householder transformation.

2. Obtain the QR decomposition for the matrix $A = \begin{bmatrix} 1 & 2 & 1 \\ 2 & 2 & 3 \\ 1 & 3 & 3 \end{bmatrix}$ using plane rotations.

4.8 Eigen-values: Bounds and Power Method

Finding eigen-values of a matrix is important but it is difficult to find them for matrices of a large order. This is because the problem reduces to finding the roots of characteristic polynomials and finding roots of a polynomial of degree ≥ 4 is not easy. Hence numerical approach becomes essential. This calls for choosing the initial approximation. Gershgorin gave a simple technique to find the bounds of the eigen-values of a matrix **A** by using the definition of an eigen-value. This method is described below.

Gershgorin Method

Let **A** be a square matrix of order n and λ be an eigen-value of **A** and $\mathbf{x} = (x_1 x_2 x_n)^T$ be the eigen-vector corresponding to λ then $\mathbf{Ax} = \lambda\mathbf{x}$, $\mathbf{x} \neq \mathbf{0}$.

Writing down the system of equations yields

$$\begin{aligned}
&(a_{11} - \lambda)x_1 + a_{12}x_2 + \ldots + a_{1n}x_n = 0 \\
&a_{21}x_1 + (a_{22} - \lambda)x_2 + \ldots + a_{2n}x_n = 0 \\
&\vdots \\
&a_{l1}x_1 + a_{l2}x_2 + \ldots + (a_{ll} - \lambda)x_l + \ldots + a_{ln}x_n = 0 \\
&\vdots \\
&a_{n1}x_1 + a_{n2}x_2 + \ldots + a_{nl}x_l + \ldots + (a_{nn} - \lambda)x_n = 0
\end{aligned} \tag{4.23}$$

without loss of generality, assume that

$$|x_l| = Max\{|x_1|, |x_2|, ..., |x_n|\}$$

Then considering the lth equation in the system (4.23) and taking the absolute value gives

$$|\lambda||x_l| = |a_{l1}x_1 + a_{l2}x_2 + ... + a_{ln}x_n| \leq |a_{l1}||x_1| + |a_{l2}||x_2| + ... + |a_{ln}||x_n|$$

dividing by $|x_l|$ and observing that $|x_i|/|x_l| \leq 1$, yields,

$$|\lambda| \leq |a_{l1}| + |a_{l2}| + ...|a_{ln}| = \sum_{j=1}^{n} |a_{lj}|$$

Thus an eigen-value λ of a matrix **A** is bounded by the sum of the absolute values of its row elements. This holds for all rows $l = 1, 2, ..., n$.

Since **A** and $\mathbf{A^T}$ have the same eigen-values $|\lambda| \leq \sum_{i=1}^{n} |a_{ik}|$ for each $k = 1, 2, ..., n$.

By rewriting lth equation in the system (4.23) in a slightly different fashion one can obtain Gershgorin circles or discs as follows.

$$(a_{ll}-\lambda)|\mathbf{x}_l| = |a_{l1}||\mathbf{x}_1| + |a_{l2}||\mathbf{x}_2| + ... + |a_{l(l-1)}||\mathbf{x}_{l-1}| + |a_{l(l+1)}||\mathbf{x}_{l+1}| + ... + |a_{ln}||\mathbf{x}_n|$$

since $|x_i|/|x_l| < 1$,

$$|a_{ll} - \lambda| \leq \sum_{j=1, j\neq l}^{n} |a_{lj}| = r_l$$

Thus an eigen-value λ lies in a circle of radius r_l and center a_{ll}.

As the λ is arbitrary and both **A** and $\mathbf{A^T}$ have the same eigen-values the following result can be safely concluded. This result is given by Braver.

Result 4.8.1 (1) The eigen-values of a matrix A are always less than or equal to the sum of its row elements in absolute values or the sum of its column elements in absolute values.

(2) The eigen-values of a matrix **A** lie in discs with center a_{ll} and radius $r_l = \sum_{j=1, j\neq l}^{n} |a_{lj}|$, $l = 1, 2, ..., n$ and radius $r_k = \sum_{i=1, i\neq k}^{n} |a_{ik}|$, $k = 1, 2, ..., n$.

Let C_l be the circle with radius r_l and center a_{ll}, $l = 1, 2, ..., n$ and C_k be the circle with radius r_k and center a_{kk}, $k = 1, 2, ..., n$. Write $C_R = \bigcup_{l=1}^{n} C_l$ and

$$C_C = \bigcup_{k=1}^{n} C_k$$

Then all eigen-values of all the matrix **A** will lie in the circle $C = C_C \cap C_R$.

Example 4.8.1 *(i) Find the bounds for the eigen-values of the matrix*

$$\mathbf{A} = \begin{bmatrix} -1 & 1 & 3 \\ 1 & 2 & 1 \\ -1 & 1 & 2 \end{bmatrix}$$

(ii) Find the Gershgorin circles. (iii) Find the region of existence of the eigen-values of **A**.

Solution.

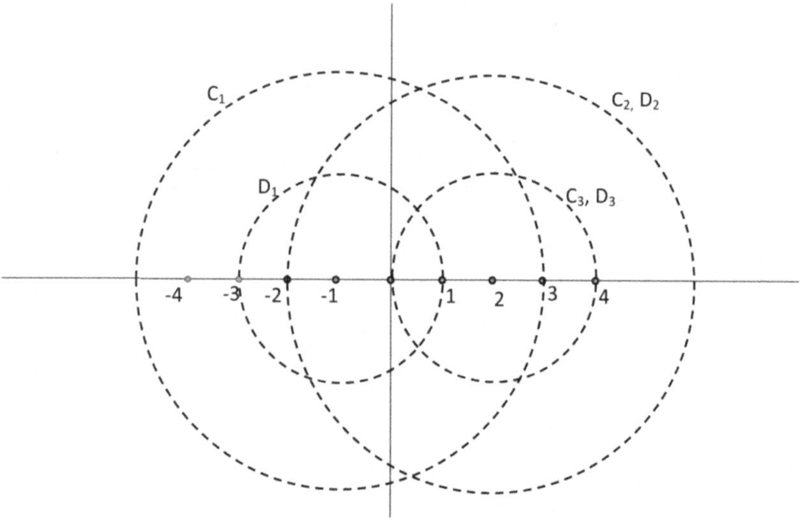

Fig. 4.1 The union of Gershgorin discs of $\mathbf{A}(C_R = C_1 \cup C_2 \cup C_3)$ and the union of Gershgorin discs of $\mathbf{A}^{\mathbf{T}}(C_C = D_1 \cup D_2 \cup D_3)$ and eigen-values of **A** lie inside $C_R \cap C_C$

(i) (a) by considering the row sums of **A**

$$|\lambda| \leq Max\{5, 4, 4\} = 5$$

(b) by considering the column sums of **A**

$$|\lambda| = Max\{3, 4, 6\} = 6$$

(ii) (a) The Gershgorin circles taking rows are

$$C_1 = \{\lambda : |\lambda + 1| \leq 4\}, C_2 = \{\lambda : |\lambda - 2| \leq 2\}, C_3 = \{\lambda : |\lambda - 2| \leq 2\}.$$

(b) The Gershgorin circles taking columns are

$$D_1 = \{\lambda : |\lambda + 1| \leq 2\}, D_2 = \{\lambda : |\lambda - 2| \leq 2\}, D_3 = \{\lambda : |\lambda - 2| \leq 4\}.$$

(iii) $\mathbf{C}_R = C_1 \cup C_2 \cup C_3$ *and* $\mathbf{C}_C = D_1 \cup D_2 \cup D_3$

$\mathbf{C}_R \cap \mathbf{C}_C$ *(see figure 4.1) is the region in which the eigen-values of* **A** *lie. The eigen-values of* **A** *are 3,* $\pm i$.

Power method. Power method is used to find the largest eigen-value of a square matrix of order n. In order to use this method certain criteria are necessary. These are assumed and are as follows.

(i) All the 'n' eigen-vectors can be determined even if some of the eigen-values are repeated.

(ii) All the 'n' eigen-vectors are linearly independent.

(iii) Let $\mathbf{v_i}$ be the eigen-vector corresponding to the eigen-value λ_i. The above assumptions guarantee that the 'n' eigen-vectors $S = \{v_1, v_2, ..., v_n\}$ form a basis.

Without loss of generality assume that $|\lambda_1| > |\lambda_2| \geq ... \geq |\lambda_n|$.

Let $\mathbf{v} = \alpha_1 \mathbf{v}_1 + \alpha_2 \mathbf{v}_2 + ... + \alpha_n \mathbf{v}_n$ be any vector in $\mathbb{R}^n$. Then

$$\begin{aligned}\mathbf{Av} &= \alpha_1 \mathbf{Av_1} + \alpha_2 \mathbf{Av_2} + ... + \alpha_n \mathbf{Av_n} \\ &= \alpha_1 \lambda_1 \mathbf{v_1} + \alpha_2 \lambda_2 \mathbf{v_2} + ... + \alpha_n \lambda_n \mathbf{v_n} \\ &= \lambda_1(\alpha_1 \mathbf{v_1} + \alpha_2(\lambda_2/\lambda_1)\mathbf{v_2} + ... + \alpha_n(\lambda_n/\lambda_1)\mathbf{v_n})\end{aligned}$$

Also,

$$\mathbf{A^2 v} = \lambda_1^2(\alpha_1 \mathbf{v_1} + \alpha_2(\lambda_2/\lambda_1)^2 \mathbf{v_2} + ... + \alpha_n(\lambda_n/\lambda_1)^2 \mathbf{v_n})$$

Proceeding so on,

$$\mathbf{A^k v} = \lambda_1^k(\alpha_1 \mathbf{v_1} + \alpha_2(\lambda_2/\lambda_1)^k \mathbf{v_2} + ... + \alpha_n(\lambda_n/\lambda_1)^k \mathbf{v_n})$$

and

$$\mathbf{A^{k+1} v} = \lambda_1^{k+1}(\alpha_1 \mathbf{v_1} + \alpha_2(\lambda_2/\lambda_1)^{k+1} \mathbf{v_2} + ... + \alpha_n(\lambda_n/\lambda_1)^{k+1} \mathbf{v_n})$$

Now as $k \to \infty$

$$\begin{aligned}\mathbf{A^k v} &\to \lambda_1^k \alpha_1 \mathbf{v_1} \text{ as } k \to \infty \\ \mathbf{A^{k+1} v} &\to \lambda_1^{k+1} \alpha_1 \mathbf{v_1} \text{ as } k \to \infty\end{aligned}$$

$$\lambda_1 = \lim_{k\to\infty} (\mathbf{A^{k+1} v})_i / (\mathbf{A^k v})_i, \quad i = 1, 2, ...n$$

Observation. $\mathbf{A}^k\mathbf{v}$, $\mathbf{A}^{k+1}\mathbf{v}$ are vectors, hence $(\mathbf{Av})_i, (\mathbf{A}^{k+1}\mathbf{v})_i$ are the ith components of the vectors.

The iteration procedure is stopped when the error of the magnitude of successive ratios is very small or within the prescribed limits.

The algorithm of the power method is as follows.

Step 1. Given a square matrix $\mathbf{A}$ of order n, choose any vector v_0 (which is not in the standard basis).

Step 2. Consider $\mathbf{Av_0} = \mathbf{v_1} = \lambda_1^{(1)} \mathbf{y^{(1)}}$ where $\lambda_1^{(1)} = max(v_{11}, v_{12}, ...v_{1n})$

Step 3. Calculate $\mathbf{Ay^{(1)}} = \mathbf{v_2} = \lambda_1^{(2)}\mathbf{y^{(2)}}$ where $\lambda_1^{(2)} = max(v_{21}, v_{22}, ...v_{2n})$

Repeat Step 3, k times till $\lambda_1^{(k)} - \lambda_1^{(k-1)} < \epsilon$ and $||\mathbf{y}^{(k)} - \mathbf{y}^{(k-1)}|| < \epsilon$, the given tolerance limit.

Example 4.8.2 *Using power method find the largest eigen-value of the following matrix*

$$\mathbf{A} = \begin{bmatrix} 2 & 4 & 8 \\ 1 & 3 & 5 \\ 9 & 4 & 3 \end{bmatrix}.$$

Solution. *Let the initial approximation be*

$$\mathbf{v}_0 = \begin{bmatrix} 1 \\ 1 \\ 1 \end{bmatrix}.$$

Then calculating

$$\mathbf{Av}_0 = \begin{bmatrix} 2 & 4 & 8 \\ 1 & 3 & 5 \\ 9 & 4 & 3 \end{bmatrix} \begin{bmatrix} 1 \\ 1 \\ 1 \end{bmatrix} = \begin{bmatrix} 14 \\ 9 \\ 16 \end{bmatrix}$$

Now taking maximum of components of $\mathbf{Av}_0$, *and dividing the vector,*

$$\mathbf{v}_1 = \frac{1}{16} \begin{bmatrix} 14 \\ 9 \\ 16 \end{bmatrix} = \begin{bmatrix} 0.875 \\ 0.5625 \\ 1 \end{bmatrix}$$

proceeding similarly,

$$\mathbf{Av}_1 = \begin{bmatrix} 2 & 4 & 8 \\ 1 & 3 & 5 \\ 9 & 4 & 3 \end{bmatrix} \begin{bmatrix} 0.875 \\ 0.5625 \\ 1 \end{bmatrix} = \begin{bmatrix} 12 \\ 7.5625 \\ 13.125 \end{bmatrix}.$$

Now,

$$\mathbf{v}_2 = \frac{1}{13.125} \begin{bmatrix} 12 \\ 7.5625 \\ 13.125 \end{bmatrix} = \begin{bmatrix} 0.9143 \\ 0.5762 \\ 1 \end{bmatrix}$$

and

$$\mathbf{Av}_2 = \begin{bmatrix} 2 & 4 & 8 \\ 1 & 3 & 5 \\ 9 & 4 & 3 \end{bmatrix} \begin{bmatrix} 0.9143 \\ 0.5762 \\ 1 \end{bmatrix} = \begin{bmatrix} 12.1333 \\ 7.6429 \\ 13.5333 \end{bmatrix}.$$

again,

$$\mathbf{v}_3 = \frac{1}{13.5333} \begin{bmatrix} 12.1333 \\ 7.6429 \\ 13.5333 \end{bmatrix} = \begin{bmatrix} 0.8966 \\ 0.5647 \\ 1 \end{bmatrix}$$

$$\mathbf{Av}_3 = \begin{bmatrix} 2 & 4 & 8 \\ 1 & 3 & 5 \\ 9 & 4 & 3 \end{bmatrix} \begin{bmatrix} 0.8966 \\ 0.5647 \\ 1 \end{bmatrix} = \begin{bmatrix} 12.0521 \\ 7.5908 \\ 13.3279 \end{bmatrix}$$

$$\mathbf{v}_4 = \frac{1}{13.3279} \begin{bmatrix} 12.0521 \\ 7.5908 \\ 13.3279 \end{bmatrix} = \begin{bmatrix} 0.9043 \\ 0.5695 \\ 1 \end{bmatrix}$$

$$\mathbf{Av}_4 = \begin{bmatrix} 2 & 4 & 8 \\ 1 & 3 & 5 \\ 9 & 4 & 3 \end{bmatrix} \begin{bmatrix} 0.9043 \\ 0.5695 \\ 1 \end{bmatrix} = \begin{bmatrix} 12.0867 \\ 7.6129 \\ 13.4166 \end{bmatrix}.$$

Again,

$$\mathbf{v}_5 = \frac{1}{13.4166} \begin{bmatrix} 12.0867 \\ 7.6129 \\ 13.4166 \end{bmatrix} = \begin{bmatrix} 0.9009 \\ 0.5674 \\ 1 \end{bmatrix}$$

and

$$\mathbf{Av}_5 = \begin{bmatrix} 2 & 4 & 8 \\ 1 & 3 & 5 \\ 9 & 4 & 3 \end{bmatrix} \begin{bmatrix} 0.9009 \\ 0.5674 \\ 1 \end{bmatrix} \begin{bmatrix} 12.0714 \\ 7.6031 \\ 13.3776 \end{bmatrix}.$$

Next,

$$\mathbf{v}_6 = \frac{1}{13.3776} \begin{bmatrix} 12.0714 \\ 7.6031 \\ 13.3776 \end{bmatrix} = \begin{bmatrix} 0.9024 \\ 0.5683 \\ 1 \end{bmatrix}$$

and

$$\mathbf{Av}_6 = \begin{bmatrix} 2 & 4 & 8 \\ 1 & 3 & 5 \\ 9 & 4 & 3 \end{bmatrix} \begin{bmatrix} 0.9024 \\ 0.5683 \\ 1 \end{bmatrix} = \begin{bmatrix} 12.0781 \\ 7.6074 \\ 13.3947 \end{bmatrix}$$

$$\mathbf{v}_7 = \frac{1}{13.3947} \begin{bmatrix} 12.0781 \\ 7.6074 \\ 13.3947 \end{bmatrix} = \begin{bmatrix} 0.9017 \\ 0.5679 \\ 1 \end{bmatrix}$$

$$\mathbf{Av}_7 = \begin{bmatrix} 2 & 4 & 8 \\ 1 & 3 & 5 \\ 9 & 4 & 3 \end{bmatrix} \begin{bmatrix} 0.9017 \\ 0.5679 \\ 1 \end{bmatrix} = \begin{bmatrix} 12.0752 \\ 7.6055 \\ 13.3872 \end{bmatrix}.$$

$$\mathbf{v}_8 = \frac{1}{13.3872} \begin{bmatrix} 12.0752 \\ 7.6055 \\ 13.3872 \end{bmatrix} = \begin{bmatrix} 0.902 \\ 0.5681 \\ 1 \end{bmatrix}$$

$$\mathbf{Av}_8 = \begin{bmatrix} 2 & 4 & 8 \\ 1 & 3 & 5 \\ 9 & 4 & 3 \end{bmatrix} \begin{bmatrix} 0.902 \\ 0.5681 \\ 1 \end{bmatrix} = \begin{bmatrix} 12.0765 \\ 7.6064 \\ 13.3905 \end{bmatrix}.$$

Again

$$\mathbf{v}_9 = \frac{1}{13.3905} \begin{bmatrix} 12.0765 \\ 7.6064 \\ 13.3905 \end{bmatrix} = \begin{bmatrix} 0.9019 \\ 0.568 \\ 1 \end{bmatrix}$$

and

$$\mathbf{A}\mathbf{v}_9 = \begin{bmatrix} 2 & 4 & 8 \\ 1 & 3 & 5 \\ 9 & 4 & 3 \end{bmatrix} \begin{bmatrix} 0.9019 \\ 0.568 \\ 1 \end{bmatrix} = \begin{bmatrix} 12.0759 \\ 7.606 \\ 13.389 \end{bmatrix}$$

$$\mathbf{v}_{10} = \frac{1}{13.389} \begin{bmatrix} 12.0759 \\ 7.606 \\ 13.389 \end{bmatrix} = \begin{bmatrix} 0.9019 \\ 0.5681 \\ 1 \end{bmatrix}.$$

Since $\mathbf{v}_9$ *and* $\mathbf{v}_{10}$ *are same the procedure is stopped.*
The maximum eigen-value is 13.39 *and the corresponding eigen-vector is*

$$\begin{bmatrix} 0.9 \\ 0.57 \\ 1 \end{bmatrix}$$

Exercise 4.8

1. Obtain the bounds for the eigen-values for the matrices

$$(i) \begin{bmatrix} 2 & 5 & 2 \\ 2 & 2 & 3 \\ 3 & 2 & 2 \end{bmatrix} \quad \begin{bmatrix} 2 & 4 & 1 \\ 2 & 2 & 1 \\ 1 & 3 & 1 \end{bmatrix}$$

2. Obtain the Gershgorin circles for the matrices.

$$(i) \begin{bmatrix} -2 & 1 & 5 \\ 4 & -3 & 1 \\ -2 & 1 & 4 \end{bmatrix} \quad (ii) \begin{bmatrix} 1 & 3 & 2 \\ -1 & -2 & 3 \\ 1 & 0 & -1 \end{bmatrix}$$

3. Using power method find the largest eigen-value for the matrices. Find the corresponding eigen-vector.

$$(i) \begin{bmatrix} 3 & 2 & 1 \\ 4 & 2 & 1 \\ 2 & 1 & 2 \end{bmatrix} \quad (ii) \begin{bmatrix} 2 & 4 & 1 & 2 \\ 0 & 0 & 2 & 3 \\ 1 & 0 & 2 & 1 \\ 0 & 4 & 2 & 2 \end{bmatrix}$$

4.9 Krylov Subspace Methods

It has been described in literature that one of the ten most important classes of numerical methods is Krylov Space methods. They are used to solve large sparse systems of linear equations and also eigen-value problems of large sparse matrices. Following is a definition of a Krylov subspace.

Definition 4.9.1 *Given a nonsingular matrix* $\mathbf{A} \in C^{n\times n}$ *and a non zero vector* $\mathbf{y} \in C^n$*, the* n^{th} *Krylov subspace* $K_n(\mathbf{A}, \mathbf{y})$ *generated by* $\mathbf{A}$ *from* $\mathbf{y}$ *is*

$$K_n = K_n(\mathbf{A}, \mathbf{y}) = span(\mathbf{y}, \mathbf{A}\mathbf{y}, ..., \mathbf{A}^{\mathbf{n-1}}\mathbf{y}).$$

Observation. $K_1 \subseteq K_2 \subseteq K_3 \ldots \subseteq K_n$.

Idea behind Krylov Space Methods

Consider a linear system $\mathbf{Ax} = \mathbf{b}$.

Choose an initial approximation, so that $\mathbf{r}_0 = \mathbf{b} - \mathbf{Ax_0} - \mathbf{b}$ is the initial error vector or the initial residual vector.

The sequence of approximations are generated from r_0 by considering the nth Krylov subspace $\mathbf{K}_n(\mathbf{A}, \mathbf{r}_0)$ and finding $\mathbf{x}_n$ such that $\mathbf{x}_n \in \mathbf{x}_0 + \mathbf{K}_n(\mathbf{A}, \mathbf{r}_0)$ and the residual $\mathbf{r}_n \in \mathbf{K}_{n+1}(\mathbf{A}, \mathbf{r}_0)$ and $r_n \to 0$ as $n \to \infty$. Thus, it is expected that in a finite number of steps $\mathbf{r}_n = 0$ and $\mathbf{x}_n = \mathbf{x}^*$ is the solution of the linear system $\mathbf{Ax} = \mathbf{b}$.

In general, the definition of Krylov subspace solver that covers all cases is difficult. The following definition is given in a simplified form.

Definition 4.9.2 *A Krylov space solver is an iterative method for solving a linear system* $\mathbf{Ax} = \mathbf{b}$ *starting from an initial approximation* $\mathbf{x}_0$ *and the corresponding residual vector* $\mathbf{r}_0 = \mathbf{b} - \mathbf{Ax_0}$ *and generating for all, or at least most 'n', till it possibly finds the exact solutions, iteration* $\mathbf{x}_n$ *such that*

$$\mathbf{x}_n - \mathbf{x}_0 = p_{n-1}(\mathbf{A})\mathbf{r}_0 \in \mathbf{K}_n(\mathbf{A}, \mathbf{r}_0)$$

with a polynomial p_{n-1} *of exact degree* $n - 1$.

For some n, $\mathbf{x}_n$ *may not exist or* $\mathbf{p}_{n-1}$ *may be of lower degree.*

Krylov space solvers converge very slowly(if they do) for very large real world problems. One of the approaches to handle this situation is by pre-conditioning the given system, that is, by pre-multiplying with an approximate inverse $\mathbf{C}$ and considering the system

$$\mathbf{CAx} = \mathbf{Cb} \text{ in place of } \mathbf{Ax} = \mathbf{b}.$$

There are many methods that come under this umbrella. Some of them are

(i) Conjugate gradient (CG) method applied when $\mathbf{A}$ is a symmetric positive definite matrix.

(ii) Conjugate direction method, which is a refined version of CG method and is used when $\mathbf{A}$ is a symmetric positive definite matrix.

(iii) Orthomin form of the CG method which is again a refinement of CG method and is used when $\mathbf{A}$ is a symmetric positive definite matrix.

(iv) Biconjugate gradient (BiCG) method for nonsymmetric matrices.

4.10 Conclusion

Elements of computational errors are introduced in this chapter. Care has been taken to introduce all the direct and iterative methods to solve a linear system of equations. As an orthogonal transformation mitigates errors, Householder transformation and plane rotation method are given to transform ordinary and symmetric matrices respectively to diagonal or tridiagonal form. The QR method is described and the bounds for eigen-values and power method for finding the largest eigen-value and its corresponding eigen-vector are given. Further, Krylov subspace solvers are introduced. For further reading the readers are suggested to refer to [5], [10], [13], [14] and [18].

Chapter 5

Applications

5.1 Introduction

The focus of this chapter is to use the tools developed in earlier chapters and apply them to diverse topics. Section 5.2 highlights the use of determinants in finding equations of conics. Section 5.3 deals with Markov chains that deal with a mathematical system having a finite number of states. In Section 5.4 economic models are discussed using a linear system of equations. The concept of congruent inverse introduced in Section 1.9 is used for encoding and decoding messages in the field of cryptography in Section 5.5. Linear transformations described in Chapter 3 are used to introduce the basics of computer graphics in Section 5.6. The notion of change of basis matrix is widely used in robotics to pass on information from one frame of reference to another and this forms the content of Section 5.7. Matrices play a key role in bioinformatics and this is showcased in Section 5.8. Section 5.9 deals with principal component analysis wherein the eigen-value decomposition and singular value decomposition theorems are well utilized. In Section 5.10, the concept of big data is introduced and an example is given to illustrate it.

5.2 Finding Curves through Given Points

This section deals with application of determinants. It is shown that equation of curves and surfaces can be found using the theorems of analytic geometry along with determinants. The main theorem that is used to obtain our results is from Chapter 2, which states that 'A homogeneous system of linear equations has a nontrivial solution if and only if the coefficient matrix is singular, that is, its determinant is zero'.

Result 1

Finding the equation of a circle passing through 3 points.

Let $(x_1 \quad y_1)$, $(x_2 \quad y_2)$, $(x_3 \quad y_3)$ be any three points through which the

DOI: 10.1201/9781351014953-5

circle must pass. Then clearly these 3 points must satisfy the equation of the circle

$$a(x^2 + y^2) + bx + cy + d = 0.$$

Hence,

$$a(x_1^2 + y_1^2) + bx_1 + cy_1 + d = 0$$
$$a(x_2^2 + y_2^2) + bx_2 + cy_2 + d = 0$$
$$a(x_3^2 + y_3^2) + bx_3 + cy_3 + d = 0$$

All the above equations form a homogeneous linear system and possess a non trivial solution if

$$\begin{vmatrix} x^2 + y^2 & x & y & 1 \\ x_1^2 + y_1^2 & x_1 & y_1 & 1 \\ x_2^2 + y_2^2 & x_2 & y_2 & 1 \\ x_3^2 + y_3^2 & x_3 & y_3 & 1 \end{vmatrix} = 0,$$

which is also the equation of a circle passing through the points $(x_1\ \ y_1)$, $(x_2\ \ y_2)$, $(x_3\ \ y_3)$.

Example 5.2.1 *Construct a circle passing through the points* $(1\ \ 0), (0\ \ 1), (2\ \ 3)$.
Then the equation of the circle passing through the given points is

$$\begin{vmatrix} x^2 + y^2 & x & y & 1 \\ 1 & 1 & 0 & 1 \\ 1 & 0 & 1 & 1 \\ 13 & 2 & 3 & 1 \end{vmatrix} = 0$$

expanding the determinant gives $4(x^2 + y^2) - 12(x + y) + 8 = 0$, *which is the equation of a circle passing through the points* $(1\ \ 0)$, $(0\ \ 1)$, $(2\ \ 3)$.

Result 2

Finding the equation of a conic section.

From analytic geometry it is known that the general equation of a conic section is

$$ax^2 + by^2 + cxy + dx + ey + f = 0.$$

Since there are six unknowns given 5 points $(x_1\ \ y_1)$, $(x_2\ \ y_2)$, $(x_3\ \ y_3)$, $(x_4\ \ y_4)$, $(x_5\ \ y_5)$, the equation of the conic section can be determined by forming a homogeneous linear system as before (one can eliminate the coefficient of x^2).

The system possesses a nontrivial solution if the determinant of the system is zero, that is,

$$\begin{vmatrix} x^2 & y^2 & xy & x & y & 1 \\ x_1^2 & y_1^2 & x_1y_1 & x_1 & y_1 & 1 \\ x_2^2 & y_2^2 & x_2y_2 & x_2 & y_2 & 1 \\ x_3^2 & y_3^2 & x_3y_3 & x_3 & y_3 & 1 \\ x_4^2 & y_4^2 & x_4y_4 & x_4 & y_4 & 1 \\ x_5^2 & y_5^2 & x_5y_5 & x_5 & y_5 & 1 \end{vmatrix} = 0,$$

which also gives the equation of a conic section.

Using this determinant the equations of curves like ellipse, parabola and hyperbola can be determined. This has applications in determining specified orbits of asteroid, etc.

EXERCISE 5.2

1. Construct a conic passing through the points $(0\ \ 0)$, $(1\ \ 1)$, $(-1\ \ 1)$, $(2\ \ 0), (3\ \ -2)$.

2. Find the determinant to find the equation of a sphere passing through four points. and hence find the equation of the sphere passing through the points, $(0\ \ 0\ \ 0)$, $(0\ \ 1\ \ 1)$, $(1\ \ 1\ \ 0)$, $(0\ \ 2\ \ 0)$.

3. Find the determinant to find the equation of a plane in a 3-D space and hence find the equation of the plane passing through the points, $(1\ \ 1\ \ 1)$, $(0\ \ 0\ \ 3)$, $(2\ \ 5\ \ 2)$.

5.3 Markov Chains

Markov chain technique was developed by the Russian mathematician Andrei Markov. This technique has a wide range of applications which include forecasting weather patterns, understanding commodity price movements, administration of plantations, studying behavior of animals, analysis of experiments in physics, etc. Andrei Markov was the first one to study matrices involving stochastic matrices and Markov chains.

Consider a physical or a mathematical system having a finite number of states and is such that at any moment of time the system occupies only one of the finite states. The behavior of the system as it changes from one state to another leads to the following definitions.

Definition 5.3.1 *Stochastic process is a process in which changing of the system from one state to another is not predetermined but can only be specified in terms of certain probabilities which are obtained from the past behavior or history of the system.*

Definition 5.3.2 *First-order Markov process or a Markov Chain is a stochastic process where in the transition probability of a state depends exclusively on its immediately preceding state.*

A more explicit definition can be given as follows.

Definition 5.3.3 *A first-order Markov chain is a sequence of trials of an experiment such that*

(1) The outcome of the nth *trial depends only on the outcome of the* $(n-1)$th *trial and not on the outcome of the earlier trials; and*

(2) In two successive trials of the experiment, the probability of going from state S_i *to state* S_j *remains the same or does not change.*

To understand a system described by a Markov process or a Markov chain the following concepts are necessary.

Definition 5.3.4 *Probability vector is a column vector,* $\mathbf{p}$*, whose entries are nonnegative and their sum is one, that is,*

$$\sum_{i=1}^{n} p_i = 1.$$

Definition 5.3.5 *Stochastic matrix is a matrix* $\mathbf{P}=[p_{ij}]$ *of order* $n \times n$ *where each of its columns is a probability vector.*

Definition 5.3.6 *Transition probability is the probability of a system moving from one state to another.*

Definition 5.3.7 *Transition matrix of a Markov chain is a probability matrix* $\mathbf{T} = [t_{ij}]$ *of order* $n \times n$ *where* t_{ij} *is the transition probability of the system going from state* S_i *to state* S_j *on successive trials of the experiment.*

Observation. The sum of entries in each column is one. In other words

$$\sum_{i=1}^{n} p_{ij} = 1, for\ j = 1, 2, ..., n,$$

which implies that a transition matrix is a stochastic matrix. The following problem involves these concepts.

Example 5.3.1 *Best Furniture Decorators (BFD) has 40% of the business in a certain city. It's only competitor Excellent Enterprises (EE) has the other 60%. To become more competitive BFD advertised to improve its image. During the advertisement campaign monthly sales figures were compiled. It was found that 80% of BFD's customers returned to BFD the following month, while 25% of EE's customers switched to BFD.*

(a) What percentage of customers use each service after one month?

(b) What percentage of customers use each service after two months?

(c) What is the long-term effect in case of each service?

Solution. *From the given information the decorating services in a city is a system in which there are two states BFD and EE and the probability of the customers choosing each state is given by*

$$P(BFD/BFD) = 0.8 \qquad P(EE/BFD) = 0.2$$

$$P(BFD/EE) = 0.25 \qquad P(EE/EE) = 0.75$$

From the given information, the initial probability vector $\mathbf{p_0} = [0.4\ 0.6]^T$.

Observe that the initial vector is a probability vector and the transition matrix $\mathbf{T}$ *is a stochastic matrix given by,*

$$\mathbf{T} = \begin{bmatrix} 0.8 & 0.25 \\ 0.2 & 0.75 \end{bmatrix}$$

Observation. The matrix $\mathbf{T}$ gives the probabilities of passing from one state to another. For example, the element a_{21} of *the matrix* $\mathbf{T}$ is the probability of moving from state 2 to state 1 in one month.

To find the number of customers availing each service, consider

$$\mathbf{Tp_0} = \begin{bmatrix} 0.8 & 0.25 \\ 0.2 & 0.75 \end{bmatrix} \begin{bmatrix} 0.4 \\ 0.6 \end{bmatrix} = \begin{bmatrix} 0.47 \\ 0.53 \end{bmatrix} = \mathbf{p_1},$$

where $\mathbf{p_1}$ *is the vector of proportions after a month. It can be easily seen that the vector of proportions after two months,* $\mathbf{p_2}$ *is given by*

$$\mathbf{p_2} = \mathbf{Tp_1} = \begin{bmatrix} 0.8 & 0.25 \\ 0.2 & 0.75 \end{bmatrix} \begin{bmatrix} 0.47 \\ 0.53 \end{bmatrix} = \begin{bmatrix} 0.5085 \\ 0.4915 \end{bmatrix}$$

Note that 0.5085 + 0.4915 = 1 and $\mathbf{p_1} = \mathbf{Tp_0}$,

$$\mathbf{p_2} = \mathbf{Tp_1} = \mathbf{T}(\mathbf{Tp_0}) = \mathbf{T^2p_0}, \textit{ similarly,}$$
$$\mathbf{p_3} = \mathbf{Tp_2} = \mathbf{T}(\mathbf{T^2p_0}) = \mathbf{T^3p_0}.$$
$$\textit{Thus, } \mathbf{p_3} = \begin{bmatrix} 0.8 & 0.25 \\ 0.2 & 0.75 \end{bmatrix} \begin{bmatrix} 0.5085 \\ 0.4915 \end{bmatrix} = \begin{bmatrix} 0.529675 \\ 0.470325 \end{bmatrix}$$

It can be seen that as the number of months increase, the proportions tend to a fixed probability vector,

$$\mathbf{p} = \begin{bmatrix} 0.555... \\ 0.444... \end{bmatrix} = \begin{bmatrix} \frac{5}{9} \\ \frac{4}{9} \end{bmatrix}.$$

$\mathbf{p}$ *is called fixed vector for Transition matrix* $\mathbf{T}$ *because*

$$\mathbf{Tp} = \begin{bmatrix} 0.8 & 0.25 \\ 0.2 & 0.75 \end{bmatrix} \begin{bmatrix} \frac{5}{9} \\ \frac{4}{9} \end{bmatrix}$$
$$= \begin{bmatrix} \frac{5}{9} & \frac{4}{9} \end{bmatrix}^T = \mathbf{p}$$

In the long run, 56% of the market is favorable to BFD and 44% of the market to EE.

The following table provides information regarding the tendency of customers' choice after n months.

n	*Probability vector after n months*
0	$(0.4, 0.6)^T$
1	$(0.47, 0.53)^T$
2	$(0.5085, 0.4915)^T$
3	$(0.529675, 0.470325)^T$
4	$(0.54132125, 0.45867875)^T$
5	$(0.547726688, 0.452273313)^T$
10	$(0.555161541, 0.444838489)^T$
15	$(0.555535725, 0.444464275)^T$
20	$(0.555554558, 0.444445442)^T$
25	$(0.555555505, 0.444444495)^T$

In order to understand Markov Processes the following concepts are necessary.

Definition 5.3.8 *A transition matrix* $\mathbf{T}$ *is said to be regular if all the entries of at least one of its powers* $\mathbf{T^n}$ *are strictly positive. A Markov chain is regular if its transition matrix is regular.*

Result 5.3.1 If $\mathbf{T}$ is a regular probability matrix then there exists a unique probability vector $\mathbf{t}$ such that

$$\mathbf{Tt} = \mathbf{t}$$

Observation. 1. This unique probability vector is also called as a steady state vector.
2. If $\mathbf{p}$ is any probability vector, then, $\mathbf{T^n p}$ tends to the fixed vector $\mathbf{t}$ as n increases.

The following example illustrates the concepts given above.

Example 5.3.2 *In state elections, it has been determined that voters will switch votes according to the following probabilities.*

From To	Socialist	Capitalist	Independent
Socialist	0.7	0.35	0.4
Capitalist	0.2	0.6	0.3
Independent	0.1	0.05	0.3

Over the long run, what percentages of voters will vote for a socialist, a capitalist and an independent?

Solution. *The transition matrix for the Markov chain is*

$$\mathbf{T} = \begin{bmatrix} 0.7 & 0.35 & 0.4 \\ 0.2 & 0.6 & 0.3 \\ 0.1 & 0.05 & 0.3 \end{bmatrix}$$

Since $\mathbf{T}$ *is regular from the above result there exists a unique fixed probability vector* $\mathbf{t}$ *such that* $\mathbf{T}\mathbf{t} = \mathbf{t}$.

Let $\mathbf{t} = (x\,y\,z)^T$. *Since* $\mathbf{t}$ *is a probability vector,* $x + y + z = 1$ *and further,*

$$\begin{bmatrix} 0.7 & 0.35 & 0.4 \\ 0.2 & 0.6 & 0.3 \\ 0.1 & 0.05 & 0.3 \end{bmatrix} \begin{bmatrix} x \\ y \\ z \end{bmatrix} = \begin{bmatrix} x \\ y \\ z \end{bmatrix}.$$

yielding a system of equations given by,

$$\begin{aligned} x + y + z &= 1 \\ -0.3x + 0.35y + 0.4z &= 0 \\ 0.2x - 0.4y + 0.3z &= 0 \\ 0.1x + 0.05y - 0.7z &= 0. \end{aligned}$$

To solve the system of equations, consider the augmented matrix given by

$$\left[\begin{array}{ccc|c} 1 & 1 & 1 & 1 \\ -0.3 & 0.35 & 0.4 & 0 \\ 0.2 & -0.4 & 0.3 & 0 \\ 0.1 & 0.05 & -0.7 & 0 \end{array}\right]$$

Now $R_2 \to R_2 + 0.3R_1$, $R_3 \to R_3 - 0.2R_1$, $R_4 \to R_4 - 0.1R_1$ *gives*

$$\sim \left[\begin{array}{ccc|c} 1 & 1 & 1 & 1 \\ 0 & 0.65 & 0.7 & 0.3 \\ 0 & -0.6 & 0.1 & -0.2 \\ 0 & -0.05 & -0.8 & -0.1 \end{array}\right]$$

$R_2 \to (1/0.65)R_2$ *gives*

$$\sim \left[\begin{array}{ccc|c} 1 & 1 & 1 & 1 \\ 0 & 1 & 1.076 & 0.4615 \\ 0 & -0.6 & 0.1 & -0.2 \\ 0 & -0.05 & -0.8 & -0.1 \end{array}\right]$$

$R_1 \to R_1 - R_2$, $R_3 \to R_3 + 0.6R_2$, $R_4 \to R_4 + 0.05R_2$ *gives*

$$\sim \left[\begin{array}{ccc|c} 1 & 0 & -0.076 & 0.5385 \\ 0 & 1 & 1.076 & 0.4615 \\ 0 & 0 & 0.7456 & 0.0769 \\ 0 & 0 & -0.7462 & -0.0769 \end{array}\right]$$

$R_3 \to (1/0.7456)R_3$ *gives*

$$\sim \left[\begin{array}{ccc|c} 1 & 0 & -0.076 & 0.5385 \\ 0 & 1 & 1.076 & 0.4615 \\ 0 & 0 & 1 & 0.1031 \\ 0 & 0 & -0.7462 & -0.0769 \end{array}\right]$$

and $R_1 \to R_1 + 0.076R_3$, $R_2 \to R_2 - 1.076R_3$, $R_4 \to R_4 + 0.7462R_3$ *gives*

$$\sim \left[\begin{array}{ccc|c} 1 & 0 & 0 & 0.5463 \\ 0 & 1 & 0 & 0.3506 \\ 0 & 0 & 1 & 0.1031 \\ 0 & 0 & 0 & 0 \end{array}\right]$$

Thus, $x = 0.5463$, $y = 0.3506$, $z = 0.1031$ *and therefore*

$$t = (0.5463 \;\; 0.3506 \;\; 0.1031)^T.$$

Hence, the percentages of voters that will vote for socialist is 54.63%, capitalist is 35.06% and an independent is 10.31%.

Note that if the given information is close to reality, then the conclusion will also be close to reality.

Age of trees in a forest. *Assume that trees in a forest are divided into four age-groups. Let* $F_B(k)$ *denote the number of baby trees in the forest with age – group 0–20 years at a given time period k. Similarly,* $F_Y(k)$, $F_M(k)$ *and* $F_O(k)$ *denote the number of young trees (21–40 years of age), middle-aged trees (41–60 years) and old trees (older than 60 years of age), respectively. The length of one time period is 20 years.*

This simple model needs the following assumptions:

(1) In each group, certain percentage of trees die.

(2) Surviving trees are placed in the next age group and old trees remain old.

(3) Trees lost are replaced by baby trees.

Note that total tree population remains constant over a given period of time. This problem is expressed as a system of difference equations as follows: Let the loss rates in percentage in each age group be L_B, L_Y, L_M, L_O, *each being a nonnegative quantity less than 1, then*

$$\begin{aligned} F_B(k+1) &= L_B F_B(k) + L_Y F_Y(k) + L_M F_M(k) + L_O F_O(k) \\ F_Y(k+1) &= (1 - L_B) F_B(k) \\ F_M(k+1) &= (1 - L_Y) F_Y(k) \\ F_O(k+1) &= (1 - L_M) F_M(k) + (1 - L_O)\, F_O(k). \end{aligned}$$

Age distribution vector **t** *is given by*

$$\mathbf{t}(k) = (F_B(k) \;\; F_Y(k) \;\; F_M(k) \;\; F_O(k))^T.$$

Transition matrix **A** *from the above system of difference equations is given by*

$$\mathbf{A} = \begin{bmatrix} L_B & L_Y & L_M & L_O \\ 1-L_B & 0 & 0 & 0 \\ 0 & 1-L_Y & 0 & 0 \\ 0 & 0 & 1-L_M & 1-L_O \end{bmatrix}.$$

Then we have, $\mathbf{At(k)} = \mathbf{t}(k+1)$

Now, choose $L_B = 0.2,\ \ L_Y = 0.1,\ \ L_M = 0.4,\ L_O = 0.3$. *Then,*

$$\mathbf{A} = \begin{bmatrix} 0.2 & 0.1 & 0.4 & 0.3 \\ 0.8 & 0 & 0 & 0 \\ 0 & 0.9 & 0 & 0 \\ 0 & 0 & 0.6 & 0.7 \end{bmatrix}.$$

Then, the fixed probability vector **t** *is given by*

$$\mathbf{t} = \frac{1}{3.96}(1\ \ 0.8\ \ 0.72\ \ 1.44)^T.$$

Suppose that a forest is newly planted with a total tree population of 70,000 trees. Initial age distribution of the forest $\mathbf{f}(0)$ *is given by*

$$\mathbf{f}(0) = (70000\ \ 0\ \ 0\ \ 0)^T$$

After 20 years, $\mathbf{f}(1) = \mathbf{At}(0) =$ *70,000* $(0.2\ \ 0.8\ \ 0\ \ 0]^T$ =$(14000\ \ 56000\ \ 0\ \ 0)^T$

After 40 years, $\mathbf{f}(2) = \mathbf{At}(1) =$ *70,000* $(0.12\ \ 0.16\ \ 0.72\ \ 0)^T = (8400\ \ 11200\ \ 50400\ \ 0)^T$

After 60 years, $\mathbf{f}(3) = \mathbf{At}(2) =$ *70,000* $(0.328\ \ 0.096\ \ 0.144\ \ 0.432)^T = (22960\ \ 6720\ \ 10080\ \ 30240)^T$

Clearly for nth *period, the number of trees in a forest is given by*

$$\mathbf{f}(n) = \mathbf{Af}(n-1) = \mathbf{A^2f}(n-2) = ... = \mathbf{A^nf}(0).$$

EXERCISE 5.3

1. Let the transition matrix be $\mathbf{T}= \begin{bmatrix} 0.55 & 0.4 \\ 0.45 & 0.6 \end{bmatrix}$

   Calculate $\mathbf{p}_n$ for n= 1,2,3,4 given $\mathbf{p}_0 = [1\ \ 0]^T$.

2. Given the transition matrix $\mathbf{T} = \begin{bmatrix} 0.6 & 0.5 & 0.7 \\ 0.2 & 0.4 & 0.1 \\ 0.2 & 0.1 & 0.2 \end{bmatrix}$ and the initial vector $\mathbf{p}_0 = \begin{bmatrix} 0 \\ 1 \\ 0 \end{bmatrix}$. Find $\mathbf{p}_1$, $\mathbf{p}_2$ and $\mathbf{p}_3$.

3. Check whether the following transition matrices are regular and find the steady state vector **t**.

   (a) $\begin{bmatrix} 1/2 & 1/4 \\ 1/2 & 3/4 \end{bmatrix}$ (b) $\begin{bmatrix} 0.78 & 0.67 \\ 0.12 & 0.33 \end{bmatrix}$

4. A person either works from home or goes to office. If he works from home one day, he works from home the following day, 1 time out of three. If he goes to office one day, he goes to office the next day 4 times out of five. Over the long run what are the chances that a person goes to office.

5. A city administration wanted to plan its services for next few years. So they have divided the city onto 3 parts: old city, residential area and suburbs. It is found that 15% of old city people move to suburbs and 10% move to residential area. 5% of people in residential area move to old city and 10% move to suburbs. 5% of residents of suburbs move to old city and 5% of residents move to residential area after a long time what percentage of people live in each region.

5.4 Leontief's Models

In this section, two economic models are discussed. The Nobel-Laureate Wassily Leontief proposed two different mathematical models describing interrelations between the constituents of an economic system. These mathematical models are linear in nature and hence the theory of matrices can be utilized to analyze them. The mathematical models are: (1) the closed or input-output model and (2) the open or production model.

In a comprehensive economic system, all industries are related to each other. A mathematical model is used for studying the interrelations between industries through some defined parameters. Using this information certain other parameters must be obtained such that they satisfy a given economic objective.

For example, suppose that for its production of goods each industry buys products of other industries as input. In some cases, buying can be only from primary producing sector. Thus industries, in general, are interdependent and study of this interdependence of industries in any economy is described in the following models.

Closed Model.

A closed or an input-output model deals with an economic system consisting of 'n' industries having an 'output' that is completely utilized in a predetermined manner by all the 'n' industries themselves. The objective of the closed or the input-output model is to find prices for the outputs so as to balance the income and expenditure of each industry during a given time period.

Let $x_i, i = 1, 2, 3, \ldots, n$ be the price of the total output of the ith industry and $a_{ij}, i, j = 1, 2, \ldots, n$ be the fraction of the total output of the jth industry purchased by the ith industry.

Then clearly (i) $x_i \geqslant 0$, (ii) $a_{ij} \geqslant 0$ and (iii) $\sum_{i=1}^{n} a_{ij} = 1,\ j = 1, 2, ..., n$. Let $\mathbf{x} = (x_1 \ \ x_2 \ \ ... \ \ x_n)$ be the cost vector and the input-output matrix be $\mathbf{A} = (a_{ij})_{n \times n}$.

According to the model, since expenditure of each industry must be equal to its income,

$$x_1 = a_{11}x_1 + ... + a_{1n}x_n$$
$$\cdot$$
$$\cdot$$
$$\cdot$$
$$x_n = a_{n1}x_1 + ... + a_{nn}x_n$$
or
$$\mathbf{x} = \mathbf{A}\mathbf{x} \ or \ (\mathbf{I} - \mathbf{A})\mathbf{x} = 0.$$

This is a linear system of homogeneous equations and hence has a nontrivial solution, if $(\mathbf{I} - \mathbf{A})$ is a singular matrix.

The following example illustrates the model.

Example 5.4.1 *Three factories X, Y, Z manufacture products and whatever they produce is consumed among themselves. Let X consume 20 % of what it produces and Y 60% of X's production and remaining is consumed by Z. The production of Y is shared by X, Y, Z in the ratio* 3 : 4 : 3. *Also X consumes 10% of Z's production, Y consumes 20% of Z's production and Z the remaining 70%. Find the ratios of incomes of X, Y, and Z.*

Solution.

Assume that the total production is unity and let x, y, and z be the incomes of X, Y and Z respectively. Let x be the amount obtained from the sale of X's product. Since X spends $\frac{x}{5}$ of his product and spends $\frac{3y}{10}$ on Y's product and $\frac{z}{10}$ on Z's product, the total expenditure of X is $\frac{x}{5} + \frac{3y}{10} + \frac{z}{10}$. The system being closed, total expenditure equals total income. Using the same reasoning for the incomes of Y and Z, yields,

$$x = \frac{x}{5} + \frac{3y}{10} + \frac{z}{10}$$
$$y = \frac{3x}{5} + \frac{2y}{5} + \frac{z}{5}$$
$$z = \frac{x}{5} + \frac{3y}{10} + \frac{7z}{10}.$$

The above mathematical model can be expressed as a linear system,

$$\begin{bmatrix} \frac{4}{5} & -\frac{3}{10} & -\frac{1}{10} \\ -\frac{3}{5} & \frac{3}{5} & -\frac{1}{5} \\ -\frac{1}{5} & -\frac{3}{10} & \frac{3}{10} \end{bmatrix} \begin{bmatrix} x \\ y \\ z \end{bmatrix} = \begin{bmatrix} 0 \\ 0 \\ 0 \end{bmatrix}.$$

Performing a sequence of elementary row operations: (i) $R_1 \to 10R_1$, *(ii)* $R_2 \to 5R_2$, *(iii)* $R_3 \to 10R_3$ *(iv)* $R_2 \to R_2 - R_1$, *(v)* $R_3 \to R_3 + 3R_1$ *and (vi)* $R_3 \to R_3 + 2R_2$ *to get*

$$\begin{bmatrix} 8 & -3 & -1 \\ -11 & 6 & 0 \\ 0 & 0 & 0 \end{bmatrix} \begin{bmatrix} x \\ y \\ z \end{bmatrix} = \begin{bmatrix} 0 \\ 0 \\ 0 \end{bmatrix}$$

This gives, $8x - 3y - z = 0, \quad -11x + 6y = 0$ *and*

$$\frac{x}{y} = \frac{6}{11}, \quad \frac{y}{z} = \frac{11}{15} \text{ or } x : y : z = 6 : 11 : 15,$$

which are the required ratio of incomes of X, Y and Z, respectively.

Open Model

In an open or a production model, the industries satisfy demand from the consumers in addition to the requirement among themselves. Thus in this model, the prices are fixed and the output to be produced by each industry to meet a given demand is to be determined.

Making an assumption that the output is measured in terms of the price of the product, the model is developed as follows.

Let x_i be financial value of the total output of the ith industry and d_i be the financial value of the output of the ith industry to satisfy the consumer demand. Further, let a_{ij} be the financial value of the output of the ith industry needed by the jth industry to produce one unit of financial value of its own output.

This gives rise to the production model.

$$\begin{aligned} x_1 &= a_{11}x_1 + a_{12}x_2 + \dots + a_{1n}x_n + d_1 \\ x_2 &= a_{21}x_1 + a_{22}x_2 + \dots + a_{2n}x_n + d_2 \\ &\vdots \\ x_n &= a_{n1}x_1 + a_{n2}x_2 + \dots + a_{nn}x_n + d_n. \end{aligned} \tag{5.1}$$

In matrix notation, $\mathbf{x}=\mathbf{A}\mathbf{x} + \mathbf{d}$ *or* $(\mathbf{I} - \mathbf{A})\mathbf{x} = \mathbf{d}$, where

$$\mathbf{A} = \begin{bmatrix} a_{11} & a_{12} & \cdots & a_{1n} \\ a_{21} & a_{22} & \cdots & a_{2n} \\ \vdots & \vdots & \vdots & \vdots \\ a_{n1} & a_{n2} & \cdots & a_{nn} \end{bmatrix},$$

$\mathbf{d} = (d_1 \ d_2 \ ... \ d_n)^T$ and $\mathbf{x} = (x_1 \ x_2 \ ... \ x_n)^T$. The matrix $\mathbf{A}$ is called as a transaction matrix.

To have a realistic model it is assumed that $\mathbf{x} \geq 0$, $\mathbf{d} \geq 0$ *and* $\mathbf{A} \geq 0$ and that $(\mathbf{I} - \mathbf{A})$ is invertible, then the output vector $\mathbf{x}$ is given by

$$\mathbf{x} = (\mathbf{I} - \mathbf{A})^{-1}\mathbf{d}.$$

The following example illustrates the model.

Example 5.4.2 *An economy produces products A and B. The two commodities serve as intermediate input in each other's production. 0.4 units of product B and 0.7 units of product A are needed to produce one unit of product B. 0.1 unit of product B and 0.6 units of product A are required to produce one unit of product A. No capital inputs are required. Verify whether the system is viable. 5 and 4 labor days are required to produce one unit of product A and B respectively. If the economy needs 100 units of product A and 50 units of product B, calculate the gross output of the two commodities and the total labor required.*

Consumer Producer	Product A	Product B	Final Demand
Product A(units)	0.6	0.7	100
Product B(units)	0.1	0.4	50

$$\textit{Given, } \mathbf{A} = \begin{bmatrix} 0.6 & 0.7 \\ 0.1 & 0.4 \end{bmatrix}, \ \mathbf{d} = \begin{bmatrix} 100 \\ 50 \end{bmatrix}, \ \mathbf{x} = \begin{bmatrix} x1 \\ x2 \end{bmatrix}.$$

Under the product B, $0.7 + 0.4 = 1.1 > 1$ *and therefore the economy is not viable. The balance equation is*

$$\mathbf{x} = \mathbf{A}\mathbf{x} + \mathbf{d} \ \ \textit{i.e.} \ \ (\mathbf{I} - \mathbf{A})\mathbf{x} = \mathbf{d} \ \ \textit{or}$$

$$\mathbf{x} = (\mathbf{I} - \mathbf{A})^{-1}\mathbf{d}.$$

$$|\mathbf{I} - \mathbf{A}| = \begin{vmatrix} 0.4 & -0.7 \\ -0.1 & 0.6 \end{vmatrix} = 0.17 \neq 0 \ \ \textit{and}$$

$$(\mathbf{I} - \mathbf{A})^{-1} = \frac{1}{0.17} \begin{bmatrix} 0.6 & 0.7 \\ 0.1 & 0.4 \end{bmatrix}. \ \ \textit{Hence}$$

$$\mathbf{x} = \frac{1}{0.17} \begin{bmatrix} 0.6 & 0.7 \\ 0.1 & 0.4 \end{bmatrix} \begin{bmatrix} 100 \\ 50 \end{bmatrix} = \frac{1}{0.17} \begin{bmatrix} 95 \\ 30 \end{bmatrix} = \begin{bmatrix} 558.8 \\ 176.5 \end{bmatrix}.$$

The gross production required is 558.8 units of product A and 176.5 units of product B.

$$\textit{Total labor required} = [5 \, 4] \begin{bmatrix} 558.8 \\ 176.5 \end{bmatrix} = 3500 \ \textit{man days}$$

Example 5.4.3 *The interrelationship between the production of three industries A, B, C in a year is given below (in crores of rupees).*

Users / Producers	A	B	C	Final demand	Output
A	40	50	50	40	160
B	40	100	30	60	200
C	40	50	50	20	150

Determine the output if the final demand changes are 60 for A, 40 for B and 60 for C.

Solution. *Input coefficients of A, B, C are given by*

A	B	C
40/160	50/200	50/150
40/160	100/200	30/150
40/160	50/200	50/150

The input-coefficient matrix $\mathbf{A}$ *is given by*

$$\mathbf{A} = \begin{bmatrix} \frac{1}{4} & \frac{1}{4} & \frac{1}{3} \\ \frac{1}{4} & \frac{1}{2} & \frac{1}{5} \\ \frac{1}{4} & \frac{1}{4} & \frac{1}{3} \end{bmatrix}$$

Further, $\mathbf{x}^T = [x\ y\ z]$, *and* $\mathbf{d}^T = [60\ 40\ 60]$

The Input-Output balance equation is $\mathbf{x} = \mathbf{A}\mathbf{x} + \mathbf{d}$ *or* $(\mathbf{I} - \mathbf{A})\mathbf{x} = \mathbf{d}$ *i.e.* $\mathbf{x}=(\mathbf{I} - \mathbf{A})^{-1}\mathbf{d}$. *Hence*

$$\mathbf{I} - \mathbf{A} = \begin{bmatrix} \frac{3}{4} & -\frac{1}{4} & -\frac{1}{3} \\ -\frac{1}{4} & \frac{1}{2} & -\frac{1}{5} \\ -\frac{1}{4} & -\frac{1}{4} & \frac{2}{3} \end{bmatrix} \quad \text{and } |\mathbf{I} - \mathbf{A}| = \frac{23}{240} \neq 0.$$

It follows that

$$\mathbf{x}=(\mathbf{I}-\mathbf{A})^{-1}\mathbf{d} = \begin{bmatrix} \frac{68}{23} & \frac{60}{23} & \frac{52}{23} \\ \frac{52}{23} & \frac{100}{23} & \frac{56}{23} \\ \frac{45}{23} & \frac{60}{23} & \frac{75}{23} \end{bmatrix} \begin{bmatrix} 60 \\ 40 \\ 60 \end{bmatrix}$$

$$= \begin{bmatrix} 417.39 \\ 455.65 \\ 401.74 \end{bmatrix}$$

Thus the final output for A, B and C are $x = 417.39$, $y = 455.65$, $z = 401.74$.

EXERCISE 5.4

1. A medical expert advised a patient to consume 75 g protein, 90 g fats and 100 g of carbohydrates. The patient consumes three types of food items A, B, C which when mixed will provide him desired food values. The table below presents the food analysis of A, B, C.

Type of food	Food value 100 g		
	Proteins	Fats	Carbohydrates
A	8	5	5
B	4	20	2.5
C	25	5	40

   What quantities of items A, B, C patient should consume so that he will receive needed food values?

2. One unit of commodity A is produced by combining 5 units of land, 3 units of labor and 1 unit of capital. One unit of B is produced by 2 units of land, 1 unit of labor and 3 units of capital. One unit of C is produced by 1 unit of land, 2 units of labor and 4 units of capital. If the cost price of A, B, C are Rs.16, Rs. 19 and Rs. 25, respectively, find the rent R, wages W and rate of interest I.

3. An amount of Rs. 5000/- is put into three investments at rates of 6, 7 and 8 percent per annum, respectively. the total annual income is Rs.358/- The combined income from the first two investments is Rs.70/-

more than income from the third investment. Find the amount of each investment.

4. A Salesman has the following record of sales during three months for three items X, Y and Z which have different rates of commission.

Months	Sale of units			Total commission (in rupees)
	X	Y	Z	
October	60	100	30	850
November	130	50	40	900
December	90	100	20	800

Find the rates of commission on items X, Y, Z.

5. A farmer feeds cattle on a matrix of three types of food namely A, B and C. Type A contains 10% proteins, 10% of the calories and 5% of carbohydrates which needs each day. Similarly, type B contains 5% proteins, 10% of calories but no carbohydrates and type C has 5% proteins, 5% of the calories and 10% of carbohydrates. How many units of each type of food should the farmer give a steer each day so that it gives 100% of the amount of proteins, calories and carbohydrates it requires?

5.5 Cryptology

Cryptography which is both an art and a science deals with techniques for encoding and decoding messages which are exchanged between two parties – one of whom is the sender and the other is the receiver. The message created by the sender is called 'plain text'. The sender encodes the plain text ('encrypted text') using one or more 'keys' and sends the encrypted text to the recipient. The recipient uses one or more keys to decrypt the text ('decrypted text') so as to obtain the original message, i.e. plain text.

In symmetric cryptography, both the sender and the recipient use the same key for encoding and decoding purposes. This obviously requires the sender to share the key with the recipient. A third party can obtain the plain text by intercepting both the encrypted text and the shared key transmitted by the sender.

In asymmetric cryptography, both sender and receiver employ different keys for encoding and decoding purposes. An example of asymmetric cryptography is the RSA [Rivest–Shamir–Adleman] algorithm where two sets of keys are used. One key is the public key known and shared by both sender and receiver. The third party can access the public key which is commonly available.

There are two other keys – a private key used by the sender and known only to him/her, and a second private key used by the receiver and known only by him/her. If a third party is able to intercept the encrypted text, they will need to compute the private keys in order to decode the encrypted text. The private keys are two very large prime numbers. Computation of these values is extremely time intensive.

Irrespective of which cryptographic techniques are employed, in order to foil hackers all of these techniques must meet two basic requirements which are as follows: (a) the time spent by a hacker in trying to find the key(s) far exceeds the time during which the plain text is seen as useful, i.e. the message becomes useless after a certain time period, and (b) the costs involved in trying to decrypt a message far exceeds the value of the plain text.

In this section, we focus on two symmetric cryptographic techniques which employ matrices for encryption/decryption purposes. In the first technique, the plain text is encoded and multiplied by any nonsingular matrix of appropriate dimensions to generate the encrypted text. The encrypted text is then multiplied by the inverse of the nonsingular matrix to yield the original text. The second technique, called the Hill method, is also like the first one but here modular arithmetic and modular matrices are involved. The following two examples show how the two techniques are employed.

Assume that the plain text is created from an alphabet drawn from letters A through S (a total of 19 letters) where the letters are encoded by the numerals 0 through 18 as shown below:

A	B	C	D	E	F	G	H	I	J	K	L	M	N	O	P	Q	R	S
0	1	2	3	4	5	6	7	8	9	10	11	12	13	14	15	16	17	18

Example 5.5.1 *Using matrix inverse.*

Consider a message line 'CLEAR PLAN' to be sent. Then from association between alphabet and numbers, 'CLEAR PLAN' is equivalent to '2 11 4 0 17 15 11 0 13'. Writing the above nine numbers as three vectors in serial order gives

$$\begin{bmatrix} 2 \\ 11 \\ 4 \end{bmatrix} \quad \begin{bmatrix} 0 \\ 17 \\ 15 \end{bmatrix} \quad \begin{bmatrix} 11 \\ 0 \\ 13 \end{bmatrix}.$$

Next consider any arbitrary nonsingular matrix $\mathbf{A} = \begin{bmatrix} 1 & 2 & 3 \\ 1 & 0 & 2 \\ 0 & 1 & 2 \end{bmatrix}$ $(det(\mathbf{A}) \neq 0)$

and send the message as 3 vectors $\mathbf{x}, \mathbf{y}, \mathbf{z}$ given by

$$\mathbf{x} = \mathbf{A} \begin{bmatrix} 2 \\ 11 \\ 4 \end{bmatrix}$$

$$\mathbf{y} = \mathbf{A} \begin{bmatrix} 0 \\ 17 \\ 15 \end{bmatrix}$$

$$\mathbf{z} = \mathbf{A} \begin{bmatrix} 11 \\ 0 \\ 13 \end{bmatrix}$$

The receiver on obtaining the vectors $\mathbf{x}, \mathbf{y}, \mathbf{z}$ multiplies them with

$$\mathbf{A}^{-1} = \begin{bmatrix} \frac{2}{3} & \frac{1}{3} & -\frac{4}{3} \\ \frac{2}{3} & -\frac{2}{3} & -\frac{1}{3} \\ -\frac{1}{3} & \frac{1}{3} & \frac{2}{3} \end{bmatrix}$$

to get the original vectors and hence retrieves the message.

This method can be easily hacked and hence there is a better technique using modular inverse of a matrix from Section 1.9 which is as follows.

Example 5.5.2 *Using modular inverse of a matrix.*

Suppose a plain text message 'EI' must be sent. The letters reduce to the vector $x = (4\ \ 8)^T$. *Now consider the matrix* $M = \begin{bmatrix} 8 & 1 \\ 13 & 3 \end{bmatrix}$ *in Example 1.9.4.*

Then $Mx = \begin{bmatrix} 8 & 1 \\ 13 & 3 \end{bmatrix} \begin{bmatrix} 4 \\ 8 \end{bmatrix} = \begin{bmatrix} 40 \\ 76 \end{bmatrix} = \begin{bmatrix} 2 \\ 0 \end{bmatrix}$, *which is obtained using* $a \bmod b \equiv 19$. *The vector* $\begin{bmatrix} 2 \\ 0 \end{bmatrix}$ *in alphabet reduces to* $\begin{bmatrix} C \\ A \end{bmatrix}$ *and this message is sent to the receiver. The receiver picks the message CA transforms it to vector* $\mathbf{z} = \begin{bmatrix} 2 \\ 0 \end{bmatrix}$ *uses the modular inverse of M (from Example 1.9.4),* $\mathbf{M}^{-1} = \begin{bmatrix} 2 & 12 \\ 4 & 18 \end{bmatrix}$ *and finds* $\mathbf{M}^{-1}\mathbf{z} = \begin{bmatrix} 2 & 12 \\ 4 & 18 \end{bmatrix} \begin{bmatrix} 2 \\ 0 \end{bmatrix} = \begin{bmatrix} 4 \\ 8 \end{bmatrix}$ *and gets the original message.*

Observation. A 2×2 matrix is required to send an information containing 2 letters. Thus a 3×3 matrix is essential to send information containing 3 letters. So sending big messages becomes difficult if one were to use large matrices, but the messages can be sent in segments of 3 using a 3×3 matrix.

EXERCISE 5.5

1. Consider $\mathbf{N} = \begin{bmatrix} 8 & -1 \\ 13 & 18 \end{bmatrix}$ (Problem 4, Exercise 1.9) send message 'OM' by encrypting using modular matrix $\mathbf{N}_c$ decrypt p=19 it using its modular inverse.

2. Send the message 'CLEAR PLAN' by using a 3×3 modular matrix. Verify the result using the modular inverse of the matrix.

5.6 Application to Computer Graphics

In this section, the basic concepts in computer graphics are discussed. Visualization is an important aspect of this area, as 3D images have to be projected onto a plane. These images are essentially a bunch of vectors. Moving an image from one point to another point, magnifying the image, scaling the image, rotating the image and projecting an image – all mean multiplying all the vectors in the image by a suitable matrix. All the mentioned objectives can be achieved by the following concepts involving matrices: (1) translation, (2) scaling, (3) rotation and (4) projection.

The following concept is necessary before describing the four matrices. While working with vectors in a plane, that is, 2D or with vectors in a space, that is, 3D, sometimes it becomes necessary to consider them as vectors, in a higher dimension. This leads to the following definition.

Definition 5.6.1 *Homogeneous coordinates of a 2D vector* $(a \quad b)^T$ *are* $(a \quad b \quad 1)^T$ *and of a 3D vector* $(a \quad b \quad c)^T$ *are* $(a \quad b \quad c \quad 1)^T$

1. Translation

As translation is not a linear transformation, to translate an image placed with reference to the origin $(0 \quad 0 \quad 0)^T$ to a new frame of reference $(a \quad b \quad c)^T$, the homogeneous coordinates must be used. Every vector $(x \quad y \quad z)^T$ must be considered in its homogeneous coordinates form and multiplied by the following matrix using homogeneous coordinates.

(i) Translation matrix in 3D:

$$\mathbf{T} = \begin{bmatrix} 1 & 0 & 0 & 0 \\ 0 & 1 & 0 & 0 \\ 0 & 0 & 1 & 0 \\ a & b & c & 1 \end{bmatrix}$$

and the vector in its homogeneous coordinates is $(x \quad y \quad z \quad 1)^T$

(ii) Translation matrix in 2D:

$$\mathbf{T} = \begin{bmatrix} 1 & 0 & 0 \\ 0 & 1 & 0 \\ a & b & 1 \end{bmatrix}$$

and the vector in its homogeneous coordinates is $(x \;\; y \;\; 1)^T$. The new position vector in 3D is $(x+a \;\; y+b \;\; z+c)^T$, and in homogeneous coordinates is $(x{+}a \;\; y{+}b \;\; z{+}c \;\; 1)^T$. Similarly, in 2D the new position vector is $(x{+}a \;\; y{+}b)^T$ and in homogeneous coordinates, $(x+a \;\; y+b \;\; 1)^T$.

Observation. To change the frame of reference from origin $(0 \;\; 0 \;\; 0)^T$ to $(a \;\; b \;\; c)^T$, the origin must be taken in homogeneous coordinates as $\mathbf{x} = (0 \;\; 0 \;\; 0 \;\; 1)^T$ then the product $\mathbf{xT} = (a \;\; b \;\; c \;\; 1)^T$ is the new origin in 3D. Similar is the case for 2D.

Example 5.6.1 *Translate a vector* $\mathbf{v} = (6 \;\; 3)^T$ *to a new point with reference to the vector* $\mathbf{v} = (2 \;\; 1)^T$.
Solution. *The translation matrix is*

$$\mathbf{T} = \begin{bmatrix} 1 & 0 & 0 \\ 0 & 1 & 0 \\ 2 & 1 & 1 \end{bmatrix}$$

and

$$\mathbf{vT} = \begin{bmatrix} 6 & 3 & 1 \end{bmatrix} \begin{bmatrix} 1 & 0 & 0 \\ 0 & 1 & 0 \\ 2 & 1 & 1 \end{bmatrix} = \begin{bmatrix} 8 & 4 & 1 \end{bmatrix}.$$

New position of the vector is $(8 \;\; 4)^T$.

2. Scaling

Scaling is an important notion used to magnify or diminish a picture. Depending on how much the picture must be reduced in x, y and z directions the constants k_1, k_2 and k_3 are chosen and the scaling matrix is denoted by **S**.

(i) Scaling matrix in 3D:

$$S = \begin{bmatrix} k_1 & 0 & 0 & 0 \\ 0 & k_2 & 0 & 0 \\ 0 & 0 & k_3 & 0 \\ 0 & 0 & 0 & 1 \end{bmatrix}.$$

(ii) Scaling matrix in 2D:

$$S = \begin{bmatrix} k_1 & 0 & 0 \\ 0 & k_2 & 0 \\ 0 & 0 & 1 \end{bmatrix}.$$

Example 5.6.2 *Scale the vector* $\mathbf{w} = (9 \quad 0.4 \quad 8)$ *by* 0.3 *units in x-direction, 2 units in y-direction and 0.2 units in z-direction.*
Solution. *The scaling matrix*

$$\mathbf{S} = \begin{bmatrix} 0.3 & 0 & 0 & 0 \\ 0 & 2 & 0 & 0 \\ 0 & 0 & 0.2 & 0 \\ 0 & 0 & 0 & 1 \end{bmatrix}.$$

The new position of vector $\mathbf{w}$ *in homogeneous coordinates is*

$$\mathbf{w}_S = \begin{bmatrix} 9 & 0.4 & 8 & 1 \end{bmatrix} \begin{bmatrix} 0.3 & 0 & 0 & 0 \\ 0 & 2 & 0 & 0 \\ 0 & 0 & 0.2 & 0 \\ 0 & 0 & 0 & 1 \end{bmatrix} = \begin{bmatrix} 2.7 & 0.8 & 1.6 & 1 \end{bmatrix}$$

3. Rotation

Rotating the picture or an image involves considering a rotation matrix. In 2D the rotation of a vector is about the origin. In order to turn a vector $\mathbf{w} = (a \quad b)$ through an angle α, it is necessary to consider a 3×3 matrix of the type,

$$\mathbf{R} = \begin{bmatrix} cos\alpha & sin\alpha & 0 \\ -sin\alpha & cos\alpha & 0 \\ 0 & 0 & 1 \end{bmatrix}.$$

Example 5.6.3 *Rotate the vector* $\mathbf{w} = (3 \quad 3)$ *through an angle* $\alpha = \pi/6$ *about another vector* $\mathbf{v} = (2 \quad 1)$. *What is the new position of* $\mathbf{w}$?
Solution. *In order to rotate the vector* $\mathbf{w} = (3 \quad 3)$ *through an angle* $\alpha = \pi/6$ *about another vector* $\mathbf{v} = (2 \quad 1)$ *the notion of translation along with rotation must be used. First the vector about which the rotation must be taken is translated to the origin by the translation matrix* $\mathbf{T}$ *in Example 5.6.1.*

Next the rotation matrix

$$\mathbf{R} = \begin{bmatrix} cos\pi/6 & sin\pi/6 & 0 \\ -sin\pi/6 & cos\pi/6 & 0 \\ 0 & 0 & 1 \end{bmatrix}.$$

is applied and the vector is transformed to the position vector $(2, 1)$ *by the translation matrix,*

$$\mathbf{T}_- = \begin{bmatrix} 1 & 0 & 0 \\ 0 & 1 & 0 \\ -2 & -1 & 1 \end{bmatrix}.$$

Now the vector $\mathbf{w} = (3 \quad 3)$ *rotated through an angle* $\alpha = \pi/6$ *about the vector* $(2 \quad 1)$ *is the vector given by the product,*

$$\mathbf{wT\_RT} = (1.866 \quad 3.232 \quad 1).$$

Hence the new position of $\mathbf{w} = (1.866 \;\; 3.232)$.

In 3D, the rotation of a vector is about an axis. For example, the rotation matrix about the y-axis in 3D is

$$T = \begin{bmatrix} cos\beta & 0 & sin\beta & 0 \\ -sin\beta & 0 & cos\beta & 0 \\ 0 & 1 & 0 & 0 \\ 0 & 0 & 0 & 1 \end{bmatrix}.$$

Example 5.6.4 *Suppose a camera is focused on the point* $\mathbf{w} = (5 \;\; 3)$ *of an image in the xy plane. It is moved by an angle* $\frac{\pi}{4}$ *in the* **xy** *plane. On what point will the camera focus on?*

Solution. *Taking* $\alpha = \pi/4$ *in the rotation matrix* $\mathbf{R}$, *the camera will focus on the point given by the product*

$$\begin{bmatrix} 5 & 3 & 1 \end{bmatrix} \begin{bmatrix} cos\frac{\pi}{4} & sin\frac{\pi}{4} & 0 \\ -sin\frac{\pi}{4} & cos\frac{\pi}{4} & 0 \\ 0 & 0 & 1 \end{bmatrix} = \begin{bmatrix} \frac{5}{\sqrt{2}} - \frac{3}{\sqrt{2}} & \frac{5}{\sqrt{2}} + \frac{3}{\sqrt{2}} & 1 \end{bmatrix}.$$

Thus the camera will focus on the point $(1.4 \;\; 5.6)$.

4. Projection

To project 3D vectors onto a plane there are two cases (i) planes passing through the origin and (ii) planes not passing through the origin.

Case (i) Let K be a plane passing through the origin and $\mathbf{n}$ be a unit vector normal to the plane.

Then the projection matrix onto the plane is given by $\mathbf{I} - \mathbf{n}\mathbf{n}^{\mathbf{T}}$ and in homogeneous coordinates the projection matrix is given by

$$\mathbf{P}_M = \begin{bmatrix} & & & 0 \\ \mathbf{I} - \mathbf{n}\mathbf{n}^{\mathbf{T}} & & & 0 \\ & & & 0 \\ 0 & & 0 \;\; 0 & 1 \end{bmatrix}$$

Case (ii) Let K be a plane not passing through the origin.

Then the projection onto the affine plane (or flat as it is called) is done in 3 steps.

Step 1. Consider any point on the plane represented by the vector $\mathbf{v}_0$ and translate $\mathbf{v}_0$ to the origin by the matrix $\mathbf{T}_-$.

Step 2. Apply projection matrix along the direction of $\mathbf{n}$.

Step 3. Translate back along the vector $\mathbf{v}_0$.

In this case, the projection matrix is

$$\mathbf{T\_P}_M\mathbf{T} = \begin{bmatrix} \mathbf{I} & 0 \\ -\mathbf{v}_0 & 1 \end{bmatrix} \begin{bmatrix} \mathbf{I}-\mathbf{nn}^\mathbf{T} & 0 \\ 0 & 1 \end{bmatrix} \begin{bmatrix} \mathbf{I} & 0 \\ \mathbf{v}_0 & 1 \end{bmatrix}$$

$$= \begin{bmatrix} \mathbf{T}_{3\times3} & & & 0 \\ & & & 0 \\ & & & 0 \\ -v_{10} & -v_{20} & -v_{30} & 1 \end{bmatrix} \begin{bmatrix} & & & 0 \\ \mathbf{I}-\mathbf{nn}^\mathbf{T} & & & 0 \\ & & & 0 \\ 0 & 0 & 0 & 1 \end{bmatrix} \begin{bmatrix} T_{3\times3} & & & 0 \\ & & & 0 \\ & & & 0 \\ v_{10} & v_{20} & v_{30} & 1 \end{bmatrix}.$$

Observation.

1. The projection matrix $\mathbf{P}_M\mathbf{n}= (\mathbf{I}-\mathbf{nn}^\mathbf{T})\mathbf{n} = 0$.
2. The projection matrix $\mathbf{P}_M\mathbf{v} = \mathbf{v}$, $\mathbf{v}$ is any vector on the plane K.
3. A reflection moves each point twice as far as projection.

The reflection goes through the plane on to the other side. So for a mirror matrix the projection is $\mathbf{I} - 2\mathbf{nn}^T$.

EXERCISE 5.6

1. Consider a quadrilateral with coordinates $A(2\ \ 3)$, $B(3\ \ 3)$, $C(2\ \ 0)$ and $D(0\ \ 0)$. Apply the translation with a distance of 1 unit from the origin on the x-axis and 1 unit from the origin on the y-axis. Obtain the new coordinates of the quadrilateral.
2. Consider the coordinates $A(1\ \ 4\ \ 2)$, $B(3\ \ 3\ \ 1)$ and $C(5\ \ 2\ \ 3)$ of a 3-D object. Apply the translation with a distance of 1 unit from the origin on the x-axis, 1 unit from the origin on the y-axis and 2 units from the origin on the z-axis. Obtain the new coordinates of the object.
3. A picture represented by a matrix $\mathbf{A} = (a_{ij})_{3\times3}$ must be reflected onto the x-axis and then projected onto the y-axis. Find the reflection and projection matrices and the matrix obtained after the operations are performed.
4. Rotate the vectors $(2\ \ 1\ \ 2)$, $(1\ \ 3\ \ 4)$ and $(5\ \ 4\ \ 1)$ by an angle $\pi/4$ in the yz-plane.
5. Project the vectors $(x_1\ \ y_1\ \ z_1)$, $(x_2\ \ y_2\ \ z_2)$ and $(x_3\ \ y_3\ \ z_3)$ onto the xy-plane by using a suitable projection matrix.

5.7 Application to Robotics

Robotics is a discipline that includes design and development of robots. The robots are expected to replicate human working patterns, especially in areas that are dangerous, difficult, dirty and dull for humans.

A scientific definition of a robot is as follows.

Definition 5.7.1 *A robot is a programmed mechanical device that is software controllable and uses sensors to guide one or more end-effectors to manipulate physical objects in a workspace.*

There are many types of robots. In this section, an industrial robot is considered. This is also called as a robotic manipulator or a robotic arm. It is modeled by a chain of rigid links interconnected by flexible joints. There is an end effector, called a tool, gripper or a hand, connected to a robotic arm which is used to handle physical objects. It is roughly similar to a human arm. The tool of a robotic manipulator is mounted on a plate and secured to the wrist of a robot.

The joints of a robot are of six types. Of these following two are widely used:

(i) Revolute joint (R), which allows a rotary motion about an axis;

(ii) Prismatic joint (P), which has a sliding or a linear motion along an axis.

A robot needs a frame of reference to work with. The axes of the first three joints of a robot, which dictate the movement of the wrist, are considered as the major axes and these determine the coordinate system adopted by the robot.

Types of Robots

1. Cartesian coordinate robot consists of all prismatic joints as shown in Fig. 5.1. The 3 major axes correspond to 3 sliding joints allowing the movement of wrist up and down, in and out, and back and forth.

2. Cylindrical coordinate robot is obtained by replacing the 1st prismatic joint in Cartesian coordinate robot with a revolute joint. The other joints are left untouched. The first revolute joint swings the arm back and forth about a vertical axis. The two prismatic joints then move the wrist up and down along the vertical axis and in and out along the radial axis as can be seen from, Fig. 5.2.

3. Spherical coordinate robot is the result of replacing both the 1st and 2nd

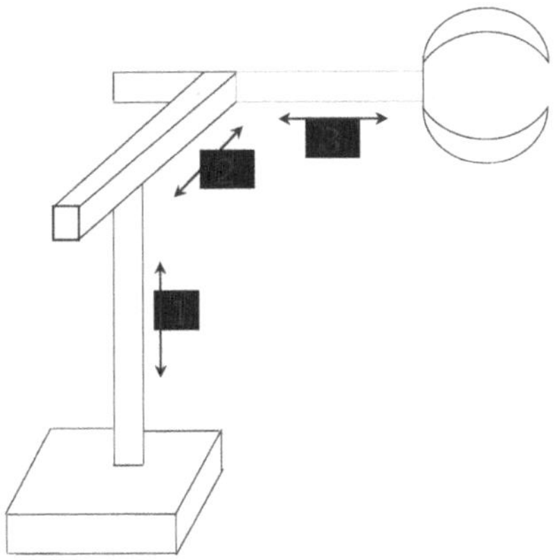

Fig. 5.1 Cartesian robotic arm

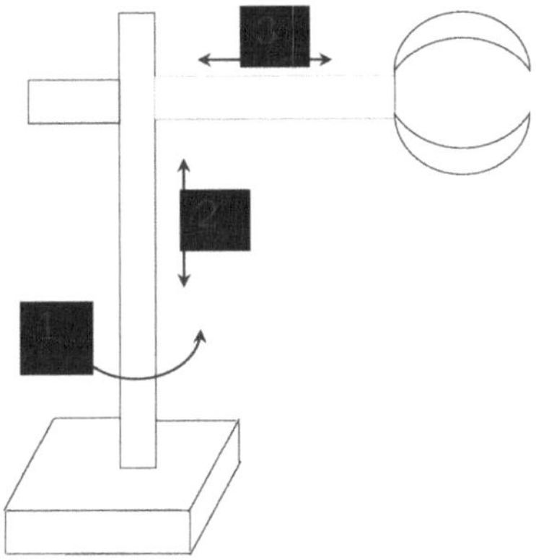

Fig. 5.2 Cylindrical robotic arm

prismatic joints in Cartesian coordinate robot with revolute joints. The arm of this robot swings back and forth about a vertical base axis due to the first revolute joint. By the second revolute joint the arm moves up and down about a horizontal shoulder axis and the wrist moves in and out due to the prismatic joint, Fig. 5.3.

The other important part of the robot is the tool. The tool is fixed to the wrist and has its own movements. These movements are of many types and they specify the tool orientation. The tool orientation called yaw-pitch-roll (YPR) system adopted by the aeronautics industry is described here.

The tool orientation consists of two reference frames

(i) wrist coordinate frame $B_1 = [\mathbf{v_1}, \mathbf{v_2}, \mathbf{v_3}]$ and

(ii) tool coordinate frame $B_2 = [\mathbf{w_1}, \mathbf{w_2}, \mathbf{w_3}]$.

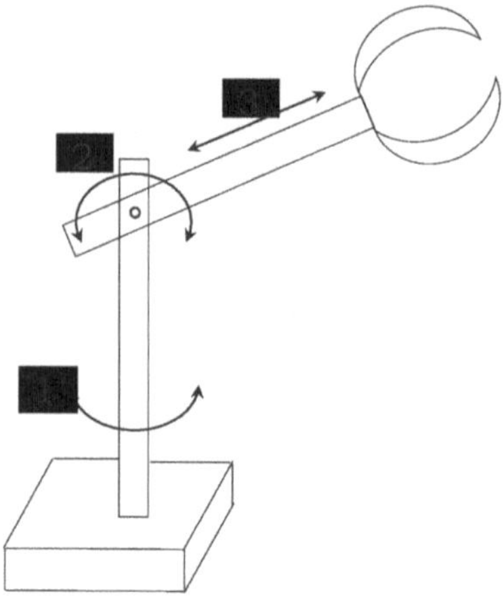

Fig. 5.3 Spherical or polar robotic arm

The wrist frame is attached to the end of the forearm and the tool frame is attached to the tool and moves with it. At the outset both the coordinate frames B_1 and B_2 are coincident. The yaw motion is performed first. This is done by rotating the tool about wrist axis $\mathbf{v_1}$. Then the pitch motion is done by rotating the tool about wrist axis $\mathbf{v_2}$ and finally, the roll motion is performed by rotating the tool about wrist axis $\mathbf{v_3}$. These operations can be specified with reference to the tool frame B_2 also.

With the frame of reference in place, consider the direct kinematics problem, which will help in understanding the movements of a robot and the application of matrices to this topic.

Definition 5.7.2 *Direct kinematics problem is to determine the position and orientation of the tool with respect to (w.r.t.) a coordinate frame attached to the robot base when the joint variables of a robotic manipulator are given.*

Consider a single-axis robot given in Fig. 5.4. The robot consists of a single revolute joint connecting a fixed link to a mobile link, whose frames of references are given by $B_1 = [\mathbf{v_1}, \mathbf{v_2}, \mathbf{v_3}]$ and $B_2 = [\mathbf{w_1}, \mathbf{w_2}, \mathbf{w_3}]$, respectively. Both B_1 and B_2 form orthonormal bases. Now consider any point P on the body and the coordinates of P w.r.t. the basis B_1 be given by $[P]_{B_1}=(a_1, a_2, a_3)$ and let the coordinate of P w.r.t. the basis B_2 be $[P]_{B_2} =(b_1, b_2, b_3)$.

Then the vectors in basis B_1 can be written as a linear combination of the vectors in B_2 as,

$$\mathbf{v_j} = \sum_{i=1}^{3} c_{ij}\mathbf{w_i}, \ j = 1, 2 \ldots 3. \tag{5.2}$$

Now using the change of basis matrix (or coordinate transformation matrix),

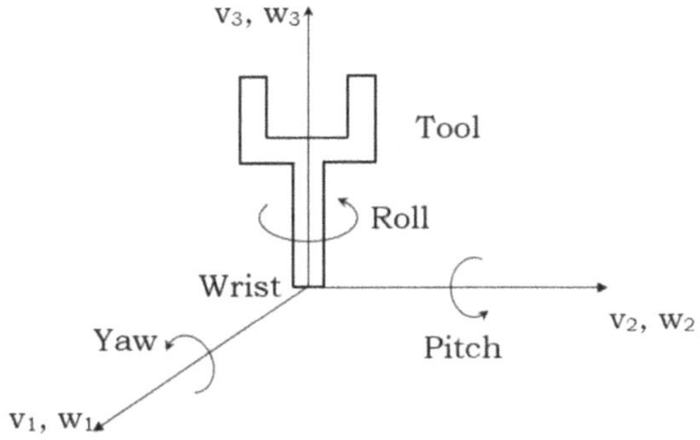

Fig. 5.4 Roll, pitch, yaw motions

that is, using the relation (3.5) gives

$$[P]_{B_1} = \boldsymbol{C}[P]_{B_2}, \text{ where } \boldsymbol{C} = [c_{ij}]_{3\times3}$$

Observation. If the coordinates of P w.r.t. B_2 are known and the coordinates of P w.r.t. B_1 are to be found

Then $[P]_{B_2} = \boldsymbol{C}^T\ [P]_{B_1}$

This follows from the fact that B_1 and B_2 are orthonormal bases and thus $\boldsymbol{C}^{-1} = \boldsymbol{C}^T$.

Representation of fundamental rotations in terms of matrices is essential in using the robot manipulator. These are described below.

Suppose the mobile frame (basis) B_1 is rotated about one of the unit vectors of the fixed frame(basis) B_2, then the change of basis matrix (or the coordinate transformation matrix) is called a fundamental rotation matrix.

The 3 fundamental rotation matrices that arise are as follows.

(i) Let the mobile frame B_1 move about the unit vector $\mathbf{v_1}$ of the fixed frame B_2 by an angle θ. Then the vectors $\mathbf{v_1}$ and $\mathbf{w_1}$ coincide and the vectors $\mathbf{v_2}$ and $\mathbf{w_2}$, $\mathbf{v_3}$ and $\mathbf{w_3}$ vary by θ as in Fig. 5.5

Then the resulting change of basis matrix (or the coordinate transformation matrix) is given by

$$\boldsymbol{M}_{1\theta} = \begin{bmatrix} v_1.w_1 & v_1.w_2 & v_1.w_3 \\ v_2.w_1 & v_2.w_2 & v_2.w_3 \\ v_3.w_1 & v_3.w_2 & v_3.w_3 \end{bmatrix}$$

$$= \begin{bmatrix} 1 & 0 & 0 \\ 0 & cos\theta & -sin\theta \\ 0 & sin\theta & cos\theta \end{bmatrix}$$

Similarly, the fundamental rotation matrix about $\mathbf{v_2}$ by an angle ϕ is given by

$$\boldsymbol{M}_{2\phi} = \begin{bmatrix} cos\phi & 0 & sin\phi \\ 0 & 1 & 0 \\ -sin\phi & 0 & cos\phi \end{bmatrix}$$

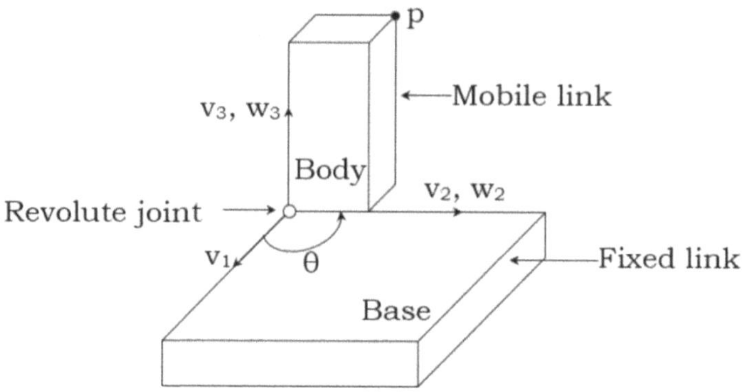

Fig. 5.5 Rotation of a body

The fundamental rotation matrix about $\mathbf{v_3}$ by an angle ψ yields

$$\text{and } \mathbf{M}_{3\psi} = \begin{bmatrix} cos\psi & -sin\psi & 0 \\ sin\psi & cos\psi & 0 \\ 0 & 0 & 1 \end{bmatrix}$$

These three fundamental rotations correspond to the yaw, pitch and roll of the tool and can be represented by matrices. Since matrix multiplication is not commutative, the order in which the rotations are done play an important role.

In order to avoid ambiguity the following algorithm is used.

Algorithm for Multiple Rotation

(1) Set the mobile frame basis and the fixed frame basis to coincide then the rotation matrix $\mathbf{M} = \mathrm{I}$.

(2) Suppose the mobile frame (B_2) is rotated by an angle θ about the $\mathbf{v_i}$, unit vector of the fixed frame(B_1), then premultiply $\mathbf{M}$ with $\mathbf{M_i}$ $(\boldsymbol{\theta})$.

(3) Suppose the mobile frame (B_2) is rotated by an angle θ about its own unit vector $(\mathbf{w_i})$ then post multiply $\mathbf{M}$ with $\mathbf{M_i}$ $(\boldsymbol{\theta})$.

(4) Repeat the process by going to step 2, if there are more rotations. Else the process is complete.

The end result is a matrix $\mathbf{M}$ that maps mobile frame (B_2) coordinates into fixed frame (B_1) coordinates.

The composite yaw, pitch, roll transformation matrix is as follows.

Suppose the mobile frame B_2 is rotated about the unit vector $\mathbf{v_1}$ with a yaw of θ, next about the unit vector $\mathbf{v_2}$ with a pitch of ϕ and then about the unit vector $\mathbf{v_3}$ with a roll of ψ. Then

$$\text{YPR}(\theta) = \begin{bmatrix} cos\psi & -sin\psi & 0 \\ sin\psi & cos\psi & 0 \\ 0 & 0 & 1 \end{bmatrix} \begin{bmatrix} cos\phi & 0 & sin\phi \\ 0 & 1 & 0 \\ -sin\phi & 0 & cos\phi \end{bmatrix} \begin{bmatrix} 1 & 0 & 0 \\ 0 & cos\theta & -sin\theta \\ 0 & sin\theta & cos\theta \end{bmatrix}$$

Homogeneous coordinates

The basic movements of a robot are rotation and translation. While rotation is with respect to the given origin, translation implies a change of origin. In order to accommodate both these transformations in one mathematical operation, just as in computer graphics, the homogeneous coordinates are introduced. A brief description is as below.

Let $\mathbf{p}$ be a vector (point) in $\mathbb{R}^3$ and $\mathcal{B}$ be an orthonormal coordinate frame (basis) in $\mathbb{R}^3$. Further, let γ be any nonzero scaling factor.

Definition. The homogeneous coordinates of $\mathbf{p}$ with respect to $\mathcal{B}$ are defined by $[\mathbf{p}]_B = \gamma\ (p_1, p_2, p_3, 1)^T$.

The original vector $\mathbf{p}$ can be obtained by

$$\mathbf{p} = \mathbf{H}_\gamma\ [\mathbf{p}]_\mathcal{B},$$

where $\mathbf{H}_\gamma$ is a 3×4 matrix called homogeneous coordinate conversion matrix and is given by $\mathbf{H}_\gamma = \frac{1}{\gamma}\ [\mathbf{I}, 0]$.

Observation.

1. The homogeneous coordinates of a point $\mathbf{p}$ are not unique because of the scale factor γ.

2. In robotics, normally, the scale factor γ=1.

The transformations of rotation and translation are performed by changing from one coordinate frame to another giving rise a transformation matrix. Thus a homogeneous transformation matrix $\mathbf{T}$ is a 4×4 matrix and can be partitioned into 4 separate submatrices as

$$\mathbf{T} = \begin{bmatrix} \mathbf{R} & : & \mathbf{p} \\ \cdots & \cdots & \cdots \\ \eta^T & : & \gamma \end{bmatrix},$$

where $\mathbf{R}$ is a 3×3 rotation matrix, $\mathbf{p}$ is a 3×1 perspective vector such that η =0. The situation $\eta \neq 0$ deals with situation where an overhead camera is used and is beyond the scope of this section.

Example 5.7.1 *A vector* $\mathbf{v} = 3i + 2j + 7k$ *is rotated by* $60°$ *about the z-axis of the reference frame. It is then rotated by* $60°$ *about the x-axis of the reference* *frame. Find the rotation transformation.*

Solution.

$$\mathbf{H}_1 = \mathbf{M}_3(60^\circ) = \begin{bmatrix} cos60^\circ & -sin60^\circ & 0 & 0 \\ sin60^\circ & cos60^\circ & 0 & 0 \\ 0 & 0 & 1 & 0 \\ 0 & 0 & 0 & 1 \end{bmatrix}$$

$$\mathbf{H}_2 = \mathbf{M}_1(60^\circ) = \begin{bmatrix} 1 & 0 & 0 & 0 \\ 0 & cos60^\circ & -sin60^\circ & 0 \\ 0 & sin60^\circ & cos60^\circ & 0 \\ 0 & 0 & 0 & 1 \end{bmatrix}$$

$$\mathbf{H}_2\mathbf{H}_1 = \begin{bmatrix} 1 & 0 & 0 & 0 \\ 0 & 0.5 & -0.866 & 0 \\ 0 & 0.866 & 0.5 & 0 \\ 0 & 0 & 0 & 1 \end{bmatrix} \begin{bmatrix} 0.5 & -0.866 & 0 & 0 \\ 0.866 & 0.5 & 0 & 0 \\ 0 & 0 & 1 & 0 \\ 0 & 0 & 0 & 1 \end{bmatrix}$$

$$= \begin{bmatrix} 0.5 & -0.866 & 0 & 0 \\ 0.433 & 0.25 & -0.866 & 0 \\ 0.75 & 0.433 & 0.5 & 0 \\ 0 & 0 & 0 & 1 \end{bmatrix},$$

thus,

$$\mathbf{v}' = \mathbf{H}_2\mathbf{H}_1\mathbf{v} = \begin{bmatrix} 0.5 & -0.866 & 0 & 0 \\ 0.433 & 0.25 & -0.866 & 0 \\ 0.75 & 0.433 & 0.5 & 0 \\ 0 & 0 & 0 & 1 \end{bmatrix} \begin{bmatrix} 3 \\ 2 \\ 7 \\ 1 \end{bmatrix} = \begin{bmatrix} -0.23 \\ -4.26 \\ 6.62 \\ 1 \end{bmatrix}$$

EXERCISE 5.7

1. Find the transformation matrices for the following operations on the vector $2\mathbf{i} + 5\mathbf{j} + 4\mathbf{k}$
   (i) Rotate 60° about x-axis and then translate 4 units along y-axis.
   (ii) Translate -5 units along y-axis and rotate 30° about y-axis.

2. Find the matrix in homogeneous coordinates that is required to move a robot from the $(1\ \ 1\ \ 3)$ position to the position $(4\ \ -5\ \ 2)$.

3. If $B_1 = \{(0\ \ 1\ \ 1), (1\ \ 0\ \ 1), (1\ \ 1\ \ 0)\}$ is the frame of reference of the rigid axis of the robot and $B_2 = \{(1\ 0\ 2),\ (2\ 3\ 4),\ (1\ 4\ 1)\}$ is the frame of reference of the robotic arms. Find the transformation matrix that is, change of basis matrix B_2 to B_1 and represent a vector $(4\ \ 4\ \ 1)$ in basis B_1 and B_2.

4. A point P $(2\ \ 3\ \ -4)^T$ is attached to a rotating frame. The frame rotates an angle 'θ' about the x-axis of the frame of reference. The

coordinates of the point relative to the frame of reference after the rotation is $P'(2 \quad -4 \quad 3)$. Find the rotation matrix.

5. Find the axis of rotation and angle of rotation about the same axis for the matrix $\mathbf{R} = \begin{bmatrix} \frac{\sqrt{3}}{2} & 0 & 0.5 \\ 0 & 1 & 0 \\ -0.5 & 0 & \frac{\sqrt{3}}{2} \end{bmatrix}$.

5.8 Bioinformatics

Bioinformatics deals with the sequencing of DNA molecules, protein molecules and genes, etc. The protein molecule and the DNA molecule consist of amino acids and nucleic acids respectively which are represented by the alphabet and are called characters.

To study the similarity and comparison between different protein sequences, gene sequences or DNA sequences is a vital component of bioinformatics. The investigation deals with finding the alignment between any two given sequences.

The given sequences can be considered as vectors (called strings) and the problem reduces to a mathematical one. Placing one string vertically and another string row wise immediately gives rise to a matrix.

To obtain a best possible alignment between the two sequences Needleman–Wunsch algorithm is used. The Needleman–Wunsch algorithm is a global alignment technique versus the Smith–Waterman algorithm which is a local alignment technique.

The following concepts are necessary to introduce the Needleman–Wunsch algorithm.

Definition 5.8.1 *Scoring matrix is a matrix consisting of a preassigned value (score) given to each pair of characters depending on their structure. The value can be positive or negative.*

Let n be the size of the alphabet. Then the matrix consists of n^2 entries. But due to symmetry the number of entries sufficient for our study is $\frac{n(n-1)}{2}$.

Example 5.8.1 *(a) BLOSUM 50 is the scoring matrix for proteins, whose values come from biology. The protein molecule consists of 21-letter alphabet and the matrix consists of 210 entries.*
(b) The DNA molecule consists of 4 letter alphabet and the number of entries in the scoring matrix is 6.

The sequences under study may have undergone some insertions and deletions. These are taken care of by introducing gaps which are defined below.

Definition 5.8.2 *Gaps are blank characters.*

Given two strings $\mathbf{x} = (x_1 x_2 \ldots x_n)$ and $\mathbf{y} = (y_1 y_2 \ldots y_m)$ blank characters called gaps are introduced into the strings.

These gaps can be individual blanks or a string of blanks.

Definition 5.8.3 *(i) Gap penalty (gp) is a numerical constant 'a' given for a single blank.*
(ii) Gap extension penalty is a numerical constant 'b' given for a string of blanks. If l is the length of the string

$$gp(l) = -a - (l-1)b,$$

for some constants a and b with $b < a$.

Definition 5.8.4 *Optimal match between two sequences occurs when the score is maximum. This match is achieved by adding blanks to the given sequences in a systematic fashion so that the resulting strings are aligned optimally.*

The procedure for optimal alignment known as Needleman–Wunsch algorithm is given below.

Needleman–Wunsch Algorithm

Given two strings $\mathbf{x} = (x_1, x_2, ..., x_n)$ and $\mathbf{y} = (y_1, y_2, ..., y_m)$, let $\mathbf{x}$ be placed in the row and $\mathbf{y}$ be placed in the column giving rise to an $m \times n$ relational matrix $\mathbf{R}$. This matrix gives scores between x_i and y_j, denoted by $r(x_i, y_j)$. These scores are preassigned values which have been determined from the relation that exists between the elements of the sequences. [See Example 5.8.1.]

Now the alignment matrix $\mathbf{A}_{(m+1)/(n+1)}$ is constructed as follows. A zero row and a zero column are added. Set $a_{00} = 0$, $a_{i0} = ia$, $i = 1, 2 \cdots m$, and $a_{j0} = ja$, $j = 1, 2 \cdots, n$. Assume that the information of $a_{i-1\,j}$, $a_{i\,j-1}$, and $a_{i-1\,j-1}$ is known. [I.e. in order to find the value of a_{ij} the values of up element, left element and upper left diagonal element must be given.]

Thus the (i, j)th element of alignment matrix is given by

$$a_{ij} = Max \begin{cases} a_{i-1\,j-1} + r(x_i, y_j), & ....\text{(i)} \\ a_{i-1\,j} - a, & ....\text{(ii)} \\ a_{i\,j-1} - a, & ....\text{(iii)} \end{cases}$$

In order to understand the computation, a pointer indicates the path followed in filling the alignment matrix. The choice taken to obtain the entry a_{ij} is denoted by a diagonal arrow if (i) is chosen, left to right arrow if (ii) is chosen and down arrow if (iii) is chosen. Note that one of the options will surely be taken. Starting at the top this procedure is adopted to reach the right-most corner a_{mn}.

Thus, the matrix **A** along with the arrows gives the path for optimal alignment. If any two of the paths yield the same value for a_{ij}, then one of the paths can be chosen. But to get optimal value, all paths must be considered.

Once the alignment matrix is completed the optimal alignment is given by two new strings **u** and **v** that are traced using the arrows and along with the following choice.

If $a_{ij} = (i)$, then x_i is aligned with y_j.
If $a_{ij} = (ii)$, then x_i is aligned with a new gap.
If $a_{ij} = (iii)$, then y_j is aligned with a new gap.

Proceeding in this fashion both sequences constructed are in alignment with one another. The following example gives a better understanding of the Needleman–Wunsch algorithm.

Example 5.8.2 *Using the scoring matrix, fig 5.6 given below find the optimal alignment between the strings TAPES and TAMPS.*

	A	E	M	P	S	T
A	4					
E	-1	5				
M	-1	-2	5			
P	0	-1	-2	7		
S	1	0	-1	-1	4	
T	0	-1	-1	-1	1	5

Fig. 5.6 Scoring matrix

Then the relational matrix is given by

$R = (x_i, y_j) =$

	T	A	M	P	S
T	5	0	−1	−1	1
A	0	4	−1	0	1
P	−1	0	−2	7	−1
E	−1	−1	−2	−1	0
S	1	1	−1	−1	4

Now the alignment matrix is obtained by using the formula

$$a_{ij} = Max \begin{cases} a_{i-1\,j-1} + r(x_i, y_j) & (i) \\ a_{i-1\,j} - a & (ii) \\ a_{i\,j-1} - a & (iii) \end{cases}$$

Thus the alignment matrix begins with first row and first column consisting of gap penalty as $-ai$ *and* $-aj$*, respectively.*

Then the first row and column are respectively

$$[0\ \ -3\ \ -6\ \ -9\ \ -12\ \ -15]\ and\ [0\ \ -3\ \ -6\ \ -9\ \ -12\ \ -15]^T$$

Then

$$a_{22} = Max\{\ 5,\ \ -6,\ \ -6\ \} = 5$$

Next

$$a_{23} = Max\{\ -3,\ \ -9,\ \ 2\ \} = 2$$

proceeding

$$a_{33} = Max\{\ 9,\ \ -1,\ \ -5\ \} = 9$$

proceeding in this fashion following matrix is obtained.

		T	A	M	P	S
	0	-3	-6	-9	-12	-15
T	-3	5	2	-1	-4	-7
A	-6	2	9	6	3	-3
P	-9	-1	6	7	13	10
E	-12	-4	3	4	10	13
S	-15	-7	0	2	7	14

Fig. 5.7 The global alignment by traceback

Now starting from the last corner position $6-6$ position $a_{66} = 14$ the path is chosen by taking the maximum element in upward, in diagonal or to the left sideward of the path. The optimal path is given by dark color arrows as

$$14 \to 10 \to 13 \to 6 \to 9 \to 5 \to 0.$$

Thus

$$
\begin{aligned}
a_{66} &= (i) && \text{thus } x_6 \text{ and } y_6 \text{ are aligned,}\\
a_{55} &= (iii) && \text{thus } y_5 \text{ is aligned with } a\ gap,\\
a_{45} &= (i) && \text{thus } x_4 \text{ is aligned with } y_4,\\
a_{34} &= (ii) && \text{thus } x_3 \text{ is aligned with } a\ gap,\\
a_{33} &= (i) && \text{thus } x_2 \text{ is aligned with } y_2,\\
a_{22} &= (i) && \text{thus } x_1 \text{ is aligned with } y_1.
\end{aligned}
$$

Thus the new alignment sequences are
T A M P _ S
T A _ P E S.

EXERCISE 5.8

1. Using the scoring matrix given in Fig. 5.6. Find the optimal alignment between the strings APES and AMPS.

2. Using the scoring matrix given in Fig. 5.6. Find the optimal alignment between the strings SPAM and PAAM.

5.9 Principal Component Analysis (PCA)

PCA is considered as one of the most valuable results derived from linear algebra. It is used to extract relevant information from a confusing data set. Further, it is used to reduce the dimension of the data set and also to obtain hidden and simplified dynamics of the data set.

In order to understand any system, some parameters (say, n) are chosen and a data set is constructed through experimentation. A relation between the data is obtained by finding the covariance matrix, (say $\mathbf{X}$), which is a symmetric matrix. The problem is to

(i) simplify the data set by finding parameters that contain the complete information;

(ii) uncorrelate the important parameters;

(iii) filter out noise and reveal the hidden dynamics;

(iv) identify the parameters (or directions) in which most of the data is concentrated.

This problem is handled using tools from linear algebra.

As the data set is a matrix $\mathbf{X}_{n\times n}$, the rank of the matrix (r, say) gives the number of linearly independent parameters in the system.

The system can be considered as an n-dimensional space and there are r linearly independent vectors (since rank $\mathbf{X} = r$). These r vectors can be extended to form a basis containing n vectors.

As there are infinitely many bases, the choice of the basis is vital.

The basis consisting of the unit eigen-vectors corresponding to the eigen-values of the covariance matrix $\mathbf{X}$ is considered.

An orthogonal matrix $\mathbf{P}$ of order $n \times n$ is constructed using the r unit eigen-vectors and adding $n - r$ orthonormal vectors.

The idea is to introduce a new matrix $\mathbf{Y}$, such that $\mathbf{X} = \mathbf{PY}$ so that the matrix $\mathbf{Y}$ contains columns that are uncorrelated and are arranged in decreasing order of variance.

$$\mathbf{x}_k = \mathbf{P}\mathbf{y}_k \text{ or } \mathbf{y}_k = \mathbf{P}^T\mathbf{x}_k \quad (\text{as P is an orthogonal matrix, } \mathbf{P}^{-1} = \mathbf{P}^T).$$

From Chapter 2, the matrix $\mathbf{P}$ diagonalizes $\mathbf{X}$ so that $\mathbf{X} = \mathbf{PDP^T}$ or $\mathbf{D} = \mathbf{P}^T\mathbf{XP}$, where $\mathbf{D}$ is the diagonal matrix consisting of eigen-values $\lambda_1, \lambda_2, \lambda_3, \ldots, \lambda_r$ of $\mathbf{X}$ arranged in decreasing order, that is, $\lambda_1 \geq \lambda_2 \geq \lambda_3 \geq \ldots \geq \lambda_r$. Suppose $\mathbf{u}_1, \mathbf{u}_2, \mathbf{u}_3, \ldots, \mathbf{u}_r$ are the eigen-vectors corresponding to the eigen-values $\lambda_1, \lambda_2, \ldots, \lambda_r$. Then

$$\mathbf{X}\mathbf{u}_i = \lambda_i \mathbf{u}_i, \quad i = 1, 2, 3, \ldots, r.$$

This means that the data concentration is in the direction of $\mathbf{u}_i$ with λ_i as magnitude, $i = 1, 2, 3, \ldots, r$.

As $\lambda_1 = max\{\lambda_i : i = 1, 2, \ldots, r]$, thus one can conclude most of the data is concentrated in the direction of $\mathbf{u}_1$ with magnitude λ_1. Thus $\mathbf{u}_1$ is called the first principal component. Consider the next maximum eigen-value λ_2, excluding λ_1, then λ_2 gives the magnitude of data concentrated in the direction $\mathbf{u}_2$, which is the second principal component.

Depending on the required accuracy of the problem the number of principal components are considered and the required information is obtained. The new matrix $\mathbf{Y}$ consisting of the columns $\mathbf{y}_i$ is obtained by

$$\mathbf{y}_i = \mathbf{u_i^T}\,\mathbf{X} = c_1\mathbf{x}_1 + c_2\mathbf{x}_2 + \ldots + c_n\mathbf{x}_n,$$

where $\mathbf{u}_i = (c_i)$.

Thus $\mathbf{y}_i$ is the linear combination of the original variables with weights as the entries of $\mathbf{u}_i$.

Similarly, all the significant 'r' columns can be obtained. The $\mathbf{y}_i's$, $i = 1, 2, \ldots, r$ are the new uncorrelated variables.

Procedure to find the principal components of a data set

Step 1. From the given dataset find the covariance matrix (or consider the given matrix).

Step 2. Find the unit eigen-vectors corresponding to the eigen-values.

Step 3. Arrange the eigen-values in descending order.

Step 4. Then $\mathbf{u}_1$ is the first principal component and $\mathbf{u}_2$ is the second principal component and so on. This process can be continued till the required degree of accuracy is reached.

Step 5. $\mathbf{y}_i = \mathbf{u}_i^{\mathbf{T}}\ \mathbf{X}$, $i = 1, 2, ..., r$ gives the necessary uncorrelated variables, which is the new data set.

Another procedure to obtain PCA is the singular value decomposition. The singular value decomposition (SVD) discussed in Chapter 3 is used in many applications. It is considered as a practical tool for PCA as SVD caters to any $m \times n$ matrix.

Procedure for PCA using SVD

Suppose $\mathbf{A}$ is an $m \times n$ matrix of observations having rank r.

Step 1. Calculate $\mathbf{A}^T\mathbf{A}$ and find its eigen-values. Let $\lambda_1, ..., \lambda_n$ be the eigen-values of $\mathbf{A}^T\mathbf{A}$ and $\mathbf{v}_1, ..., \mathbf{v}_n$ be the orthonormal vectors corresponding to $\lambda_1 \geq \lambda_2 \geq ... \geq \lambda_n$.

Step 2. Since rank $\mathbf{A} = r$, there exists $\sigma_i > 0$, $i = 1, 2, ..., r$ such that

$$\sigma_i = \sqrt{\lambda_i}, \quad i = 1, 2, ..., r.$$

Set $\mathbf{u}_i = \frac{1}{\sigma_i}\mathbf{A}\mathbf{v}_i$
Then $\sigma_i \mathbf{u}_i = \mathbf{A}\mathbf{v}_i$
$\mathbf{u}_i$, $i = 1, 2, ..., r$, are called left singular vectors of $\mathbf{A}$
$\mathbf{v}_i$, $i = 1, 2, ..., r$, are called right singular vectors of $\mathbf{A}$

$\mathbf{v}_i$: the right singular vectors of $\mathbf{A}$ are called the principal components of the matrix $\mathbf{A}$.

Observation. Compared to eigen-value decomposition of $\mathbf{A}$, SVD is faster in calculation and is more accurate.

Example 5.9.1 *Consider the Iris flower data set of samples from each of 3 species of Iris, wherein four features were measured from each sample.*

The notation used is:
SL: sepal length, SW: sepal width, PL: petal length, PW: petal width

The problem is to reduce the dimension of the given data.

Solution.

Step 1. Consider the data set which is given by

$$\begin{array}{c} \\ 1\\2\\3\\4\\5\\6\\7\\8 \end{array}\begin{array}{c} \begin{array}{cccc} SL & SW & PL & PW \end{array} \\ \begin{bmatrix} 5.5 & 3.5 & 1.4 & 0.2\\ 4.9 & 3 & 1.4 & 0.2\\ 4.7 & 3.2 & 1.3 & 0.2\\ 4.6 & 3.1 & 1.5 & 0.2\\ 7 & 3.2 & 4.7 & 1.4\\ 6.4 & 3.2 & 4.5 & 1.5\\ 6.9 & 3.1 & 4.9 & 1.5\\ 5.5 & 2.3 & 4 & 1.3 \end{bmatrix}\end{array}$$

Step 2. Calculate the covariance matrix using the formula

$$cov(x,y) = \begin{cases} \dfrac{\sum(x_i-\bar{x})(y_i-\bar{y})}{N-1} \quad \text{[N is number of rows]} & x \neq y \\ \text{Variance } (x) = \dfrac{\sum(x_i-\bar{x})^2}{N} & x = y \end{cases}$$

The covariance matrix is

$$X = \begin{bmatrix} 0.9355 & 0.0368 & 1.4694 & 0.5530 \\ 0.0367 & 0.1193 & -0.1625 & -0.0725 \\ 1.4694 & -0.1625 & 2.857 & 1.1062 \\ 0.5530 & -0.0725 & 1.1062 & 0.4327 \end{bmatrix}$$

Step 3. Calculate eigen-values and eigen-vectors considering the characteristic equation $det(X - \lambda\mathbf{I}) = 0$.

The eigen-values are $\lambda_1 = 4.077$, $\lambda_2 = 0.2468$, $\lambda_3 = 0.0184$, $\lambda_4 = 0.0023$. Since the eigen-vector satisfies the equation

$$Xx = \lambda x \text{ or } (X - \lambda I)x = 0$$

on substituting λ values in the above equation yields the following eigen-vectors corresponding to the eigen-value λ_i written in the descending order

$$\begin{aligned} x_1 &= [0.4464 \quad 0.6893 \quad 0.5571 \quad 0.1233]^T \\ x_2 &= [-0.036 \quad 0.6473 \quad -0.7591 \quad -0.0584]^T \\ x_3 &= [0.8342 \quad -0.2722 \quad -0.2397 \quad -0.4153]^T \\ x_4 &= [0.3217 \quad -0.1782 \quad -0.2364 \quad 0.8994]^T \end{aligned}$$

Step 4. Select the number of principal components. One can select any value from 1 to n [n is number of λ values].

In this for example choose $n = 2$. Then take the top 2 eigen-values corresponding and top 2 eigen-vectors.

Values arranged in descending order must be taken. Next transpose the resultant values.

Thus, considering eigen-vectors corresponding to the top 2 eigen-values, x_1 and x_2, the resultant matrix is

$$\mathbf{A} = \begin{bmatrix} 0.4464 & 0.6893 & 0.5571 & 0.1233 \\ -0.036 & 0.6473 & -0.7591 & -0.0584 \end{bmatrix}$$

Next transposing the above matrix yields

$$\mathbf{A}^T = \begin{bmatrix} 0.4464 & -0.036 \\ 0.6893 & 0.6473 \\ 0.5571 & -0.7591 \\ 0.1233 & -0.0584 \end{bmatrix}_{4\times 2}$$

Step 5. The required new dataset is given by the product of covariance matrix and the transpose matrix of the chosen eigen-vectors, and is.

$$= \begin{bmatrix} 1.3298 & -1.1577 \\ -0.0008 & 0.2035 \\ 2.272 & -2.3915 \\ 0.8666 & -0.9319 \end{bmatrix}_{4\times 2}$$

Thus the resultant dataset has 2 principal components.

EXERCISE 5.9

1. Consider the matrix

$$A = \begin{bmatrix} 70 & 120 & 30 & 50 \\ 4 & 5 & 1 & 4 \\ 1 & 5 & 1 & 0 \\ 5 & 8 & 7 & 8 \end{bmatrix}$$

and find the first three principal components and build the new dataset.

2. Consider the matrix

$$A = \begin{bmatrix} 4 & -3 & 4 \\ 5 & -7 & 7 \\ 4 & -1 & 2 \end{bmatrix}$$

and find the first two principal components and build the new dataset.

5.10 Big Data

Big data mainly deals with unstructured and semi-structured data in terms of pictures, messages, etc., created by people all over the world. It is characterized by the 4 *Vs*: Volume, Variety, Veracity and Velocity–which are briefly described below.

Volume

The main characteristic that makes data "big" is its sheer volume. The total amount of information is growing exponentially every year. For example, Google has a 30 billion x 30 billion matrix and everyday 250,000 pages of new data is being added to it.

Variety

Variety is one of the most interesting developments in technology as more and more information is digitized. A picture, a voice recording, a tweet – they all can be different but express ideas and thoughts based on human understanding and all these are identified as unstructured data, a fundamental concept in big data. One of the goals of big data is to use technology to consider this unstructured data and make sense of it.

Veracity

Veracity refers to the trustworthiness of the data. There will be inherent discrepancies in all the data collected.

Velocity

Velocity is the frequency of the incoming data that needs to be processed. The many SMS messages, Facebook status updates or credit card swipes that are being sent on a particular telecom carrier every minute of every day indicate the high velocity of data that is being transmitted daily.

Analyzing this data and coming up with meaningful interpretations, that is required by companies and institutions for their benefit and profit, is the big challenge.

In terms of understanding the data and making it structured so as to come up with useful interpretations it is essential to think in terms of mathematical language. A brief description of the math used is given below for a better understanding of the data.

In order to analyze the data, mathematicians use curve fitting in one dimension, surface fitting in two dimensions and volume fitting in three dimensions. Big data can be characterized as data having too many data points in n-dimensions where the dimension of the data crops up from the number of parameters involved. Thus attempts have to be made to do n-manifold fitting by extending the least squares method used for curve fitting to n-dimensions. Further it may be necessary to use nonlinear optimization for analyzing the data.

Big data is by nature very dynamic. The number of parameters and their character keeps on changing. In essence there are two types of parameters-static and dynamic- for both of which the values keep on varying. Further a parameter involving probability must be included so as to make predictions using the data. Since parameters represent dimension a big issue is which parameters to keep and which to toss so as to reduce the dimension.

As stated earlier, big data consists of mainly unstructured data, which must be transformed to structured data, so that it can be understood and meaningful interpretations could be made. The principal component analysis (PCA) and singular value decomposition are used for data compression. In what follows a simple example of groceries is considered and changed to structured data so as to analyze it.

Example 5.10.1 *Consider a store which keeps track of 4 customers [A, B, C, D] and which/how many of 16 items the store carries that a customer purchases during any given month. Assume the 16 items are eggs, butter, milk, soda, candy, beer, utensils, soap, paper towels, fruits, vegetables, rice, lentils, coffee, tea and incense.*
Suppose the store has collected the following data for each of the four customers during a given month: what did the customer buy, in what quantities, total bill, top two items purchased, and how many times the customer has shopped at the store. The data can be represented using a 4 by 35 matrix, one row per customer. Thirty-five columns are needed because we need 16 columns to store which items were purchased, 16 columns to indicate which two of the 16 items are a customer's top items (these columns will have only two non-zero values), one column for the date, one column for the total bill, and one column indicating how many times the customer has shopped at the store during that month.
However, one can reduce the size of the matrix by introducing, say, 4 categories to group the 16 items. For example, eggs, butter and milk can be placed in the dairy category, the beverage category would contain beer, soda, coffee and tea, the non-food category will contain soap, utensils, incense and paper towels, and the 'other' category will account for the remaining items. Under this approach, only 4 by 11 matrix is required, since there are only four categories to represent items purchased, four category columns to indicate top choices, plus one column each for the date, total bill and store visit frequency. A simple illustration is as follows.

Representation in a matrix form using categories is as follows (food, beverages, non-food, and other)

Customer	Date	Items	Top 2 items	Total bill	Store visits
A	18-Jan	Beer, eggs, milk, coffee, rice	eggs, milk	150	2
B	13-Jan	Butter, fruits, soap	fruits, soap	85	4
C	5-Jan	Milk, soda, beer	soda, beer	62	1
D	26-Jan	Coffee, tea, incense, utensils, lentils	tea, incense	212	2

$$\begin{bmatrix} 18 & 1 & 1 & 0 & 1 & 1 & 0 & 0 & 0 & 150 & 2 \\ 13 & 1 & 0 & 1 & 1 & 0 & 0 & 1 & 1 & 85 & 4 \\ 5 & 1 & 1 & 0 & 0 & 0 & 1 & 0 & 0 & 62 & 1 \\ 26 & 0 & 1 & 0 & 1 & 0 & 1 & 1 & 0 & 212 & 2 \end{bmatrix}$$

Based on the above matrix, we can now compute the pseudo inverse, eigen-values, singular values, eigen-vectors, singular vectors, etc.

$$P = \begin{bmatrix} -0.53515 & 0.09607 & 0.12532 & 0.82986 \\ -0.30462 & -0.87462 & 0.34705 & 0.34705 \\ -0.22017 & 0.44976 & 0.80593 & -0.31576 \\ -0.75653 & 0.15332 & -0.46295 & -0.43569 \end{bmatrix}.$$

$$Q = \begin{bmatrix} 282.34 & 0 & 0 & 0 & 0 & 0 & 0 & 0 & 0 & 0 & 0 \\ 0 & 4.66 & 0 & 0 & 0 & 0 & 0 & 0 & 0 & 0 & 0 \\ 0 & 0 & 2.46 & 0 & 0 & 0 & 0 & 0 & 0 & 0 & 0 \\ 0 & 0 & 0 & 1.36 & 0 & 0 & 0 & 0 & 0 & 0 & 0 \end{bmatrix}.$$

$$R = \begin{bmatrix} -0.121 & -0.730 & -0.501 & 0.081 & -0.118 & -0.013 & 0.180 & -0.104 & -.0965 & -.260 & -0.240 \\ -0.003 & -0.0705 & 0.517 & 0.268 & -0.144 & -0.223 & 0.174 & 0.078 & -0.169 & 0.163 & -0.701 \\ -0.005 & 0.150 & 0.189 & 0.057 & 0.116 & 0.080 & -0.318 & 0.036 & -0.019 & -0.884 & -0.179 \\ -0.001 & -0.187 & 0.140 & -0.108 & 0.124 & 0.595 & -0.460 & -0.470 & -0.154 & 0.252 & -0.204 \\ -0.005 & -0.134 & 0.003 & 0.180 & 0.956 & -0.109 & 0.112 & 0.065 & 0.005 & 0.073 & -.0253 \\ -0.001 & 0.020 & 0.050 & 0.607 & -0.085 & 0.653 & 0.303 & 0.261 & 0.093 & -0.029 & 0.152 \\ -0.003 & 0.129 & 0.138 & -0.550 & 0.097 & 0.291 & 0.69 & -0.194 & -0.063 & -0.178 & -0.108 \\ -0.003 & -0.154 & -0.046 & -0.427 & 0.0411 & 0.236 & -0.190 & 0.804 & -0.088 & 0.102 & -0.178 \\ -0.001 & -0.187 & 0.140 & -0.108 & -0.008 & 0.024 & -0.031 & -0.032 & 0.950 & 0.014 & -0.162 \\ -0.992 & 0.098 & 0.049 & -0.008 & 0.009 & 0.000 & -0.023 & 0.009 & 0.013 & 0.036 & 0.026 \\ -0.014 & -0.547 & 0.615 & -0.086 & -0.055 & -0.090 & 0.015 & 0.034 & -0.112 & -0.091 & 0.529 \end{bmatrix}.$$

In order to understand and analyze the patterns, the analyst changes one category and studies the corresponding effect on the matrices P, Q and R.

Once the eigen-values, eigen-vectors, singular values and singular vectors have been computed, the big data analyst will now have to perform a perturbation/sensitivity analysis. Such an analysis is done by changing one or more values in the original data matrix and recomputing the eigen-values, singular values, etc. This is a tedious and an iterative process but needs to be carried out in order to extract any patterns that may emerge so as to be able to predict consumer behavior. Once these patterns have been extracted, the interesting challenge becomes how best to interpret the patterns and give them real-world significance which will be of use to the store. Different stores may have different criteria for pattern interpretation.

EXERCISE 5.10

1. Find the first two principal components of the data matrix obtained by considering the first 4 columns given in the above example using eigen-value decomposition.

2. Consider from the 7th column to 11th column of the matrix in example 5.10.1. Form a matrix and find the first two principal components using eigen-value decomposition and SVD.

5.11 Conclusion

This chapter showcased some of the applications of various aspects of matrices to diverse areas ranging from mathematics and economics to computer graphics, robotics and big data. This chapter just touches the tip of the iceberg of the vast range of applications of matrices in the field of knowledge in mathematical, financial, physical, social and biological sciences, engineering and technology. For further reading the readers are suggested to refer to [1], [7], [9],[11], [15] and [17]

Chapter 6

Kronecker Product

6.1 Introduction

The present chapter is devoted to a concept which has raised the theory of matrices to new heights. Further, it enriched the scope of matrix applications in basic sciences, engineering, technology and social sciences. Very few text books on matrix theory of linear algebra include the concept of Kronecker product. The present text includes Kronecker products and interlinks all the subsequent chapters by using it as a tool.

Kronecker product (also known as **'direct product'** or **'tensor product'**) has its origin in group theory and stability analysis. Conversion of a matrix into a vector form and vice versa using a Kronecker product provides a way for solving several problems in matrices. Further, the knowledge of eigen-values and their related eigen-vectors of matrices of lower orders such as $n = 2, 3, 4$ were utilized to obtain the eigen-values and the corresponding eigen-vectors of matrices of higher order, thus enhancing the scope of applications of matrices of higher order. These concepts are presented in this chapter. Following is the plan which sequentially brings home the ideas concerning Kronecker product.

In Section 6.2 the concepts of primary matrices are introduced and converting a matrix into a vector form is discussed. Kronecker products are introduced in Section 6.3 and some basic properties are given. Section 6.4 deals with some more useful properties of Kronecker product. In Section 6.5 the Kronecker product of the linear transformations is given. Section 6.6 deals with the utility of combining Kronecker products and vector operators. Permutation matrices and Kronecker products form the content of Section 6.7. In Section 6.8 analytic functions of Kronecker products are studied. Kronecker sum is defined in Section 6.9 and a few important properties are obtained.

6.2 Primary Matrices

In this section, the topics considered in the first few chapters are revisited and are formulated in a manner useful for the development of the material in the upcoming sections. The following is a beginning in this direction.

DOI: 10.1201/9781351014953-6

Definition 6.2.1 *A primary matrix* $\mathbf{P}_{kl}$ *is an* $m \times n$ *matrix having '1' in the* (k, l) *entry and zero elsewhere, that is*

$$\mathbf{P}_{kl} = (p_{ij})_{m \times n}, \quad \text{where} \quad p_{ij} = \begin{cases} 1 & \text{if } i = k, j = l \\ 0 & \text{otherwise.} \end{cases}$$

Example 6.2.1 *Primary matrices* $\mathbf{P}_{kl}$ *of order* 2×2 *are*

$$\mathbf{P}_{11} = \begin{bmatrix} 1 & 0 \\ 0 & 0 \end{bmatrix}, \mathbf{P}_{12} = \begin{bmatrix} 0 & 1 \\ 0 & 0 \end{bmatrix}, \mathbf{P}_{21} = \begin{bmatrix} 0 & 0 \\ 1 & 0 \end{bmatrix} \text{ and } \mathbf{P}_{22} = \begin{bmatrix} 0 & 0 \\ 0 & 1 \end{bmatrix}.$$

Observation.

1. $I_2 = \mathbf{P}_{11} + \mathbf{P}_{22}$
2. $S = \{\mathbf{P}_{11}, \mathbf{P}_{12}, \mathbf{P}_{21}, \mathbf{P}_{22}\}$ is a basis for all 2×2 real matrices, $\mathbb{R}^{2\times2}$.

Example 6.2.2 *Show that any matrix* $\mathbf{A}_{m \times n}$ *can be written as a linear combination of primary matrices* $\mathbf{P}_{kl}$ *of order* $m \times n$ *(i.e. of same order as* $\mathbf{A}$*).*
Solution.
Let $\mathbf{P}_{kl} = (p_{ij})_{m \times n}$, *where* $p_{ij} = \begin{cases} 1 & \text{if } i = k, j = l \\ 0 & \text{otherwise} \end{cases}$. *Consider the set* $S = \{\mathbf{P}_{kl} : k = 1, 2, \ldots, m; l = 1, 2, \ldots, n\}$ *of* mn *matrices. Then any matrix* $\mathbf{A} = (a_{ij})_{m \times n}$ *can be written as a linear combination of elements* S *as follows:*

$$\mathbf{A} = (a_{ij})_{m \times n} = \sum_{j=1}^{n} \sum_{i=1}^{m} a_{ij} \mathbf{P}_{ij}.$$

The concept of unit vectors introduced in Chapter 1 is linked to primary matrices as follows.

Definition 6.2.2 *Primary column matrix,* $\mathbf{P}_{k!}$ *is a unit column vector having* m *rows and 1 column, such that* $\mathbf{P}_{k!} = (p_{i1})_{m \times 1}$ *with* $p_{i1} = \begin{cases} 1 & \text{if } i = k \\ 0 & \text{otherwise.} \end{cases}$

Definition 6.2.3 *Primary row matrix,* $\mathbf{P}_{!l}$ *is a unit row vector having 1 row and* n *columns, such that,* $\mathbf{P}_{!l} = (p_{1j})_{1 \times n}$ *with* $p_{1j} = \begin{cases} 1 & \text{if } j = l \\ 0 & \text{otherwise.} \end{cases}$

Example 6.2.3

$$\mathbf{P}_{1!} = \mathbf{e}_1 \quad \mathbf{P}_{2!} = \mathbf{e}_2 \ldots \mathbf{P}_{n!} = \mathbf{e}_n$$

$$\mathbf{P}_{!1} = \mathbf{e}_1^T \quad \mathbf{P}_{!2} = \mathbf{e}_2^T \cdots \mathbf{P}_{!m} = \mathbf{e}_m^T.$$

Relation between unit vectors and primary matrices

The primary matrix $\mathbf{P}_{kl}$ of order $m \times n$ is the product of the m-unit vector $\mathbf{e}_k$ and the transpose of the n-unit vector $\mathbf{e}_l$, e_l^T,

$$(\mathbf{P}_{kl})_{m\times n} = (\mathbf{e}_k)_{m\times 1}\left(\mathbf{e}_l^T\right)_{1\times n}$$

which in terms of primary matrices can be written as

$$(\mathbf{P}_{kl})_{m\times n} = (\mathbf{P}_{k!})_{m\times 1}\left(\mathbf{P}_{!l}^T\right)_{1\times n} \tag{6.1}$$

$$\text{Thus the matrix } \mathbf{P}_{kl} = P_{k!}P_{!l} = \begin{pmatrix} 0 \\ 0 \\ . \\ . \\ 1 \\ . \\ . \\ 0 \\ 0 \end{pmatrix} \begin{pmatrix} 0 & 0 & . & . & 1 & . & . & 0 & 0 \end{pmatrix}$$

k^{th} *place* ↗ (the 1 in the column) $\qquad$ ↑ l^{th} *place* (the 1 in the row)

$$= \begin{pmatrix} 0 & \cdots & 0 & \cdots & 0 \\ 0 & \cdots & 0 & \cdots & 0 \\ 0 & \cdots & 1_{kl} & \cdots & 0 \\ 0 & \cdots & 0 & \cdots & 0 \\ 0 & \cdots & 0 & \cdots & 0 \end{pmatrix}$$

Example 6.2.4 *The primary matrices of order* 2×2 *can be written using primary column and row matrices of order* 2×2 *as follows*

$$(\mathbf{P}_{11})_{2\times 2} = (\mathbf{P}_{1!})_{2\times 1}(\mathbf{P}_{!1})_{1\times 2}, \quad (\mathbf{P}_{12})_{2\times 2} = (\mathbf{P}_{1!})_{2\times 1}(\mathbf{P}_{!2})_{1\times 2}$$
$$(\mathbf{P}_{21})_{2\times 2} = (\mathbf{P}_{!1})_{2\times 1}(\mathbf{P}_{!1})_{1\times 2} \quad \textit{and} \quad (\mathbf{P}_{22})_{2\times 2} = (\mathbf{P}_{2!})_{2\times 1}(\mathbf{P}_{!2})_{1\times 2}.$$

Partitions of a matrix

Consider an $m \times n$ matrix $\mathbf{A}$. Represent its 'n' columns by $\mathbf{A}_{*1}, \mathbf{A}_{*2}, \ldots, \mathbf{A}_{*n}$ and its 'm' rows by $\mathbf{A}_{1*}, \mathbf{A}_{2*}, \ldots, \mathbf{A}_{m*}$.

Definition 6.2.4 *Column partition of matrix* $\mathbf{A}$ *is the matrix written using its columns as submatrices and is as*

$$\mathbf{A} = \begin{bmatrix}\mathbf{A}_{*1} & \mathbf{A}_{*2} & \cdots & \mathbf{A}_{*n}\end{bmatrix}$$

Definition 6.2.5 *Row partition of a matrix* $\mathbf{A}$ *is the matrix consisting of its rows as submatrices and is written as*

$$\mathbf{A} = \begin{bmatrix}\mathbf{A}_{1*} & \mathbf{A}_{2*} & \mathbf{A}_{3*} & \cdots & \mathbf{A}_{m*}\end{bmatrix}^T$$

Observation. $\left[\mathbf{A}_{i*}\right]^T$ is the ith row of $\mathbf{A}$ and is simply represented by $\mathbf{A}_{i*}^T$. Here $\mathbf{A}_{i*}$ is the column of the ith row of $\mathbf{A}$ for $i = 1, 2, \ldots, m$.

Example 6.2.5 *Let* $\mathbf{A} = \begin{bmatrix} 1 & 3 \\ 2 & 4 \end{bmatrix}$

Then $\mathbf{A}_{*1} = \begin{bmatrix} 1 \\ 2 \end{bmatrix}$ *and* $\mathbf{A}_{*2} = \begin{bmatrix} 3 \\ 4 \end{bmatrix}$ *and* $\mathbf{A}_{1*} = \begin{bmatrix} 1 \\ 3 \end{bmatrix}$ *which gives* $\mathbf{A}_{1*}^T = \begin{bmatrix} 1 & 3 \end{bmatrix}$ *and* $\mathbf{A}_{2*} = \begin{bmatrix} 2 \\ 4 \end{bmatrix}$ *this gives* $\mathbf{A}_{2*}^T = \begin{bmatrix} 2 & 4 \end{bmatrix}$.

Relation between partitioned matrices and primary matrices

(1) The lth column of $\mathbf{A}$, $\mathbf{A}_{*l}$ is given by

$$\mathbf{AP}_{l!} = \begin{bmatrix} \mathbf{A}_{*1} & \mathbf{A}_{*2} & \cdots & \mathbf{A}_{*n} \end{bmatrix} \begin{bmatrix} 0 \\ 0 \\ \cdot \\ \cdot \\ 0 \\ 1 \\ 0 \\ \cdot \\ \cdot \\ 0 \\ 0 \end{bmatrix} \begin{matrix} \\ \\ \\ \\ \\ \rightarrow l\text{th } \textit{place} \\ \\ \\ \\ \\ \\ \end{matrix}$$

$$= \mathbf{A}_{*l}$$

Thus

$$\mathbf{A}_{*l} = \mathbf{AP}_{l!} \tag{6.2}$$

(2) the kth row of $\mathbf{A}$, $\mathbf{A}_{k*}$ is represented as a column by nomenclature. So to obtain the kth row of $\mathbf{A}$, the kth row of $\mathbf{A}^T$ is used.

$$\begin{bmatrix} 0 & . & . & . & 0 & 1 & 0 & . & . & . & 0 \end{bmatrix} \begin{bmatrix} \mathbf{A}_{1*} & \mathbf{A}_{2*} & \mathbf{A}_{3*} & \cdots & \mathbf{A}_{m*} \end{bmatrix}^T$$

$\uparrow k$th place

$$= \mathbf{A}_{k*}^T$$

thus, the kth row of $\mathbf{A}$ is

$$\mathbf{A}_{k*}^T = \mathbf{P}_{!k}\mathbf{A} \tag{6.3}$$

Example 6.2.6 *Let* $\mathbf{A} = \begin{bmatrix} 4 & 5 \\ 2 & 3 \\ 1 & 8 \end{bmatrix}$. *Express the* 2nd *column of* $\mathbf{A}$ *and* 3rd *row of* $\mathbf{A}$ *using primary matrices.*

Solution.

$$\mathbf{A}_{*2} = \mathbf{AP}_{2!} = \begin{bmatrix} 4 & 5 \\ 2 & 3 \\ 1 & 8 \end{bmatrix} \begin{bmatrix} 0 \\ 1 \end{bmatrix} = \begin{bmatrix} 5 \\ 3 \\ 8 \end{bmatrix}$$

$$\mathbf{A}_{3*} = \textit{the column matrix of the } 3^{rd} \textit{ row of } \mathbf{A}.$$

To get the 3rd row of $\mathbf{A}$ consider $\mathbf{A}_{3*}^T$

$$\mathbf{A}_{3*}^T = \mathbf{P}_{13}\mathbf{A} = \begin{bmatrix} 0 & 0 & 1 \end{bmatrix} \begin{bmatrix} 4 & 5 \\ 2 & 3 \\ 1 & 8 \end{bmatrix} = \begin{bmatrix} 1 & 8 \end{bmatrix}.$$

To obtain the matrix $\mathbf{A}$ using row partitioned matrices and column partitioned matrices

(1) From (6.2)

$$\mathbf{A} = \sum_{l=1}^{n} \mathbf{A}_{*l}\mathbf{P}_{!l} \tag{6.4}$$

From (6.3)

$$\mathbf{A} = \sum_{k=1}^{m} \mathbf{P}_{k!}\mathbf{A}_{k*}^T \tag{6.5}$$

Now in Example 6.2.6, using R.H.S of the relation (6.4) *gives*

$$\begin{aligned} \sum_{l=1}^{2} \mathbf{A}_{*l}\mathbf{P}_{!l} &= \begin{bmatrix} 4 \\ 2 \\ 1 \end{bmatrix} \begin{bmatrix} 1 & 0 \end{bmatrix} + \begin{bmatrix} 5 \\ 3 \\ 8 \end{bmatrix} \begin{bmatrix} 0 & 1 \end{bmatrix} \\ &= \begin{bmatrix} 4 & 0 \\ 2 & 0 \\ 1 & 0 \end{bmatrix} + \begin{bmatrix} 0 & 5 \\ 0 & 3 \\ 0 & 8 \end{bmatrix} \\ &= \begin{bmatrix} 4 & 5 \\ 2 & 3 \\ 1 & 8 \end{bmatrix} = \mathbf{A}. \end{aligned}$$

To obtain $\mathbf{A}$ using the rows of $\mathbf{A}$, consider the R.H.S of (6.5),

$$\begin{aligned} \sum_{k=1}^{3} \mathbf{P}_{k!}\mathbf{A}_{k*}^T &= \begin{bmatrix} 1 \\ 0 \\ 0 \end{bmatrix} \begin{bmatrix} 4 & 5 \end{bmatrix} + \begin{bmatrix} 0 \\ 1 \\ 0 \end{bmatrix} \begin{bmatrix} 2 & 3 \end{bmatrix} + \begin{bmatrix} 0 \\ 0 \\ 1 \end{bmatrix} \begin{bmatrix} 1 & 8 \end{bmatrix} \\ &= \begin{bmatrix} 4 & 5 \\ 0 & 0 \\ 0 & 0 \end{bmatrix} + \begin{bmatrix} 0 & 0 \\ 2 & 3 \\ 0 & 0 \end{bmatrix} + \begin{bmatrix} 0 & 0 \\ 0 & 0 \\ 1 & 8 \end{bmatrix} \\ &= \begin{bmatrix} 4 & 5 \\ 2 & 3 \\ 1 & 8 \end{bmatrix} = \mathbf{A}. \end{aligned}$$

Example 6.2.7 *Let* $\mathbf{A} = \begin{bmatrix} a_1 & a_2 & a_3 \\ a_4 & a_5 & a_6 \end{bmatrix}$. *Verify the relations* (6.4) *and* (6.5)

Solution. *Verification of relation* (6.4)

Consider

$$\sum_{l=1}^{3} \mathbf{A}_{*l}\mathbf{P}_{!l} = \begin{bmatrix} a_1 \\ a_4 \end{bmatrix} \begin{bmatrix} 1 & 0 & 0 \end{bmatrix} + \begin{bmatrix} a_2 \\ a_5 \end{bmatrix} \begin{bmatrix} 0 & 1 & 0 \end{bmatrix} + \begin{bmatrix} a_3 \\ a_6 \end{bmatrix} \begin{bmatrix} 0 & 0 & 1 \end{bmatrix}$$

$$= \begin{bmatrix} a_1 & 0 & 0 \\ a_4 & 0 & 0 \end{bmatrix} + \begin{bmatrix} 0 & a_2 & 0 \\ 0 & a_5 & 0 \end{bmatrix} + \begin{bmatrix} 0 & 0 & a_3 \\ 0 & 0 & a_6 \end{bmatrix}$$

$$= \begin{bmatrix} a_1 & a_2 & a_3 \\ a_4 & a_5 & a_6 \end{bmatrix}, \text{ hence (6.4) holds}$$

Consider

$$\sum_{k=1}^{2} \mathbf{P}_{k!}\mathbf{A}_{k*}^{T} = \begin{bmatrix} 1 \\ 0 \end{bmatrix} \begin{bmatrix} a_1 & a_2 & a_3 \end{bmatrix} + \begin{bmatrix} 0 \\ 1 \end{bmatrix} \begin{bmatrix} a_4 & a_5 & a_6 \end{bmatrix}$$

$$= \begin{bmatrix} a_1 & a_2 & a_3 \\ 0 & 0 & 0 \end{bmatrix} + \begin{bmatrix} 0 & 0 & 0 \\ a_4 & a_5 & a_6 \end{bmatrix} = \begin{bmatrix} a_1 & a_2 & a_3 \\ a_4 & a_5 & a_6 \end{bmatrix}, \text{ hence (6.5) holds.}$$

Further, from the relation (6.1) *one can obtain that the* (i, j)th *entry of*

$$\mathbf{A} = a_{ij} = \mathbf{P}_{!i}A\mathbf{P}_{j!} \tag{6.6}$$

and that

$$\mathbf{A} = \sum_{j=1}^{n}\sum_{i=1}^{m} a_{ij}\mathbf{P}_{ij}, \tag{6.7}$$

where $\mathbf{P}_{ij}$ *is an* $m \times n$ *matrix (same order as* $\mathbf{A}$*) such that the* (i, j)th *entry of* $\mathbf{P}_{ij} = 1$ *and elsewhere '0'.*

Example 6.2.8 *Given* $\mathbf{A} = \begin{bmatrix} 1 & 2 & 3 \\ 3 & 4 & 5 \\ 5 & 6 & 7 \end{bmatrix}$

(i) *Obtain the* 2nd *column of* **A** *using the required primary matrix*

(ii) *Obtain the* 3rd *row of* **A** *using the required primary matrix*

(iii) *Obtain the* $(1, 3)$ *entry of* **A** *using the required primary matrices.*

(iv) *Express* **A** *as a sum of (a) column vectors of* **A** *(b) row vectors of* **A** *(c) entries of* **A**.

Solution.

(i) Consider $\mathbf{AP}_{2!} = \begin{bmatrix} 1 & 2 & 3 \\ 3 & 4 & 5 \\ 5 & 6 & 7 \end{bmatrix} \begin{bmatrix} 0 \\ 1 \\ 0 \end{bmatrix} = \begin{bmatrix} 2 \\ 4 \\ 6 \end{bmatrix}$

(ii) *Consider* $\mathbf{P}_{!3}A = \begin{bmatrix} 0 & 0 & 1 \end{bmatrix} \begin{bmatrix} 1 & 2 & 3 \\ 3 & 4 & 5 \\ 5 & 6 & 7 \end{bmatrix} = \begin{bmatrix} 5 & 6 & 7 \end{bmatrix}$

(iii) *Consider* $\mathbf{P}_{!1}\mathbf{A}\mathbf{P}_{3!} = \begin{bmatrix} 1 & 0 & 0 \end{bmatrix} \begin{bmatrix} 1 & 2 & 3 \\ 3 & 4 & 5 \\ 5 & 6 & 7 \end{bmatrix} \begin{bmatrix} 0 \\ 0 \\ 1 \end{bmatrix} = \begin{bmatrix} 1 & 2 & 3 \end{bmatrix} \begin{bmatrix} 0 \\ 0 \\ 1 \end{bmatrix}$

$= [3]$

(iv) *(a)* *Consider* $\sum_{l=1}^{3} \mathbf{A}_{*l}\mathbf{P}_{!l} = \mathbf{A}_{*1}\mathbf{P}_{!1} + \mathbf{A}_{*2}\mathbf{P}_{!2} + \mathbf{A}_{*3}\mathbf{P}_{!3}$

$$= \begin{bmatrix} 1 \\ 3 \\ 5 \end{bmatrix} \begin{bmatrix} 1 & 0 & 0 \end{bmatrix} + \begin{bmatrix} 2 \\ 4 \\ 6 \end{bmatrix} \begin{bmatrix} 0 & 1 & 0 \end{bmatrix} + \begin{bmatrix} 3 \\ 5 \\ 7 \end{bmatrix} \begin{bmatrix} 0 & 0 & 1 \end{bmatrix}$$

$$= \begin{bmatrix} 1 & 2 & 3 \\ 3 & 4 & 5 \\ 5 & 6 & 7 \end{bmatrix} = \mathbf{A}$$

(b) *Consider*

$$\sum_{k=1}^{3} \mathbf{P}_{k1}\mathbf{A}_{k*}^{T} = \begin{bmatrix} 1 \\ 0 \\ 0 \end{bmatrix} \begin{bmatrix} 1 & 2 & 3 \end{bmatrix} + \begin{bmatrix} 0 \\ 1 \\ 0 \end{bmatrix} \begin{bmatrix} 3 & 4 & 5 \end{bmatrix} + \begin{bmatrix} 0 \\ 0 \\ 1 \end{bmatrix} \begin{bmatrix} 5 & 6 & 7 \end{bmatrix}$$

$$= \begin{bmatrix} 1 & 2 & 3 \\ 3 & 4 & 5 \\ 5 & 6 & 7 \end{bmatrix} = \mathbf{A}$$

(c) *Consider*

$$\sum_{l=1}^{3} \sum_{k=1}^{3} a_{kl}\mathbf{P}_{kl}$$

$$= \begin{matrix} a_{11}\mathbf{P}_{11} + a_{21}\mathbf{P}_{21} + a_{31}\mathbf{P}_{31} \\ a_{12}\mathbf{P}_{12} + a_{22}\mathbf{P}_{22} + a_{32}\mathbf{P}_{32} \\ a_{13}\mathbf{P}_{13} + a_{23}\mathbf{P}_{23} + a_{33}\mathbf{P}_{33} \end{matrix}$$

$$= 1 \begin{bmatrix} 1 & 0 & 0 \\ 0 & 0 & 0 \\ 0 & 0 & 0 \end{bmatrix} + 3 \begin{bmatrix} 0 & 0 & 0 \\ 1 & 0 & 0 \\ 0 & 0 & 0 \end{bmatrix} + 5 \begin{bmatrix} 0 & 0 & 0 \\ 0 & 0 & 0 \\ 1 & 0 & 0 \end{bmatrix}$$

$$+ 2 \begin{bmatrix} 0 & 1 & 0 \\ 0 & 0 & 0 \\ 0 & 0 & 0 \end{bmatrix} + 4 \begin{bmatrix} 0 & 0 & 0 \\ 0 & 1 & 0 \\ 0 & 0 & 0 \end{bmatrix} + 6 \begin{bmatrix} 0 & 0 & 0 \\ 0 & 0 & 0 \\ 0 & 1 & 0 \end{bmatrix}$$

$$+ 3 \begin{bmatrix} 0 & 0 & 1 \\ 0 & 0 & 0 \\ 0 & 0 & 0 \end{bmatrix} + 5 \begin{bmatrix} 0 & 0 & 0 \\ 0 & 0 & 1 \\ 0 & 0 & 0 \end{bmatrix} + 7 \begin{bmatrix} 0 & 0 & 0 \\ 0 & 0 & 0 \\ 0 & 0 & 1 \end{bmatrix}$$

$$= \begin{bmatrix} 1 & 2 & 3 \\ 3 & 4 & 5 \\ 5 & 6 & 7 \end{bmatrix} = \mathbf{A}$$

Next the concept of writing a matrix in terms of a vector is introduced.

Converting a matrix to a vector

Let $\mathbf{P}$ be an $m \times n$ matrix represented by its columns as

$$\mathbf{P} = \begin{bmatrix} \mathbf{P}_{*1} & \cdots & \mathbf{P}_{*l} & \cdots & \mathbf{P}_{*n} \end{bmatrix},$$

where $\mathbf{P}_{*l}$ is a column with m rows, for $l = 1, 2, \ldots, n$.

Definition 6.2.6 *The vector operator of a matrix $\mathbf{P}$, denoted by $Vec\,\mathbf{P}$, is defined as a column vector consisting of columns of $\mathbf{P}$. Thus,*

$$Vec\ \mathbf{P} = \begin{bmatrix} \mathbf{P}_{*1} \\ \mathbf{P}_{*2} \\ \vdots \\ \mathbf{P}_{*n} \end{bmatrix}$$

Example 6.2.9 *Let* $\mathbf{P} = \begin{bmatrix} p_1 & p_2 & p_5 \\ p_3 & p_4 & p_6 \end{bmatrix}$

Then,

$$Vec\ \mathbf{P} = \begin{bmatrix} p_1 & p_3 & p_2 & p_4 & p_5 & p_6 \end{bmatrix}^T$$

Example 6.2.10 *Let* $\mathbf{Q} = \begin{bmatrix} 2 & 1 \\ 4 & 8 \\ 5 & 3 \end{bmatrix}$

Then,

$$Vec\ \mathbf{Q} = \begin{bmatrix} 2 & 4 & 5 & 1 & 8 & 3 \end{bmatrix}^T$$

EXERCISE 6.2

1. Write the primary matrices $\mathbf{P}_{11}, \mathbf{P}_{12}, \mathbf{P}_{31}$ of order 3 using primary column matrices (unit vectors) of order 3.

2. Write I_3 in terms of primary matrices

3. Define Kronecker delta $\delta_{ij} = \begin{cases} 1 & \text{if } i = j \\ 0 & \text{if } i \neq j \end{cases}$. Show that

   (i) $\delta_{ij} = \mathbf{P}_{1i}\mathbf{P}_{j1} = \mathbf{P}_{1j}\mathbf{P}_{i1}$

   (ii) $\mathbf{P}_{ij}\mathbf{P}_{kl} = \delta_{jk}\mathbf{P}_{il}$

   (iii) $\mathbf{P}_{ij}\mathbf{P}_{jl}\mathbf{P}_{kl} = \mathbf{P}_{il}\mathbf{P}_{kl} = \mathbf{P}_{il}$

4. Let $\mathbf{A} = (a_{ij})_{3\times3}$. Write $\mathbf{A}$ as a sum of
   (i) column partitions of $\mathbf{A}$ (ii) row partitions of $\mathbf{A}$.
   (iii) as entry sum of $\mathbf{A}$ using primary matrices (iv) express a_{32} using primary column 1 matrices.

5. Let $\mathbf{P}$ and $\mathbf{Q}$ be matrices of order n. Then $tr(\mathbf{PQ}) = (Vec\mathbf{P}^T)^T\ Vec\mathbf{Q}$. Verify this property for $\mathbf{P} = (p_{ij})_{2\times2}$ and $\mathbf{Q} = (q_{ij})_{2\times2}$.

6.3 Kronecker Products

In this section, Kronecker product is introduced and some of its properties are discussed. This concept is also known as a direct product or a tensor product.

Consider two matrices $\mathbf{P} = (p_{ij})_{m\times n}$ and $\mathbf{Q} = (q_{ij})_{r\times s}$

Definition 6.3.1 *The Kronecker product of two matrices* $\mathbf{P}$ *and* $\mathbf{Q}$*, denoted by* $\mathbf{P}\otimes\mathbf{Q}$ *is defined as a matrix,*

$$\mathbf{P}\otimes\mathbf{Q} = \begin{bmatrix} p_{11}\mathbf{Q} & p_{12}\mathbf{Q} & \cdots & p_{1n}\mathbf{Q} \\ p_{21}\mathbf{Q} & p_{22}\mathbf{Q} & \cdots & p_{2n}\mathbf{Q} \\ \vdots & & & \\ p_{m1}\mathbf{Q} & \cdots & \cdots & p_{mn}\mathbf{Q} \end{bmatrix}$$

Observation.

(i) the (i, j)th entry of $\mathbf{P}\otimes\mathbf{Q}$ is the matrix $p_{ij}\mathbf{Q}$ of order $r\times s$.

(ii) The matrix $\mathbf{P}\otimes\mathbf{Q}$ has mn blocks and each block is a matrix of the form $p_{ij}\mathbf{Q}$.

(iii) The order of the matrix $\mathbf{P}\otimes\mathbf{Q}$ is $mr\times ns$. This follows as each block has r rows and s columns and there are m rows and n columns each of such blocks.

Example 6.3.1

$$\textit{Let}\ \mathbf{A} = \begin{bmatrix} 2 & 4 \\ 3 & 5 \end{bmatrix},\quad \mathbf{B} = \begin{bmatrix} 6 & 8 \\ 7 & 9 \end{bmatrix}\ \text{then}$$

$$\mathbf{A}\otimes\mathbf{B} = \begin{bmatrix} 2\mathbf{B} & 4\mathbf{B} \\ 3\mathbf{B} & 5\mathbf{B} \end{bmatrix} = \begin{bmatrix} 12 & 16 & 24 & 32 \\ 14 & 18 & 28 & 36 \\ 18 & 24 & 30 & 40 \\ 21 & 27 & 35 & 45 \end{bmatrix},$$

Observation. $\mathbf{A}\otimes\mathbf{B}\ \neq\ \mathbf{B}\otimes\mathbf{A}$.

Example 6.3.2 *Let*

$$\mathbf{A} = \begin{bmatrix} a_{11} & a_{12} \\ a_{21} & a_{22} \end{bmatrix} \quad \text{and} \quad \mathbf{B} = \begin{bmatrix} b_{11} & b_{12} & b_{13} \\ b_{21} & b_{22} & b_{23} \end{bmatrix}, \quad \text{then}$$

$$\mathbf{A} \otimes \mathbf{B} = \begin{bmatrix} a_{11}\mathbf{B} & a_{12}\mathbf{B} \\ a_{21}\mathbf{B} & a_{22}\mathbf{B} \end{bmatrix} = \begin{bmatrix} a_{11}b_{11} & a_{11}b_{12} & a_{11}b_{13} & a_{12}b_{11} & a_{12}b_{12} & a_{12}b_{13} \\ a_{11}b_{21} & a_{11}b_{22} & a_{11}b_{23} & a_{12}b_{21} & a_{12}b_{22} & a_{12}b_{23} \\ a_{21}b_{11} & a_{21}b_{12} & a_{21}b_{13} & a_{22}b_{11} & a_{22}b_{12} & a_{22}b_{13} \\ a_{21}b_{21} & a_{21}b_{22} & a_{21}b_{23} & a_{22}b_{21} & a_{22}b_{22} & a_{22}b_{23} \end{bmatrix}.$$

Some of the properties of Kronecker products that are useful in later are stated and proved below.

(i) $a\mathbf{P} \otimes \mathbf{R} = \mathbf{P} \otimes a\mathbf{R} = a(\mathbf{P} \otimes \mathbf{R})$

(ii) Kronecker product is associative $\mathbf{P} \otimes (\mathbf{Q} \otimes \mathbf{R}) = (\mathbf{P} \otimes \mathbf{Q}) \otimes \mathbf{R}$

(iii) Zero element involving a Kronecker product is given by

$$\mathbf{O}_{mn} = \mathbf{O}_m \otimes \mathbf{O}_n,$$

where $\mathbf{O}_m$ is a zero square matrix of order m.

(iv) The unit matrix associated with a Kronecker product is given by

$$\mathbf{I}_{mn} = \mathbf{I}_m \otimes \mathbf{I}_n,$$

where $\mathbf{I}_m$ is an Identity matrix of order m.

(v) Kronecker product is distributive with respect to addition

(a) $(\mathbf{P} + \mathbf{Q}) \otimes \mathbf{R} = \mathbf{P} \otimes \mathbf{R} + \mathbf{Q} \otimes \mathbf{R}$

(b) $\mathbf{P} \otimes (\mathbf{Q} + \mathbf{R}) = \mathbf{P} \otimes \mathbf{Q} + \mathbf{P} \otimes \mathbf{R}$

Proof.

(i) By the definition of the Kronecker product

$$\begin{aligned} a\mathbf{P} \otimes \mathbf{R} &= \begin{bmatrix} ap_{11}\mathbf{R} & ap_{12}\mathbf{R} & \cdots & ap_{1n}\mathbf{R} \\ ap_{21}\mathbf{R} & ap_{22}\mathbf{R} & \cdots & ap_{2n}\mathbf{R} \\ \vdots & \vdots & & \vdots \\ ap_{m1}\mathbf{R} & ap_{m2}\mathbf{R} & \cdots & ap_{mn}\mathbf{R} \end{bmatrix} \\ &= a \begin{bmatrix} p_{11}\mathbf{R} & p_{12}\mathbf{R} & \cdots & p_{1n}\mathbf{R} \\ \\ p_{m1}\mathbf{R} & p_{m2}\mathbf{R} & \cdots & p_{mn}\mathbf{R} \end{bmatrix} \\ &= a(\mathbf{P} \otimes \mathbf{R}) \end{aligned}$$

again by definition of Kronecker product.

Also

$$\mathbf{P} \otimes a\mathbf{R} = \begin{bmatrix} p_{11}a\mathbf{R} & p_{12}a\mathbf{R} & \cdots & p_{1n}a\mathbf{R} \\ p_{21}a\mathbf{R} & p_{22}a\mathbf{R} & \cdots & p_{2n}a\mathbf{R} \\ \vdots & \vdots & & \vdots \\ p_{m1}a\mathbf{R} & p_{m2}a\mathbf{R} & \cdots & p_{mn}a\mathbf{R} \end{bmatrix}$$

$$= a \begin{bmatrix} p_{11}\mathbf{R} & p_{12}\mathbf{R} & \cdots & p_{1n}\mathbf{R} \\ p_{21}\mathbf{R} & p_{22}\mathbf{R} & \cdots & p_{2n}\mathbf{R} \\ \vdots & \vdots & & \vdots \\ p_{m1}\mathbf{R} & p_{m2}\mathbf{R} & \cdots & p_{mn}\mathbf{R} \end{bmatrix}$$

$$= a(\mathbf{P} \otimes \mathbf{R})$$

(ii) This proof is done by considering the (i, j)th block of the Kronecker products.

Consider L.H.S $\mathbf{P} \otimes (\mathbf{Q} \otimes \mathbf{R})$

the (i, j)th block of $\mathbf{P} \otimes (\mathbf{Q} \otimes \mathbf{R})$

$$= p_{ij}(\mathbf{Q} \otimes \mathbf{R}) = p_{ij} \begin{bmatrix} q_{11}\mathbf{R} & q_{12}\mathbf{R} & \cdots & q_{1n}\mathbf{R} \\ q_{21}\mathbf{R} & q_{22}\mathbf{R} & \cdots & q_{2n}\mathbf{R} \\ \vdots & \vdots & & \vdots \\ q_{m1}\mathbf{R} & q_{m2}\mathbf{R} & \cdots & q_{mn}\mathbf{R} \end{bmatrix}$$

$$= \begin{bmatrix} p_{ij}q_{11}\mathbf{R} & p_{ij}q_{12}\mathbf{R} & \cdots & p_{ij}q_{1n}\mathbf{R} \\ p_{ij}q_{21}\mathbf{R} & p_{ij}q_{22}\mathbf{R} & \cdots & p_{ij}q_{2n}\mathbf{R} \\ \vdots & \vdots & & \vdots \\ p_{ij}q_{m1}\mathbf{R} & p_{ij}q_{m2}\mathbf{R} & \cdots & p_{ij}q_{mn}\mathbf{R} \end{bmatrix} \tag{6.8}$$

Consider R.H.S $(\mathbf{P} \otimes \mathbf{Q}) \otimes \mathbf{R}$

the (i, j)th block of $(\mathbf{P} \otimes \mathbf{Q}) \otimes \mathbf{R} = (i, j)$th element of $(\mathbf{P} \otimes \mathbf{Q}) \otimes \mathbf{R}$

$$= \begin{bmatrix} p_{ij}q_{11} & p_{ij}q_{12} & \cdots & p_{ij}q_{1n} \\ p_{ij}q_{21} & p_{ij}q_{22} & \cdots & p_{ij}q_{2n} \\ \vdots & \vdots & & \vdots \\ p_{ij}q_{m1} & p_{ij}q_{m2} & \cdots & p_{ij}q_{mn} \end{bmatrix} \otimes \mathbf{R}$$

$$= \begin{bmatrix} p_{ij}q_{11}\mathbf{R} & p_{ij}q_{12}\mathbf{R} & \cdots & p_{ij}q_{1n}\mathbf{R} \\ p_{ij}q_{21}\mathbf{R} & p_{ij}q_{22}\mathbf{R} & \cdots & p_{ij}q_{2n}\mathbf{R} \\ \vdots & \vdots & & \vdots \\ p_{ij}q_{m1}\mathbf{R} & p_{ij}q_{m2}\mathbf{R} & \cdots & p_{ij}q_{mn}\mathbf{R} \end{bmatrix} \tag{6.9}$$

Since (6.8) and (6.9) are equal the result follows.

(iii) Property (iii) is clear and so the proof is not given.

(iv) Property (iv) can be easily proved and hence is avoided.

(v) Consider the (i,j)th block of $(\mathbf{P}+\mathbf{Q})\otimes\mathbf{R}$

$$= (p_{ij}+q_{ij})\mathbf{R} = p_{ij}\mathbf{R}+q_{ij}\mathbf{R}$$
$$= (i,j)\text{th block of } \mathbf{P}\otimes\mathbf{R} + (i,j)\text{th block of } \mathbf{Q}\otimes\mathbf{R}$$

This holds for every $i=1,2,\ldots,m$ and $j=1,2,\ldots,n$. Hence the proof. The other distributive property can be proved in a similar fashion.

Example 6.3.3 *Let* $\mathbf{P}=\begin{bmatrix}p_1 & p_2\\ p_3 & p_4\end{bmatrix}$, $\mathbf{Q}=\begin{bmatrix}q_1 & q_2\\ q_3 & q_4\end{bmatrix}$ *and* $\mathbf{R}=\begin{bmatrix}r_1\\ r_2\end{bmatrix}$
Verify the properties (i), (ii), (iv)

(i) Let $a\in\mathbb{R}$ $\quad a\mathbf{P}\otimes\mathbf{R}=\begin{bmatrix}ap_1 & ap_2\\ ap_3 & ap_4\end{bmatrix}\otimes\begin{bmatrix}r_1\\ r_2\end{bmatrix}=\begin{bmatrix}ap_1r_1 & ap_2r_1\\ ap_1r_2 & ap_2r_2\\ ap_3r_1 & ap_4r_1\\ ap_3r_2 & ap_4r_2\end{bmatrix}$

$$= a\begin{bmatrix}p_1\mathbf{R} & p_2\mathbf{R}\\ p_3\mathbf{R} & p_4\mathbf{R}\end{bmatrix} = a\,(\mathbf{P}\otimes\mathbf{R})$$

Similarly

$$\mathbf{P}\otimes a\mathbf{R}=\begin{bmatrix}p_1 & p_2\\ p_3 & p_4\end{bmatrix}\otimes\begin{bmatrix}ar_1\\ ar_2\end{bmatrix}$$
$$=\begin{bmatrix}p_1ar_1 & p_2ar_1\\ p_1ar_2 & p_2ar_2\\ p_3ar_1 & p_4ar_1\\ p_3ar_2 & p_4ar_2\end{bmatrix}$$
$$= a\begin{bmatrix}p_1\mathbf{R} & p_2\mathbf{R}\\ p_3\mathbf{R} & p_4\mathbf{R}\end{bmatrix} = a\,(\mathbf{P}\otimes\mathbf{R})$$

(ii) $\quad \text{L.H.S} = \mathbf{P}\otimes(\mathbf{Q}\otimes\mathbf{R}) = \begin{bmatrix}p_1 & p_2\\ p_3 & p_4\end{bmatrix}\otimes\left[\begin{bmatrix}q_1 & q_2\\ q_3 & q_4\end{bmatrix}\otimes\begin{bmatrix}r_1\\ r_2\end{bmatrix}\right]$

$$=\begin{bmatrix}p_1 & p_2\\ p_3 & p_4\end{bmatrix}\otimes\begin{bmatrix}q_1r_1 & q_2r_1\\ q_1r_2 & q_2r_2\\ q_3r_1 & q_4r_1\\ q_3r_2 & q_4r_2\end{bmatrix}$$

$$=\begin{bmatrix}p_1q_1r_1 & p_1q_2r_1 & p_2q_1r_1 & p_2q_2r_1\\ p_1q_1r_2 & p_1q_2r_2 & p_2q_1r_2 & p_2q_2r_2\\ p_1q_3r_1 & p_1q_4r_1 & p_2q_3r_1 & p_2q_4r_1\\ p_1q_3r_2 & p_1q_4r_2 & p_2q_3r_2 & p_2q_4r_2\\ p_3q_1r_1 & p_3q_2r_1 & p_4q_1r_1 & p_4q_2r_1\\ p_3q_1r_2 & p_3q_2r_2 & p_4q_1r_2 & p_4q_2r_2\\ p_3q_3r_1 & p_3q_4r_1 & p_4q_3r_1 & p_4q_4r_1\\ p_3q_3r_2 & p_3q_4r_2 & p_4q_3r_2 & p_4q_4r_2\end{bmatrix}$$

$$
\begin{aligned}
\text{R.H.S} &= (\mathbf{P} \otimes \mathbf{Q}) \otimes \mathbf{R} \\
&= \left(\begin{bmatrix} p_1 & p_2 \\ p_3 & p_4 \end{bmatrix} \otimes \begin{bmatrix} q_1 & q_2 \\ q_3 & q_4 \end{bmatrix} \right) \otimes \begin{bmatrix} r_1 \\ r_2 \end{bmatrix} \\
&= \begin{bmatrix} p_1q_1 & p_1q_2 & p_2q_1 & p_2q_2 \\ p_1q_3 & p_1q_4 & p_2q_3 & p_2q_4 \\ p_3q_1 & p_3q_2 & p_4q_1 & p_4q_2 \\ p_3q_3 & p_3q_4 & p_4q_3 & p_4q_4 \end{bmatrix} \otimes \begin{bmatrix} r_1 \\ \\ r_2 \end{bmatrix} \\
&= \begin{bmatrix} p_1q_1r_1 & p_1q_2r_1 & p_2q_1r_1 & p_2q_2r_1 \\ p_1q_1r_2 & p_1q_2r_2 & p_2q_1r_2 & p_2q_2r_2 \\ p_1q_3r_1 & p_1q_4r_1 & p_2q_3r_1 & p_2q_4r_1 \\ p_1q_3r_2 & p_1q_4r_2 & p_2q_3r_2 & p_2q_4r_2 \\ p_3q_1r_1 & p_3q_2r_1 & p_4q_1r_1 & p_4q_2r_1 \\ p_3q_1r_2 & p_3q_2r_2 & p_4q_1r_2 & p_4q_2r_2 \\ p_3q_3r_1 & p_3q_4r_1 & p_4q_3r_1 & p_4q_4r_1 \\ p_3q_3r_2 & p_3q_4r_2 & p_4q_3r_2 & p_4q_4r_2 \end{bmatrix}
\end{aligned}
$$

Since L.H.S = R.H.S, the associative law $\mathbf{P} \otimes (\mathbf{Q} \otimes \mathbf{R}) = (\mathbf{P} \otimes \mathbf{Q}) \otimes \mathbf{R}$ is verified.

(v) (a) $(\mathbf{P} + \mathbf{Q}) \otimes \mathbf{R} = (\mathbf{P} \otimes \mathbf{R}) + (\mathbf{Q} \otimes \mathbf{R})$
Since $\mathbf{P}$ and $\mathbf{Q}$ are of same order the addition on both sides holds.

$$
\begin{aligned}
\mathbf{P} + \mathbf{Q} &= \begin{bmatrix} p_1 + q_1 & p_2 + q_2 \\ p_3 + q_3 & p_4 + q_4 \end{bmatrix} \\
(\mathbf{P} + \mathbf{Q}) \otimes \mathbf{R} &= \begin{bmatrix} p_1 + q_1 & p_2 + q_2 \\ p_3 + q_3 & p_4 + q_4 \end{bmatrix} \otimes \begin{bmatrix} r_1 \\ r_2 \end{bmatrix} \\
&= \begin{bmatrix} (p_1 + q_1)r_1 & (p_2 + q_2)r_1 \\ (p_1 + q_1)r_2 & (p_2 + q_2)r_2 \\ (p_3 + q_3)r_1 & (p_4 + q_4)r_1 \\ (p_3 + q_3)r_2 & (p_4 + q_4)r_2 \end{bmatrix} \\
&= \begin{bmatrix} p_1r_1 & p_2r_1 \\ p_1r_2 & p_2r_2 \\ p_3r_1 & p_4r_1 \\ p_3r_2 & p_4r_2 \end{bmatrix} + \begin{bmatrix} q_1r_1 & q_2r_1 \\ q_1r_2 & q_2r_2 \\ q_3r_1 & q_4r_1 \\ q_3r_2 & q_4r_2 \end{bmatrix} \\
&= \begin{bmatrix} p_1\mathbf{R} & p_2\mathbf{R} \\ p_3\mathbf{R} & p_4\mathbf{R} \end{bmatrix} + \begin{bmatrix} q_1\mathbf{R} & q_2\mathbf{R} \\ q_3\mathbf{R} & q_4\mathbf{R} \end{bmatrix} \\
&= (\mathbf{P} \otimes \mathbf{R}) + (\mathbf{Q} \otimes \mathbf{R})
\end{aligned}
$$

(b) The order of the matrices $\mathbf{Q}$ and $\mathbf{R}$ is not compatible for addition. Hence, the second distributive law does not exist.

EXERCISE 6.3

1. Let $\mathbf{A} = \begin{bmatrix} 2 & 3 \\ -1 & 4 \end{bmatrix}$, $\mathbf{B} = \begin{bmatrix} 3 & 2 \\ -4 & 1 \end{bmatrix}$ then verify $\mathbf{A} \otimes \alpha\mathbf{B} = \alpha(\mathbf{A} \otimes \mathbf{B})$, where $\alpha \in \mathbb{R}$.

2. If $\mathbf{A} = \begin{bmatrix} 1 & 4 \\ 3 & 2 \end{bmatrix}$, $\mathbf{B} = \begin{bmatrix} 2 & 3 \\ -1 & 5 \end{bmatrix}$ then

   (i) calculate $3\mathbf{A} \otimes 4\mathbf{B}$ and verify that

   (ii) $3\mathbf{A} \otimes 4\mathbf{B} = 12(\mathbf{A} \otimes \mathbf{B})$.

3. Given $\mathbf{A} = \begin{bmatrix} 1 & 4 \\ 2 & 3 \end{bmatrix}$, $\mathbf{B} = \begin{bmatrix} 2 & 1 \\ 3 & 2 \end{bmatrix}$ find the Kronecker product of $\mathbf{A} \otimes \mathbf{B}$ and $\mathbf{B} \otimes \mathbf{A}$

4. If $\mathbf{A} = \begin{bmatrix} 2 & 1 \\ 0 & 1 \end{bmatrix}$, $\mathbf{B} = \begin{bmatrix} -1 & 1 \\ 2 & 0 \end{bmatrix}$ then calculate $\mathbf{A} \otimes \mathbf{B}$ and $\mathbf{B} \otimes \mathbf{A}$;

5. Let $\mathbf{A} = \begin{bmatrix} 1 & -1 \\ 2 & 3 \end{bmatrix}$, $\mathbf{B} = \begin{bmatrix} 2 & 3 \\ -1 & 2 \end{bmatrix}$, $\mathbf{C} = \begin{bmatrix} 3 & 4 \\ 2 & 1 \end{bmatrix}$. Verify that

   (i) $(\mathbf{A} \otimes \mathbf{B}) \otimes \mathbf{C} = \mathbf{A} \otimes (\mathbf{B} \otimes \mathbf{C})$;

   (ii) $(\mathbf{A} + \mathbf{B}) \otimes \mathbf{C} = (\mathbf{A} \otimes \mathbf{C}) + (\mathbf{B} \otimes \mathbf{C})$;

   (iii) $\mathbf{A} \otimes (\mathbf{B} + \mathbf{C}) = (\mathbf{A} \otimes \mathbf{B}) + (\mathbf{A} \otimes \mathbf{C})$.

6.4 Further Properties of Kronecker Products

In this section, the transpose of a Kronecker product and the mixed product rule will be obtained. Further, using the mixed product rule relations between eigen-values and eigen-vectors of a Kronecker product and the inverse of a Kronecker product will be established.

Consider the matrices $\mathbf{P} = (p_{ij})_{m\times n}$, $\mathbf{Q} = (q_{ij})_{p\times q}$, $\mathbf{R} = (r_{ij})_{n\times k}$ and $\mathbf{S} = (s_{ij})_{q\times l}$ in the given order.

(1) Transpose property of Kronecker product

$$(\mathbf{P} \otimes \mathbf{Q})^T = \mathbf{P}^T \otimes \mathbf{Q}^T.$$

Proof.

$$\mathbf{P} \otimes \mathbf{Q} = \begin{bmatrix} p_{11}\mathbf{Q} & p_{12}\mathbf{Q} & \cdots & p_{1n}\mathbf{Q} \\ p_{21}\mathbf{Q} & p_{22}\mathbf{Q} & \cdots & p_{2n}\mathbf{Q} \\ \vdots & \vdots & \cdots & \vdots \\ p_{m1}\mathbf{Q} & p_{m2}\mathbf{Q} & \cdots & p_{mn}\mathbf{Q} \end{bmatrix}$$

$$\text{hence } (\mathbf{P} \otimes \mathbf{Q})^T = \begin{bmatrix} p_{11}\mathbf{Q}^T & p_{21}\mathbf{Q}^T & \cdots & p_{m1}\mathbf{Q}^T \\ p_{12}\mathbf{Q}^T & p_{22}\mathbf{Q}^T & \cdots & p_{m2}\mathbf{Q}^T \\ \vdots & \vdots & \cdots & \vdots \\ p_{n1}\mathbf{Q}^T & p_{n2}\mathbf{Q}^T & \cdots & p_{mn}\mathbf{Q}^T \end{bmatrix}$$
$$= \mathbf{P}^T \otimes \mathbf{Q}^T,$$

by definition of the Kronecker product.

Observation. 1. The order of $\mathbf{P} \otimes \mathbf{Q}$ is $mp \times nq$. The order of $(\mathbf{P} \otimes \mathbf{Q})^T$ is $nq \times mp$. The order of $\mathbf{P}^T$ is $n \times m$, $\mathbf{Q}^T$ is $q \times p$ and $\mathbf{P}^T \otimes \mathbf{Q}^T$ is $nq \times mp$.
2. In ordinary multiplication of matrices, $(\mathbf{P}\,\mathbf{Q})^T = \mathbf{Q}^T\,\mathbf{P}^T$.

Example 6.4.1 *Verify the transpose property for*

$$\mathbf{P} = \begin{bmatrix} p_1 & p_2 \\ p_3 & p_4 \end{bmatrix} \quad \textit{and} \quad \mathbf{Q} = \begin{bmatrix} q_1 \\ q_2 \end{bmatrix}$$

Solution. *Given*

$$\mathbf{P} \otimes \mathbf{Q} = \begin{bmatrix} p_1q_1 & p_2q_1 \\ p_1q_2 & p_2q_2 \\ p_3q_1 & p_4q_1 \\ p_3q_2 & p_4q_2 \end{bmatrix}$$

$$(\mathbf{P} \otimes \mathbf{Q})^T = \begin{bmatrix} p_1q_1 & p_1q_2 & p_3q_1 & p_3q_2 \\ p_2q_1 & p_2q_2 & p_4q_1 & p_4q_4 \end{bmatrix}$$

$$\mathbf{P}^T = \begin{bmatrix} p_1 & p_3 \\ p_2 & p_4 \end{bmatrix}, \quad \mathbf{Q}^T = [q_1, q_2]$$

$$\mathbf{P}^T \otimes \mathbf{Q}^T = \begin{bmatrix} p_1q_1 & p_1q_2 & p_3q_1 & p_3q_2 \\ p_2q_1 & p_2q_2 & p_4q_1 & p_4q_2 \end{bmatrix}$$

The result is verified.

Example 6.4.2 *Verify the transpose property for*

$$\mathbf{A} = \begin{bmatrix} 2 & 3 \\ 1 & 4 \end{bmatrix} \qquad \mathbf{B} = \begin{bmatrix} 1 & 2 & 1 \\ 2 & 3 & 2 \end{bmatrix}$$

Solution:

$$\mathbf{A}\otimes\mathbf{B} = \begin{bmatrix} 2 & 4 & 2 & 3 & 6 & 3 \\ 4 & 6 & 4 & 6 & 9 & 6 \\ 1 & 2 & 1 & 4 & 8 & 4 \\ 2 & 3 & 2 & 8 & 12 & 8 \end{bmatrix}$$

$$(\mathbf{A}\otimes\mathbf{B})^T = \begin{bmatrix} 2 & 4 & 1 & 2 \\ 4 & 6 & 2 & 3 \\ 2 & 4 & 1 & 2 \\ 3 & 6 & 4 & 8 \\ 6 & 9 & 8 & 12 \\ 3 & 6 & 4 & 8 \end{bmatrix}$$

$$\mathbf{A}^T = \begin{bmatrix} 2 & 1 \\ 3 & 4 \end{bmatrix} \qquad \mathbf{B}^T = \begin{bmatrix} 1 & 2 \\ 2 & 3 \\ 1 & 2 \end{bmatrix} \quad \text{and} \quad \mathbf{A}^T\otimes B^T = \begin{bmatrix} 2 & 4 & 1 & 2 \\ 4 & 6 & 2 & 3 \\ 2 & 4 & 1 & 2 \\ 3 & 6 & 4 & 8 \\ 6 & 9 & 8 & 12 \\ 3 & 6 & 4 & 8 \end{bmatrix}$$

The result is verified.

(2) Mixed product rule.

$$(\mathbf{P}\otimes\mathbf{Q})(\mathbf{R}\otimes\mathbf{S}) = \mathbf{PR}\otimes\mathbf{QS}.$$

Proof. First observe that the matrix products are compatible and L.H.S and R.H.S gives rise to a matrix of the same order.

$$\begin{aligned} \text{L.H.S} &= (\mathbf{P}\otimes\mathbf{Q})_{mp\times nq}(\mathbf{R}\otimes\mathbf{S})_{nq\times kl} \\ &= ((\mathbf{P}\otimes\mathbf{Q})(\mathbf{R}\otimes\mathbf{S}))_{mp\times kl} \end{aligned}$$

R.H.S: $(\mathbf{PR})_{m\times k}\otimes(\mathbf{QS})_{p\times l} = (\mathbf{PR}\otimes\mathbf{QS})_{mp\times kl}$.

The proof of the rule is as follows.
Consider the L.H.S. $(\mathbf{P}\otimes\mathbf{Q})(\mathbf{R}\otimes\mathbf{S})$

$$= \begin{bmatrix} p_{11}\mathbf{Q} & p_{12}\mathbf{Q} & \cdots & p_{1n}\mathbf{Q} \\ p_{21}\mathbf{Q} & p_{22}\mathbf{Q} & \cdots & p_{2n}\mathbf{Q} \\ \vdots & \vdots & \cdots & \vdots \\ p_{m1}\mathbf{Q} & p_{m2}\mathbf{Q} & \cdots & p_{mn}\mathbf{Q} \end{bmatrix} \begin{bmatrix} r_{11}\mathbf{S} & r_{12}\mathbf{S} & \cdots & r_{1k}\mathbf{S} \\ r_{21}\mathbf{S} & r_{22}\mathbf{S} & \cdots & r_{2k}\mathbf{S} \\ \vdots & \vdots & \cdots & \vdots \\ r_{n1}\mathbf{S} & r_{n2}\mathbf{S} & \cdots & r_{nk}\mathbf{S} \end{bmatrix}$$

Now the (i,j)th entry of $(\mathbf{P}\otimes\mathbf{Q})(\mathbf{R}\otimes\mathbf{S})$

$$\begin{aligned} &= p_{i1}\mathbf{Q}r_{1j}\mathbf{S} + p_{i2}\mathbf{Q}r_{2j}\mathbf{S} + \cdots + p_{in}\mathbf{Q}r_{nj}\mathbf{S} \\ &= \sum_{t=1}^{n} p_{it}r_{tj}\mathbf{QS}. \end{aligned} \tag{6.10}$$

Next consider the L.H.S product $\mathbf{PR}\otimes\mathbf{QS}$.

The (i, j)th element of the product $\mathbf{PR}$ is given by

$$\sum_{i=1}^{n} p_{it} r_{tj}$$

and the (i, j)th element of the Kronecker product $\mathbf{PR} \otimes \mathbf{QS}$ is

$$= \sum_{i=1}^{n} p_{it} r_{tj} \mathbf{QS} \tag{6.11}$$

The result follows from (6.10) and (6.11).

Example 6.4.3 *Verify the mixed product rule for*

$$\mathbf{P} = \begin{bmatrix} 2 \\ 3 \end{bmatrix}_{2\times 1} \quad \mathbf{Q} = \begin{bmatrix} 1 & 2 & 3 \end{bmatrix}_{1\times 3} \quad \mathbf{R} = \begin{bmatrix} 1 & 4 \end{bmatrix}_{1\times 2} \quad \mathbf{S} = \begin{bmatrix} 1 & 0 & 1 \end{bmatrix}_{3\times 1}^{T}$$

Solution.

$$\begin{aligned} \mathbf{P} \otimes \mathbf{Q} &= \begin{bmatrix} 2 \\ 3 \end{bmatrix} \otimes \begin{bmatrix} 1 & 2 & 3 \end{bmatrix} \\ &= \begin{bmatrix} 2 & 4 & 6 \\ 3 & 6 & 9 \end{bmatrix} \end{aligned}$$

$$\mathbf{R} \otimes \mathbf{S} = \begin{bmatrix} 1 & 4 \end{bmatrix} \otimes \begin{bmatrix} 1 \\ 0 \\ 1 \end{bmatrix} = \begin{bmatrix} 1 & 4 \\ 0 & 0 \\ 1 & 4 \end{bmatrix}$$

$$\begin{aligned} \text{L.H.S} &= (\mathbf{P} \otimes \mathbf{Q})(\mathbf{R} \otimes \mathbf{S}) = \begin{bmatrix} 2 & 4 & 6 \\ 3 & 6 & 9 \end{bmatrix} \begin{bmatrix} 1 & 4 \\ 0 & 0 \\ 1 & 4 \end{bmatrix} \\ &= \begin{bmatrix} 8 & 32 \\ 12 & 48 \end{bmatrix} \end{aligned}$$

Now

$$\mathbf{PR} = \begin{bmatrix} 2 \\ 3 \end{bmatrix} \begin{bmatrix} 1 & 4 \end{bmatrix} = \begin{bmatrix} 2 & 8 \\ 3 & 12 \end{bmatrix}$$

$$\mathbf{QS} = \begin{bmatrix} 1 & 2 & 3 \end{bmatrix} \begin{bmatrix} 1 \\ 0 \\ 1 \end{bmatrix} = \begin{bmatrix} 4 \end{bmatrix}$$

$$\begin{aligned} \mathbf{PR} \otimes \mathbf{QS} &= \begin{bmatrix} 2 & 8 \\ 3 & 12 \end{bmatrix} \otimes \begin{bmatrix} 4 \end{bmatrix} \\ &= \begin{bmatrix} 8 & 32 \\ 12 & 48 \end{bmatrix}. \end{aligned}$$

The result is verified.

(3) The inverse of the Kronecker product $\mathbf{P}_{m\times m}$ and $\mathbf{Q}_{n\times n}$ is $(\mathbf{P}\otimes\mathbf{Q})^{-1} = \mathbf{P}^{-1}\otimes\mathbf{Q}^{-1}$.

The result follows from the mixed product rule by considering

$$\begin{aligned}(\mathbf{P}\otimes\mathbf{Q})(\mathbf{P}^{-1}\otimes\mathbf{Q}^{-1}) &= \mathbf{P}\mathbf{P}^{-1}\otimes\mathbf{Q}\mathbf{Q}^{-1}\\ &= \mathbf{I}_m\otimes\mathbf{I}_n.\end{aligned}$$

Example 6.4.4 *Let* $\mathbf{P} = \begin{bmatrix}1 & 2\\ 2 & 3\end{bmatrix}$ *and* $\mathbf{Q} = \begin{bmatrix}1 & 1\\ 3 & 4\end{bmatrix}$. *Verify the inverse property.*

Solution.

$$\mathbf{P}^{-1} = \begin{bmatrix}-3 & 2\\ 2 & -1\end{bmatrix} \quad \textit{and} \quad \mathbf{Q}^{-1} = \begin{bmatrix}4 & -1\\ -3 & 1\end{bmatrix}$$

Now

$$(\mathbf{P}\otimes\mathbf{Q}) = \begin{bmatrix}1 & 1 & 2 & 2\\ 3 & 4 & 6 & 8\\ 2 & 2 & 3 & 3\\ 6 & 8 & 9 & 12\end{bmatrix}$$

$$\mathbf{P}^{-1}\otimes\mathbf{Q}^{-1} = \begin{bmatrix}-12 & 3 & 8 & -2\\ 9 & -3 & -6 & 2\\ 8 & -2 & -4 & 1\\ -6 & 2 & 3 & -1\end{bmatrix}$$

$$\begin{aligned}(\mathbf{P}\otimes\mathbf{Q})(\mathbf{P}^{-1}\otimes\mathbf{Q}^{-1}) &= \begin{bmatrix}1 & 1 & 2 & 2\\ 3 & 4 & 6 & 8\\ 2 & 2 & 3 & 3\\ 6 & 8 & 9 & 12\end{bmatrix}\begin{bmatrix}-12 & 3 & 8 & -2\\ 9 & -3 & -6 & 2\\ 8 & -2 & -4 & 1\\ -6 & 2 & 3 & -1\end{bmatrix}\\ &= \begin{bmatrix}1 & 0 & 0 & 0\\ 0 & 1 & 0 & 0\\ 0 & 0 & 1 & 0\\ 0 & 0 & 0 & 1\end{bmatrix}\\ &= \mathbf{I}_2\otimes\mathbf{I}_2.\end{aligned}$$

Hence the result is verified.

(4) The eigen-values and eigen-vectors of a Kronecker product.

Let $\mathbf{P}_{m\times m}$ and $\mathbf{Q}_{n\times n}$ be given. Further, suppose that $\{\lambda_i\}$ $i = 1, 2, \ldots, m$ and $\{\mu_j\}$ $j = 1, 2, \ldots, n$ are the corresponding given values of $\mathbf{P}$ and $\mathbf{Q}$, respectively. Then the eigen-values and eigen-vectors of the Kronecker product $\mathbf{P}\otimes\mathbf{Q}$ are $\{\lambda_i\mu_j\}$ and $\mathbf{x}_i\otimes\mathbf{y}_j$, $i = 1, 2, \ldots, m$ and $j = 1, 2, \ldots, n$, respectively.

Proof. Consider $(\mathbf{P}\otimes\mathbf{Q})(\mathbf{x}_i\otimes\mathbf{y}_j)$ for some arbitrary fixed i and j. Using the mixed product rule

$$\begin{aligned}(\mathbf{P}\otimes\mathbf{Q})(\mathbf{x}_i\otimes\mathbf{y}_j) &= \mathbf{P}\mathbf{x}_i\otimes\mathbf{Q}\mathbf{y}_j\\ &= \lambda_i\mathbf{x}_i\otimes\mu_j\mathbf{y}_j\\ &= (\lambda_i\mu_j)\mathbf{x}_i\otimes\mathbf{y}_j,\end{aligned}$$

which implies that $\lambda_i \mu_j$ is the eigen-value of $\mathbf{P} \otimes \mathbf{Q}$ and its corresponding eigen-vector is $\mathbf{x}_i \otimes \mathbf{y}_j$.

Since the result holds for all i and all j the property is proved.

Example 6.4.5 *Let* $\mathbf{P} = \begin{bmatrix} 5 & 2 \\ 2 & 2 \end{bmatrix}$ *and* $\mathbf{Q} = \begin{bmatrix} 2 & 1 \\ 0 & 1 \end{bmatrix}$. *Verify the property 4.*

Solution. The eigen-values and the eigen-vectors of $\mathbf{P}$ are $\lambda_1 = 1$ and $\lambda_2 = 6$ and $\mathbf{x}_1 = \binom{-1}{2}$ and $\mathbf{x}_2 = \binom{2}{1}$.

The eigen-values and the eigen-vectors of $\mathbf{Q}$ are $\mu_1 = 1$ and $\mu_2 = 2$ and $\mathbf{y}_1 = \binom{-1}{1}$ and $\mathbf{y}_2 = \binom{1}{0}$.

Now

$$\mathbf{P} \otimes \mathbf{Q} = \begin{bmatrix} 10 & 5 & 4 & 2 \\ 0 & 5 & 0 & 2 \\ 4 & 2 & 4 & 2 \\ 0 & 2 & 0 & 2 \end{bmatrix}$$

whose eigen-values are 1, 2, 6, 12 and the corresponding eigen-vectors are

$$\begin{pmatrix} 1 \\ -1 \\ -2 \\ 2 \end{pmatrix}, \begin{pmatrix} -1 \\ 0 \\ 2 \\ 0 \end{pmatrix}, \begin{pmatrix} -2 \\ 2 \\ -1 \\ 1 \end{pmatrix}, \begin{pmatrix} 2 \\ 0 \\ 1 \\ 0 \end{pmatrix}$$

This property 4 can be easily verified as follows.

$$\lambda_1 \mu_1 = 1 \quad \text{and} \quad \mathbf{x}_1 \otimes \mathbf{y}_1 = \begin{pmatrix} -1 \\ 2 \end{pmatrix} \otimes \begin{pmatrix} -1 \\ 1 \end{pmatrix} = \begin{pmatrix} 1 \\ -1 \\ -2 \\ 2 \end{pmatrix}$$

$$\lambda_1 \mu_2 = 2 \quad \text{and} \quad \mathbf{x}_1 \otimes \mathbf{y}_2 = \begin{pmatrix} -1 \\ 2 \end{pmatrix} \otimes \begin{pmatrix} 1 \\ 0 \end{pmatrix} = \begin{pmatrix} -1 \\ 0 \\ 2 \\ 0 \end{pmatrix}$$

$$\lambda_2 \mu_1 = 6 \quad \text{and} \quad \mathbf{x}_2 \otimes \mathbf{y}_1 = \begin{pmatrix} 2 \\ 1 \end{pmatrix} \otimes \begin{pmatrix} -1 \\ 1 \end{pmatrix} = \begin{pmatrix} -2 \\ 2 \\ -1 \\ 1 \end{pmatrix}$$

$$\lambda_2 \mu_2 = 12 \quad \text{and} \quad \mathbf{x}_2 \otimes \mathbf{y}_2 = \begin{pmatrix} 2 \\ 1 \end{pmatrix} \otimes \begin{pmatrix} 1 \\ 0 \end{pmatrix} = \begin{pmatrix} 2 \\ 0 \\ 1 \\ 0 \end{pmatrix}$$

(5) Suppose $\mathbf{P}$ and $\mathbf{Q}$ are arbitrary square matrices of sizes m and n, respectively. Then $det(\mathbf{P} \otimes \mathbf{Q}) = (det\ \mathbf{P})^n (det\ \mathbf{Q})^m$.

Proof. Let $\lambda_1, \lambda_2, \ldots, \lambda_m$ and $\mu_1, \mu_2, \ldots, \mu_n$ be the eigen-values of $\mathbf{P}$ and $\mathbf{Q}$, respectively.

Then $det\ \mathbf{P} = \lambda_1\ \lambda_2\ \ldots\ \lambda_m$ and $det\ \mathbf{Q} = \mu_1\ \mu_2\ \ldots\ \mu_n$. Further

$$\begin{aligned}
det\ (\mathbf{P} \otimes \mathbf{Q}) &= \prod_{i,j} \lambda_i \mu_j \\
&= (\lambda_1\mu_1)(\lambda_1\mu_2)\cdots(\lambda_1\mu_n) \\
&\quad (\lambda_2\mu_1)(\lambda_2\mu_2)\cdots(\lambda_2\mu_n) \\
&\quad \vdots \\
&\quad (\lambda_m\mu_1)(\lambda_m\mu_2)\cdots(\lambda_m\mu_n) \\
&= (\lambda_1)^n(\mu_1\mu_2\cdots\mu_n) \\
&\quad (\lambda_2)^n(\mu_1\mu_2\cdots\mu_n) \\
&\quad (\lambda_m)^n(\mu_1\mu_2\cdots\mu_n) \\
&= (\lambda_1\cdots\lambda_m)^n(\mu_1\cdots\mu_n)^m \\
&= (det\ \mathbf{P})^n(det\ \mathbf{Q})^m, \text{ proving the result.}
\end{aligned}$$

Example 6.4.6 *Verify the property 5 for the matrices* **P** *and* **Q** *given below*

$$\mathbf{P} = \begin{bmatrix} 2 & 1 \\ 1 & 2 \end{bmatrix} \quad \mathbf{Q} = \begin{bmatrix} 3 & 3 \\ 1 & 2 \end{bmatrix}$$

$$det\ \mathbf{P} = 3 \qquad det\ \mathbf{Q} = 3.$$

$$\mathbf{P} \otimes \mathbf{Q} = \begin{bmatrix} 6 & 6 & 3 & 3 \\ 2 & 4 & 1 & 2 \\ 3 & 3 & 6 & 6 \\ 1 & 2 & 2 & 4 \end{bmatrix}$$

$$R_1 \longleftrightarrow R_4 = \begin{bmatrix} 1 & 2 & 2 & 4 \\ 2 & 4 & 1 & 2 \\ 3 & 3 & 6 & 6 \\ 6 & 6 & 3 & 3 \end{bmatrix}$$

$R_2 \longleftrightarrow R_2 - 2R1, R_3 \longleftrightarrow R_3 - 3R_1, R_4 \longleftrightarrow R_4 - 6R_1$

$$\approx \begin{bmatrix} 1 & 2 & 2 & 4 \\ 0 & 0 & -3 & -6 \\ 0 & -3 & 0 & -6 \\ 0 & -6 & -9 & -21 \end{bmatrix}$$

$R_4 \longleftrightarrow R_4 - 2R_3, R_4 \longleftrightarrow R_4 + 3R_2$

$$\approx \begin{bmatrix} 1 & 2 & 2 & 4 \\ 0 & 0 & -3 & -6 \\ 0 & -3 & 0 & -6 \\ 0 & 0 & 0 & -9 \end{bmatrix} = 1 \begin{bmatrix} 0 & -3 & -6 \\ -3 & 0 & -6 \\ 0 & 0 & -9 \end{bmatrix} = 81.$$

$det(\mathbf{P} \otimes \mathbf{Q}) = 81 = 3^2 \cdot 3^2 = (det\ \mathbf{P})^2(det\ \mathbf{Q})^2$. *Hence verified.*

EXERCISE 6.4

1. Let $\mathbf{A} = \begin{bmatrix} 2 & 1 \\ 0 & 1 \end{bmatrix}$ and $\mathbf{B} = \begin{bmatrix} -1 & 1 \\ 2 & 0 \end{bmatrix}$

   Verify (i) $(\mathbf{A} \otimes \mathbf{B})^T = \mathbf{A}^T \otimes \mathbf{B}^T$
   (ii) $(\mathbf{A} \otimes \mathbf{B})^{-1} = \mathbf{A}^{-1} \otimes \mathbf{B}^{-1}$

2. Verify the mixed product rule using the following matrices

   $\mathbf{A} = \begin{bmatrix} 5 & 4 \\ 3 & 2 \end{bmatrix}$, $\mathbf{B} = \begin{bmatrix} 3 & 2 \\ -4 & -1 \end{bmatrix}$, $\mathbf{C} = \begin{bmatrix} 1 & 2 \\ 4 & 3 \end{bmatrix}$ and $\mathbf{D} = \begin{bmatrix} 3 & 5 \\ 6 & 1 \end{bmatrix}$.

3. Verify the transpose property of Kronecker product for

   $\mathbf{A} = \begin{bmatrix} 1 & 4 \\ 3 & 2 \end{bmatrix}$ and $\mathbf{B} = \begin{bmatrix} 2 & -1 \\ 3 & 5 \end{bmatrix}$

4. Verify the inverse property of Kronecker product using

   $\mathbf{A} = \begin{bmatrix} 2 & 5 \\ 1 & 3 \end{bmatrix}$ $\mathbf{B} = \begin{bmatrix} 2 & 1 \\ 3 & 2 \end{bmatrix}$

5. Let $\mathbf{A} = \begin{bmatrix} 1 & 2 \\ 3 & 4 \end{bmatrix}$ and $\mathbf{B} = \begin{bmatrix} 2 & 1 \\ -4 & 3 \end{bmatrix}$ Verify the determinant property.

6. $\mathbf{A} = \begin{bmatrix} 1 & 0 \\ 1 & 2 \end{bmatrix}$ and $\mathbf{B} = \begin{bmatrix} 2 & 1 \\ 0 & 3 \end{bmatrix}$ Verify the eigen-value property for $\mathbf{A} \otimes \mathbf{B}$.

6.5 Kronecker Product of Two Linear Transformations

In this section, two linear transformations are considered and their Kronecker product is described.

Let $\mathbf{P} = (p_{ij})_{m\times n}$, $\mathbf{Q} = (q_{ij})_{k\times l}$ be two rectangular matrices and let $\mathbf{x} = \mathbf{Py}$ and $\mathbf{z} = \mathbf{Qw}$ be two linear transformations defined on them.

Then one can conclude that $\mathbf{x} = (x_1 \quad x_2 \quad \dots \quad x_m)^T$ and $\mathbf{y} = (y_1 \ y_2 \ \dots \ y_n)^T$.

Also $\mathbf{z} = (z_1 \ z_2 \ \dots \ z_k)^T$ and $\mathbf{w} = (w_1 \ w_2 \ \dots \ w_l)^T$.

To obtain the Kronecker product of the linear transformations

$$\begin{aligned} \mathbf{x} \otimes \mathbf{z} &= \mathbf{Py} \otimes \mathbf{Qw} \\ &= (\mathbf{P} \otimes \mathbf{Q})(\mathbf{y} \otimes \mathbf{w}) \end{aligned}$$

by the mixed product rule.

Observation. $\mathbf{x} \otimes \mathbf{z}$ is an $mk \times 1$ vectors,

$\mathbf{P} \otimes \mathbf{Q}$ is an $mk \times nl$ matrix,
and $\mathbf{y} \otimes \mathbf{w}$ is an $nl \times 1$ vectors. Hence the Kronecker product of the linear transformations is compatible.

Example 6.5.1 *Find a Kronecker product of the linear transformations involving the matrices*

$$\mathbf{P} = \begin{bmatrix} p_1 & p_2 \\ p_3 & p_4 \end{bmatrix} \quad \textit{and} \quad \mathbf{Q} = \begin{bmatrix} q_1 & q_2 \\ q_3 & q_4 \end{bmatrix}$$

given by $\mathbf{x} = \mathbf{Py}$ *and* $\mathbf{z} = \mathbf{Qw}$, *where* $\mathbf{x}, \mathbf{y}, \mathbf{z}, \mathbf{w}$ *are suitable vectors.*

Solution. The Kronecker product of the linear transformations is

$$\mathbf{x} \otimes \mathbf{z} = \mathbf{Py} \otimes \mathbf{Qw},$$

where $\mathbf{x} = (x_1 \;\; x_2)^T$, $\mathbf{y} = (y_1 \;\; y_2)^T$, $\mathbf{z} = (z_1 \;\; z_2)^T$ and $\mathbf{w} = (w_1 \;\; w_2)^T$

Now using the mixed product rule

$$\mathbf{x} \otimes \mathbf{z} = \mathbf{Py} \otimes \mathbf{Qw} = (\mathbf{P} \otimes \mathbf{Q})(\mathbf{y} \otimes \mathbf{w})$$

Thus

$$\begin{pmatrix} x_1 \\ x_2 \end{pmatrix} \otimes \begin{pmatrix} z_1 \\ z_2 \end{pmatrix} = \left(\begin{bmatrix} p_1 & p_2 \\ p_3 & p_4 \end{bmatrix} \otimes \begin{bmatrix} q_1 & q_2 \\ q_3 & q_4 \end{bmatrix} \right) \left(\begin{pmatrix} y_1 \\ y_2 \end{pmatrix} \otimes \begin{pmatrix} w_1 \\ w_2 \end{pmatrix} \right)$$

That is

$$\begin{pmatrix} x_1 z_1 \\ x_1 z_2 \\ x_2 z_1 \\ x_2 z_2 \end{pmatrix} = \begin{bmatrix} p_1 q_1 & p_1 q_2 & p_2 q_1 & p_2 q_2 \\ p_1 q_3 & p_1 q_4 & p_2 q_3 & p_2 q_4 \\ p_3 q_1 & p_3 q_2 & p_4 q_1 & p_4 q_2 \\ p_3 q_3 & p_3 q_4 & p_4 q_3 & p_2 q_4 \end{bmatrix} \begin{pmatrix} y_1 w_1 \\ y_1 w_2 \\ y_2 w_1 \\ y_2 w_2 \end{pmatrix}$$

Example 6.5.2 *Let* $\begin{bmatrix} x_1 \\ x_2 \\ x_3 \end{bmatrix} = \begin{bmatrix} 3 & 0 & 0 \\ 1 & 2 & 0 \\ 2 & 0 & 1 \end{bmatrix} \begin{pmatrix} y_1 \\ y_2 \\ y_3 \end{pmatrix}$ *and* $\begin{pmatrix} z_1 \\ z_2 \end{pmatrix} = \begin{bmatrix} 1 & 2 \\ 2 & 3 \end{bmatrix} \begin{pmatrix} w_1 \\ w_2 \end{pmatrix}$.
Find the Kronecker product of the above linear transformations.

Solution. *The product of the linear transformations is* $\begin{pmatrix} x_1 \\ x_2 \\ x_3 \end{pmatrix} \otimes \begin{pmatrix} z_1 \\ z_2 \end{pmatrix} =$
$\begin{bmatrix} 3 & 0 & 0 \\ 1 & 2 & 0 \\ 2 & 0 & 1 \end{bmatrix} \begin{pmatrix} y_1 \\ y_2 \\ y_3 \end{pmatrix} \otimes \begin{bmatrix} 1 & 2 \\ 2 & 3 \end{bmatrix} \begin{pmatrix} w_1 \\ w_2 \end{pmatrix}$

By mixed product rule $= \left(\begin{bmatrix} 3 & 0 & 0 \\ 1 & 2 & 0 \\ 2 & 0 & 1 \end{bmatrix} \otimes \begin{bmatrix} 1 & 2 \\ 2 & 3 \end{bmatrix} \right) \begin{pmatrix} y_1 \\ y_2 \\ y_3 \end{pmatrix} \otimes \begin{pmatrix} w_1 \\ w_2 \end{pmatrix}$

Thus $\begin{bmatrix} x_1 z_1 \\ x_1 z_2 \\ x_2 z_1 \\ x_2 z_2 \\ x_3 z_1 \\ x_3 z_2 \end{bmatrix} \begin{bmatrix} 3 & 6 & 0 & 0 & 0 & 0 \\ 6 & 9 & 0 & 0 & 0 & 0 \\ 1 & 2 & 2 & 4 & 0 & 0 \\ 2 & 3 & 4 & 6 & 0 & 0 \\ 2 & 4 & 0 & 0 & 1 & 2 \\ 4 & 6 & 0 & 0 & 2 & 3 \end{bmatrix} \begin{bmatrix} y_1 w_1 \\ y_1 w_2 \\ y_2 w_1 \\ y_2 w_2 \\ y_3 w_1 \\ y_3 w_2 \end{bmatrix}$

EXERCISE 6.5

1. Consider the linear transformations $\mathbf{x} = \mathbf{A}\mathbf{y}$, $\mathbf{z} = \mathbf{B}\mathbf{w}$ with $\mathbf{A} = (a_{ij})_{2\times 2}$ and $\mathbf{B} = (b_{ij})_{2\times 2}$ and the vectors $\mathbf{x} = [x_1, x_2]^T$, $\mathbf{y} = [y_1, y_2]^T$, $\mathbf{z} = [z_1, z_2]^T$ and $\mathbf{w} = [w_1, w_2]^T$. Find the Kronecker product of the two linear transformations x and z.

2. Let $\mathbf{x} = \begin{bmatrix} 1 & 0 \\ 0 & 1 \end{bmatrix} \mathbf{y}$ and $\mathbf{z} = \begin{bmatrix} -1 & 0 \\ 2 & -1 \end{bmatrix} \mathbf{w}$. Set the vectors $\mathbf{x}, \mathbf{y}, \mathbf{z}$ and $\mathbf{w}$ and then find the Kronecker product of the two linear transformations $\mathbf{x}$ and $\mathbf{z}$.

3. Consider $\mathbf{A} = \begin{bmatrix} 1 & 4 \\ 2 & 3 \end{bmatrix}$ and $\mathbf{B} = \begin{bmatrix} 2 & 1 \\ 3 & 2 \end{bmatrix}$. Define linear transformations using $\mathbf{A}$ and $\mathbf{B}$ as in exercise 1 and find the Kronecker product of the two linear transformations so obtained.

6.6 Kronecker Product and Vector Operators

An important result in this section deals with expressing a vector operator of a product of 3 matrices as a Kronecker product involving the concerned matrices. This requires the kth column of the product of the 3 matrices. The following sequence of simple problems are useful in reaching the goal. Let $\mathbf{P}_{kl}$, $\mathbf{P}_{!k}$, $\mathbf{P}_{l!}$ be primary matrices defined in Section 6.2.

Example 6.6.1 *Given any matrix* $\mathbf{P}_{m\times n}$*, show that* $\mathbf{P}\mathbf{P}_{kl} = \mathbf{P}_{*k}\mathbf{P}_{!l}$ *and hence deduce* $\mathbf{P}\mathbf{P}_{ll} = \mathbf{P}_{*l}\mathbf{P}_{!l}$.

Solution. Consider $\mathbf{P}\mathbf{P}_{kl} = \mathbf{P}\mathbf{P}_{k1}\mathbf{P}_{1l} = \mathbf{P}_{*k}\mathbf{P}_{!l}$, using (6.2). Set $k = l$ Then $\mathbf{P}\mathbf{P}_{!l} = \mathbf{P}_{*l}\mathbf{P}_{!l}$.

Example 6.6.2 *Show that* $\mathbf{P}\mathbf{P}_{kl}\mathbf{Q} = \mathbf{P}_{*k}\mathbf{Q}_{l*}^T$.

Solution. Consider

$$\begin{aligned}
\mathbf{P}\mathbf{P}_{kl}\mathbf{Q} &= (\mathbf{P}\mathbf{P}_{k1})(\mathbf{P}_{!l}\mathbf{Q}) \\
&= \mathbf{P}_{*k}\mathbf{Q}_{l*}^T, \quad \text{using (6.2) and (6.3).}
\end{aligned}$$

Example 6.6.3 *Show that* $\mathbf{P}_{kl}\mathbf{P}\mathbf{P}_{rs} = p_{lr}\mathbf{P}_{ks}$
and hence deduce $\mathbf{P}_{ll}\mathbf{P}\mathbf{P}_{rr} = P_{lr}\mathbf{P}_{lr}$.

Solution. Consider

$$\begin{aligned}\mathbf{P}_{kl}\mathbf{P}\mathbf{P}_{rs} &= \mathbf{P}_{k1}\left(\mathbf{P}_{1l}\mathbf{P}\mathbf{P}_{r1}\right)\mathbf{P}_{1s} \\ &= \mathbf{P}_{k1}\left(p_{lr}\right)\mathbf{P}_{1s}, \\ &= p_{lr}\mathbf{P}_{k1}\mathbf{P}_{1s} = p_{lr}\mathbf{P}_{ks}\end{aligned}$$

on using (6.6) and (6.1), respectively.

Setting $k = l$ and $s = r$ gives

$$\mathbf{P}_{ll}\mathbf{P}\mathbf{P}_{rr} = p_{lr}\mathbf{P}_{lr}.$$

Example 6.6.4 *Express the product* **PQ** *using primary matrices.*

Solution. Assuming that **P** and **Q** are compatible so that the product **PQ** is well defined, using (6.4) and relation in Example 6.6.1,

$$\mathbf{P} = \sum \mathbf{P}_{*j}\mathbf{P}_{1j} = \sum \mathbf{P}\mathbf{P}_{jj}$$

Thus

$$\begin{aligned}\mathbf{PQ} &= \sum_j (\mathbf{P}\mathbf{P}_{jj})\mathbf{Q} \\ &= \sum_j (\mathbf{P}\mathbf{P}_{j!})(\mathbf{P}_{!j}\mathbf{Q}) \\ &= \sum_j \mathbf{P}_{*j}\mathbf{Q}_{j*}^T.\end{aligned}$$

Example 6.6.5 *Find the* kth *column of the product* **PQ**.

Solution. Consider

$$\begin{aligned}(\mathbf{PQ})_{*k} &= \mathbf{PQP}_{k1} \\ &= \mathbf{P}(\mathbf{QP}_{k1}) = \mathbf{PQ}_{*k} \\ &= \sum_j (\mathbf{P}\mathbf{P}_{jj})\mathbf{Q}_{*k} \\ &= \sum_j \mathbf{P}\mathbf{P}_{j1}\mathbf{P}_{1j}\mathbf{Q}_{*k} \\ &= \sum_j \mathbf{P}_{*j}q_{jk} \\ &= \sum_j (q_{jk})\mathbf{P}_{*j},\end{aligned}$$

on using the definition of kth column of a matrix, Example 6.6.1 and (6.1).

Example 6.6.6 *Find the* kth *column of the product of 3 matrices* $\mathbf{P}$, $\mathbf{Y}$ *and* $\mathbf{Q}$.

Solution. The result can be obtained using results from the above examples. Consider

$$\begin{aligned}(\mathbf{PYQ})_{*k} &= (\mathbf{PYQ})\mathbf{P}_{k1} \\ &= (\mathbf{PY})\mathbf{Q}_{*k} \\ &= \sum_j (\mathbf{PY})_{*j} q_{jk} \\ &= \sum_j (q_{jk}\mathbf{P})\mathbf{Y}_{*j}\end{aligned}$$

Now the main result of this section is as follows.

Result 6.6.1 $Vec(\mathbf{PYQ}) = \mathbf{Q}^T \otimes \mathbf{P}\, Vec\ \mathbf{Y}$.

Proof. Suppose that $\mathbf{P}_{m\times n}\mathbf{Y}_{n\times r}\mathbf{Q}_{r\times s}$ matrices, so that the product $(\mathbf{PYQ})_{m\times s}$ matrix exists.

Now the kth column of $(\mathbf{PYQ})$ is obtained using the Example 6.6.6 as

$$\begin{aligned}(\mathbf{PYQ})_{*k} &= \sum_j (q_{jk}\mathbf{P})\mathbf{Y}_{*j} \\ &= \begin{bmatrix} q_{1k}\mathbf{P} & q_{2k}\mathbf{P} & \cdots & q_{rk}\mathbf{P}\end{bmatrix}\begin{bmatrix}\mathbf{Y}_{*1}\\ \mathbf{Y}_{*2}\\ \vdots \\ \mathbf{Y}_{*r}\end{bmatrix} \\ &= \mathbf{Q}_{*k}^T \otimes \mathbf{P}\begin{bmatrix}\mathbf{Y}_{*1}\\ \mathbf{Y}_{*2}\\ \vdots \\ \mathbf{Y}_{*r}\end{bmatrix} \\ &= \left[[\mathbf{Q}_{*k}]^T \otimes \mathbf{P}\right]\ Vec\ \mathbf{Y}\end{aligned}$$

Hence

$$Vec(\mathbf{PYQ}) = \begin{bmatrix}[\mathbf{Q}_{*1}]^T \otimes \mathbf{P} & Vec\ \mathbf{Y}\\ [\mathbf{Q}_{*2}]^T \otimes \mathbf{P} & Vec\ \mathbf{Y}\\ \vdots & \\ [\mathbf{Q}_{*s}]^T \otimes \mathbf{P} & Vec\ \mathbf{Y}\end{bmatrix}.$$

Example 6.6.7 *Verify Result 6.6.1 for the following matrices*

$$\mathbf{P} = \begin{bmatrix} p_1 & p_2 \\ p_3 & p_4\end{bmatrix} \quad \mathbf{Y} = \begin{bmatrix} y_1 \\ y_2\end{bmatrix} \quad \mathbf{Q} = \begin{bmatrix} q_1 & q_2\end{bmatrix}$$

Solution. The matrix multiplication is compatible and

$$\begin{aligned}\mathbf{PYQ} &= \begin{bmatrix} p_1 & p_2 \\ p_3 & p_4 \end{bmatrix} \begin{bmatrix} y_1 \\ y_2 \end{bmatrix} \begin{bmatrix} q_1 & q_2 \end{bmatrix} \\ &= \begin{bmatrix} p_1y_1 + p_2y_2 \\ p_3y_1 + p_4y_2 \end{bmatrix} \begin{bmatrix} q_1 & q_2 \end{bmatrix} \\ &= \begin{bmatrix} (p_1y_1 + p_2y_2)\, q_1 & (p_1y_1 + p_2y_2)\, q_2 \\ (p_3y_1 + p_4y_2)\, q_1 & (p_3y_1 + p_4y_2)\, q_2 \end{bmatrix}\end{aligned}$$

hence

$$\begin{aligned}Vec(\mathbf{PYQ}) &= \begin{bmatrix} (p_1y_1 + p_2y_2)\, q_1 \\ (p_3y_1 + p_4y_2)\, q_1 \\ (p_1y_1 + p_2y_2)\, q_2 \\ (p_3y_1 + p_4y_2)\, q_2 \end{bmatrix}. \\ &= \text{L.H.S of Result 6.6.1}\end{aligned}$$

Now $\mathbf{Q}^T = \begin{bmatrix} q_1 \\ q_2 \end{bmatrix}$

$$\begin{aligned}\text{1st column of R.H.S} &= \left[[\mathbf{Q}_{*1}]^T \otimes \mathbf{P}\right] \,Vec\ \mathbf{Y} \\ &= \left[q_1 \otimes \mathbf{P}\right] \begin{bmatrix} y_1 \\ y_2 \end{bmatrix} \\ &= \begin{bmatrix} q_1p_1 & q_1p_2 \\ q_1p_3 & q_1p_4 \end{bmatrix} \begin{bmatrix} y_1 \\ y_2 \end{bmatrix} \\ &= \begin{bmatrix} q_1p_1y_1 + q_1p_2y_2 \\ q_1p_3y_1 + q_1p_4y_2 \end{bmatrix}\end{aligned}$$

$$\begin{aligned}\text{2nd column of R.H.S} &= \left[[\mathbf{Q}_{*2}]^T \otimes \mathbf{P}\right] \,Vec\ \mathbf{Y} \\ &= \left[q_2 \otimes \mathbf{P}\right] \begin{bmatrix} y_1 \\ y_2 \end{bmatrix} \\ &= \begin{bmatrix} q_2p_1 & q_2p_2 \\ q_2p_3 & q_2p_4 \end{bmatrix} \begin{bmatrix} y_1 \\ y_2 \end{bmatrix} \\ &= \begin{bmatrix} q_2p_1y_1 + q_2p_2y_2 \\ q_2p_3y_1 + q_2p_4y_2 \end{bmatrix}\end{aligned}$$

$$\text{Hence the matrix on R.H.S} \ = \begin{bmatrix} q_1\,(p_1y_1 + p_2y_2) \\ q_1\,(p_3y_1 + p_4y_2) \\ q_2\,(p_1y_1 + p_2y_2) \\ q_2\,(p_3y_1 + p_4y_2) \end{bmatrix}$$

Since L.H.S = R.H.S, the Result 1 is verified.

EXERCISE 6.6

1. Write the equation $\mathbf{AY} = \mathbf{B}$,

$$\begin{bmatrix} a_1 & a_3 \\ a_2 & a_4 \end{bmatrix} \begin{bmatrix} y_1 & y_3 \\ y_2 & y_4 \end{bmatrix} = \begin{bmatrix} b_1 & b_3 \\ b_2 & b_4 \end{bmatrix}$$

   in a vector-matrix form.

   Hint. Consider $\mathbf{AYI}_2 = \mathbf{B}$.

2. Show that $tr(\mathbf{AB}) = (\mathit{Vec}\ \mathbf{A}^T)^T\ \mathit{Vec}\ \mathbf{B}$.

3. Let $\mathbf{A} = \begin{bmatrix} 2 & -1 \\ 3 & 4 \end{bmatrix}$, $\mathbf{B} = \begin{bmatrix} 3 & -4 \\ 2 & 1 \end{bmatrix}$, $Y = \begin{bmatrix} y_1 & y_3 \\ y_2 & y_4 \end{bmatrix}$
   Verify that $\mathit{Vec}\ (\mathbf{AYB}) = (\mathbf{B}^T \otimes \mathbf{A})\ \mathit{Vec}\ \mathbf{Y}$.

4. Let $\mathbf{A} = \begin{bmatrix} 1 & 1 \\ -2 & 3 \end{bmatrix}$, $\mathbf{B} = \begin{bmatrix} 2 & 7 \\ 11 & 9 \end{bmatrix}$, $\mathbf{Z} = \begin{bmatrix} z_1 & z_3 \\ z_2 & z_4 \end{bmatrix}$
   Verify that $\mathit{Vec}\ (\mathbf{AZB}) = (\mathbf{B}^T \otimes \mathbf{A})\ \mathit{Vec}\ \mathbf{Z}$.

6.7 Permutation Matrices and Kronecker Products

In this section, the concept of a permutation matrix is introduced and their usefulness in expressing Kronecker products is given.

The primary column matrices are the basic ingredients of a permutation matrix.

Definition 6.7.1 *Permutation matrix of order n is an $n \times n$ square matrix consisting of 'n' primary column matrices of order $n \times 1$.*

Example 6.7.1 *The permutation matrices using primary column matrices of order 2, $\begin{bmatrix} 1 & 0 \end{bmatrix}^T$ and $\begin{bmatrix} 0 & 1 \end{bmatrix}^T$ are* $\begin{bmatrix} 1 & 0 \\ 0 & 1 \end{bmatrix}, \begin{bmatrix} 1 & 1 \\ 0 & 0 \end{bmatrix}, \begin{bmatrix} 0 & 0 \\ 1 & 1 \end{bmatrix}, \begin{bmatrix} 0 & 1 \\ 1 & 0 \end{bmatrix}$.

Example 6.7.2 *Let $\mathbf{P}_{ij}$ be a primary matrix for order 2 and $i, j = 1, 2$. Find*

$$\mathbf{U} = \sum_{j=1}^{2} \sum_{i=1}^{2} \mathbf{P}_{ij} \otimes \mathbf{P}_{ji}$$

Solution.

$$\mathbf{U} = \mathbf{P}_{11} \otimes \mathbf{P}_{11} + \mathbf{P}_{12} \otimes \mathbf{P}_{21} + \mathbf{P}_{21} \otimes \mathbf{P}_{12} + \mathbf{P}_{22} \otimes \mathbf{P}_{22}$$
$$= \begin{bmatrix} 1 & 0 \\ 0 & 0 \end{bmatrix} \otimes \begin{bmatrix} 1 & 0 \\ 0 & 0 \end{bmatrix} + \begin{bmatrix} 0 & 1 \\ 0 & 0 \end{bmatrix} \otimes \begin{bmatrix} 0 & 0 \\ 1 & 0 \end{bmatrix} + \begin{bmatrix} 0 & 0 \\ 1 & 0 \end{bmatrix} \otimes \begin{bmatrix} 0 & 1 \\ 0 & 0 \end{bmatrix}$$
$$+ \begin{bmatrix} 0 & 0 \\ 0 & 1 \end{bmatrix} \otimes \begin{bmatrix} 0 & 0 \\ 0 & 1 \end{bmatrix} = \begin{bmatrix} 1 & 0 & 0 & 0 \\ 0 & 0 & 1 & 0 \\ 0 & 1 & 0 & 0 \\ 0 & 0 & 0 & 1 \end{bmatrix}$$

Example 6.7.3 *Find the permutation matrix obtained by* $\mathbf{I}_3 \otimes \mathbf{I}_2$.

Solution.

$$\mathbf{I}_3 \otimes \mathbf{I}_2 = \begin{bmatrix} 1 & 0 & 0 \\ 0 & 1 & 0 \\ 0 & 0 & 1 \end{bmatrix} \otimes \begin{bmatrix} 1 & 0 \\ 0 & 1 \end{bmatrix}$$
$$= \begin{bmatrix} 1 & 0 & 0 & 0 & 0 & 0 \\ 0 & 1 & 0 & 0 & 0 & 0 \\ 0 & 0 & 1 & 0 & 0 & 0 \\ 0 & 0 & 0 & 1 & 0 & 0 \\ 0 & 0 & 0 & 0 & 1 & 0 \\ 0 & 0 & 0 & 0 & 0 & 1 \end{bmatrix}.$$

Result 6.7.1 Given a matrix $\mathbf{A}$ there exists a permutation matrix $\mathbf{U}$ such that $Vec\ \mathbf{A}^T = \mathbf{U}\, Vec\ \mathbf{A}$.

Proof. Let $\mathbf{A} = (a_{ij})_{m \times n}$. Then the permutation matrix can be obtained either by considering primary column matrices of order $mn \times 1$ or by considering the primary matrices of order $m \times n$ and using the corresponding Vec operator. The proof using primary matrices of order $m \times n$ is as follows.

$$\text{The matrix } \mathbf{A} = \sum\sum a_{ij}\mathbf{P}_{ij}$$
$$\text{hence } \mathbf{A}^T = \sum_j \sum_i a_{ij}\mathbf{P}_{ij}^T$$
$$Vec\ \mathbf{A}^T = \left[Vec\ \mathbf{P}_{11}^T \;\; Vec\ \mathbf{P}_{21}^T \;\; \cdots \;\; Vec\ \mathbf{P}_{m1}^T \;\; Vec\ \mathbf{P}_{12}^T \;\; \cdots \right.$$
$$\left. Vec\ \mathbf{P}_{m2}^T \;\; \cdots \;\; Vec\ \mathbf{P}_{mn}^T\right] Vec\ \mathbf{A}$$
$$\text{Thus } \mathbf{U} = \left[Vec\ \mathbf{P}_{11}^T \;\; Vec\ \mathbf{P}_{21}^T \;\; \cdots \;\; Vec\ \mathbf{P}_{m1}^T \;\; Vec\ \mathbf{P}_{12}^T \;\; \cdots \right.$$
$$\left. Vec\ \mathbf{P}_{m2}^T \;\; \cdots \;\; Vec\ \mathbf{P}_{mn}^T\right]$$

In case primary column matrices are used. Each matrix is of order $mn \times 1$ and

$$\mathbf{U} = \left[\mathbf{P}_{1\,1} \;\; \mathbf{P}_{m+1\,1} \;\; \mathbf{P}_{2m+1\,1} \;\; \cdots \mathbf{P}_{(n-1)m+1\,1} \;\; \mathbf{P}_{2\,1} \;\; \mathbf{P}_{m+2\,1} \;\; \mathbf{P}_{2m+2\,1} \;\; \cdots \right.$$
$$\left. \mathbf{P}_{(n-1)m+2\,1}\mathbf{P}_{m\,1} \;\; \mathbf{P}_{m+m\,1} \;\; \mathbf{P}_{2m+m\,1} \;\; \cdots \mathbf{P}_{mn\,1}\right]$$

Then $Vec\ \mathbf{A} = \mathbf{U}^T\ Vec\ \mathbf{A}^T$.

Example 6.7.4 *Find the permutation matrix using primary column matrices for the matrix* $\mathbf{A} = \begin{bmatrix} a_{11} & a_{12} & a_{13} \\ a_{21} & a_{22} & a_{23} \\ a_{31} & a_{32} & a_{33} \end{bmatrix}$ *so that* $Vec\ \mathbf{A} = \mathbf{U}^T\ Vec\ \mathbf{A}^T$.

Solution. Using the above formula with $m = 3, n = 3$

$$\mathbf{U} = \begin{bmatrix} \mathbf{P}_{11} & \mathbf{P}_{41} & \mathbf{P}_{71} & \mathbf{P}_{21} & \mathbf{P}_{51} & \mathbf{P}_{81} & \mathbf{P}_{31} & \mathbf{P}_{61} & \mathbf{P}_{91} \end{bmatrix}^T$$

$$\mathbf{U} = \begin{bmatrix} 1 & 0 & 0 & 0 & 0 & 0 & 0 & 0 & 0 \\ 0 & 0 & 0 & 1 & 0 & 0 & 0 & 0 & 0 \\ 0 & 0 & 0 & 0 & 0 & 0 & 1 & 0 & 0 \\ 0 & 1 & 0 & 0 & 0 & 0 & 0 & 0 & 0 \\ 0 & 0 & 0 & 0 & 1 & 0 & 0 & 0 & 0 \\ 0 & 0 & 0 & 0 & 0 & 0 & 0 & 1 & 0 \\ 0 & 0 & 1 & 0 & 0 & 0 & 0 & 0 & 0 \\ 0 & 0 & 0 & 0 & 0 & 1 & 0 & 0 & 0 \\ 0 & 0 & 0 & 0 & 0 & 0 & 0 & 0 & 1 \end{bmatrix}^T$$

$$\text{Consider}\quad \mathbf{U}^T\, Vec\ \mathbf{A}^T = \begin{bmatrix} 1 & 0 & 0 & 0 & 0 & 0 & 0 & 0 & 0 \\ 0 & 0 & 0 & 1 & 0 & 0 & 0 & 0 & 0 \\ 0 & 0 & 0 & 0 & 0 & 0 & 1 & 0 & 0 \\ 0 & 1 & 0 & 0 & 0 & 0 & 0 & 0 & 0 \\ 0 & 0 & 0 & 0 & 1 & 0 & 0 & 0 & 0 \\ 0 & 0 & 0 & 0 & 0 & 0 & 0 & 1 & 0 \\ 0 & 0 & 1 & 0 & 0 & 0 & 0 & 0 & 0 \\ 0 & 0 & 0 & 0 & 0 & 1 & 0 & 0 & 0 \\ 0 & 0 & 0 & 0 & 0 & 0 & 0 & 0 & 1 \end{bmatrix} \begin{bmatrix} a_{11} \\ a_{12} \\ a_{13} \\ a_{21} \\ a_{22} \\ a_{23} \\ a_{31} \\ a_{32} \\ a_{33} \end{bmatrix} = \begin{bmatrix} a_{11} \\ a_{21} \\ a_{31} \\ a_{12} \\ a_{22} \\ a_{32} \\ a_{13} \\ a_{23} \\ a_{33} \end{bmatrix} = Vec\,\mathbf{A}.$$

Example 6.7.5 *Verify the Result 6.7.1 for the matrix*

$$\mathbf{A} = \begin{bmatrix} a_{11} & a_{12} \\ a_{21} & a_{22} \end{bmatrix}$$

Consider the vector formed by using the primary matrices $\mathbf{P}_{11}\ \ \mathbf{P}_{12}\ \ \mathbf{P}_{21}\ \ \mathbf{P}_{22}$

$$Vec\ \mathbf{P}_{11} = \begin{bmatrix} 1 \\ 0 \\ 0 \\ 0 \end{bmatrix} \quad Vec\ \mathbf{P}_{12} = \begin{bmatrix} 0 \\ 0 \\ 1 \\ 0 \end{bmatrix} \quad Vec\ \mathbf{P}_{21} = \begin{bmatrix} 0 \\ 1 \\ 0 \\ 0 \end{bmatrix} \quad Vec\ \mathbf{P}_{22} = \begin{bmatrix} 0 \\ 0 \\ 0 \\ 1 \end{bmatrix}$$

$$\mathbf{U} = \begin{bmatrix} 1 & 0 & 0 & 0 \\ 0 & 0 & 1 & 0 \\ 0 & 1 & 0 & 0 \\ 0 & 0 & 0 & 1 \end{bmatrix}$$

$$\mathbf{U}^T \ Vec\ \mathbf{A} = \begin{bmatrix} 1 & 0 & 0 & 0 \\ 0 & 0 & 1 & 0 \\ 0 & 1 & 0 & 0 \\ 0 & 0 & 0 & 1 \end{bmatrix} \begin{bmatrix} a_{11} \\ a_{21} \\ a_{12} \\ a_{22} \end{bmatrix} = \begin{bmatrix} a_{11} \\ a_{12} \\ a_{21} \\ a_{22} \end{bmatrix} = Vec\ \mathbf{A}^T.$$

Observation. The permutation matrix is not an identity matrix.

Result 6.7.2 Let $\mathbf{P} = [p_{ij}]_{m\times n}$ and $\mathbf{Q} = [q_{ij}]_{k\times l}$ be matrices of order $m \times n$ and $k \times l$, respectively. Then there exist permutation matrices $\mathbf{U}_1$ and $\mathbf{U}_2$ such that

$$\mathbf{P} \otimes \mathbf{Q} = \mathbf{U}_1(\mathbf{Q} \otimes \mathbf{P})\mathbf{U}_2.$$

Proof. The proof follows from the Result 6.6.1, which gives $Vec\ (\mathbf{PYQ}) = (\mathbf{Q}^T \otimes \mathbf{P})\ Vec\ \mathbf{Y}$.

$$\begin{aligned} \text{Set} \qquad \mathbf{PYQ}^T &= \mathbf{X} \\ \text{Then } Vec\ (\mathbf{PYQ}^T) &= Vec\ \mathbf{X} \\ \Longrightarrow (\mathbf{Q} \otimes \mathbf{P})\ Vec\ \mathbf{Y} &= Vec\ \mathbf{X}. \end{aligned}$$

Using $(\mathbf{PYQ}^T)^T = \mathbf{QY}^T\mathbf{P}^T = \mathbf{X}^T$
applying the Result 6.6.1, gives

$$(\mathbf{P} \otimes \mathbf{Q})\ Vec\ \mathbf{Y}^T = Vec\ \mathbf{X}^T$$

Now from Result 6.7.1, there exists permutation matrices $\mathbf{U}_1$ and $\mathbf{U}_2$ such that

$$Vec\ \mathbf{X}^T = \mathbf{U}_1\ Vec\ \mathbf{X} \quad \text{and} \quad Vec\ \mathbf{Y} = \mathbf{U}_2\ Vec\ \mathbf{Y}^T$$

multiplying the relation with $\mathbf{U}_1$ both sides gives

$$\mathbf{U}_1(\mathbf{Q} \otimes \mathbf{P})\mathbf{U}_2\ Vec\ \mathbf{Y}^T = \mathbf{U}_1\ Vec\mathbf{X}.$$

Also, $(\mathbf{P} \otimes \mathbf{Q})\ Vec\ \mathbf{Y}^T = \mathbf{U}_1\ Vec\ \mathbf{X}$
comparing the above relations give

$$\mathbf{P} \otimes \mathbf{Q} = \mathbf{U}_1(\mathbf{Q} \otimes \mathbf{P})\mathbf{U}_2,$$

where $\mathbf{U}_1$ and $\mathbf{U}_2$ are permutation matrices.

Example 6.7.6 *Given* $\mathbf{P} = \begin{bmatrix} 1 & 2 \\ 3 & 4 \end{bmatrix}$ *and* $\mathbf{Q} = \begin{bmatrix} 2 & 1 \\ -4 & 3 \end{bmatrix}$. *Find permutation matrices* $\mathbf{U}_1$ *and* $\mathbf{U}_2$ *such that* $\mathbf{P} \otimes \mathbf{Q} = \mathbf{U}_1(\mathbf{Q} \otimes \mathbf{P})\mathbf{U}_2$

Solution.

$$\text{From Example 6.7.5, } \mathbf{U}_1 = \mathbf{U}_2 = \begin{bmatrix} 1 & 0 & 0 & 0 \\ 0 & 0 & 1 & 0 \\ 0 & 1 & 0 & 0 \\ 0 & 0 & 0 & 1 \end{bmatrix}$$

Verification

$$L.H.S = \mathbf{P} \otimes \mathbf{Q} = \begin{bmatrix} 1 & 2 \\ 3 & 4 \end{bmatrix} \otimes \begin{bmatrix} 2 & 1 \\ -4 & 3 \end{bmatrix}$$

$$= \begin{bmatrix} 2 & 1 & 4 & 2 \\ -4 & 3 & -8 & 6 \\ 6 & 3 & 8 & 4 \\ -12 & 9 & -16 & 12 \end{bmatrix}$$

$$\mathbf{Q} \otimes \mathbf{P} = \begin{bmatrix} 2 & 1 \\ -4 & 3 \end{bmatrix} \otimes \begin{bmatrix} 1 & 2 \\ 3 & 4 \end{bmatrix} = \begin{bmatrix} 2 & 4 & 1 & 2 \\ 6 & 8 & 3 & 4 \\ -4 & -8 & 3 & 6 \\ -12 & -16 & 9 & 12 \end{bmatrix}$$

$$\mathbf{U}_1(\mathbf{Q} \otimes \mathbf{P}) = \begin{bmatrix} 1 & 0 & 0 & 0 \\ 0 & 0 & 1 & 0 \\ 0 & 1 & 0 & 0 \\ 0 & 0 & 0 & 1 \end{bmatrix} \begin{bmatrix} 2 & 4 & 1 & 2 \\ 6 & 8 & 3 & 4 \\ -4 & -8 & 3 & 6 \\ -12 & -16 & 9 & 12 \end{bmatrix}$$

$$= \begin{bmatrix} 2 & 4 & 1 & 2 \\ -4 & -8 & 3 & 6 \\ 6 & 8 & 3 & 4 \\ -12 & -16 & 9 & 12 \end{bmatrix}$$

$$\mathbf{U}_1(\mathbf{Q} \otimes \mathbf{P})\mathbf{U}_2 = \begin{bmatrix} 2 & 4 & 1 & 2 \\ -4 & -8 & 3 & 6 \\ 6 & 8 & 3 & 4 \\ -12 & -16 & 9 & 12 \end{bmatrix} \begin{bmatrix} 1 & 0 & 0 & 0 \\ 0 & 0 & 1 & 0 \\ 0 & 1 & 0 & 0 \\ 0 & 0 & 0 & 1 \end{bmatrix}$$

$$= \begin{bmatrix} 2 & 1 & 4 & 2 \\ -4 & 3 & -8 & 6 \\ 6 & 3 & 8 & 4 \\ -12 & 9 & -16 & 12 \end{bmatrix}$$

$$= \mathbf{P} \otimes \mathbf{Q}$$

Hence the example verifies the Result 6.7.2.

EXERCISE 6.7

1. Write down the primary matrices of order 3 and find the permutation matrix
$$\mathbf{U} = \sum\sum \mathbf{P}_{ij} \otimes \mathbf{P}_{ji}$$
What is the order of the matrix $\mathbf{U}$?

2. Given $\mathbf{A} = \begin{bmatrix} 2 & 5 \\ 4 & 3 \end{bmatrix}$, find a permutation matrix, $\mathbf{U}$ such that $\mathit{Vec}\ \mathbf{A}^T = \mathbf{U}\ \mathit{Vec}\ \mathbf{A}$.

3. Let $\mathbf{A} = \begin{bmatrix} 2 & 3 \\ 4 & 2 \end{bmatrix}$ and $\mathbf{B} = \begin{bmatrix} 2 & 5 \\ 3 & 3 \end{bmatrix}$. Find permutation matrices $\mathbf{U}_1$ and $\mathbf{U}_2$ such that $\mathbf{U}_1(\mathbf{B} \otimes \mathbf{A})\mathbf{U}_2$.

6.8 Analytical Functions and Kronecker Product

In this section, an analytical function is considered and a relation is obtained using the Kronecker product. Let h be an analytical function of z. Then from the theory of analytical functions, h can be expressed as a convergent series,

$$h(z) = c_0 + c_1 z + c_2 z^2 + \cdots + c_n z^n + \cdots,$$

where $c_0, c_1, c_2, \ldots, c_n, \cdots$ are all real constants.

The following result involves Kronecker products.

Result 6.8.1 Let $\mathbf{P}$ be a square matrix of order m and h be an analytical function. Then,

$$\text{(a)} \quad h(\mathbf{I}_n \otimes \mathbf{P}) = \mathbf{I}_n \otimes h(\mathbf{P})$$

and

$$\text{(b)} \quad h(\mathbf{P} \otimes \mathbf{I}_n) = h(\mathbf{P}) \otimes \mathbf{I}_n.$$

Proof. Consider $h(\mathbf{I}_n \otimes \mathbf{P})$, since h is an analytical function

$$h(\mathbf{I}_n \otimes \mathbf{P}) = \sum c_k (\mathbf{I}_n \otimes \mathbf{P})^k.$$

Now using the mixed product rule $(\mathbf{I}_n \otimes \mathbf{P})^2$

$$= (\mathbf{I}_n \otimes \mathbf{P})(\mathbf{I}_n \otimes \mathbf{P}) = \mathbf{I}_n \otimes \mathbf{P}^2$$
$$(\mathbf{I}_n \otimes \mathbf{P})^3 = (\mathbf{I}_n \otimes \mathbf{P}^2)(I \otimes \mathbf{P}) = \mathbf{I}_n \otimes \mathbf{P}^3$$

In general, $(\mathbf{I}_n \otimes \mathbf{P})^k = \mathbf{I}_n \otimes \mathbf{P}^k$

$$\begin{aligned}
\text{Thus} \quad h(\mathbf{I}_n \otimes \mathbf{P}) &= \sum c_k(\mathbf{I}_n \otimes \mathbf{P})^k \\
&= \sum c_k(\mathbf{I}_n \otimes \mathbf{P}^k) \\
&= \sum \mathbf{I}_n \otimes c_k\mathbf{P}^k \\
&= \mathbf{I}_n \otimes \sum c_k\mathbf{P}^k = \mathbf{I}_n \otimes h(\mathbf{P})^k.
\end{aligned}$$

Similarly, (b) can be proved by considering

$$h(\mathbf{P} \otimes \mathbf{I}_n) = \sum c_n(\mathbf{P} \otimes \mathbf{I})^n.$$

Example 6.8.1 *Verify Result 6.8.1 (a) and (b) for the polynomial* $h(x) = 3x + 1$ *for* $\mathbf{I}_2$ *and* $\mathbf{P} = \begin{bmatrix} 1 & 2 & 0 \\ 0 & 1 & 2 \\ 0 & 0 & 1 \end{bmatrix}$

$$\textit{Let} \quad \mathbf{I}_n = \mathbf{I}_2 = \begin{bmatrix} 1 & 0 \\ 0 & 1 \end{bmatrix}$$

$$\begin{aligned}
\textit{Then} \quad \mathbf{I}_2 \otimes \mathbf{P} &= \begin{bmatrix} 1 & 0 \\ 0 & 1 \end{bmatrix} \otimes \begin{bmatrix} 1 & 2 & 0 \\ 0 & 1 & 2 \\ 0 & 0 & 1 \end{bmatrix} \\
&= \begin{bmatrix} 1 & 2 & 0 & 0 & 0 & 0 \\ 0 & 1 & 2 & 0 & 0 & 0 \\ 0 & 0 & 1 & 0 & 0 & 0 \\ 0 & 0 & 0 & 1 & 2 & 0 \\ 0 & 0 & 0 & 0 & 1 & 2 \\ 0 & 0 & 0 & 0 & 0 & 1 \end{bmatrix}
\end{aligned}$$

$$\begin{aligned}
h(\mathbf{I}_2 \otimes \mathbf{P}) &= 3(\mathbf{I}_2 \otimes \mathbf{P}) + \mathbf{I} \\
&= 3\begin{bmatrix} 1 & 2 & 0 & 0 & 0 & 0 \\ 0 & 1 & 2 & 0 & 0 & 0 \\ 0 & 0 & 1 & 0 & 0 & 0 \\ 0 & 0 & 0 & 1 & 2 & 0 \\ 0 & 0 & 0 & 0 & 1 & 2 \\ 0 & 0 & 0 & 0 & 0 & 1 \end{bmatrix} + \begin{bmatrix} 1 & 0 & 0 & 0 & 0 & 0 \\ 0 & 1 & 0 & 0 & 0 & 0 \\ 0 & 0 & 1 & 0 & 0 & 0 \\ 0 & 0 & 0 & 1 & 0 & 0 \\ 0 & 0 & 0 & 0 & 1 & 0 \\ 0 & 0 & 0 & 0 & 0 & 1 \end{bmatrix} \\
&= \begin{bmatrix} 4 & 6 & 0 & 0 & 0 & 0 \\ 0 & 4 & 6 & 0 & 0 & 0 \\ 0 & 0 & 4 & 0 & 0 & 0 \\ 0 & 0 & 0 & 4 & 6 & 0 \\ 0 & 0 & 0 & 0 & 4 & 6 \\ 0 & 0 & 0 & 0 & 0 & 4 \end{bmatrix}
\end{aligned}$$

$$\textit{Further,}\ h(\mathbf{P}) = 3\begin{bmatrix}1&2&0\\0&1&2\\0&0&1\end{bmatrix} + \begin{bmatrix}1&0&0\\0&1&0\\0&0&1\end{bmatrix} = \begin{bmatrix}4&6&0\\0&4&6\\0&0&4\end{bmatrix}$$

$$\textit{Now,}\ \mathbf{I}_2 \otimes h(\mathbf{P}) = \begin{bmatrix}1&0\\0&1\end{bmatrix} \otimes \begin{bmatrix}4&6&0\\0&4&6\\0&0&4\end{bmatrix} = \begin{bmatrix}4&6&0&0&0&0\\0&4&6&0&0&0\\0&0&4&0&0&0\\0&0&0&4&6&0\\0&0&0&0&4&6\\0&0&0&0&0&4\end{bmatrix}$$

hence the Result 6.8.1(a) is verified. 6.8.1(b) is left as an exercise.

Example 6.8.2 *For $h(z) = e^z$ verify the Result 6.8.1(a) and 6.8.1(b) with $\mathbf{I}_2$ and $\mathbf{P}$ as in above example.*

Solution. Consider $h(z) = e^z$. Since e^z is an analytical function from Result 6.8.1(a) to show $e^{\mathbf{I}_2 \otimes \mathbf{P}} = \mathbf{I}_2 \otimes e^P$

Consider $e^{\mathbf{I}_2 \otimes \mathbf{P}} = \mathbf{I}_6 + \mathbf{I}_2 \otimes \mathbf{P} + \frac{1}{2!}(\mathbf{I}_2 \otimes \mathbf{P})^2 + \cdots$

$$= \begin{bmatrix}1&0&0&0&0&0\\0&1&0&0&0&0\\0&0&1&0&0&0\\0&0&0&1&0&0\\0&0&0&0&1&0\\0&0&0&0&0&1\end{bmatrix} + \begin{bmatrix}1&2&0&0&0&0\\0&1&2&0&0&0\\0&0&1&0&0&0\\0&0&0&1&2&0\\0&0&0&0&1&2\\0&0&0&0&0&1\end{bmatrix} + \frac{1}{2}\begin{bmatrix}1&4&4&0&0&0\\0&1&4&0&0&0\\0&0&1&0&0&0\\0&0&0&1&4&4\\0&0&0&0&1&4\\0&0&0&0&0&1\end{bmatrix}$$

$$= \begin{bmatrix}5/2&4&2&0&0&0\\0&5/2&4&0&0&0\\0&0&5/2&0&0&0\\0&0&0&5/2&4&2\\0&0&0&0&5/2&4\\0&0&0&0&0&5/2\end{bmatrix} + \cdots$$

Now $e^{\mathbf{P}} = \mathbf{I}_3 + \mathbf{P} + \mathbf{P}^2/2! + \cdots$

$$= \begin{bmatrix}1&0&0\\0&1&0\\0&0&1\end{bmatrix} + \begin{bmatrix}1&2&0\\0&1&2\\0&0&1\end{bmatrix} + \frac{1}{2}\begin{bmatrix}1&4&4\\0&1&4\\0&0&1\end{bmatrix} + \cdots$$

$$= \begin{bmatrix}5/2&4&2\\0&5/2&4\\0&0&5/2\end{bmatrix}.$$

$$\text{Now} \quad \mathbf{I}_2 \otimes e^{\mathbf{P}} = \begin{bmatrix} 1 & 0 \\ 0 & 1 \end{bmatrix} \otimes \begin{bmatrix} 5/2 & 4 & 2 \\ 0 & 5/2 & 4 \\ 0 & 0 & 5/2 \end{bmatrix}$$

$$= \begin{bmatrix} 5/2 & 4 & 2 & 0 & 0 & 0 \\ 0 & 5/2 & 4 & 0 & 0 & 0 \\ 0 & 0 & 5/2 & 0 & 0 & 0 \\ 0 & 0 & 0 & 5/2 & 4 & 2 \\ 0 & 0 & 0 & 0 & 5/2 & 4 \\ 0 & 0 & 0 & 0 & 0 & 5/2 \end{bmatrix}$$

Hence the Result 6.8.1(a) is verified.

Observation.

1. Note that in $e^{\mathbf{I}_2 \otimes \mathbf{P}}$ the 1st three terms have been considered and also in e^P, the 1st three terms were considered. The same number of terms in $e^{\mathbf{I}_2 \otimes \mathbf{P}}$ and e^P must be chosen.

2. Result 6.8.1(b) is left as exercise.

   The following example deals with using the tools of linear algebra to verify the Result 6.8.1(a).

Example 6.8.3 *Given* $\mathbf{A} = \begin{bmatrix} 3 & 4 \\ -2 & -3 \end{bmatrix}$ *calculate*

(i) $exp(\mathbf{A})$ *(ii)* $exp(\mathbf{A} \otimes \mathbf{I})$

(iii) verify the result $exp(\mathbf{A}) \otimes \mathbf{I} = exp(\mathbf{A} \otimes \mathbf{I})$.

(i) Clearly the eigen-values of $\mathbf{A}$ are 1 and -1 and eigen-vectors are $\mathbf{x}_1 = [2 \quad -1]^T$ and $\mathbf{x}_2 = [1 \quad -1]^T$, respectively.

Next matrix $\mathbf{P} = [\mathbf{x}_1 \quad \mathbf{x}_2]$ is given by,

$$\mathbf{P} = \begin{bmatrix} 2 & 1 \\ -1 & -1 \end{bmatrix} \text{ and hence } \mathbf{P}^{-1} = \begin{bmatrix} 1 & 1 \\ -1 & -2 \end{bmatrix}$$

Observe that $\mathbf{P}^{-1}\mathbf{A}\mathbf{P} = \begin{bmatrix} 1 & 0 \\ 0 & -1 \end{bmatrix} = \mathbf{D}$ and

$$exp\,\mathbf{A} = \mathbf{P}e^{\mathbf{D}}\mathbf{P}^{-1} = \begin{bmatrix} 2 & 1 \\ -1 & -1 \end{bmatrix} \begin{bmatrix} e & 0 \\ 0 & e^{-1} \end{bmatrix} \begin{bmatrix} 1 & 1 \\ -1 & -2 \end{bmatrix},$$

$$\text{or } exp\,\mathbf{A} = \begin{bmatrix} 2e - e^{-1} & 2(e - e^{-1}) \\ e^{-1} - e & 2e^{-1} - e \end{bmatrix}$$

(ii) Clearly,

$$\mathbf{A} \otimes \mathbf{I}_2 = \begin{bmatrix} 3 & 0 & 4 & 0 \\ 0 & 3 & 0 & 4 \\ -2 & 0 & -3 & 0 \\ 0 & -2 & 0 & -3 \end{bmatrix},$$

The characteristic equation for matrix $\mathbf{A} \otimes \mathbf{I}_2$ is

$$(\mathbf{A} \otimes \mathbf{I}_2) - \lambda \mathbf{I}_4 = \begin{bmatrix} 3-\lambda & 0 & 4 & 0 \\ 0 & 3-\lambda & 0 & 4 \\ -2 & 0 & -3-\lambda & 0 \\ 0 & -2 & 0 & -3-\lambda \end{bmatrix}$$

Verify that the eigen-values are 1 and -1 with multiplicity 2 in each case. The eigen-vectors associated to 1, are given by the system

$$\begin{aligned} 2x + 4z &= 0, \\ 2y + 4w &= 0, \\ -2x - 4z &= 0, \\ -2y - 4w &= 0. \end{aligned}$$

This gives

$$\mathbf{x}_1 = \alpha[0 \quad -2 \ \ 0 \ \ 1]^T + \beta[-2 \ \ 0 \ \ 1 \ \ 0]^T.$$

The eigen-vectors associated to -1 are given by the system

$$\begin{aligned} 4x + 4z &= 0 \\ 4y + 4w &= 0 \\ -2x - 2z &= 0 \\ -2y - 2w &= 0 \end{aligned}$$

This gives

$$\mathbf{x}_2 = \gamma[1 \ \ 0 \quad -1 \ \ 0]^T + \delta[0 \ \ 1 \ \ 0 \quad -1]^T.$$

Note that the geometric multiplicity of 1 and -1 is 2, the same as its algebraic multiplicity. Now, set

$$\mathbf{P} = \begin{bmatrix} 0 & -2 & 1 & 0 \\ -2 & 0 & 0 & 1 \\ 0 & 1 & -1 & 0 \\ 1 & 0 & 0 & -1 \end{bmatrix} \text{ then } \mathbf{P}^{-1} = \begin{bmatrix} 0 & -1 & 0 & -1 \\ -1 & 0 & -1 & 0 \\ -1 & 0 & -2 & 0 \\ 0 & -1 & 0 & -2 \end{bmatrix}$$

$$\text{and } \mathbf{P}^{-1}\mathbf{AP} = \begin{bmatrix} 1 & 0 & 0 & 0 \\ 0 & 1 & 0 & 0 \\ 0 & 0 & -1 & 0 \\ 0 & 0 & 0 & -1 \end{bmatrix}$$

Hence

$$exp(\mathbf{A} \otimes \mathbf{I}_2) = \mathbf{P}e^{\mathbf{D}}\mathbf{P}^{-1}$$
$$= \begin{bmatrix} 2e - e^{-1} & 0 & 2e - 2e^{-1} & 0 \\ 0 & 2e - e^{-1} & 0 & 2e - 2e^{-1} \\ e^{-1} - e & 0 & 2e^{-1} - e & 0 \\ 0 & e^{-1} - e & 0 & 2e^{-1} - e \end{bmatrix}.$$

(iii) It follows that $(exp\mathbf{A}) \otimes \mathbf{I}_2 = exp(\mathbf{A} \otimes \mathbf{I}_2)$.

EXERCISE 6.8

1. If $\mathbf{A} = \begin{bmatrix} 3 & 2 \\ 1 & 4 \end{bmatrix}$, calculate

   (i) $exp\mathbf{A}$

   (ii) $exp(\mathbf{A} \otimes \mathbf{I})$

   (iii) verify the property $(exp\mathbf{A}) \otimes \mathbf{I} = exp(\mathbf{A} \otimes \mathbf{I})$

2. Let $\mathbf{A} = \begin{bmatrix} 1 & 1 \\ -2 & 4 \end{bmatrix}$, $B = \begin{bmatrix} 5 & 4 \\ 1 & 2 \end{bmatrix}$

   (i) find the eigen-values and eigen-vectors of $\mathbf{A}$ and $\mathbf{B}$.

   (ii) find $exp\mathbf{A}$ and $exp\mathbf{B}$.

   (iii) verify the property $(exp\mathbf{A}) \otimes \mathbf{I} = exp(\mathbf{A} \otimes \mathbf{I})$

   (iv) verify the property $(exp\mathbf{B}) \otimes \mathbf{I} = exp(\mathbf{B} \otimes \mathbf{I})$

3. If $\mathbf{A} = \begin{bmatrix} 0 & 1 \\ 1 & 0 \end{bmatrix}$, show that $exp\mathbf{A} = \begin{bmatrix} \cosh 1 & \sinh 1 \\ \sinh 1 & \cosh 1 \end{bmatrix}$.

4. If $\mathbf{A} = \begin{bmatrix} 0 & x \\ x & 0 \end{bmatrix}$, show that $exp\mathbf{A}x = \begin{bmatrix} \cosh x & \sinh x \\ \sinh x & \cosh x \end{bmatrix}$.

5. If $\mathbf{A} = \begin{bmatrix} 0 & x \\ -x & 0 \end{bmatrix}$, show that $exp\mathbf{A}x = \begin{bmatrix} \cos x & \sin x \\ \sin x & \cos x \end{bmatrix}$.

6.9 Kronecker Sum

In this section, the Kronecker sum ($\oplus$) of two matrices is defined and some of its properties are obtained. The symbol $\oplus$ is different from the symbol for Kronecker product, $\otimes$. Let $\mathbf{P}$ and $\mathbf{Q}$ be square matrices of orders m and n, respectively.

Definition 6.9.1 *The Kronecker sum of two matrices* $\mathbf{P}$ *and* $\mathbf{Q}$*, denoted by* $\mathbf{P} \oplus \mathbf{Q}$*, defined as*

$$\mathbf{P} \oplus \mathbf{Q} = \mathbf{P} \otimes \mathbf{I}_n + \mathbf{I}_m \otimes \mathbf{Q}.$$

Example 6.9.1 *Let* $\mathbf{P} = \begin{bmatrix} 1 & 1 \\ 2 & 1 \end{bmatrix}$ *and* $\mathbf{Q} = \begin{bmatrix} 2 & 1 \\ -1 & 3 \end{bmatrix}$. *Find* $\mathbf{P} \oplus \mathbf{Q}$.

Solution. By definition

$$\begin{aligned}\mathbf{P} \oplus \mathbf{Q} &= \begin{bmatrix} 1 & 1 \\ 2 & 1 \end{bmatrix} \otimes \begin{bmatrix} 1 & 0 \\ 0 & 1 \end{bmatrix} + \begin{bmatrix} 1 & 0 \\ 0 & 1 \end{bmatrix} \times \begin{bmatrix} 2 & 1 \\ -1 & 3 \end{bmatrix} \\ &= \begin{bmatrix} 1 & 0 & 1 & 0 \\ 0 & 1 & 0 & 1 \\ 2 & 0 & 1 & 0 \\ 0 & 2 & 0 & 1 \end{bmatrix} + \begin{bmatrix} 2 & 1 & 0 & 0 \\ -1 & 3 & 0 & 0 \\ 0 & 0 & 2 & 1 \\ 0 & 0 & -1 & 3 \end{bmatrix} = \begin{bmatrix} 3 & 1 & 1 & 0 \\ -1 & 4 & 0 & 1 \\ 2 & 0 & 3 & 1 \\ 0 & 2 & -1 & 4 \end{bmatrix}.\end{aligned}$$

Result 6.9.1 The following are some of the properties of a Kronecker sum

(i) $\mathbf{P} \oplus \mathbf{Q} = \mathbf{O}_{mn} \implies \mathbf{P} = \mathbf{O}_m$ and $\mathbf{Q} = \mathbf{O}_n$,
where $\mathbf{O}_m$ and $\mathbf{O}_n$ are zero matrices of order m and n, respectively

(ii) $\mathbf{I}_m \oplus \mathbf{I}_n = 2\mathbf{I}_{mn}$

(iii) Mixed addition rule: $(\mathbf{P} \oplus \mathbf{Q}) + (\mathbf{R} \oplus \mathbf{S}) = (\mathbf{P} + \mathbf{R}) \oplus (\mathbf{Q} + \mathbf{S})$, where $\mathbf{P}$ and $\mathbf{R}$ are square matrices of order m and $\mathbf{Q}$ and $\mathbf{S}$ are square matrices of order n

(iv) Associative law holds $\mathbf{P} \oplus (\mathbf{Q} \oplus \mathbf{R}) = (\mathbf{P} \oplus \mathbf{Q}) \oplus \mathbf{R}$

(v) If $\{\lambda_k\}$ and $\{\mu_l\}$ are the eigen-values and $\{\mathbf{x}_k\}$ and $\{\mathbf{y}_l\}$ are the corresponding eigen-vectors of $\mathbf{P}$ and $\mathbf{Q}$, respectively then $\lambda_k + \mu_l$ are the eigen-values and $\mathbf{x}_k \otimes \mathbf{y}_l$ are the corresponding eigen-vectors of $\mathbf{P} \oplus \mathbf{Q}$.

Proof.

(i) Clearly if $\mathbf{P} \neq \mathbf{O}$ and $\mathbf{Q} \neq \mathbf{O}$ then

$$\mathbf{P} \oplus \mathbf{Q} = \mathbf{P} \oplus \mathbf{I}_n + \mathbf{I}_m \oplus \mathbf{Q} \neq \mathbf{O}.$$

Suppose if possible $\mathbf{P} \neq \mathbf{O}$ and $\mathbf{Q} = \mathbf{O}$, then

$$\begin{aligned}\mathbf{P} \oplus \mathbf{Q} = \mathbf{O} &\implies \mathbf{P} \otimes \mathbf{I}_n + \mathbf{I}_m \otimes \mathbf{Q} = \mathbf{P} \otimes \mathbf{I}_n + \mathbf{O} = \mathbf{O} \implies \mathbf{P} \otimes \mathbf{I}_n = \mathbf{O} \\ &\implies \mathbf{P} = \mathbf{O}.\end{aligned}$$

Similarly, if $\mathbf{P} = \mathbf{O}$ and if possible $\mathbf{Q} \neq \mathbf{O}$, then it can be shown that $\mathbf{Q} = \mathbf{O}$.

(ii) $\mathbf{I}_m \oplus \mathbf{I}_n = \mathbf{I}_m \otimes \mathbf{I}_n + \mathbf{I}_m \otimes \mathbf{I}_n = 2(\mathbf{I}_m \otimes \mathbf{I}_n) = 2\mathbf{I}_{mn}.$

(iii) Consider $\mathbf{P}, \mathbf{R}$ and $\mathbf{Q}, \mathbf{S}$ square matrices of order m and n, respectively. Then

$$\begin{aligned}
&(\mathbf{P} \oplus \mathbf{Q}) + (\mathbf{R} \oplus \mathbf{S}) \\
&\quad = (\mathbf{P} \otimes \mathbf{I}_n + \mathbf{I}_m \otimes \mathbf{Q}) + (\mathbf{R} \otimes \mathbf{I}_n + \mathbf{I}_m \otimes \mathbf{S}) \\
&\quad = (\mathbf{P} \otimes \mathbf{I}_n + \mathbf{R} \otimes \mathbf{I}_n) + (\mathbf{I}_m \otimes \mathbf{Q} + \mathbf{I}_m \otimes \mathbf{S}) \\
&\quad = (\mathbf{P} + \mathbf{R}) \otimes \mathbf{I}_n + \mathbf{I}_m \otimes (\mathbf{Q} + \mathbf{S}) \\
&\quad = (\mathbf{P} + \mathbf{R}) \oplus (\mathbf{Q} + \mathbf{S}), \text{by definition of Kronecker sum.}
\end{aligned}$$

(iv) Let $\mathbf{P}, \mathbf{Q}$ and $\mathbf{R}$ be square matrices of order m. Then

$$\begin{aligned}
\text{L.H.S} &= \mathbf{P} \oplus (\mathbf{Q} \oplus \mathbf{R}) \\
&= \mathbf{P} \oplus (\mathbf{Q} \otimes \mathbf{I}_m + \mathbf{I}_m \otimes \mathbf{R}) \\
&= \mathbf{P} \otimes \mathbf{I}_{m^2} + \mathbf{I}_m \otimes \mathbf{Q} \otimes \mathbf{I}_m + \mathbf{I}_{m^2} \otimes \mathbf{R}
\end{aligned}$$

$$\begin{aligned}
\text{R.H.S} &= (\mathbf{P} \oplus \mathbf{Q}) \oplus \mathbf{R} \\
&= (\mathbf{P} \otimes \mathbf{I}_m + \mathbf{I}_m \otimes \mathbf{Q}) \oplus \mathbf{R} \\
&= (\mathbf{P} \otimes \mathbf{I}_m + \mathbf{I}_m \otimes \mathbf{Q}) \otimes \mathbf{I}_m + \mathbf{I}_{m^2} \otimes \mathbf{R} \\
&= \mathbf{P} \otimes \mathbf{I}_{m^2} + \mathbf{I}_m \otimes \mathbf{Q} \otimes \mathbf{I}_m + \mathbf{I}_{m^2} \otimes \mathbf{R}
\end{aligned}$$

Since L.H.S = R.H.S the result holds.

(v) Let $\mathbf{P}_{m\times m}$ and $\mathbf{Q}_{n\times n}$ be square matrices. Consider

$$\begin{aligned}
&(\mathbf{P} \oplus \mathbf{Q})(\mathbf{x}_k \otimes \mathbf{y}_l) \\
&\quad = (\mathbf{P} \otimes \mathbf{I}_n + \mathbf{I}_m \otimes \mathbf{Q})(\mathbf{x}_k \otimes \mathbf{y}_l) \\
&\quad = (\mathbf{P} \otimes \mathbf{I}_n)(\mathbf{x}_k \otimes \mathbf{y}_l) + (\mathbf{I}_m \otimes \mathbf{Q})(\mathbf{x}_k \otimes \mathbf{y}_l) \\
&\quad = (\mathbf{P}\mathbf{x}_k \otimes \mathbf{y}_l) + (\mathbf{x}_k \otimes \mathbf{Q}\mathbf{y}_l) \\
&\quad = \lambda_k \mathbf{x}_k \otimes \mathbf{y}_l + \mathbf{x}_k \otimes \mu_l \mathbf{y}_l \\
&\quad = (\lambda_k + \mu_l)(\mathbf{x}_k \otimes \mathbf{y}_l),
\end{aligned}$$

which implies by the definition of eigen-values and eigen-vectors that $(\lambda_k + \mu_l)$ is an eigen-value of $(\mathbf{P} \oplus \mathbf{Q})$ with corresponding eigen-vector $\mathbf{x}_k \otimes \mathbf{y}_l$.

Example 6.9.2 *Let* $\mathbf{P} = \begin{bmatrix} 1 & 0 \\ 0 & 1 \end{bmatrix}$, $\mathbf{Q} = \begin{bmatrix} -1 & 0 \\ 0 & 0 \end{bmatrix}$ *and* $\mathbf{R} = \begin{bmatrix} 0 & 0 \\ 0 & 0 \end{bmatrix}$. *Find (i)* $\mathbf{P} \oplus \mathbf{Q}$ *(ii)* $\mathbf{P} \oplus \mathbf{R}$

(i) By definition

$$\mathbf{P} \oplus \mathbf{Q} = \begin{bmatrix} 1 & 0 \\ 0 & 0 \end{bmatrix} \otimes \begin{bmatrix} 1 & 0 \\ 0 & 1 \end{bmatrix} + \begin{bmatrix} 1 & 0 \\ 0 & 1 \end{bmatrix} \begin{bmatrix} -1 & 0 \\ 0 & 0 \end{bmatrix}$$

$$= \begin{bmatrix} 1 & 0 & 0 & 0 \\ 0 & 1 & 0 & 0 \\ 0 & 0 & 0 & 0 \\ 0 & 0 & 0 & 0 \end{bmatrix} + \begin{bmatrix} -1 & 0 & 0 & 0 \\ 0 & 0 & 0 & 0 \\ 0 & 0 & -1 & 0 \\ 0 & 0 & 0 & 0 \end{bmatrix}$$

$$= \begin{bmatrix} 0 & 0 & 0 & 0 \\ 0 & 1 & 0 & 0 \\ 0 & 0 & -1 & 0 \\ 0 & 0 & 0 & 0 \end{bmatrix}$$

(ii)

$$\mathbf{P} \oplus \mathbf{R} = \begin{bmatrix} 1 & 0 \\ 0 & 0 \end{bmatrix} \otimes \begin{bmatrix} 1 & 0 \\ 0 & 1 \end{bmatrix} + \begin{bmatrix} 1 & 0 \\ 0 & 1 \end{bmatrix} \otimes \begin{bmatrix} 0 & 0 \\ 0 & 0 \end{bmatrix}$$

$$= \begin{bmatrix} 1 & 0 & 0 & 0 \\ 0 & 1 & 0 & 0 \\ 0 & 0 & 0 & 0 \\ 0 & 0 & 0 & 0 \end{bmatrix} + \begin{bmatrix} 0 & 0 & 0 & 0 \\ 0 & 0 & 0 & 0 \\ 0 & 0 & 0 & 0 \\ 0 & 0 & 0 & 0 \end{bmatrix}$$

$$= \begin{bmatrix} 1 & 0 & 0 & 0 \\ 0 & 1 & 0 & 0 \\ 0 & 0 & 0 & 0 \\ 0 & 0 & 0 & 0 \end{bmatrix}$$

Observation.

(1) Though $\mathbf{P} + \mathbf{Q} = \mathbf{O}$, $\mathbf{P} \oplus \mathbf{Q} \neq \mathbf{O}$

(2) Even if at least one entry of $\mathbf{P} \neq \mathbf{O}$ and $\mathbf{Q} \neq \mathbf{O}$, $\mathbf{P} \oplus \mathbf{Q} \neq \mathbf{O}$

Example 6.9.3 *Let $m = 3$ and $n = 2$. Find $\mathbf{I}_m \oplus \mathbf{I}_n$*
Consider $\mathbf{I}_3 \oplus \mathbf{I}_2 = \mathbf{I}_3 \otimes \mathbf{I}_2 + \mathbf{I}_3 \otimes \mathbf{I}_2$
$= 2\mathbf{I}_3 \otimes \mathbf{I}_2 = 2\mathbf{I}_6.$

Example 6.9.4 *Verify associative law for Kronecker sum by taking*

$$\mathbf{P} = \begin{bmatrix} p_1 & p_2 \\ p_3 & p_4 \end{bmatrix} \quad \mathbf{Q} = \begin{bmatrix} q_1 & q_2 \\ q_3 & q_4 \end{bmatrix} \quad \text{and} \quad \mathbf{R} = \begin{bmatrix} r_1 & r_2 \\ r_3 & r_4 \end{bmatrix}$$

Consider $\mathbf{P} \oplus (\mathbf{Q} \oplus \mathbf{R}) = \mathbf{P} \oplus [\mathbf{Q} \otimes \mathbf{I}_2 + \mathbf{I}_2 \otimes \mathbf{R}]$

$$= (\mathbf{P} \otimes \mathbf{I}_4) + \mathbf{I}_2 \otimes [\mathbf{Q} \otimes \mathbf{I}_2 + \mathbf{I}_2 \otimes \mathbf{R}]$$

$$= \begin{bmatrix} p_1 & p_2 \\ p_3 & p_4 \end{bmatrix} \otimes \begin{bmatrix} 1 & 0 & 0 & 0 \\ 0 & 1 & 0 & 0 \\ 0 & 0 & 1 & 0 \\ 0 & 0 & 0 & 1 \end{bmatrix} + \begin{bmatrix} 1 & 0 \\ 0 & 1 \end{bmatrix} \otimes \left\{ \begin{bmatrix} q_1 & 0 & q_2 & 0 \\ 0 & q_1 & 0 & q_2 \\ q_3 & 0 & q_4 & 0 \\ 0 & q_3 & 0 & q_4 \end{bmatrix} + \begin{bmatrix} r_1 & r_2 & 0 & 0 \\ r_3 & r_4 & 0 & 0 \\ 0 & 0 & r_1 & r_2 \\ 0 & 0 & r_3 & r_4 \end{bmatrix} \right\}$$

$$= \begin{bmatrix} p_1 & 0 & 0 & 0 & p_2 & 0 & 0 & 0 \\ 0 & p_1 & 0 & 0 & 0 & p_2 & 0 & 0 \\ 0 & 0 & p_1 & 0 & 0 & 0 & p_2 & 0 \\ 0 & 0 & 0 & p_1 & 0 & 0 & 0 & p_2 \\ p_3 & 0 & 0 & 0 & p_4 & 0 & 0 & 0 \\ 0 & p_3 & 0 & 0 & 0 & p_4 & 0 & 0 \\ 0 & 0 & p_3 & 0 & 0 & 0 & p_4 & 0 \\ 0 & 0 & 0 & p_3 & 0 & 0 & 0 & p_4 \end{bmatrix}$$

$$+ \begin{bmatrix} q_1 + r_1 & r_2 & q_2 & 0 & 0 & 0 & 0 & 0 \\ r_3 & q_1 + r_4 & 0 & q_2 & 0 & 0 & 0 & 0 \\ q_3 & 0 & q_4 + r_1 & r_2 & 0 & 0 & 0 & 0 \\ 0 & q_3 & r_3 & q_4 + r_4 & 0 & 0 & 0 & 0 \\ 0 & 0 & 0 & 0 & q_1 + r_1 & r_2 & q_2 & 0 \\ 0 & 0 & 0 & 0 & r_3 & q_1 + r_4 & 0 & q_2 \\ 0 & 0 & 0 & 0 & q_3 & 0 & q_4 + r_1 & r_2 \\ 0 & 0 & 0 & 0 & 0 & q_3 & r_3 & q_4 + r_4 \end{bmatrix}$$

$$= \begin{bmatrix} p_1 + q_1 + r_1 & r_2 & q_2 & 0 & p_2 & 0 & 0 & 0 \\ r_3 & p_1 + q_1 + r_4 & 0 & q_2 & 0 & p_2 & 0 & 0 \\ q_3 & 0 & p_1 + q_4 + r_1 & r_2 & 0 & 0 & p_2 & 0 \\ 0 & q_3 & r_3 & p_1 + q_4 + r_4 & 0 & 0 & 0 & p_2 \\ p_3 & 0 & 0 & 0 & p_4 + q_1 + r_1 & r_2 & q_2 & 0 \\ 0 & p_3 & 0 & 0 & r_3 & p_4 + q_1 + r_4 & 0 & q_2 \\ 0 & 0 & p_3 & 0 & q_3 & 0 & p_4 + q_4 + r_1 & r_2 \\ 0 & 0 & 0 & p_3 & 0 & q_3 & r_3 & p_4 + q_4 + r_4 \end{bmatrix}$$

Consider R.H.S $= (\mathbf{P} \oplus \mathbf{Q}) \oplus \mathbf{R} = (\mathbf{P} \oplus \mathbf{Q}) \otimes \mathbf{I}_2 + \mathbf{I}_4 \otimes \mathbf{R}$

$$= \left[\begin{bmatrix} p_1 & p_2 \\ p_3 & p_4 \end{bmatrix} \otimes \mathbf{I}_2 + \mathbf{I}_2 \times \begin{bmatrix} q_1 & q_2 \\ q_3 & q_4 \end{bmatrix} \right] \otimes \mathbf{I}_2 + \mathbf{I}_4 \otimes \mathbf{R}$$

$$= \left[\begin{bmatrix} p_1 & 0 & p_2 & 0 \\ 0 & p_1 & 0 & p_2 \\ p_3 & 0 & p_4 & 0 \\ 0 & p_3 & 0 & p_4 \end{bmatrix} + \begin{bmatrix} q_1 & q_2 & 0 & 0 \\ q_3 & q_4 & 0 & 0 \\ 0 & 0 & q_1 & q_2 \\ 0 & 0 & q_3 & q_4 \end{bmatrix} \right] \otimes \mathbf{I}_2$$

$$+ \begin{bmatrix} 1 & 0 & 0 & 0 \\ 0 & 1 & 0 & 0 \\ 0 & 0 & 1 & 0 \\ 0 & 0 & 0 & 1 \end{bmatrix} \otimes \begin{bmatrix} r_1 & r_2 \\ r_3 & r_4 \end{bmatrix}$$

$$= \begin{bmatrix} p_1+q_1 & 0 & q_2 & 0 & p_2 & 0 & 0 & 0 \\ 0 & p_1+q_1 & 0 & q_2 & 0 & p_2 & 0 & 0 \\ q_3 & 0 & p_1+q_4 & 0 & 0 & 0 & p_2 & 0 \\ 0 & q_3 & 0 & p_1+q_4 & 0 & 0 & 0 & p_2 \\ p_3 & 0 & 0 & 0 & p_4+q_1 & 0 & q_2 & 0 \\ 0 & p_3 & 0 & 0 & 0 & p_4+q_1 & 0 & q_2 \\ 0 & 0 & p_3 & 0 & q_3 & 0 & p_1+q_4 & 0 \\ 0 & 0 & 0 & p_3 & 0 & q_3 & 0 & p_4+q_4 \end{bmatrix}$$

$$+ \begin{bmatrix} r_1 & r_2 & 0 & 0 & 0 & 0 & 0 & 0 \\ r_3 & r_4 & 0 & 0 & 0 & 0 & 0 & 0 \\ 0 & 0 & r_1 & r_2 & 0 & 0 & 0 & 0 \\ 0 & 0 & r_3 & r_4 & 0 & 0 & 0 & 0 \\ 0 & 0 & 0 & 0 & r_1 & r_2 & 0 & 0 \\ 0 & 0 & 0 & 0 & r_3 & r_4 & 0 & 0 \\ 0 & 0 & 0 & 0 & 0 & 0 & r_1 & r_2 \\ 0 & 0 & 0 & 0 & 0 & 0 & r_3 & r_4 \end{bmatrix}$$

$$= \begin{bmatrix} p_1+q_1+r_1 & r_2 & q_2 & 0 & p_2 & 0 & 0 & 0 \\ r_3 & p_1+q_1+r_4 & 0 & q_2 & 0 & p_2 & 0 & 0 \\ q_3 & 0 & p_1+q_4+r_1 & r_2 & 0 & 0 & p_2 & 0 \\ 0 & q_3 & r_3 & p_1+q_4+r_4 & 0 & 0 & 0 & p_2 \\ p_3 & 0 & 0 & 0 & p_4+q_1+r_1 & r_2 & q_2 & 0 \\ 0 & p_3 & 0 & 0 & r_3 & p_4+q_1+r_4 & 0 & q_2 \\ 0 & 0 & p_3 & 0 & q_3 & 0 & p_4+q_4+r_1 & r_2 \\ 0 & 0 & 0 & p_3 & 0 & q_3 & r_3 & p_4+q_4+r_4 \end{bmatrix}$$

Thus L.H.S = R.H.S.

Hence $\mathbf{P} \oplus (\mathbf{Q} \oplus R) = (\mathbf{P} \oplus \mathbf{Q}) \oplus \mathbf{R}$ *is verified.*

Example 6.9.5 *Given* $\mathbf{P} = \begin{bmatrix} 1 & 2 \\ 2 & 1 \end{bmatrix}$, and $\mathbf{Q} = \begin{bmatrix} 1 & -1 \\ 0 & 2 \end{bmatrix}$, *Find*

(i) the eigen-values and eigen-vectors of $\mathbf{P}$ *and* $\mathbf{Q}$.

(ii) the eigen-values and eigen-vectors of $\mathbf{P} \oplus \mathbf{Q}$.

(iii) Verify the property concerning the eigen-values, eigen-vectors and Kronecker sum.

Solution: The characteristic equation for the matrix $\mathbf{P}$ is $|\mathbf{P} - \lambda \mathbf{I}_2| = 0$. It is seen that the eigen-values of $\mathbf{P}$ are -1 and 3 and corresponding eigen-vectors are respectively $\mathbf{x}_1 = [1 \quad -1]^T$ and $\mathbf{x}_2 = [1 \ \ 1]^T$.

The characteristic equation for matrix $\mathbf{Q}$ is $|\mathbf{Q} - \mu \mathbf{I}_2| = 0$. It can be verified that the eigen-values of $\mathbf{Q}$ are 1 and 2 and corresponding eigen-vectors are $\mathbf{y}_1 = [1 \ \ 0]^T$ and $\mathbf{y}_2 = [1 \quad -1]^T$. Now,

$$\mathbf{P} \oplus \mathbf{Q} = \begin{bmatrix} 1 & 2 \\ 2 & 1 \end{bmatrix} \oplus \begin{bmatrix} 1 & -2 \\ 0 & 1 \end{bmatrix}$$

$$= \begin{bmatrix} 2 & -1 & 2 & 0 \\ 0 & 3 & 0 & 2 \\ 2 & 0 & 2 & -1 \\ 0 & 2 & 0 & 3 \end{bmatrix}.$$

The characteristic equation for matrix $\mathbf{P} \oplus \mathbf{Q}$ is $|\mathbf{P} \oplus \mathbf{Q} - \alpha \mathbf{I}_4| = 0$ or

$$\begin{vmatrix} 2-\alpha & -1 & 2 & 0 \\ 0 & 3-\alpha & 0 & 2 \\ 2 & 0 & 2-\alpha & -1 \\ 0 & 2 & 0 & 3-\alpha \end{vmatrix} = 0$$

Expanding this determinant, we obtain

$$\alpha\,(\alpha-1)(\alpha-4)(\alpha-5) = 0$$

It follows that eigen-values of $\mathbf{P} \oplus \mathbf{Q}$ are $0, 1, 4, 5$. The associated eigen-vectors are given by the linear system

$$(\mathbf{P} \oplus \mathbf{Q} = -\lambda \mathbf{I})\mathbf{x} = 0$$

Corresponding to eigen-value 0, the associated eigen-vectors are given by the linear system

$$\begin{aligned} 2x - y + 2z &= 0 \\ 3y + 2w &= 0 \\ 2x + 2z - w &= 0 \\ 2y + 3w &= 0 \end{aligned}$$

An eigen-vector is $[1 \ \ 0 \ \ -1 \ \ 0]^T$. The following table is illustrative.

Eigen-value	System of equations	Eigen-vector
1	$x - y + 2z = 0$	
	$2y + 2w = 0$	
	$2x + z - w = 0$	
	$2y + 2w = 0$	$[\,1 \ \ -1 \ \ -1 \ \ 1\,]^T$.
4	$-2x - y + 2z = 0$	
	$-y + 2w = 0$	
	$2x - 2z - w = 0$	
	$2y - w = 0$	$[\,1 \ \ 0 \ \ 1 \ \ 0\,]^T$.
5	$-3x - y + 2z = 0$	
	$-2y + 2w = 0$	
	$2x - 3z - w = 0$	
	$2y - 2w = 0$	$[\,1 \ \ -1 \ \ 1 \ \ -1\,]^T$.

Observe that,

$\alpha = 0 = \lambda_1 + \mu_1$	and	$x_1 \oplus y_1 = [\,1 \ \ 0 \ \ -1 \ \ 0\,]^T$
$\alpha = 1 = \lambda_1 + \mu_2$	and	$x_1 \oplus y_2 = [\,1 \ \ -1 \ \ -1 \ \ 1\,]^T$
$\alpha = 4 = \lambda_2 + \mu_1$	and	$x_2 \oplus y_1 = [\,1 \ \ 0 \ \ 1 \ \ 0\,]^T$
$\alpha = 5 = \lambda_2 + \mu_2$	and	$x_2 \oplus y_2 = [\,1 \ \ -1 \ \ 1 \ \ -1\,]^T$

This example suggests that Kronecker sum helps to get eigen-values and eigen-vectors of matrices of higher order.

EXERCISE 6.9

1. Find the Kronecker sum of the matrices $A = \begin{bmatrix} 2 & 4 \\ 1 & 3 \end{bmatrix}$ and $B = \begin{bmatrix} 8 & 1 \\ 5 & 6 \end{bmatrix}$.

2. Find the eigen-values and eigen-vectors corresponding to the Kronecker sum of the matrices $A = \begin{bmatrix} 1 & 3 \\ 2 & 2 \end{bmatrix}$ and $B = \begin{bmatrix} 2 & 3 \\ 4 & 3 \end{bmatrix}$.

6.10 Lyapunov Function

In this section. Kronecker products are used to reduce a specific matrix equation to a matrix-vector form and solve it.

Definition 6.10.1 *Lyapunov Equation is a linear matrix equation of the form*

$$\mathbf{PX} + \mathbf{XQ} = \mathbf{R}, \tag{6.12}$$

where $\mathbf{P}, \mathbf{Q}$ *and* $\mathbf{R}$ *are given matrices of order* n *and* $\mathbf{X}$ *is an unknown matrix of order* n.

The following are particular cases of (6.12) are of interest.

(i) $\mathbf{Q} = \mathbf{O}$, *then* (6.12) *reduces to* $\mathbf{PX} = \mathbf{R}$

(ii) When $\mathbf{P} = \mathbf{Q}$ *and* $\mathbf{R} = \mathbf{O}$, *i.e.* $\mathbf{PX} \pm \mathbf{XP} = \mathbf{O}$

(iii) $\mathbf{PX} \pm \mathbf{XP^T} = \mathbf{O}$

(iv) $\mathbf{PX} \pm \mathbf{XP^{-1}} = \mathbf{O}$

The solution for the equation (6.12) is given by

$$\mathbf{X} = -\int_0^\infty e^{\mathbf{P}t}\mathbf{R}e^{\mathbf{Q}t}\,dt$$

provided the integral exists for an arbitrary constant matrix $\mathbf{C}$ [see 4].

Result. A necessary and a sufficient condition for the equation (6.12) to have a unique solution for each value of $\mathbf{R}$ is that $\lambda_i + \mu_j \neq 0$ for all values of i and j, where λ_i is an eigen-value of $\mathbf{P}$ and μ_j is an eigen-value of $\mathbf{Q}$ for each i and j.

The aim in this section is to solve the Lyapunov equation using Kronecker products.

The equation (6.12) can be written in vector matrix form as follows using the relation (6.1) and Kronecker sum.

$$\begin{aligned} Vec(\mathbf{PX}+\mathbf{XQ}) &= Vec(\mathbf{PXI}_n+\mathbf{I_n XQ}) \\ &= Vec(\mathbf{PXI}_n)+Vec(\mathbf{I_n XQ}) \\ &= \left((\mathbf{I}_n\otimes\mathbf{P})+\mathbf{Q}^T\otimes\mathbf{I}_n\right) Vec\mathbf{X} \\ &= (\mathbf{Q}^T\oplus\mathbf{P})\, Vec\mathbf{X} \end{aligned}$$

or the equation (6.12) reduces to the form

$$\mathbf{Hx}=\mathbf{c}, \tag{6.13}$$

where $\mathbf{H}=\mathbf{Q}^T\oplus\mathbf{P}$, $\mathbf{x}=Vec\mathbf{X}$ and $\mathbf{c}=Vec\mathbf{R}$.

The equation (6.13) has a unique solution iff $\mathbf{H}$ is nonsingular, that is, the eigen-values of $\mathbf{Q}^T\oplus\mathbf{P}$ are all nonzero. Thus if $\lambda_i+\mu_j\neq 0$ for all i and j then the equation (6.12) has a unique solution.

Example 6.10.1 *A medical expert advised two patients Usha and Nisha of different ages to consume* $(150,145)$ *g of proteins and* $(330,270)$ *g of carbohydrates, respectively. The patients consume the two types of food A and B and when mixed provide them approximately the desired food values. Noticing that in course of time food values of A and B decrease, the expert advised them to add vitamins C and D containing proteins and carbohydrates as boosters. The following tables present food analysis of A and B and vitamins C and D.*

Type of food	*Food value 100 g*	
	Proteins	*Carbohydrates*
A	*5*	*15*
B	*30*	*25*

The food items are boosted by vitamins C and D having potency as follows

Type of food	*Food value 100 g*	
	Proteins	*Carbohydrates*
C	*5*	*20*
D	*10*	*5*

What amount of items A and B Usha and Nisha consume so that they receive needed food values.

Solution. Let x_{11} and x_{21} be the quantities in units in 100 g of A and B for Usha and x_{12}, x_{22} the quantities for Nisha. These quantities are then subjected to foods containing vitamins C and D. The resulting model is then given by a matrix equation

$$\begin{bmatrix}5 & 15\\30 & 25\end{bmatrix}\begin{bmatrix}x_{11} & x_{12}\\x_{21} & x_{22}\end{bmatrix}+\begin{bmatrix}x_{11} & x_{12}\\x_{21} & x_{22}\end{bmatrix}\begin{bmatrix}5 & 20\\10 & 5\end{bmatrix}=\begin{bmatrix}150 & 145\\330 & 270\end{bmatrix}$$

i.e. $\mathbf{PX} + \mathbf{XQ} = \mathbf{R}$. Then

$$\begin{aligned}(\mathbf{Q}^T \oplus \mathbf{P}) &= [(\mathbf{Q}^T \otimes \mathbf{I}_2) + (\mathbf{I}_2 \otimes \mathbf{P})]\\ &= \begin{bmatrix} 5 & 10 \\ 20 & 5 \end{bmatrix} \otimes \begin{bmatrix} 1 & 0 \\ 0 & 1 \end{bmatrix} + \begin{bmatrix} 1 & 0 \\ 0 & 1 \end{bmatrix} \otimes \begin{bmatrix} 5 & 15 \\ 30 & 25 \end{bmatrix}\\ &= \begin{bmatrix} 5 & 0 & 10 & 0 \\ 0 & 5 & 0 & 10 \\ 20 & 0 & 5 & 0 \\ 0 & 20 & 0 & 50 \end{bmatrix} + \begin{bmatrix} 5 & 15 & 0 & 0 \\ 30 & 25 & 0 & 0 \\ 0 & 0 & 5 & 15 \\ 0 & 0 & 30 & 25 \end{bmatrix}\\ &= \begin{bmatrix} 10 & 15 & 10 & 0 \\ 30 & 30 & 0 & 10 \\ 20 & 0 & 10 & 15 \\ 0 & 20 & 30 & 30 \end{bmatrix}\end{aligned}$$

Using equation (6.2),

$$\begin{bmatrix} 10 & 15 & 10 & 0 \\ 30 & 30 & 0 & 10 \\ 20 & 0 & 10 & 15 \\ 0 & 20 & 30 & 30 \end{bmatrix} \begin{bmatrix} x_{11} \\ x_{21} \\ x_{12} \\ x_{22} \end{bmatrix} = \begin{bmatrix} 150 \\ 330 \\ 145 \\ 270 \end{bmatrix}.$$

Thus the system of equations obtained are

$$\begin{aligned} 10x_{11} + 15x_{21} + 10x_{12} &= 150\\ 30x_{11} + 30x_{21} + 10x_{22} &= 330\\ 20x_{11} + 10x_{12} + 15x_{22} &= 145\\ 20x_{21} + 30x_{12} + 30x_{22} &= 270 \end{aligned}$$

Thus, solving the above system yields that Usha consumes 4 units and 6 units and Nisha consumes 2 units and 3 units of food **A** and B, respectively.

EXERCISE 6.10

1. Solve the system
$$\mathbf{AX} + \mathbf{XB} = \mathbf{C},$$
where $\mathbf{A} = \begin{bmatrix} 1 & -1 \\ 0 & 2 \end{bmatrix}$, $\mathbf{B} = \begin{bmatrix} -3 & 4 \\ 1 & 0 \end{bmatrix}$, $\mathbf{C} = \begin{bmatrix} 1 & 3 \\ -2 & 2 \end{bmatrix}$ and $\mathbf{X}$ is an unknown matrix of order 2.

2. Solve the system
$$\mathbf{AX} + \mathbf{XB} = \mathbf{C},$$
where $\mathbf{A} = \begin{bmatrix} 1 & -1 \\ 0 & 2 \end{bmatrix}$, $\mathbf{B} = \begin{bmatrix} -3 & 4 \\ 1 & 0 \end{bmatrix}$, $\mathbf{C} = \begin{bmatrix} 0 & 5 \\ 2 & -9 \end{bmatrix}$ and $\mathbf{X}$ is an unknown matrix of order 2.

3. Determine the solution of the system

$$\mathbf{AX} + \mathbf{XA} = \mu\mathbf{X},$$

where $\mathbf{A} = \begin{bmatrix} 1 & 0 \\ 2 & 3 \end{bmatrix}$. $\mu = -2$ and $\mathbf{X}$ is an unknown matrix of order 2.

6.11 Conclusion

The concepts related to Kronecker product and sum have been studied in this chapter. The main feature was to convert a matrix into a vector and employ it effectively in the matrix analysis. This fact has been successfully utilized in the subsequent chapters, Chapters 8 and 9. Another useful technique given is obtaining the eigen-values and eigen-vectors of higher-order matrices through the Kronecker product and Kronecker sum of lower-order matrices. An application involving Lyapunov equation is given. For further reading the readers are suggested to refer to [4] and [8].

Chapter 7

Calculus of Matrices

7.1 Introduction

One of the interesting developments in the theory of matrices is to club it with elements of calculus and introduce differentiation and integration of matrices. As calculus deals with rate of change of functions with respect to its independent variable, matrix calculus deals with the study of rate of change of matrix functions with respect to independent variables (which may be entries of a matrix) and the study of various concepts related to integration and differentiation of matrix functions. It is natural to expect that applications of matrix calculus will be far reaching in sciences, engineering and technology.

The present chapter introduces the necessary concepts which can be extrapolated to an entire gamut of applications that arise in modeling of real-life situations using matrices.

In Section 7.2, the concept of a matrix-valued function w.r.t. a scalar, (i.e.) with respect to an independent variable is given. Section 7.3 deals with the derivative of a vector-valued function (of a vector) w.r.t. a vector. Section 7.4 introduces the notion of a derivative of a scalar-valued function (of a matrix) w.r.t. a matrix. In Section 7.5, the notion of the derivative of a matrix-valued function w.r.t. its entries and the converse are discussed. Section 7.6 deals with the matrix differential. The derivative of a matrix-valued function w.r.t. a matrix form the content of Section 7.7. Section 7.8 deals with the rules obtained for the derivative defined in Section 7.7 using Kronecker products. Section 7.9 introduces another notion of a derivative of a matrix-valued function w.r.t. a matrix.

7.2 Derivative of a Matrix Valued Function with Respect to a Scalar

Consider an object flying in a space. The position of the flying object can be determined in a 3-dimensional coordinate system w.r.t. a frame of reference. The position of the object at any time is given by a 3-vector valued function

DOI: 10.1201/9781351014953-7

$v(t) = (x(t) \;\; y(t) \;\; z(t)) \quad t \in \mathbb{R}$ and $v \in \mathbb{R}^3$. Naturally, the functions $x, y, z : \mathbb{R} \to \mathbb{R}$ are scalar-valued functions of t and are assumed to be continuous (as the object is flying in space). If the object is flying gracefully or smoothly, the derivatives of all the functions $x(t), y(t), z(t)$ exist and the derivative of a vector with respect to the time variable is given by

$$\begin{bmatrix} x(t) \\ y(t) \\ z(t) \end{bmatrix}' = \begin{bmatrix} x'(t) \\ y'(t) \\ z'(t) \end{bmatrix}.$$

The definition in precise terms is as follows.

Let $v : \mathbb{R}^n \to \mathbb{R}^n$ be an n-vector valued function $v(t) = (x_1(t) \quad x_2(t) \quad \cdots \quad x_n(t))^T$, $t \in \mathbb{R}$. Assume that the functions $x_1(t), x_2(t), \cdots, x_n(t)$ are differentiable w.r.t. t.

Definition 7.2.1 *The derivative of an n-vector valued function is a vector formed by the derivatives $x_1'(t), x_2'(t), \cdots, x_n'(t)$, as follows*

$$v'(t) = \frac{d}{dt}\begin{pmatrix} x_1(t) \\ x_2(t) \\ \vdots \\ x_n(t) \end{pmatrix} = \begin{pmatrix} x_1'(t) \\ x_2'(t) \\ \vdots \\ x_n'(t) \end{pmatrix}, \qquad t \in \mathbb{R}.$$

Example 7.2.1 *Let $v = (t \;\; t^2 \;\; 1)$, $t \in \mathbb{R}$.*
Then $v'(t) = \left(\frac{d}{dt}t \;\; \frac{d}{dt}t^2 \;\; \frac{d}{dt}1\right) = (1 \;\; 2t \;\; 0)$.

In order to define the integral of a vector valued function it is sufficient for the real-valued functions $x_1, x_2, \cdots, x_n : \mathbb{R} \to \mathbb{R}$ be bounded. Let $v : \mathbb{R}^n \to \mathbb{R}^n$ be a bounded function on an interval I.

Definition 7.2.2 *The integral of an n-vector valued function $\int_I v dt$ is as follows,*

$$\int_I v dt = \int_I \begin{pmatrix} x_1(t) \\ x_2(t) \\ \vdots \\ x_n(t) \end{pmatrix} dt = \begin{pmatrix} \int_I x_1(t)dt \\ \int_I x_2(t)dt \\ \vdots \\ \int_I x_n(t)dt \end{pmatrix}.$$

Example 7.2.2 *Let* $v = \begin{cases} t, & 0 \le t < 1 \\ t^3, & 1 < t < 2. \end{cases}$

Then,

$$\begin{aligned}\int_0^2 v(t)dt &= \int_0^1 tdt + \int_1^2 t^3 dt \\ &= \left.\frac{t^2}{2}\right|_0^1 + \left.\frac{t^4}{4}\right|_1^2 \\ &= \frac{17}{4}.\end{aligned}$$

These notations can be naturally extended to matrices. As with the components of each vector, the entries of a matrix need to be differentiable.

Consider a matrix function $\mathbf{P} : \mathbb{R}^{m\times n} \to \mathbb{R}^{m\times n}$ with entries p_{ij} and which are real-valued functions of $\mathbb{R}$ for $i = 1, 2, \ldots, m$ and $j = 1, 2, \ldots, n$. Suppose that each of the functions $p_{ij}(t)$ are differentiable w.r.t. to the scalar variable t, for $i = 1, 2, \ldots, m$ and $j = 1, 2, \ldots, n$.

Definition 7.2.3 *The derivative of the matrix function* $\mathbf{P}'(t) = \frac{d}{dt}\mathbf{P}(t)$ *is the matrix whose entries are derivative of each of its entries* $p_{ij}(t)$ *for* $i = 1, 2, \ldots, m$ *and* $j = 1, 2, \ldots, n$ *and is denoted as*

$$\mathbf{P}'(t) = \left[\frac{d}{dt}p_{ij}(t)\right]_{m\times n} = \left[p'_{ij}(t)\right]_{m\times n}.$$

Example 7.2.3 *Let* $\mathbf{P}(t) = \begin{bmatrix} 4t^3 & cos\, t \\ t^2+8 & 2 \end{bmatrix}$, $t \in \mathbb{R}$,

then $\mathbf{P}'(t) = \begin{bmatrix} 12t^2 & -sin\, t \\ 2t & 0 \end{bmatrix}$.

To define the integral it suffices for the matrix function $\mathbf{P}(t)$ *to be bounded and piecewise continuous on the interval over which it is defined.*

Definition 7.2.4 *The integral of the matrix* $\mathbf{P}(t)$ *is the matrix given as follows*

$$\int_I \mathbf{P}(t)dt = \left[\int_I p_{ij}(t)dt\right]_{m\times n}.$$

Example 7.2.4 *Let* $\mathbf{P}(t) = \begin{bmatrix} t & \frac{1}{2} & sin\, t \\ t^2 & cos\, t & e^t \end{bmatrix}$ *on* $[0, \pi/2]$.

Then

$$\int_0^{\pi/2} \mathbf{P}(t)dt = \begin{bmatrix} \displaystyle\int_0^{\pi/2} tdt & \displaystyle\int_0^{\pi/2} \frac{1}{2}dt & \displaystyle\int_0^{\pi/2} \sin t\, dt \\ \displaystyle\int_0^{\pi/2} t^2 dt & \displaystyle\int_0^{\pi/2} \cos t\, dt & \displaystyle\int_0^{\pi/2} e^t dt \end{bmatrix}$$

$$= \begin{bmatrix} \left.\dfrac{t^2}{2}\right|_0^{\pi/2} & \left.\dfrac{t}{2}\right|_0^{\pi/2} & \left.-\cos t\right|_0^{\pi/2} \\ \left.\dfrac{t^3}{3}\right|_0^{\pi/2} & \left.\sin t\right|_0^{\pi/2} & \left.e^t\right|_0^{\pi/2} \end{bmatrix}$$

$$= \begin{bmatrix} \dfrac{\pi^2}{8} & \dfrac{\pi}{4} & 1 \\ \dfrac{\pi^3}{24} & 1 & e^{\pi/2} - 1 \end{bmatrix}$$

Note that a constant matrix must be added in case of an indefinite integral.

Example 7.2.5 *Find* $\displaystyle\int \begin{bmatrix} 3t & t^2 \\ 8 & \cos t \end{bmatrix} dt$

Solution.

$$\int \begin{bmatrix} 3t & t^2 \\ 8 & \cos t \end{bmatrix} dt = \begin{bmatrix} \int 3tdt & \int t^2 dt \\ \int 8dt & \int \cos t\, dt \end{bmatrix} dt$$

$$= \begin{bmatrix} \frac{3t^2}{2} + c_1 & \frac{t^2}{3} + c_2 \\ 8t + c_3 & \sin t + c_4 \end{bmatrix}$$

$$= \begin{bmatrix} \frac{3t^2}{2} & \frac{t^3}{3} \\ 8t & \sin t \end{bmatrix} + \begin{bmatrix} c_1 & c_2 \\ c_3 & c_4 \end{bmatrix}$$

$$Hence, \quad \int \mathbf{P}(t)dt = \left[\int p_{ij}(t)\right]_{m\times n} + \mathbf{C},$$

where $\mathbf{C} = (c_{ij})_{m\times n}$ *is an arbitrary matrix.*

Next, the natural extension of these concepts to the differentiation formula are stated. Since matrix differentiation involves differentiation of each of its entries, the proof are omitted.

Differentiation formulae

Let $P(t)$ and $Q(t)$ be two matrix-valued functions whose orders are compatible for the matrix operations defined below.

1. $\dfrac{d}{dt}\,[\,k\,\mathbf{P}(t)\,] \;=\; k\,\dfrac{d}{dt}\,\mathbf{P}(t),\quad k \in \mathbb{R}$ is a constant.
2. $\dfrac{d}{dt}\,[\,\mathbf{P}(t) \;+\; \mathbf{Q}(t)\,] \;=\; \dfrac{d}{dt}\,\mathbf{P}(t) \;+\; \dfrac{d}{dt}\,\mathbf{Q}(t).$
3. $\dfrac{d}{dt}\,[\,\mathbf{P}(t)\,\mathbf{Q}(t)\,] \;=\; \left[\dfrac{d}{dt}\,\mathbf{P}(t)\right]\,\mathbf{Q}(t) \;+\; \mathbf{P}(t)\,\left[\dfrac{d}{dt}\,\mathbf{Q}(t)\right].$
4. *Let* $\mathbf{S}(t) = \mathbf{P}(t) \otimes \mathbf{Q}(t)$

   Then, $\dfrac{d}{dt}\mathbf{S}(t) = \left[\dfrac{d}{dt}\mathbf{P}(t)\right] \otimes \mathbf{Q}(t) + \mathbf{P}(t) \otimes \dfrac{d}{dt}\mathbf{Q}(t).$

Observation. The proof of (4) follows from the definition of Kronecker product and considering the (i, j)th block of $\mathbf{S}$. The proofs are left as an exercise.

Example 7.2.6 *Verify the product rule for*

$$\mathbf{P}(t) = \begin{bmatrix} t & 2 \\ t^2+1 & \cos t \end{bmatrix} \qquad \mathbf{Q}(t) = \begin{bmatrix} \sin t & t^2 \\ 2t & 1 \end{bmatrix}$$

Solution.

$$\mathbf{P}(t)\mathbf{Q}(t) = \begin{bmatrix} t\sin t + 4t & t^3 + 2 \\ (\sin t)(t^2+1) + 2t\cos t & t^2(t^2+1) + \cos t \end{bmatrix}$$

$$\frac{d}{dt}(\mathbf{P}(t)\mathbf{Q}(t)) = \begin{bmatrix} \sin t + t\cos t + 4 & 3t^2 \\ (\cos t)(t^2+1) + 2\cos t & 4t^3 + 2t - \sin t \end{bmatrix}$$

$$\frac{d}{dt}\mathbf{P}(t) = \begin{bmatrix} 1 & 0 \\ 2t & -\sin t \end{bmatrix} \qquad \frac{d}{dt}\mathbf{Q}(t) = \begin{bmatrix} \cos t & 2t \\ 2 & 0 \end{bmatrix}$$

$$\begin{aligned} \frac{d}{dt}\mathbf{P}(t)\mathbf{Q}(t) + \mathbf{P}(t)\frac{d}{dt}\mathbf{Q}(t) &= \begin{bmatrix} 1 & 0 \\ 2t & -\sin t \end{bmatrix}\begin{bmatrix} \sin t & t^2 \\ 2t & 1 \end{bmatrix} + \begin{bmatrix} t & 2 \\ t^2+1 & \cos t \end{bmatrix}\begin{bmatrix} \cos t & 2t \\ 2 & 0 \end{bmatrix} \\ &= \begin{bmatrix} \sin t + t\cos t + 4 & 3t^2 \\ (t^2+1)\cos t + 2\cos t & 4t^3 + 2t - \sin t \end{bmatrix}. \end{aligned}$$

Thus the result is verified.

Example 7.2.7 *Verify the product rule for Kronecker product* $\mathbf{P} \otimes \mathbf{Q}$*, where*

$\mathbf{P} = \begin{bmatrix} t & t^2 \\ 1 & e^t \end{bmatrix}$, $\mathbf{Q} = \begin{bmatrix} 2t & t^2 \end{bmatrix}$.

Solution.

$$\mathbf{P} \otimes \mathbf{Q} = \begin{bmatrix} 2t^2 & t^3 & 2t^3 & t^4 \\ 2t & t^2 & 2te^t & t^2e^t \end{bmatrix}$$

$$\frac{d}{dt}(\mathbf{P} \otimes \mathbf{Q}) = \begin{bmatrix} 4t & 3t^2 & 6t^2 & 4t^3 \\ 2 & 2t & 2e^t + 2te^t & 2te^t + t^2e^t \end{bmatrix}$$

$$\frac{d\mathbf{P}}{dt} = \begin{bmatrix} 1 & 2t \\ 0 & e^t \end{bmatrix} \quad \frac{d\mathbf{Q}}{dt} = \begin{bmatrix} 2 & 2t \end{bmatrix}$$

Consider $\frac{d\mathbf{P}}{dt} \otimes \mathbf{Q} + \mathbf{P} \otimes \frac{d\mathbf{Q}}{dt}$

$$= \begin{bmatrix} 1 & 2t \\ 0 & e^t \end{bmatrix} \otimes \begin{bmatrix} 2t & t^2 \end{bmatrix} + \begin{bmatrix} t & t^2 \\ 1 & e^t \end{bmatrix} \otimes \begin{bmatrix} 2 & 2t \end{bmatrix}$$

$$= \begin{bmatrix} 2t & t^2 & 4t^2 & 2t^3 \\ 0 & 0 & 2te^t & t^2e^t \end{bmatrix} + \begin{bmatrix} 2t & 2t^2 & 2t^2 & 2t^3 \\ 2 & 2t & 2e^t & 2te^t \end{bmatrix}$$

$$= \begin{bmatrix} 4t & 3t^2 & 6t^2 & 4t^3 \\ 2 & 2t & 2e^t + 2te^t & 2te^t + t^2e^t \end{bmatrix}.$$

Hence, the result is verified.

As noted in earlier chapters, the determinant of a matrix is a type of a measure of the matrix. Hence, its derivative is an interesting and useful concept which is introduced below.

Definition 7.2.5 *The derivative of the determinant of a square matrix* $\mathbf{P}$ *of order* n *is the sum of the determinants of the* n *matrices formed by taking the derivative of one row at a time.*

$$\begin{vmatrix} p_{11}(t) & p_{12}(t) & \cdots & p_{1n}(t) \\ p_{21}(t) & p_{22}(t) & \cdots & p_{2n}(t) \\ \cdots & \cdots & \cdots & \\ p_{n1}(t) & p_{n2}(t) & \cdots & p_{nn}(t) \end{vmatrix}'$$

$$= \begin{vmatrix} p'_{11}(t) & p'_{12}(t) & \cdots & p'_{1n}(t) \\ p_{21}(t) & p_{22}(t) & \cdots & p_{2n}(t) \\ \cdots & \cdots & \cdots & \\ p_{n1}(t) & p_{n2}(t) & \cdots & p_{nn}(t) \end{vmatrix} + \begin{vmatrix} p_{11}(t) & p_{12}(t) & \cdots & p_{1n}(t) \\ p'_{21}(t) & p'_{22}(t) & \cdots & p'_{2n}(t) \\ \cdots & \cdots & \cdots & \\ p_{n1}(t) & p_{n2}(t) & \cdots & p_{nn}(t) \end{vmatrix} + \cdots$$

$$+ \begin{vmatrix} p_{11}(t) & p_{12}(t) & \cdots & p_{1n}(t) \\ p_{21}(t) & p_{22}(t) & \cdots & p_{2n}(t) \\ \cdots & \cdots & \cdots & \\ p'_{n1}(t) & p'_{n2}(t) & \cdots & p'_{nn}(t) \end{vmatrix}.$$

Example 7.2.8 *Find the derivative of the determinant of the matrix* $\mathbf{P}(t) = \begin{vmatrix} e^t & te^t & t^2e^t \\ 0 & \sin t & \cos t \\ 0 & t & t\ \sin t \end{vmatrix}$

Solution.

$$\begin{vmatrix} e^t & te^t & t^2e^t \\ 0 & \sin t & \cos t \\ 0 & 1 & t\ \sin t \end{vmatrix}' = \begin{vmatrix} e^t & te^t + e^t & t^2e^t + 2te^t \\ 0 & \sin t & \cos t \\ 0 & 1 & t\ \sin t \end{vmatrix}$$

$$+ \begin{vmatrix} e^t & te^t & t^2e^t \\ 0 & \cos t & -\sin t \\ 0 & 1 & t\ \sin t \end{vmatrix} + \begin{vmatrix} e^t & te^t & t^2e^t \\ 0 & \sin t & \cos t \\ 0 & 0 & \sin t + t\ \cos t \end{vmatrix}$$

$$= e^t \left[t\ \sin^2 t - \cos t\right] + e^t \left[t\ \cos t\ \sin t + \sin t\right]$$

$$+ e^t \left[\sin^2 t + t\ \cos t\ \sin t\right]$$

$$= e^t \left[(t+1)\sin^2 t + 2t\ \cos t\ \sin t + \sin t - \cos t\right].$$

EXERCISE 7.2

1. Consider the vector $\mathbf{p}(t) = [t^3 \quad -\sin t \quad t]^T$
   Find $\frac{d}{dt}[\mathbf{p}(t)]$ and $\int \mathbf{p}(t)dt$.

2. Consider the vector $\int \mathbf{p}(t)dt = [e^t \quad \cos 2t \quad 5t]$, Find $\mathbf{p}(t)$.

3. Consider the vector $\frac{d}{dt}\mathbf{p}(t) = [te^t + e^t \quad t\sin t \quad \cos t]$, Find $\mathbf{p}(t)$.

4. Differentiate the matrix $\mathbf{P}(t)$ when

   (i) $\mathbf{P}(t) = \begin{bmatrix} 1 & t \\ t^2 & t^3 \end{bmatrix}$. (ii) $\mathbf{P}(t) = \begin{bmatrix} t\cos t & t & \sin t \\ 1 & t\sin t & e^t \\ t^2 & te^t & t\tan t \end{bmatrix}$.

5. Find $\mathbf{P}(t,)$, given

   (i) $\mathbf{P}'(t) = \begin{bmatrix} \cos t & t \\ t^2 & \sin t \end{bmatrix}$ and (ii) $\mathbf{P}'(t) = \begin{bmatrix} 1 & t & t^2/2 \\ \cos t & \sin t & 0 \\ 1 & \tan t & \sec t \end{bmatrix}$.

6. Consider $f(x, y, z) = [xy \quad x^2 + 2y + z^2 \quad xyz]$, find $\frac{\partial f}{\partial x}, \frac{\partial f}{\partial \mathbf{y}}, \frac{\partial f}{\partial z}$.

7. Given the determinant $\mathbf{P}(t)$ find its derivative, where

   (i) $\mathbf{P}(t) = \begin{vmatrix} 1 & t \\ t^2 & t^3 \end{vmatrix}$. (ii) $\mathbf{P}(t) = \begin{vmatrix} \cos t & \sin t \\ -\sin t & \cos t \end{vmatrix}$.

   (iii) $\mathbf{P}(t) = \begin{vmatrix} 1 & e^t & te^t \\ \cos t & \sin t & 0 \\ t^2e^t & t\ell nt & 1 \end{vmatrix}$.

7.3 Derivative of a Vector-Valued Function w.r.t. a Vector

The idea of considering a vector-valued function which depends on another vector is a very useful concept in applications. This notion is well utilized when performing change of variables to simplify problems in multivariable calculus.

Consider a vector-vlaued function $\mathbf{y} : \mathbb{R}^m \rightarrow \mathbb{R}^n$ given by $\mathbf{y}(x_1 \ x_2 \ \ldots \ x_m)$
$= (y_1(x_1 \ x_2 \ \ldots \ x_m) \quad y_2(x_1 \ x_2 \ \ldots \ x_m) \quad \ldots \ y_n(x_1 \ x_2 \ \ldots \ x_m))$. Assume that the partial derivatives of the functions $y_1, y_2, \ldots, y_n : \mathbb{R}^n \rightarrow \mathbb{R}$ all the variables $x_1, x_2, \ldots, x_m$ exist. Set $\mathbf{x} = (x_1 \ x_2 \ \ldots \ x_m)$. For the sake of simplicity of expression, the notation used is $\mathbf{y}(\mathbf{x}) = (y_1(\mathbf{x}) \ \ldots \ y_n(\mathbf{x}))$.

Definition 7.3.1 *The derivative of the vector-valued function* $\mathbf{y}(\mathbf{x})$ *with respect to the vector '*$\mathbf{x}$*' is denoted by* $\frac{\partial \mathbf{y}}{\partial \mathbf{x}}$ *and is defined as*

$$\frac{\partial \mathbf{y}}{\partial \mathbf{x}} = \left[\frac{\partial y_i}{\partial x_j}\right]_{m \times n}$$
$$= \begin{bmatrix} \frac{\partial y_1}{\partial x_1} & \frac{\partial y_2}{\partial x_1} & \frac{\partial y_3}{\partial x_1} & \cdots & \frac{\partial y_n}{\partial x_1} \\ \frac{\partial y_1}{\partial x_2} & \frac{\partial y_2}{\partial x_2} & \frac{\partial y_3}{\partial x_2} & \cdots & \frac{\partial y_n}{\partial x_2} \\ \vdots & \vdots & & \vdots & \vdots \\ \frac{\partial y_1}{\partial x_m} & \frac{\partial y_2}{\partial x_m} & \frac{\partial y_3}{\partial x_m} & \cdots & \frac{\partial y_n}{\partial x_m} \end{bmatrix}.$$

Observation. 1. The matrix $\frac{\partial \mathbf{y}}{\partial \mathbf{x}}$ is also called as Jacobian.

2. Case (i) if $n = 1$, then $y : \mathbb{R}^n \rightarrow \mathbb{R}$ and is a scalar-valued function. Thus $\frac{\partial y}{\partial \mathbf{x}}$ is an m-column vector and the derivative is given by

$$\frac{\partial y}{\partial \mathbf{x}} = \begin{bmatrix} \frac{\partial y}{\partial x_1} & \frac{\partial y}{\partial x_2} & \cdots & \frac{\partial y}{\partial x_m} \end{bmatrix}^T.$$

is the derivative of the scalar-valued function with respect to a vector.

Case (ii) If $m = 1$, then the function $\mathbf{y} : \mathbb{R} \rightarrow \mathbb{R}^n$ is a vector-valued function of a scalar. Thus, $\frac{\partial \mathbf{y}}{\partial x}$ is an n-row vector and the derivative is given by

$$\frac{\partial \mathbf{y}}{\partial x} = \begin{bmatrix} \frac{\partial y_1}{\partial x} & \frac{\partial y_2}{\partial x} & \cdots & \frac{\partial y_n}{\partial x} \end{bmatrix},$$

Example 7.3.1 *Let* $(y_1 \ y_2) = (x_1^3 - 3x_2 + \sin x_3 \ \ 4x_1 + x_2 + x_3^2)$. *Differentiate* $\mathbf{y} = [y_1 \ y_2]^T$ *w.r.t.* $\mathbf{x} = [x_1 \ x_2 \ x_3]^T$.

Solution.

$$\frac{\partial \mathbf{y}}{\partial \mathbf{x}} = \begin{bmatrix} \frac{\partial y_1}{\partial x_1} & \frac{\partial y_2}{\partial x_1} \\ \frac{\partial y_1}{\partial x_2} & \frac{\partial y_2}{\partial x_2} \\ \frac{\partial y_1}{\partial x_3} & \frac{\partial y_2}{\partial x_3} \end{bmatrix} = \begin{bmatrix} 3x_1^2 & 4 \\ -3 & 1 \\ \cos x_3 & 2x_3 \end{bmatrix}.$$

Example 7.3.2 *Differentiate* $\mathbf{y} = \mathbf{A}\mathbf{x}$ *w.r.t.* $\mathbf{x}$, *given* $\mathbf{y} = [y_1 \ \ y_2]^T$, $\mathbf{A} = \begin{bmatrix} a_{11} & a_{12} \\ a_{21} & a_{22} \end{bmatrix}$ *and* $\mathbf{x} = [x_1 \ \ x_2]^T$.

Solution. *Given*

$$\begin{bmatrix} y_1 \\ y_2 \end{bmatrix} = \begin{bmatrix} a_{11} & a_{12} \\ a_{21} & a_{22} \end{bmatrix} \begin{bmatrix} x_1 \\ x_2 \end{bmatrix} = \begin{bmatrix} a_{11}x_1 + a_{12}x_2 \\ a_{21}x_1 + a_{22}x_2 \end{bmatrix}$$

$$\frac{\partial \mathbf{y}}{\partial \mathbf{x}} = \begin{bmatrix} \frac{\partial y_1}{\partial x_1} & \frac{\partial y_2}{\partial x_1} \\ \frac{\partial y_1}{\partial x_2} & \frac{\partial y_2}{\partial x_2} \end{bmatrix} = \begin{bmatrix} a_{11} & a_{21} \\ a_{12} & a_{22} \end{bmatrix} = \mathbf{A}^T.$$

Example 7.3.3 *The transformation from cylindrical coordinates to Cartesian coordinates is defined by*

$$x = r \cos \theta, \ y = r \sin \theta, \ z = z,$$

where $r > 0, \ 0 < \theta < \pi$. *Obtain the Jacobian of the transformation.*

Solution.

Let $y_1 = x$, $y_2 = y$, $y_3 = z$, *and* $x_1 = r$, $x_2 = \theta$, *and* $x_3 = z$, *so that* $y_1 = x_1 \cos x_2$; $y_2 = x_1 \sin x_2$; $y_3 = x_3$; *then the Jacobian,*

$$\mathbf{J} = \left|\frac{\partial \mathbf{y}}{\partial \mathbf{x}}\right| = \begin{vmatrix} \dfrac{\partial y_1}{\partial x_1} & \dfrac{\partial y_2}{\partial x_1} & \dfrac{\partial y_3}{\partial x_1} \\ \dfrac{\partial y_1}{\partial x_2} & \dfrac{\partial y_2}{\partial x_2} & \dfrac{\partial y_3}{\partial x_2} \\ \dfrac{\partial y_1}{\partial x_3} & \dfrac{\partial y_2}{\partial x_3} & \dfrac{\partial y_3}{\partial x_3} \end{vmatrix}$$

$$= \begin{vmatrix} cos\ x_2 & sin\ x_2 & 0 \\ -\ x_1\ sin\ x_2 & x_1\ cos\ x_2 & 0 \\ 0 & 0 & 1 \end{vmatrix}$$

$$= x_1.$$

Example 7.3.4 *Let* $\mathbf{x} = (x_1 \quad x_2 \quad \ldots \quad x_n)^T$ *and* $\mathbf{A} = (a_{ij})_{n\times n}$ *and* $\alpha = \mathbf{x}^T\mathbf{A}\mathbf{x}$ *be the quadratic form. Find* $\frac{\partial \alpha}{\partial \mathbf{x}}$.
Solution. *Clearly* α *is a scalar-valued function of* $\mathbf{x}$. *Observe that from the quadratic form the quadratic form* α *is differentiable with respect to the components* $x_1, x_2, \ldots, x_n$ *of the vector* $\mathbf{x}$.

Consider the quadratic form $\alpha = \mathbf{x}^T\mathbf{A}\mathbf{x} = \sum_{j=1}^{n}\sum_{i=1}^{n} a_{ij}x_ix_j.$

Then, $\dfrac{\partial \alpha}{\partial x_k} = \sum_{i=1}^{n} a_{ik}x_i + \sum_{j=1}^{n} a_{kj}x_j.$

Thus, $\dfrac{\partial \alpha}{\partial \mathbf{x}} = \mathbf{A}^T\mathbf{x} + \mathbf{A}\mathbf{x}.$

Observation. If $\mathbf{A}$ is a symmetric matrix, then $\frac{\partial \alpha}{\partial \mathbf{x}} = 2\mathbf{A}\mathbf{x}$.

Chain rule for vector-valued function of a vector

The concept of chain rule for a vector-valued function is useful and given below

Let $\mathbf{x} = (x_1 \quad x_2 \quad \ldots \quad x_n)^T$, $\mathbf{y} = (y_1 \quad y_2 \quad \ldots \quad y_r)^T$ and $\mathbf{z} = (z_1 \quad z_2 \quad \ldots \quad z_m)^T$.

Suppose that $\mathbf{x}$ consists of independent variables $x_1, x_2, \ldots, x_n$ and $\mathbf{y} = f(\mathbf{x})$ and $\mathbf{z} = h(\mathbf{y})$. Further assume that all the required partial derivatives exist.

Differentiating the vector **z** w.r.t **x** gives

$$\frac{\partial \mathbf{z}}{\partial \mathbf{x}} = \begin{bmatrix} \frac{\partial z_1}{\partial x_1} & \frac{\partial z_2}{\partial x_1} & \cdots & \frac{\partial z_m}{\partial x_1} \\ \frac{\partial z_1}{\partial x_2} & \frac{\partial z_2}{\partial x_2} & \cdots & \frac{\partial z_m}{\partial x_2} \\ \vdots & \vdots & & \vdots \\ \frac{\partial z_1}{\partial x_n} & \frac{\partial z_2}{\partial x_n} & \cdots & \frac{\partial z_m}{\partial x_n} \end{bmatrix}_{n\times m}.$$

Now observing that each z_i is a function of $y_1, y_2, \ldots, y_r$ and the $y_1, y_2, \ldots, y_r$ are functions of $x_1, x_2, \ldots, x_n$,

$$\frac{\partial z_i}{\partial x_j} = \frac{\partial z_i}{\partial y_1}\frac{\partial y_1}{\partial x_j} + \frac{\partial z_i}{\partial y_2}\frac{\partial y_2}{\partial x_j} + \cdots + \frac{\partial z_i}{\partial y_r}\frac{\partial y_r}{\partial x_j}, \quad \begin{array}{l} i = 1, 2, \ldots, m \\ j = 1, 2, \ldots, n. \end{array}$$

This gives

$$\begin{aligned} \frac{\partial z_i}{\partial x_j} &= \begin{bmatrix} \frac{\partial z_i}{\partial y_1} & \frac{\partial z_i}{\partial y_2} & \cdots & \frac{\partial z_i}{\partial y_r} \end{bmatrix} \begin{bmatrix} \frac{\partial y_1}{\partial x_j} & \frac{\partial y_2}{\partial x_j} \cdots \frac{\partial y_r}{\partial x_j} \end{bmatrix}^T, \quad \begin{array}{l} i = 1, 2, \ldots, m \\ j = 1, 2, \ldots, n. \end{array} \\ &= \frac{\partial z_i}{\partial \mathbf{y}} \cdot \frac{\partial \mathbf{y}}{\partial x_j} = \frac{\partial \mathbf{y}}{\partial x_j} \cdot \frac{\partial z_i}{\partial \mathbf{y}}, \quad \begin{array}{l} i = 1, 2, \ldots, m \\ j = 1, 2, \ldots, n. \end{array} \end{aligned}$$

Thus,

$$\begin{aligned} \frac{\partial \mathbf{z}}{\partial \mathbf{x}} &= \begin{bmatrix} \frac{\partial \mathbf{y}}{\partial x_1}\frac{\partial z_1}{\partial \mathbf{y}} & \frac{\partial \mathbf{y}}{\partial x_1}\frac{\partial z_2}{\partial \mathbf{y}} & \cdots & \frac{\partial \mathbf{y}}{\partial x_1}\frac{\partial z_m}{\partial \mathbf{y}} \\ \frac{\partial \mathbf{y}}{\partial x_2}\frac{\partial z_1}{\partial \mathbf{y}} & \frac{\partial \mathbf{y}}{\partial x_2}\frac{\partial z_2}{\partial \mathbf{y}} & \cdots & \frac{\partial \mathbf{y}}{\partial x_2}\frac{\partial z_m}{\partial \mathbf{y}} \\ \vdots & \vdots & \cdots & \vdots \\ \frac{\partial \mathbf{y}}{\partial x_n}\frac{\partial z_1}{\partial \mathbf{y}} & \frac{\partial \mathbf{y}}{\partial x_n}\frac{\partial z_2}{\partial \mathbf{y}} & \cdots & \frac{\partial \mathbf{y}}{\partial x_n}\frac{\partial z_m}{\partial \mathbf{y}} \end{bmatrix} \\ &= \begin{bmatrix} \frac{\partial \mathbf{y}}{\partial x_1} \\ \frac{\partial \mathbf{y}}{\partial x_2} \\ \vdots \\ \frac{\partial \mathbf{y}}{\partial x_n} \end{bmatrix} \begin{bmatrix} \frac{\partial z_1}{\partial \mathbf{y}} & \frac{\partial z_2}{\partial \mathbf{y}} \cdots \frac{\partial z_m}{\partial \mathbf{y}} \end{bmatrix} \\ &= \frac{\partial \mathbf{y}}{\partial \mathbf{x}}\frac{\partial \mathbf{z}}{\partial \mathbf{y}}. \end{aligned}$$

Observation. (i) The order of the LHS matrix $\frac{\partial \mathbf{z}}{\partial \mathbf{x}}$ is $n \times m$ and on RHS the order of the matrices $\frac{\partial \mathbf{y}}{\partial \mathbf{x}}$ is $n \times r$ and $\frac{\partial \mathbf{z}}{\partial \mathbf{y}}$ is $r \times m$.

Hence, $\frac{\partial \mathbf{z}}{\partial \mathbf{x}}$ and $\frac{\partial \mathbf{y}}{\partial \mathbf{x}}\frac{\partial \mathbf{z}}{\partial \mathbf{y}}$ are of the same dimension. If x, y, z are scalars, the chain rule $\left[\frac{\partial z}{\partial x} = \frac{\partial z}{\partial y}\frac{\partial y}{\partial x}\right]$ is satisfied.

Example 7.3.5 *Let* $\mathbf{x} = (x_1 \;\; x_2)^T$, $\mathbf{y} = (y_1 \;\; y_2)^T$ *and* $\mathbf{z} = (z_1 \;\; z_2)^T$, *where* $y_1 = x_1^2 + x_2$, $y_2 = 3x_1 + x_2$, $z_1 = y_1 + y_2^2$, $z_2 = 2y_1 + y_2$. *Verify the chain rule.*

Solution.

$$\begin{aligned}\mathbf{z} = (z_1 \;\; z_2)^T &= \left(x_1^2 + x_2 + (3x_1 + x_2)^2 \;\; 2(x_1^2 + x_2) + 3x_1 + x_2\right) \\ &= \left(10x_1^2 + x_2^2 + 6x_1x_2 + x_2 \;\; 2x_1^2 + 3x_2 + 3x_1\right)\end{aligned}$$

Thus,

$$\frac{\partial \mathbf{z}}{\partial \mathbf{x}} = \begin{pmatrix} 20x_1 + 6x_2 & 4x_1 + 3 \\ 2x_2 + 6x_1 + 1 & 3 \end{pmatrix}$$

$$\frac{\partial \mathbf{z}}{\partial \mathbf{y}} = \begin{pmatrix} 1 & 2 \\ 2y_2 & 1 \end{pmatrix} \quad \frac{\partial \mathbf{y}}{\partial \mathbf{x}} = \begin{pmatrix} 2x_1 & 3 \\ 1 & 1 \end{pmatrix}$$

$$\begin{aligned}\frac{\partial \mathbf{y}}{\partial \mathbf{x}} \cdot \frac{\partial \mathbf{z}}{\partial \mathbf{y}} &= \begin{pmatrix} 2x_1 & 3 \\ 1 & 1 \end{pmatrix}\begin{pmatrix} 1 & 2 \\ 2y_2 & 1 \end{pmatrix} \\ &= \begin{pmatrix} 2x_1 + 6y_2 & 4x_1 + 3 \\ 1 + 2y_2 & 3 \end{pmatrix} \\ &= \begin{pmatrix} 20x_1 + 6x_2 & 4x_1 + 3 \\ 2x_2 + 6x_1 + 1 & 3 \end{pmatrix}.\end{aligned}$$

The chain rule is verified.

EXERCISE 7.3

1. Let $\mathbf{y} = [y_1 \;\; y_2]^T$, $\mathbf{x} = [x_1 \;\; x_2 \;\; x_3]^T$ and $y_1 = x_1^3 - 3x_1x_2x_3 + x_2^2$, $y^2 = \sin x_1 - e^{x_2} + \ell n x_3$. Find $\frac{\partial \mathbf{y}}{\partial \mathbf{x}}$.

2. Let $\mathbf{x} = (x_1 \;\; x_2)^T$, $\mathbf{y} = (y_1 \;\; y_2)^T$, $\mathbf{z} = (z_1 \;\; z_2)^T$.
   Find $\frac{\partial \mathbf{z}}{\partial \mathbf{x}}$, where $z_1 = 1 + 2y_2$, $z_2 = 3y_1 + y_2$, $y_1 = 3x_1^3 + 5$, and $y_2 = x_1 + x_2^2$

3. Let $\mathbf{x} = [x_1 \;\; x_2 \;\; x_3]^T$, $\mathbf{y} = [y_1 \;\; y_2 \;\; y_3]^T$, $\mathbf{z} = [z_1 \;\; z_2 \;\; z_3]^T$, where $\mathbf{y}$ is a function of $\mathbf{x}$ and $\mathbf{z}$ is a function of $\mathbf{y}$. Find $\frac{\partial \mathbf{y}}{\partial \mathbf{x}}, \frac{\partial \mathbf{z}}{\partial \mathbf{x}}, \frac{\partial \mathbf{z}}{\partial \mathbf{y}}$ and verify the chain rule for derivatives of vectors,

   i.e., $\frac{\partial \mathbf{z}}{\partial \mathbf{x}} = \frac{\partial \mathbf{y}}{\partial \mathbf{x}} \cdot \frac{\partial \mathbf{z}}{\partial \mathbf{y}}$.

4. Find the Jacobian of the transformation from spherical to Cartesian coordinates defined by $x = r\ cos\ \theta\ cos\ \phi$, $y = r\ cos\ \theta\ sin\ \phi$, $z = r\ sin\ \theta$, where $r > 0$, $0 < \theta < \pi$, $0 \leq \phi < 2\pi$.

7.4 Derivative of a Scalar-Valued Function w.r.t. a Matrix

There exist many situations where a matrix is transformed into a scalar by some function such as the determinant of a matrix. In such cases, it is interesting to know how this function changes w.r.t. each variable. This leads to the concept of a derivative in scalar-valued function w.r.t. a matrix.

Let y be a scalar-valued function of a matrix $\mathbf{X} = (x_{ij})_{m\times n}$ such that the partial derivatives of $\mathbf{y}$ w.r.t. each variable x_{ij}, $i = 1, 2, \ldots, m$ and $j = 1, 2, \ldots, n$ exist.

Definition 7.4.1 *The gradient matrix* $\mathbf{G}$ *of* $y(\mathbf{X})$ *is*

$$\begin{aligned}\mathbf{G} &= \frac{\partial y}{\partial \mathbf{X}} = \left[\frac{\partial y}{\partial x_{ij}}\right]_{m\times n} \\ &= \begin{bmatrix} \frac{\partial y}{\partial x_{11}} & \frac{\partial y}{\partial x_{12}} & \cdots & \frac{\partial y}{\partial x_{1n}} \\ \frac{\partial y}{\partial x_{21}} & \frac{\partial y}{\partial x_{22}} & \cdots & \frac{\partial y}{\partial x_{2n}} \\ \vdots & \vdots & \vdots & \vdots \\ \frac{\partial y}{\partial x_{m1}} & \frac{\partial y}{\partial x_{m2}} & \cdots & \frac{\partial y}{\partial x_{mn}} \end{bmatrix} \\ &= \sum_{j=1}^{n}\sum_{i=1}^{m} \mathbf{P}_{ij}\,\frac{\partial y}{\partial x_{ij}},\end{aligned}$$

where $\mathbf{P}_{ij}$ *is a primary matrix of order* $m \times n$.

Observation. The gradient matrix is the matrix consisting of partial derivatives of y w.r.t. each entry x_{ij}.

Example 7.4.1 *Find the gradient matrix for* $y = tr\ \mathbf{X} = \sum_{i=1}^{m} x_{ii}$.

Solution. *Given* $y = tr\ \mathbf{X} = x_{11} + x_{22} + \cdots + x_{mm}$,

$$\frac{\partial y}{\partial \mathbf{X}} = \begin{bmatrix} \frac{\partial y}{\partial x_{11}} & \frac{\partial y}{\partial x_{12}} & \cdots & \frac{\partial y}{\partial x_{1m}} \\ \frac{\partial y}{\partial x_{21}} & \frac{\partial y}{\partial x_{22}} & \cdots & \frac{\partial y}{\partial x_{2m}} \\ \vdots & \vdots & \vdots & \vdots \\ \frac{\partial y}{\partial x_{m1}} & \frac{\partial y}{\partial x_{m2}} & \cdots & \frac{\partial y}{\partial x_{mm}} \end{bmatrix}$$

$$= \begin{bmatrix} 1 & 0 & \cdots & 0 \\ 0 & 1 & \cdots & 0 \\ \vdots & \vdots & & \vdots \\ 0 & 0 & \cdots & 1 \end{bmatrix} = \mathbf{I}_m.$$

Example 7.4.2 *Find the gradient matrix for* $z = |\mathbf{X}|$, $\mathbf{X} = (x_{pj})_{m\times m}$.
Solution. *Let* X_{pj} *be the cofactor of* x_{pj}
Then $z = |\mathbf{X}| = \sum_j X_{pj}x_{pj}$ *for a fixed* i.
$\frac{\partial z}{\partial x_{pj}} = X_{pj}$ *since, the cofactor* X_{pj} *is independent of* x_{pj}.
Hence

$$\frac{\partial z}{\partial \mathbf{X}} = \begin{bmatrix} 0 & 0 & 0\cdots & 0 & \cdots & 0 \\ \cdots & 0 & \cdots & 0 & & \\ X_{p1} & X_{p2} & \cdots & X_{pj} & X_{pj+1} & X_{pm} \\ 0 & 0 & 0 & 0 & 0 & 0 \\ \vdots & \vdots & \vdots & \vdots & \vdots & \vdots \\ 0 & 0 & & \cdots & & 0 \end{bmatrix}$$

for a fixed i, $j = 1, \ldots, m$.

Derivative of the determinant of a matrix.

Consider a matrix $\mathbf{Z} = (z_{ij}(\mathbf{X}))_{r\times r}$ *whose entries are functions of a matrix* $\mathbf{X} = (x_{ij})_{m\times n}$.

Then, $det\mathbf{Z} = |\mathbf{Z}| = \sum_j Z_{pj}z_{pj}$ *over a fixed row* p, *where* Z_{pj} *is the cofactor of* z_{pj} *and hence is independent of* z_{pj}.

$$\textit{Now,} \quad \frac{\partial|\mathbf{Z}|}{\partial z_{ij}} = \sum_j \frac{\partial}{\partial z_{ij}}(Z_{pj}z_{pj}) = Z_{pj}$$

$$\textit{Next, for} \quad i \neq p \quad \frac{\partial|\mathbf{Z}|}{\partial z_{pj}} = \sum_{s=1}^{r} z_{ps}\frac{\partial}{\partial z_{is}}(Z_{ps})$$

$$\textit{Thus,} \ \frac{\partial|\mathbf{Z}|}{\partial x_{kl}} = \begin{cases} \sum_i \sum_j Z_{pj}\frac{\partial z_{ij}}{\partial x_{kl}} & \textit{for } i = p \\ \sum_i \sum_j (\sum_{s=1}^{r} z_{ps}\frac{\partial}{\partial z_{is}}(Z_{ps}))\frac{\partial z_{ij}}{\partial x_{kl}} & \textit{for } i \neq p \end{cases}$$

Thus, $\dfrac{\partial|\mathbf{Z}|}{\partial\mathbf{X}} = \left[\dfrac{\partial|Z|}{\partial x_{kl}}\right]_{m\times n}.$

Example 7.4.3 *Let* $\mathbf{X} = [x_{ij}]_{2\times 2}$ *and* $y = |\mathbf{X}|$. *Find* $\frac{\partial y}{\partial \mathbf{X}}$.
Solution.

$$y = |\mathbf{X}| = \begin{vmatrix} x_{11} & x_{12} \\ x_{21} & x_{22} \end{vmatrix} = x_{11}\, x_{22} - x_{21}\, x_{12}$$

$$\text{Then } \frac{\partial y}{\partial \mathbf{X}} = \begin{bmatrix} \dfrac{\partial y}{\partial x_{11}} & \dfrac{\partial y}{\partial x_{12}} \\ \dfrac{\partial y}{\partial x_{21}} & \dfrac{\partial y}{\partial x_{22}} \end{bmatrix} = \begin{bmatrix} x_{22} & -x_{21} \\ -x_{12} & x_{11} \end{bmatrix}.$$

Example 7.4.4 *Let* $y = det \begin{bmatrix} e^{x_{11}} & x_{11}x_{22}^2 \\ x_{12}^3 & \ell n x_{21} \end{bmatrix}$ *find* $\frac{\partial y}{\partial \mathbf{X}}$.

Solution. *Given* $y = e^{x_{11}}\ell n x_{21} - x_{11}x_{22}^2x_{12}^3$. *Then*

$$\frac{\partial y}{\partial \mathbf{X}} = \begin{bmatrix} e^{x_{11}}\ell n x_{21} - x_{22}^2x_{12}^3 & -3x_{11}x_{22}^2x_{12}^2 \\ \frac{e^{x_{11}}}{x_{21}} & -2x_{11}x_{22}x_{12}^3 \end{bmatrix}$$

EXERCISE 7.4

1. Let $\mathbf{X} = \begin{bmatrix} x_{11} & x_{12} \\ x_{21} & x_{22} \end{bmatrix}$, $|\mathbf{X}|$ denotes the determinant of $\mathbf{X}$ and

   $||\mathbf{X}|| = \sum_{i,j=1}^{2} |x_{ij}|$ denotes the norm of $\mathbf{X}$.

   Let (i) $y = \frac{|\mathbf{X}|}{||x||}$, $\quad (ii)$ $y = |\mathbf{X}|\ ||x||$.

   Find $\frac{\partial y}{\partial \mathbf{X}}$ when $x_{ij} > 0$ for $i, j = 1, 2$.

2. Find $\frac{\partial y}{\partial \mathbf{X}}$ where $y = det \begin{bmatrix} x_{11} & x_{12} & x_{13} \\ x_{21} & x_{22} & x_{23} \\ x_{31} & x_{32} & x_{33} \end{bmatrix}$ is found over 2nd row.

3. Find $\frac{\partial y}{\partial \mathbf{X}}$ where $y = det \begin{bmatrix} x_{11} & x_{12} & x_{13} \\ x_{21} & x_{22} & x_{23} \\ x_{31} & x_{32} & x_{33} \end{bmatrix}$ is found over 2nd column.

7.5 Derivative of a Matrix Valued Function w.r.t. its Entries and Vice versa

In Section 7.4 the derivative of a scalar valued function of a matrix w.r.t. a matrix was introduced. This idea will be extended in this section. The derivative of a matrix w.r.t. its entries and the derivative of an entry w.r.t. the matrix form the content of this section.

Let $\mathbf{X} = (x_{ij})_{m\times n}$ be any arbitrary matrix.

Definition 7.5.1 *The derivative of the matrix* $\mathbf{X}$ *w.r.t. one of its entries* x_{kl} *is denoted by* $\frac{\partial \mathbf{X}}{\partial x_{kl}}$ *and is given by*

$$\frac{\partial \mathbf{X}}{\partial x_{ij}} = (a_{ij})_{m\times n}, \text{ where } a_{ij} = \begin{cases} 1, & \text{if } i = k,\ j = l \\ 0, & \text{otherwise.} \end{cases}$$

In other words, $\frac{\partial \mathbf{X}}{\partial x_{kl}} = \mathbf{P}_{kl}$, *where* $\mathbf{P}_{kl}$ *is a primary matrix of order* $m \times n$.

Observation. The derivative of $\mathbf{X}^T$ w.r.t. x_{kl} is $\frac{\partial \mathbf{X}^T}{\partial x_{kl}} = \mathbf{P}^T_{kl}$.

Example 7.5.1 *Let* $\mathbf{X} = \begin{bmatrix} x_{11} & x_{12} \\ x_{21} & x_{22} \end{bmatrix}$

Then $\frac{\partial \mathbf{X}}{\partial x_{12}} = \begin{bmatrix} 0 & 1 \\ 0 & 0 \end{bmatrix}$ *and* $\frac{\partial \mathbf{X}}{\partial x_{21}} = \begin{bmatrix} 0 & 0 \\ 1 & 0 \end{bmatrix}$

and $\frac{\partial \mathbf{X}^T}{\partial x_{12}} = \begin{bmatrix} 0 & 0 \\ 1 & 0 \end{bmatrix}$ *and* $\frac{\partial \mathbf{X}^T}{\partial x_{21}} = \begin{bmatrix} 0 & 1 \\ 0 & 0 \end{bmatrix}$.

Definition 7.5.2 *Let* $\mathbf{Z}_{r\times s}$ *be a matrix function of* $\mathbf{X}$, *then,* $\frac{\partial \mathbf{Z}}{\partial x_{kl}} = \left[\frac{\partial z_{ij}}{\partial x_{kl}}\right]_{r\times s}$.

Example 7.5.2 *Let*

$$\mathbf{Z} = \begin{bmatrix} x_{11}^2 & x_{11}x_{12} \\ x_{11} & x_{22}^2 x_{11} \end{bmatrix}$$

$$\frac{\partial \mathbf{Z}}{\partial x_{11}} = \begin{bmatrix} 2x_{11} & x_{12} \\ 1 & x_{22}^2 \end{bmatrix}$$

$$\frac{\partial \mathbf{Z}}{\partial x_{22}} = \begin{bmatrix} 0 & 0 \\ 0 & 2x_{22}x_{11} \end{bmatrix}.$$

Product rule for the derivative of a matrix w.r.t. its Entries

Let $\mathbf{P}_{r\times m}$ and $\mathbf{Q}_{m\times s}$ be matrix functions of the matrix $\mathbf{X}_{m\times n}$ such that

(i) each entry p_{ij} and q_{ij} are functions of the matrix $\mathbf{X}$

(ii) the partial derivatives of p_{ij} and q_{ij} w.r.t. x_{kl} exist for all $i,\ j,\ k,\ l$.

Set $\mathbf{Z}_{r\times s} = \mathbf{PQ}$. Then $z_{ij} = \sum_{u=1}^{m} p_{iu}q_{uj}$, for all $i,\ j$.
Now, for each fixed $k,\ l$,

$$(i,j)^{th} \text{ entry of } \frac{\partial \mathbf{Z}}{\partial x_{kl}} = \frac{\partial z_{ij}}{\partial x_{kl}} = \sum_{u=1}^{m} \frac{\partial p_{iu}}{\partial x_{kl}} q_{uj} + \sum_{u=1}^{m} p_{iu} \frac{\partial q_{uj}}{\partial x_{kl}}.$$

Hence the natural extension

$$\frac{\partial \mathbf{PQ}}{\partial x_{kl}} = \frac{\partial \mathbf{P}}{\partial x_{kl}}\mathbf{Q} + \mathbf{P}\frac{\partial \mathbf{Q}}{\partial x_{kl}} \text{ holds.}$$

The derivative of the (k,l) entry of the product $\mathbf{Z} = \mathbf{PQ}$ w.r.t. the matrix $\mathbf{X}$ is given by

$$\frac{\partial z_{kl}}{\partial \mathbf{X}} = \sum_{u=1}^{m} \frac{\partial p_{ku}}{\partial \mathbf{X}} q_{ul} + \sum_{u=1}^{m} p_{ku} \frac{\partial q_{ul}}{\partial \mathbf{X}}$$

Example 7.5.3 *Verify the product rule for*

$$\mathbf{P} = \begin{bmatrix} x_{11}^2 \\ x_{11}x_{12} \end{bmatrix} \qquad \mathbf{Q} = \begin{bmatrix} x_{12} & x_{11}x_{22} \end{bmatrix}$$

by differentiating w.r.t. x_{11}.
Solution.

$$\mathbf{PQ} = \begin{bmatrix} x_{11}^2 x_{12} & x_{11}^3 x_{22} \\ x_{11}x_{12}^2 & x_{11}^2 x_{12}x_{22} \end{bmatrix}$$

$$\textit{Hence,} \quad \frac{\partial \mathbf{PQ}}{\partial x_{11}} = \begin{bmatrix} 2x_{11}x_{12} & 3x_{11}^2 x_{22} \\ x_{12}^2 & 2x_{11}x_{12}x_{22} \end{bmatrix}.$$

$$\textit{Now} \quad \frac{\partial \mathbf{P}}{\partial x_{11}} = \begin{bmatrix} 2x_{11} \\ x_{12} \end{bmatrix} \quad \textit{and} \quad \frac{\partial \mathbf{Q}}{\partial x_{11}} = \begin{bmatrix} 0 & x_{22} \end{bmatrix}$$

$$\textit{Hence,} \quad \frac{\partial \mathbf{P}}{\partial x_{11}}\mathbf{Q} + \mathbf{P}\frac{\partial \mathbf{Q}}{\partial x_{11}} = \begin{bmatrix} 2x_{11} \\ x_{12} \end{bmatrix} \begin{bmatrix} x_{12} & x_{11}x_{22} \end{bmatrix} + \begin{bmatrix} x_{11}^2 \\ x_{11}x_{12} \end{bmatrix} \begin{bmatrix} 0 & x_{22} \end{bmatrix}$$

$$= \begin{bmatrix} 2x_{11}x_{12} & 3x_{11}^2 x_{22} \\ x_{12}^2 & 2x_{11}x_{12}x_{22} \end{bmatrix}.$$

Hence, the result is verified.

Derivative of the product $\mathbf{Z} = \mathbf{PXQ}$

Consider the matrices $\mathbf{P}$, $\mathbf{X}$ and $\mathbf{Q}$ of orders $r\times m$, $m\times n$ and $n\times s$, respectively. Then the matrix $\mathbf{Z} = \mathbf{PXQ}$ is of order $r \times s$. Clearly, $\mathbf{Z}$ is a function of the matrix $\mathbf{X}$.

The problem is: given $\mathbf{Z} = \mathbf{PXQ}$, find $\frac{\partial z_{ij}}{\partial \mathbf{X}}$ and $\frac{\partial \mathbf{Z}}{\partial x_{kl}}$, and to solve this problem, using results from Chapter 6, set.

$$Vec(\mathbf{Z}) = Vec(\mathbf{PXQ}) = (\mathbf{Q}^T \otimes \mathbf{P})\, Vec\ \mathbf{X}.$$

Now $Vec\ \mathbf{Z}$ is an $rs \times 1$ column matrix. $\mathbf{Q}^T \otimes \mathbf{P}$ is a matrix of order $rs \times mn$ and $Vec\ \mathbf{X}$ is of order $mn \times 1$. Hence $(\mathbf{Q}^T \otimes \mathbf{P})\, Vec\ \mathbf{X}$ is of order $rs \times 1$.

Consider z_{i1} element of $Vec\ \mathbf{Z}$. Then

$$\begin{aligned}
z_{i1} &= \text{the product of } i\text{th row element of } (\mathbf{Q}^T \otimes \mathbf{P}) \text{ and } Vec\ \mathbf{X}.\\
&= \text{product of the } i\text{th row of the matrix } (q_{11}\mathbf{P}, q_{21}P, \ldots, q_{n1}\mathbf{P}) \text{ and } Vec\ \mathbf{X}\\
&= (q_{11}p_{i1}\ q_{11}p_{i2} \cdots q_{11}p_{im}\quad q_{21}p_{i1}\ q_{21}p_{i2} \cdots q_{21}p_{im}\quad q_{n1}p_{i1} \cdots q_{n1}p_{im}) \begin{bmatrix} x_{11} \\ x_{21} \\ \vdots \\ x_{m1} \\ \vdots \\ x_{mn} \end{bmatrix}\\
&= \sum_{v=1}^{n} \sum_{u=1}^{m} p_{iu} q_{v1} x_{uv}
\end{aligned}$$

Thus $\frac{\partial z_{i1}}{\partial x_{kl}} = p_{ik}q_{l1}$. Hence,

$$\begin{aligned}
\frac{\partial z_{i1}}{\partial \mathbf{X}} &= \left(\frac{\partial z_{i1}}{\partial x_{kl}}\right)_{m\times n}\\
&= \begin{bmatrix} p_{i1}q_{11} & p_{i1}q_{21} & \cdots & p_{i1}q_{n1} \\ p_{i2}q_{11} & p_{i2}q_{21} & \cdots & p_{i2}q_{n1} \\ \vdots & \vdots & \vdots & \vdots \\ p_{im}q_{i1} & p_{im}q_{21} & \cdots & p_{im}q_{n1} \end{bmatrix}\\
&= \begin{bmatrix} p_{i1} \\ p_{i2} \\ \vdots \\ p_{im} \end{bmatrix} \begin{bmatrix} q_{11} & q_{21} & \cdots & q_{n1} \end{bmatrix}\\
&= \mathbf{P}_{i*}\mathbf{Q}_{*1}^T\\
&= \mathbf{P}^T\mathbf{P}_{i1}\mathbf{P}_{11}\mathbf{Q}^T\\
&= \mathbf{P}^T\mathbf{P}_{i1}\mathbf{Q}^T.
\end{aligned}$$

where $\mathbf{P}_{ij}$ is the primary matrix compatible with $\mathbf{P}^T$ and $\mathbf{Q}^T$. Hence $\mathbf{P}_{ij}$ is of order $r \times s$. Thus $\frac{\partial z_{i1}}{\partial \mathbf{X}} = \mathbf{P}^T\mathbf{P}_{ij}\mathbf{Q}^T$.

Next,

$$\begin{aligned}
\frac{\partial \mathbf{Z}}{\partial x_{kl}} &= \left[\frac{\partial z_{i1}}{\partial x_{kl}}\right]_{r\times s} \\
&= \begin{bmatrix} p_{1k}q_{l1} & p_{1k}q_{l2} & \cdots & p_{1k}q_{ls} \\ p_{2k}q_{l1} & p_{2k}q_{l2} & \cdots & p_{2k}q_{ls} \\ \vdots & \vdots & \vdots & \vdots \\ p_{rk}q_{l1} & p_{rk}q_{l2} & \cdots & p_{rk}q_{ls} \end{bmatrix} \\
&= \begin{bmatrix} p_{1k} \\ p_{2k} \\ \vdots \\ p_{rk} \end{bmatrix} \begin{bmatrix} q_{l1} & q_{l2} & \cdots & q_{lk} \end{bmatrix} \\
&= \mathbf{P}_{*k}\mathbf{Q}_{*l} \\
&= \mathbf{P}\mathbf{P}_{k!}\mathbf{P}_{!l}\mathbf{Q} \\
&= \mathbf{P}\mathbf{P}_{kl}\mathbf{Q},
\end{aligned}$$

where $\mathbf{P}_{kl}$ is the primary matrix of order $m \times n$.

Example 7.5.4 *Let* $\mathbf{Z} = \mathbf{PXQ}$, *where*

$$\mathbf{Z} = \begin{bmatrix} z_{11} & z_{12} \\ z_{21} & z_{22} \end{bmatrix} \mathbf{P} = \begin{bmatrix} p_1 & p_3 \\ p_2 & p_4 \end{bmatrix} \mathbf{X} = \begin{bmatrix} x_{11} & x_{12} \\ x_{21} & x_{22} \end{bmatrix} \mathbf{Q} = \begin{bmatrix} q_1 & q_3 \\ q_2 & q_4 \end{bmatrix}$$

$$\text{Find} \quad (i)\ \frac{\partial \mathbf{Z}}{\partial x_{ij}} \qquad (ii)\ \frac{\partial z_{ij}}{\partial \mathbf{X}}$$

Solution. *Converting the above matrix relation into a vector form gives,*

$$\text{Vec}\,(\mathbf{Z}) = \begin{bmatrix} z_{11} \\ z_{21} \\ z_{12} \\ z_{22} \end{bmatrix} = \begin{bmatrix} p_1x_{11}q_1 + p_3x_{21}q_1 + p_1x_{12}q_2 + p_3x_{22}q_2 \\ p_2x_{11}q_1 + p_4x_{21}q_1 + p_2x_{12}q_2 + p_4x_{22}q_2 \\ p_1x_{11}q_3 + p_3x_{21}q_3 + p_1x_{12}q_4 + p_3x_{22}q_4 \\ p_2x_{11}q_3 + p_4x_{21}q_3 + p_2x_{12}q_4 + p_4x_{22}q_4 \end{bmatrix}$$

Differentiate the column vector w.r.t. x_{11}, x_{12}, x_{21} *and* x_{22} *and covert the resulting vector in each case into a matrix. Hence it follows that*

$$(i)\quad \frac{\partial \mathbf{Z}}{\partial x_{11}} = \begin{bmatrix} p_1q_1 & p_1q_3 \\ p_2q_1 & p_2q_3 \end{bmatrix}, \qquad \frac{\partial \mathbf{Z}}{\partial x_{12}} = \begin{bmatrix} p_1q_2 & p_1q_4 \\ p_2q_2 & p_2q_4 \end{bmatrix}$$

$$\frac{\partial \mathbf{Z}}{\partial x_{21}} = \begin{bmatrix} p_3q_1 & p_3q_3 \\ p_4q_1 & p_4q_3 \end{bmatrix}, \qquad \frac{\partial \mathbf{Z}}{\partial x_{22}} = \begin{bmatrix} p_3q_2 & p_3q_4 \\ p_4q_2 & p_4q_4 \end{bmatrix}$$

Further,

$$(ii)\quad \frac{\partial z_{11}}{\partial \mathbf{X}} = \begin{bmatrix} \frac{\partial z_{11}}{\partial x_{11}} & \frac{\partial z_{11}}{\partial x_{12}} \\ \frac{\partial z_{11}}{\partial x_{21}} & \frac{\partial z_{11}}{\partial x_{22}} \end{bmatrix} = \begin{bmatrix} p_1q_1 & p_1q_2 \\ p_3q_1 & p_3q_2 \end{bmatrix};$$

$$\frac{\partial z_{12}}{\partial \mathbf{X}} = \begin{bmatrix} \frac{\partial z_{12}}{\partial x_{11}} & \frac{\partial z_{12}}{\partial x_{12}} \\ \frac{\partial z_{12}}{\partial x_{21}} & \frac{\partial z_{12}}{\partial x_{22}} \end{bmatrix} = \begin{bmatrix} p_1q_3 & p_1q_4 \\ p_3q_3 & p_3q_4 \end{bmatrix}$$

$$\frac{\partial z_{21}}{\partial \mathbf{X}} = \begin{bmatrix} \frac{\partial z_{21}}{\partial x_{11}} & \frac{\partial z_{21}}{\partial x_{12}} \\ \frac{\partial z_{21}}{\partial x_{21}} & \frac{\partial z_{21}}{\partial x_{22}} \end{bmatrix} = \begin{bmatrix} p_2q_1 & p_2q_2 \\ p_4q_1 & p_4q_2 \end{bmatrix};$$

$$\frac{\partial z_{22}}{\partial \mathbf{X}} = \begin{bmatrix} \frac{\partial z_{21}}{\partial x_{11}} & \frac{\partial z_{22}}{\partial x_{12}} \\ \frac{\partial z_{22}}{\partial x_{21}} & \frac{\partial z_{22}}{\partial x_{22}} \end{bmatrix} = \begin{bmatrix} p_2q_3 & p_2q_4 \\ p_4q_3 & p_4q_4 \end{bmatrix}$$

Example 7.5.5 *Let* $\mathbf{Z} = \mathbf{P}\ \mathbf{X}^{-1}\mathbf{Q}$, *where* $\mathbf{P}, \mathbf{X}, \mathbf{Q}$ *are invertible matrices of orders* $r \times n$, $n \times n$ *and* $n \times s$, *respectively. Find* $\frac{\partial \mathbf{Z}}{\partial x_{kl}}$.

Solution. *Differentiating the identity* $\mathbf{Z}\ \mathbf{Z}^{-1} = \mathbf{I}$,

$$\frac{\partial \mathbf{Z}}{\partial x_{kl}} \mathbf{Z}^{-1} + \mathbf{Z}\ \frac{\partial \mathbf{Z}^{-1}}{\partial x_{kl}} = \mathbf{O}$$

$$\Rightarrow \quad \frac{\partial \mathbf{Z}}{\partial x_{kl}} = -\mathbf{Z}\ \frac{\partial \mathbf{Z}^{-1}}{\partial x_{kl}}\ \mathbf{Z}$$

But $\frac{\partial \mathbf{P}\ \mathbf{X}\ \mathbf{Q}}{\partial x_{kl}} = \mathbf{P}\ \mathbf{P}_{kl}\ \mathbf{Q}$ *gives*

$$\frac{\partial \mathbf{Z}^{-1}}{\partial x_{kl}} = \frac{\partial}{\partial x_{kl}} \left(\mathbf{Q}^{-1}\ \mathbf{X}\ \mathbf{P}^{-1} \right) = \mathbf{Q}^{-1}\ \mathbf{P}_{kl}\ \mathbf{P}^{-1}$$

$$\Rightarrow \quad \frac{\partial \mathbf{Z}}{\partial x_{kl}} = \frac{\partial}{\partial x_{kl}} \left(\mathbf{P}\ \mathbf{X}^{-1}\ \mathbf{Q} \right) = -\mathbf{P}\ \mathbf{X}^{-1}\ \mathbf{Q}\ \ \mathbf{Q}^{-1}\ \mathbf{P}_{kl}\ \mathbf{P}^{-1}\ \ \mathbf{P}\ \mathbf{X}^{-1}\ \mathbf{Q}$$

$$= -\mathbf{P}\ \mathbf{X}^{-1}\mathbf{P}_{kl}\ \ \mathbf{X}^{-1}\ \mathbf{Q}$$

Example 7.5.6 *Let* $\mathbf{Z} = \mathbf{P}\ \mathbf{X}^{-1}\ \mathbf{Q}$, *where* $\mathbf{P}, \mathbf{X}, \mathbf{Q}$ *are of orders* $r \times n$, $n \times n$ *and* $n \times s$, *respectively and* $\mathbf{X}$ *be an invertible matrix. Determine* $\frac{\partial z_{i1}}{\partial x_{kl}}$.

Solution. *From Example 7.5.5,*

$$\begin{aligned}
\frac{\partial \mathbf{Z}}{\partial x_{kl}} &= -\mathbf{P}\,\mathbf{X}^{-1}\,\mathbf{P}_{kl}\,\mathbf{X}^{-1}\,\mathbf{Q} \\
\textit{Let}\quad \mathbf{P}\,\mathbf{X}^{-1} &= \mathbf{P}_1\,, \quad \mathbf{X}^{-1}\,\mathbf{Q} = \mathbf{Q}_1 \\
\frac{\partial \mathbf{Z}}{\partial x_{kl}} &= -\mathbf{P}_1\,\mathbf{P}_{kl}\,\mathbf{Q}_1 \\
\frac{\partial z_{i1}}{\partial \mathbf{X}} &= -\mathbf{P}_1^T\,\mathbf{P}_{ij}\,\mathbf{Q}_1^T \\
&= -\left(\mathbf{X}^{-1}\right)^T\,\mathbf{P}^T\,\mathbf{P}_{ij}\,\mathbf{Q}^T\,\left(\mathbf{X}^{-1}\right)^T
\end{aligned}$$

Example 7.5.7 *Let* $\mathbf{Z} = \mathbf{X}^T\,\mathbf{P}\,\mathbf{X}$, *where* $\mathbf{P}, \mathbf{X}, \mathbf{Q}$ *are of orders* $r \times n$, $n \times n$ *and* $n \times s$, *respectively. Find (i)* $\frac{\partial \mathbf{Z}}{\partial x_{kl}}$, *(ii)* $\frac{\partial z_{i1}}{\partial \mathbf{X}}$.

Solution. *(i) Consider* $\mathbf{Z} = \mathbf{X}^T\mathbf{P}\mathbf{X}$ *and applying chain rule,*

$$\begin{aligned}
\frac{\partial \mathbf{Z}}{\partial x_{kl}} &= \left(\frac{\partial}{\partial x_{kl}}\,\mathbf{X}^T\right)\mathbf{P}\,\mathbf{X} + \mathbf{X}^T\,\frac{\partial(\mathbf{P}\mathbf{X})}{\partial x_{kl}} \\
&= \mathbf{P}_{kl}^T\,\mathbf{P}\,\mathbf{X} + \mathbf{X}^T\mathbf{P}\,\mathbf{P}_{kl}.
\end{aligned}$$

$$\begin{aligned}
\textit{(ii)}\ \ \textit{Let}\ \ \mathbf{P}_1 &= I,\ \ \mathbf{Q}_1 = \mathbf{P}\,\mathbf{X},\ \ \mathbf{P}_2 = \mathbf{X}^T\,\mathbf{P},\ \ \mathbf{Q}_2 = I. \\
\textit{Then,}\ \ \frac{\partial \mathbf{Z}}{\partial x_{kl}} &= \mathbf{P}_1\,\mathbf{P}_{kl}^T\,\mathbf{Q}_1 + \mathbf{P}_2\,\mathbf{P}_{kl}\,\mathbf{Q}_2 \\
\frac{\partial z_{i1}}{\partial \mathbf{X}} &= \mathbf{Q}_1\,\mathbf{P}_{ij}^T\,\mathbf{P}_1 + \mathbf{P}_2^T\,\mathbf{P}_{ij}\,\mathbf{Q}_2^T \\
&= \mathbf{P}\,\mathbf{X}\,\mathbf{P}_{ij}^T + \mathbf{P}^T\,\mathbf{X}\,\mathbf{P}_{ij}
\end{aligned}$$

Example 7.5.8 *Let* $\mathbf{Z} = \mathbf{X}^2$, *where* $\mathbf{X}$ *is a square matrix of order* n.

$$\text{Find}\ \ (i)\ \ \frac{\partial \mathbf{Z}}{\partial x_{kl}}\ \ (ii)\ \ \frac{\partial z_{i1}}{\partial \mathbf{X}}.$$

Solution. *(i) Consider* $\mathbf{Z} = \mathbf{X}^2$, *then*

$$\frac{\partial \mathbf{Z}}{\partial x_{kl}} = \frac{\partial}{\partial x_{kl}}(\mathbf{X}^2) = \frac{\partial}{\partial x_{kl}}(\mathbf{X}\,\mathbf{X}) = \frac{\partial \mathbf{X}}{\partial x_{kl}}\mathbf{X} + X\frac{\partial \mathbf{X}}{\partial x_{kl}}$$

Now,

$$\frac{\partial \mathbf{X}}{\partial x_{kl}} = \frac{\partial}{\partial x_{kl}} \begin{bmatrix} x_{11} & x_{12} & \cdots & x_{1s} & \cdots & x_{1n} \\ \cdots & \cdots & & \cdots & & \cdots \\ x_{r1} & x_{r2} & \cdots & x_{rs} & \cdots & x_{rn} \\ \vdots & \vdots & & \vdots & & \vdots \\ x_{n1} & x_{n2} & \cdots & x_{ns} & \cdots & x_{nn} \end{bmatrix}$$

$$= \begin{bmatrix} 0 & 0 & \cdots & 0 & \cdots & 0 \\ \cdots & \cdots & & \cdots & & \cdots \\ 0 & 0 & \cdots & 1 & \cdots & 0 \\ \vdots & \vdots & & \vdots & & \vdots \\ 0 & 0 & \cdots & 0 & \cdots & 0 \end{bmatrix}$$

$$= \mathbf{P}_{kl},$$

where $\mathbf{P}_{kl}$ *is the* $n \times n$ *matrix having* 1 *in the* (k, l) *place and zero elsewhere. Hence*

$$\frac{\partial \mathbf{Z}}{\partial x_{kl}} = \mathbf{P}_{kl}\mathbf{X} + \mathbf{X}\mathbf{P}_{kl}$$

Observation. If $\mathbf{Z} = \mathbf{X}^3$, then set $\mathbf{Z} = \mathbf{X}^2 \cdot \mathbf{X}$ and employ the derivative of $\mathbf{X}^2$ obtained earlier. The procedure continues for higher orders of $\mathbf{X}$.

Example 7.5.9 *Let* $\mathbf{X} = \begin{bmatrix} x_{11} & x_{12} \\ x_{21} & x_{22} \end{bmatrix}$. *Then*

$$\mathbf{Z} = \mathbf{X}^2 = \begin{bmatrix} x_{11}^2 + x_{12}x_{21} & x_{11}x_{12} + x_{12}x_{22} \\ x_{11}x_{21} + x_{22}x_{21} & x_{21}x_{12} + x_{22}^2 \end{bmatrix}$$

$$z_{11} = x_{11}^2 + x_{12}x_{21}$$

$$\frac{\partial z_{11}}{\partial \mathbf{X}} = \begin{bmatrix} \frac{\partial z_{11}}{\partial x_{11}} & \frac{\partial z_{11}}{\partial x_{12}} \\ \frac{\partial z_{11}}{\partial x_{21}} & \frac{\partial z_{11}}{\partial x_{22}} \end{bmatrix} = \begin{bmatrix} 2x_{11} & x_{21} \\ x_{12} & 0 \end{bmatrix}$$

$$= \begin{bmatrix} 1 & 0 \\ 0 & 0 \end{bmatrix} \begin{bmatrix} x_{11} & x_{21} \\ x_{12} & x_{22} \end{bmatrix} + \begin{bmatrix} x_{11} & x_{21} \\ x_{12} & x_{22} \end{bmatrix} \begin{bmatrix} 1 & 0 \\ 0 & 0 \end{bmatrix}$$

$$= \mathbf{P}_{11}\mathbf{X}^T + \mathbf{X}^T\mathbf{P}_{11}.$$

generalizing this expression for a matrix $\mathbf{X}$ *of order* $n \times n$. *gives,*

$$\frac{\partial z_{ij}}{\partial \mathbf{X}} = \mathbf{P}_{ij}X^T + \mathbf{X}^T\mathbf{P}_{ij}$$

EXERCISE 7.5

1. Let $\mathbf{P} = \begin{bmatrix} 2 & 4 \\ 5 & 6 \end{bmatrix}$, $\mathbf{Q} = \begin{bmatrix} 1 & 3 \\ 1 & 4 \end{bmatrix}$, $\mathbf{X} = \begin{bmatrix} x_1 & x_3 \\ x_2 & x_4 \end{bmatrix}$ and $\mathbf{Z} = \mathbf{PXQ}$. Find $\frac{\partial \mathbf{Z}}{\partial x_1}$, $\frac{\partial \mathbf{Z}}{\partial x_2}$, and $\frac{\partial z_{12}}{\partial \mathbf{X}}$.

2. Let $\mathbf{X} = \begin{bmatrix} x_1 & x_3 \\ x_2 & x_4 \end{bmatrix}$. Set $\mathbf{Z} = \mathbf{X}^2$ find $\frac{\partial \mathbf{Z}}{\partial x_3}$ and $\frac{\partial \mathbf{Z}}{\partial x_4}$.

3. Let $\mathbf{X} = \begin{bmatrix} x_1 & x_3 \\ x_2 & x_4 \end{bmatrix}$, $\mathbf{Y} = \begin{bmatrix} y_1(x_1, x_2, x_3, x_4) & y_3(x_1, x_2, x_3, x_4) \\ y_2(x_1, x_2, x_3, x_4) & y_4(x_1, x_2, x_3, x_4) \end{bmatrix}$. Set $Z = XY$ find $\frac{\partial \mathbf{Z}}{\partial x_1}$, $\frac{\partial \mathbf{Z}}{\partial x_2}$, $\frac{\partial \mathbf{Z}}{\partial x_3}$ and $\frac{\partial \mathbf{Z}}{\partial x_4}$.

7.6 The Matrix Differential

In this section, the concept of a matrix differential is introduced parallel to the corresponding notion of a scalar valued function $h(\mathbf{x})$, where $\mathbf{x} = (x_1 \; x_2 \; \dots \; x_n)^T$.

Definition 7.6.1 *Let* $\mathbf{Z} = [z_{ij}]_{m\times n}$. *The matrix differential* $d\mathbf{Z}$ *of the matrix* $\mathbf{Z}$ *is defined as*

$$d\mathbf{Z} = \begin{bmatrix} dz_{11} & dz_{12} & \cdots & dz_{1n} \\ dz_{21} & dz_{22} & \cdots & dz_{2n} \\ \vdots & \vdots & & \vdots \\ dz_{m1} & dz_{m2} & \cdots & dz_{mn} \end{bmatrix} = [\, dz_{ij} \,]_{m\times n}.$$

Example 7.6.1 *Let* $\mathbf{Z} = \begin{bmatrix} x_{11}^2 & x_{11}x_{12} \\ x_{21}x_{12} & x_{22}^2 \end{bmatrix}$

Then
$$d\mathbf{Z} = \begin{bmatrix} d\left(x_{11}^2\right) & d\left(x_{11}x_{12}\right) \\ d\left(x_{21}x_{12}\right) & d\left(x_{22}^2\right) \end{bmatrix} = \begin{bmatrix} 2x_{11}dx_{11} & x_{12}dx_{11} + x_{11}dx_{12} \\ x_{21}dx_{12} + x_{12}dx_{21} & 2x_{22}dx_{22} \end{bmatrix}$$

Result 7.6.1 The matrix differential is a linear transformation

1. $d(\alpha\mathbf{Z}) = \alpha d\mathbf{Z}$, where $\alpha \in \mathbb{R}$

2. $d(\mathbf{Z}_1 + \mathbf{Z}_2) = d(\mathbf{Z}_1) + d\mathbf{Z}_2$.

Example 7.6.2 *Let* $\mathbf{Z} = \begin{bmatrix} z_{11} & z_{12} \\ z_{13} & dz_{14} \end{bmatrix}$

$$d(\alpha\mathbf{Z}) = \begin{bmatrix} d(\alpha z_{11}) & d(\alpha z_{12}) \\ d(\alpha z_{13}) & d(\alpha z_{14}) \end{bmatrix} = \alpha \begin{bmatrix} dz_{11} & dz_{12} \\ dz_{13} & dz_{14} \end{bmatrix} = \alpha d\mathbf{Z}.$$

Example 7.6.3 *Let* $\mathbf{Z}_1 = \begin{bmatrix} z_{11} & z_{12} \\ z_{13} & z_{14} \end{bmatrix}$ $\mathbf{Z}_2 = \begin{bmatrix} z_{21} & z_{22} \\ z_{23} & z_{24} \end{bmatrix}$

$$\begin{aligned} d(\mathbf{Z}_1 + \mathbf{Z}_2) &= \begin{bmatrix} d(z_{11} + z_{21}) & d(z_{12} + z_{22}) \\ d(z_{13} + z_{23}) & d(z_{14} + z_{24}) \end{bmatrix} \\ &= \begin{bmatrix} dz_{11} + dz_{21} & dz_{12} + dz_{22} \\ dz_{13} + dz_{23} & dz_{14} + dz_{24} \end{bmatrix} \\ &= \begin{bmatrix} dz_{11} & dz_{12} \\ dz_{13} & dz_{14} \end{bmatrix} + \begin{bmatrix} dz_{21} & dz_{22} \\ dz_{23} & dz_{24} \end{bmatrix} \\ &= d\mathbf{Z}_1 + d\mathbf{Z}_2. \end{aligned}$$

Product rule for differentials

Let $\mathbf{Z}_1$ and $\mathbf{Z}_2$ be two matrices of orders $r \times s$ and $s \times t$, respectively.

Then the product $\mathbf{Z}_1\mathbf{Z}_2$ is a matrix of order $r \times t$ and $d(\mathbf{Z}_1\mathbf{Z}_2)=(d\mathbf{Z}_1)\mathbf{Z}_2 + \mathbf{Z}_1 d\mathbf{Z}_2$.

Proof. Follows a similar pattern to derivative of matrices w.r.t. a variable.

Example 7.6.4 *Given* $\mathbf{A} = (a_{ij})_{2\times 2}$, $\mathbf{X} = (x_{ij})_{2\times 2}$. *Find the matrix differential (i)* $\mathbf{Z} = \mathbf{A}\,\mathbf{X}$ *(ii)* $\mathbf{Z} = \mathbf{X}^T\,\mathbf{X}$.

Solution. *(i) Given*

$$\mathbf{Z} = \mathbf{A}\,\mathbf{X} = \begin{bmatrix} a_{11}\,x_{11} + a_{12}\,x_{21} & a_{11}\,x_{12} + a_{12}\,x_{22} \\ \\ a_{21}\,x_{11} + a_{22}\,x_{21} & a_{21}\,x_{12} + a_{22}\,x_{22} \end{bmatrix}$$

then,

$$
\begin{aligned}
d\mathbf{Z} &= \begin{bmatrix} a_{11}\,dx_{11} + a_{12}\,dx_{21} & a_{11}\,dx_{12} + a_{12}\,dx_{22} \\ a_{21}\,dx_{11} + a_{22}\,dx_{21} & a_{21}\,dx_{12} + a_{22}\,dx_{22} \end{bmatrix} \\
&= \begin{bmatrix} a_{11} & a_{12} \\ a_{21} & a_{22} \end{bmatrix} \begin{bmatrix} dx_{11} & dx_{12} \\ dx_{21} & dx_{22} \end{bmatrix} \\
&= \mathbf{A}d\mathbf{X}
\end{aligned}
$$

(ii) Given

$$
\begin{aligned}
\mathbf{Z} &= \mathbf{X}^T\mathbf{X} \\
&= \begin{bmatrix} x_{11} & x_{21} \\ x_{12} & x_{22} \end{bmatrix} \begin{bmatrix} x_{11} & x_{12} \\ x_{21} & x_{22} \end{bmatrix} \\
&= \begin{bmatrix} x_{11}\,x_{11} + x_{21}\,x_{21} & x_{11}\,x_{12} + x_{21}\,x_{22} \\ x_{12}\,x_{11} + x_{22}\,x_{21} & x_{12}\,x_{12} + x_{22}\,x_{22} \end{bmatrix}
\end{aligned}
$$

$$
\begin{aligned}
\textit{Then } d\mathbf{Z} &= \begin{bmatrix} \sum_{i=1}^{2} 2\,x_{i1}\,dx_{i1} & \sum_{i=1}^{2} x_{i1}\,dx_{i2} + x_{i2}\,dx_{i1} \\ \sum_{i=1}^{2} x_{i1}\,dx_{i2} + x_{i2}\,dx_{i1} & \sum_{i=1}^{2} 2\,x_{i2}\,dx_{i2} \end{bmatrix} \\
&= \begin{bmatrix} dx_{11} & dx_{21} \\ dx_{12} & dx_{22} \end{bmatrix} \begin{bmatrix} x_{11} & x_{12} \\ x_{21} & x_{22} \end{bmatrix} + \begin{bmatrix} x_{11} & x_{12} \\ x_{21} & x_{22} \end{bmatrix} \begin{bmatrix} dx_{11} & dx_{12} \\ dx_{21} & dx_{22} \end{bmatrix} \\
&= (\,d\mathbf{X}\,)^T\,\mathbf{X} + \mathbf{X}^T\,d\mathbf{X}
\end{aligned}
$$

EXERCISE 7.6

1. Let $\mathbf{Z}_1 = \begin{bmatrix} x^2 & y^2 \\ xy & x^2y \end{bmatrix}$ $\quad \mathbf{Z}_2 = \begin{bmatrix} 1 & 2x \\ x^2 & y^2 \end{bmatrix}$ find (i) $d(3\mathbf{Z}_1)$ (ii) $d(\mathbf{Z}_1 + 4\mathbf{Z}_2)$.
2. Let $\mathbf{Z}_1 = \begin{bmatrix} 3x^2 & 4x \\ 8xy & 6y^2 \end{bmatrix}$ $\quad \mathbf{Z}_2 = \begin{bmatrix} 2xy & 3x^2 \\ 1 & x^2 \end{bmatrix}$, find $d(\mathbf{Z}_1\mathbf{Z}_2)$.

7.7 Derivative of a Matrix w.r.t. a Matrix

In this section, the concepts introduced earlier are further extended to obtain a new notion, namely, the derivative of a matrix w.r.t. a matrix. From the earlier sections, derivative of a matrix $\mathbf{Z} = [z_{ij}]_{r\times s}$ w.r.t. an element x_{kl} is given by

$$\frac{\partial \mathbf{Z}}{\partial x_{kl}} = \left[\frac{\partial z_{ij}}{\partial x_{kl}}\right]_{r\times s}.$$

Definition 7.7.1 *Let* $\mathbf{Z}$ *be a function of a matrix* $\mathbf{X} = [x_{ij}]_{m\times n}$, *then the derivative of the matrix* $\mathbf{Z}$ *w.r.t the matrix* $\mathbf{X}$, $\frac{\partial \mathbf{Z}}{\partial \mathbf{X}}$ *is the partitioned matrix whose* (kl)th *partition is* $\frac{\partial \mathbf{Z}}{\partial x_{kl}}$ *and*

$$\frac{\partial \mathbf{Z}}{\partial \mathbf{X}} = \left[\frac{\partial \mathbf{Z}}{\partial x_{kl}} \right]_{mr\times ns}$$
$$= \begin{bmatrix} \frac{\partial \mathbf{Z}}{\partial x_{11}} & \frac{\partial \mathbf{Z}}{\partial x_{12}} & \cdots & \frac{\partial \mathbf{Z}}{\partial x_{1n}} \\ \frac{\partial \mathbf{Z}}{\partial x_{21}} & \frac{\partial \mathbf{Z}}{\partial x_{22}} & \cdots & \frac{\partial \mathbf{Z}}{\partial x_{2n}} \\ \vdots & \vdots & & \vdots \\ \frac{\partial \mathbf{Z}}{\partial x_{m1}} & \frac{\partial \mathbf{Z}}{\partial x_{m2}} & \cdots & \frac{\partial \mathbf{Z}}{\partial x_{mn}} \end{bmatrix}_{mr\times ns}.$$

Observation. The derivative of the matrix $\mathbf{Z}$ w.r.t. the matrix $\mathbf{X}$ can also be written as follows

$$\frac{\partial \mathbf{Z}}{\partial \mathbf{X}} = \sum_k \sum_l P_{kl} \otimes \frac{\partial \mathbf{Z}}{\partial x_{kl}}.$$

The derivative of a determinant of a matrix w.r.t a matrix is an interesting concept and is given below.

Example 7.7.1 *Let* $\mathbf{Z} = [z_{ij}]$ *be a matrix whose entriess are functions of a matrix* $\mathbf{X} = [x_{kl}]$ *or* $z_{ij} = f_{ij}(x_{kl})$. *To find* $\frac{\partial \text{ det } \mathbf{Z}}{\partial x_{kl}}$.

Solution.

For any fixed row p, *let* $\det \mathbf{Z} = \sum_j z_{pj} Z_{pj}$ *where* Z_{pj} *is the cofactor of the element* z_{pj} *in* $\det \mathbf{Z}$.

$$\text{From Chain rule, } \frac{\partial \text{ det } \mathbf{Z}}{\partial x_{kl}} = \sum_i \sum_j \frac{\partial \text{ det } \mathbf{Z}}{\partial z_{ij}} \frac{\partial z_{ij}}{\partial x_{kl}}.$$

It has been shown in Section 7.4 that,

$$\frac{\partial |\mathbf{Z}|}{\partial x_{kl}} = \begin{cases} \sum_i \sum_j Z_{pj} \frac{\partial z_{ij}}{\partial x_{kl}} & \text{for } i = p \\ \sum_i \sum_j (\sum_{s=1}^r z_{ps} \frac{\partial}{\partial z_{is}}(Z_{ps})) \frac{\partial z_{ij}}{\partial x_{kl}} & \text{for } i \neq p \end{cases}$$

Next, define, $\mathbf{A} = [a_{ij}]$, *where* $a_{ij} = \left\{ \begin{array}{ll} Z_{pj} & \text{for } i = p \\ \sum_{s=1}^r z_{ps} \frac{\partial}{\partial z_{is}}(Z_{ps}) & \text{for } i \neq p \end{array} \right\}$

and $B = (b_{ij})$ *where* $b_{ij} = \frac{\partial z_{ij}}{\partial x_{kl}}$. *Then*

$$then\ \frac{\partial\ det\ \mathbf{Z}\ |}{\partial x_{kl}} = \sum_i \sum_j a_{ij} b_{ij}$$
$$= \sum_i \sum_j a_{ij}\ \mathbf{P}_{!j}\ b_{ij}\ \mathbf{P}_{j!}$$
$$= \sum_i \mathbf{A}_i^T\ \mathbf{B}_i$$
$$= tr(\mathbf{A}\ \mathbf{B}^T) = tr(\mathbf{B}^T\mathbf{A})$$

this is alternative form of the result.

Example 7.7.2 *Let* $\mathbf{X} = \begin{bmatrix} x_{11} & x_{12} \\ x_{21} & x_{22} \end{bmatrix}$. *Find* $\frac{\partial}{\partial \mathbf{X}}(\mathbf{X}^{-1})$.

Solution. *By definition,* $\mathbf{X}^{-1} = \dfrac{1}{x_{11}x_{22} - x_{12}x_{21}} \begin{bmatrix} x_{22} & -x_{12} \\ -x_{21} & x_{11} \end{bmatrix}$.

$$\frac{\partial}{\partial \mathbf{X}}(\mathbf{X}^{-1}) = \begin{bmatrix} \frac{\partial}{\partial x_{11}}(\mathbf{X}^{-1}) & \frac{\partial}{\partial x_{12}}(\mathbf{X}^{-1}) \\ \frac{\partial}{\partial x_{21}}(\mathbf{X}^{-1}) & \frac{\partial}{\partial x_{22}}(\mathbf{X}^{-1}) \end{bmatrix}$$

$$= \frac{1}{d^2} \begin{bmatrix} -x_{22}^2 & x_{22}x_{12} & -x_{22}x_{21} & -x_{21}x_{12} - d \\ x_{21}x_{22} & -x_{11}x_{22} + d & -x_{21}^2 & x_{11}x_{21} \\ x_{22}x_{12} & -x_{12}^2 & -x_{12}x_{11} + d & x_{11}x_{12} \\ -x_{12}x_{21} - d & x_{12}x_{11} & x_{21}x_{11} & -x_{11}^2 \end{bmatrix}$$

$$where\ d = (x_{11}x_{22} - x_{12}x_{21})$$

Example 7.7.3 *Find* $\frac{\partial \mathbf{Z}}{\partial \mathbf{X}}$, *given that* $\mathbf{X} = \begin{bmatrix} x_{11} & x_{12} \\ x_{21} & x_{22} \end{bmatrix}$ *and*

$$\mathbf{Z} = \begin{bmatrix} x_{11}^2 x_{12} & \ell n(x_{11} + x_{12}) \\ \cos(x_{21} + x_{22}) & exp(x_{12}x_{22}) \end{bmatrix}$$

Solution. *By definition,*

$$\frac{\partial \mathbf{Z}}{\partial \mathbf{X}} = \begin{bmatrix} \frac{\partial \mathbf{Z}}{\partial x_{11}} & \frac{\partial \mathbf{Z}}{\partial x_{12}} \\ \frac{\partial \mathbf{Z}}{\partial x_{21}} & \frac{\partial \mathbf{Z}}{\partial x_{22}} \end{bmatrix}$$

$$= \begin{bmatrix} 2x_{11}x_{12} & \frac{1}{x_{11}+x_{12}} & x_{11}^2 & \frac{1}{x_{11}+x_{12}} \\ 0 & 0 & 0 & x_{22}\ exp(x_{12}x_{22}) \\ 0 & 0 & 0 & 0 \\ -\sin(x_{21} + x_{22}) & 0 & -\sin(x_{21} + x_{22}) & x_{12}exp(x_{12}x_{22}) \end{bmatrix}$$

Example 7.7.4 *If* $\mathbf{A} = \begin{bmatrix} a_{11} & a_{12} \\ a_{21} & a_{22} \end{bmatrix}$, $\mathbf{X} = \begin{bmatrix} x_{11} & x_{12} \\ x_{21} & x_{22} \end{bmatrix}$ *and* $\mathbf{Z} = \mathbf{A}\ \mathbf{X}^T$. *Use*

the direct method to evaluate $\frac{\partial \ Vec\ \mathbf{Z}}{\partial \ Vec\ \mathbf{X}}$.
Solution. *Given*

$$\begin{aligned}\mathbf{Z} &= \mathbf{A}\ \mathbf{X}^T \\ &= \begin{bmatrix} a_{11} & a_{12} \\ a_{21} & a_{22} \end{bmatrix} \cdot \begin{bmatrix} x_{11} & x_{21} \\ x_{12} & x_{22} \end{bmatrix} \\ &= \begin{bmatrix} a_{11}\ x_{11} + a_{12}\ x_{12} & a_{11}\ x_{21} + a_{12}\ x_{22} \\ a_{21}\ x_{11} + a_{22}\ x_{12} & a_{21}\ x_{21} + a_{22}\ x_{22} \end{bmatrix}\end{aligned}$$

then

$$\begin{bmatrix} z_{11} \\ z_{21} \\ z_{12} \\ z_{22} \end{bmatrix} = \begin{bmatrix} a_{11}\ x_{11} + a_{12}\ x_{12} \\ a_{21}\ x_{11} + a_{22}\ x_{12} \\ a_{11}\ x_{21} + a_{12}\ x_{22} \\ a_{21}\ x_{21} + a_{22}\ x_{22} \end{bmatrix}$$

$$\frac{\partial \ Vec\ \mathbf{Z}}{\partial \ Vec\ \mathbf{X}} = \begin{bmatrix} \frac{\partial\ z_{11}}{\partial\ x_{11}} & \frac{\partial\ z_{21}}{\partial\ x_{11}} & \frac{\partial\ z_{12}}{\partial\ x_{11}} & \frac{\partial\ z_{22}}{\partial\ x_{11}} \\ \frac{\partial\ z_{11}}{\partial\ x_{21}} & \frac{\partial\ z_{21}}{\partial\ x_{21}} & \frac{\partial\ z_{12}}{\partial\ x_{21}} & \frac{\partial\ z_{22}}{\partial\ x_{21}} \\ \frac{\partial\ z_{11}}{\partial\ x_{12}} & \frac{\partial\ z_{21}}{\partial\ x_{12}} & \frac{\partial\ z_{12}}{\partial\ x_{12}} & \frac{\partial\ z_{22}}{\partial\ x_{12}} \\ \frac{\partial\ z_{11}}{\partial\ x_{22}} & \frac{\partial\ z_{21}}{\partial\ x_{22}} & \frac{\partial\ z_{12}}{\partial\ x_{22}} & \frac{\partial\ z_{22}}{\partial\ x_{22}} \end{bmatrix}$$

$$= \begin{bmatrix} a_{11} & a_{21} & 0 & 0 \\ 0 & 0 & a_{11} & a_{21} \\ a_{12} & a_{22} & 0 & 0 \\ 0 & 0 & a_{12} & a_{22} \end{bmatrix}$$

EXERCISE 7.7

1. Let $\mathbf{Z} = \begin{bmatrix} x_{11}x_{12} + x_{11}^2 & sin(x_{21} + x_{22}) \\ sin(x_{12} + x_{11}) & x_{11}^2 + x_{21}x_{22} \end{bmatrix}$ and $\mathbf{X} = \begin{bmatrix} x_{11} & x_{12} \\ x_{21} & x_{22} \end{bmatrix}$. Find $\frac{\partial \mathbf{Z}}{\partial \mathbf{X}}$.

2. Let $\mathbf{Z} = \begin{bmatrix} x_{11} & x_{11}x_{12} & x_{11}x_{21} \\ x_{21}x_{11} & x_{22}^2x_{12} & x_{22}x_{11}+3 \end{bmatrix}$. Find $\dfrac{\partial \mathbf{Z}}{\partial \mathbf{X}}$ with $\mathbf{X}$ as in problem 1.

3. Let $\mathbf{Z} = \begin{bmatrix} exp\left(x_{11}^2 + x_{22}^2\right) & \ln\left(x_{22}^2 + x_{12}^2\right) \\ x_{11}^2 + x_{22}^2 & x_{22}^2 + x_{12} \end{bmatrix}$. Find $\dfrac{\partial \mathbf{Z}}{\partial \mathbf{X}}$ with the same $\mathbf{X}$ as in problem 1.

4. Find $\dfrac{\partial \mathbf{Z}}{\partial \mathbf{X}}$, where $\mathbf{Z} = \begin{bmatrix} cos(x_{11}) + sin(x_{22}) & cos(x_{12}) + sin(x_{21}) \\ x_{11}^2 cos(x_{12}) & sin(x_{22})(x_{11}^2 + x_{22}) \end{bmatrix}$ with the same $\mathbf{X}$ as in problem 1.

7.8 Derivative Formula using Kronecker Products

In this section, the concept of Kronecker product will be utilized. To develop differentiation of product of two matrices w.r.t. a matrix and the chain rule w.r.t. a matrix.

Product rule for matrix differentiation

Let $\mathbf{Y}_{p\times r}$ and $\mathbf{Z}_{r\times s}$ be functions of a matrix $\mathbf{X}_{m\times n}$.

The product rule w.r.t. an entry, x_{kl} gives

$$\frac{\partial \mathbf{YZ}}{\partial x_{kl}} = \frac{\partial \mathbf{Y}}{\partial x_{kl}}\mathbf{Z} + \mathbf{Y}\frac{\partial \mathbf{Z}}{\partial x_{kl}}$$

Now using the definition of the derivative of a matrix w.r.t. a matrix

$$\begin{aligned} \frac{\partial \mathbf{YZ}}{\partial \mathbf{X}} &= \sum_{k,l} \mathbf{P}_{kl} \otimes \left[\frac{\partial \mathbf{Y}}{\partial x_{kl}}\mathbf{Z} + \mathbf{Y}\frac{\partial \mathbf{Z}}{\partial x_{kl}}\right] \\ &= \sum_{k,l} \mathbf{P}_{kl} \otimes \frac{\partial \mathbf{Y}}{\partial x_{kl}}\mathbf{Z} + \sum_{k,l} \mathbf{P}_{kl} \otimes \mathbf{Y}\frac{\partial \mathbf{Z}}{\partial x_{kl}}, \end{aligned}$$

where $\mathbf{P}_{kl}$ is an $m \times n$ primary matrix.

Now

$$\begin{aligned} \frac{\partial (\mathbf{YZ})}{\partial \mathbf{X}} &= \sum_{k,l} \mathbf{P}_{kl}\mathbf{I}_n \otimes \frac{\partial \mathbf{Y}}{\partial x_{kl}}\mathbf{Z} + \sum_{k,l} \mathbf{I}_m\mathbf{P}_{kl} \otimes \mathbf{Y}\frac{\partial \mathbf{Z}}{\partial x_{kl}} \\ &= \sum_{k,l} \left(\mathbf{P}_{kl} \otimes \frac{\partial \mathbf{Y}}{\partial x_{kl}}\right)(\mathbf{I}_n \otimes \mathbf{Z}) + \sum_{k,l} (\mathbf{I}_m \otimes \mathbf{Y})\left(\mathbf{P}_{kl} \otimes \frac{\partial \mathbf{Z}}{\partial x_{kl}}\right) \\ &= \frac{\partial \mathbf{Y}}{\partial \mathbf{X}}(\mathbf{I}_n \otimes \mathbf{Z}) + (\mathbf{I}_m \otimes \mathbf{Y})\frac{\partial \mathbf{Z}}{\partial \mathbf{X}}. \end{aligned}$$

Observation.

1. The mixed product rule of Kronecker products is used in the second step.
2. The order of $\frac{\partial(\mathbf{YZ})}{\partial\mathbf{X}}$ is $pm \times ns$. The right-hand side also has the same dimension as $\left(\frac{\partial\mathbf{Y}}{\partial\mathbf{X}}\right)_{pm\times nr}(\mathbf{I}_n \otimes \mathbf{Z})_{nr\times ns}$, $(\mathbf{I}_m \otimes \mathbf{Y})_{pm\times mr}$ and $\left(\frac{\partial\mathbf{Z}}{\partial\mathbf{X}}\right)_{mr\times ns}$.

Example 7.8.1 *Let* $\mathbf{X} = \begin{bmatrix} x_1 & x_3 \\ x_2 & x_4 \end{bmatrix}$, $\mathbf{Y} = \begin{bmatrix} 5x_1 + x_2 & x_3 + x_4 \end{bmatrix}$ *and* $\mathbf{Z} = \begin{bmatrix} cos(x_1) & x_2 + x_3 \\ exp(x_3) & sin(x_4) \end{bmatrix}$. *Verify the product rule for differentiation w.r.t. matrices for the product* $\mathbf{YZ}$.

Solution.

$$\begin{aligned}\mathbf{YZ} &= \begin{bmatrix} 5x_1 + x_2 & x_3 + x_4 \end{bmatrix} \begin{bmatrix} cos(x_1) & x_2 + x_3 \\ exp(x_3) & sin(x_4) \end{bmatrix} \\ &= \big[(5x_1 + x_2)cos(x_1) + (x_3 + x_4)exp(x_3) \\ &\qquad\qquad (5x_1 + x_2)(x_2 + x_3) + (x_3 + x_4)sin(x_4)\big]\end{aligned}$$

$$\frac{\partial(\mathbf{YZ})_{1\times 2}}{(\partial\mathbf{X})_{2\times 2}}$$

$$= \begin{bmatrix} 5cos(x_1) - (5x_1 + x_2)sin(x_1) & 5(x_2 + x_3) & exp(x_3)[1 + x_3 + x_4] & 5x_1 + x_2 + sin(x_4) \\ cos(x_1) & 5x_1 + 2x_2 + x_3 & exp(x_3) & (x_3 + x_4)cos(x_4) + sin(x_4) \end{bmatrix}_{2\times 4}$$

$$\begin{aligned}\mathbf{I}_2 \otimes \mathbf{Y} &= \begin{bmatrix} 1 & 0 \\ 0 & 1 \end{bmatrix} \begin{bmatrix} 5x_1 + x_2 & x_3 + x_4 \end{bmatrix} \\ &= \begin{bmatrix} 5x_1 + x_2 & x_3 + x_4 & 0 & 0 \\ 0 & 0 & 5x_1 + x_2 & x_3 + x_4 \end{bmatrix}\end{aligned}$$

$$\begin{aligned}\mathbf{I}_2 \otimes \mathbf{Z} &= \begin{bmatrix} 1 & 0 \\ 0 & 1 \end{bmatrix} \begin{bmatrix} cos(x_1) & x_2 + x_3 \\ exp(x_3) & sin(x_4) \end{bmatrix} \\ &= \begin{bmatrix} cos(x_1) & x_2 + x_3 & 0 & 0 \\ sin(x_4) & sin(x_4) & 0 & 0 \\ 0 & 0 & cos(x_1) & x_2 + x_3 \\ 0 & 0 & exp(x_3) & sin(x_4) \end{bmatrix}\end{aligned}$$

Now

$$\frac{\partial(\mathbf{Y})_{1\times 2}}{(\partial\mathbf{X})_{(2\times 2)}} = \begin{bmatrix} 5 & 0 & 0 & 1 \\ 1 & 0 & 0 & 1 \end{bmatrix}_{2\times 4}$$

$$\frac{\partial\mathbf{Z}_{1\times 2}}{(\partial\mathbf{X})_{2\times 2}} = \begin{bmatrix} -sin(x_1) & 0 & 0 & 1 \\ 0 & 0 & exp(x_3) & 0 \\ 0 & 1 & 0 & 0 \\ 0 & 0 & 0 & cos(x_1) \end{bmatrix}$$

Now substituting all the above calculations in

$$
\begin{aligned}
&\frac{\partial(\mathbf{YZ})}{\partial\mathbf{X}} \\
&= \frac{\partial Y}{\partial\mathbf{X}}(\mathbf{I}_2\otimes\mathbf{Z}) + (\mathbf{I}_2\otimes\mathbf{Y})\frac{\partial Z}{\partial\mathbf{X}} \\
&= \begin{bmatrix} 5cos(x_1) & 5(x_2+x_3) & exp(x_3) & sin(x_4) \\ cos(x_1) & (x_2+x_3) & exp(x_3) & sin(x_4) \end{bmatrix} \\
&\quad + \begin{bmatrix} (5x_1+x_2)sin(x_1) & 0 & (x_3+x_4)exp(x_3) & 5x_1+x_2 \\ 0 & 5x_1+x_2 & 0 & (x_3+x_4)cos(x_4) \end{bmatrix} \\
&= \begin{bmatrix} 5cos(x_1)-(5x_1+x_2)sin(x_1) & 5(x_2+x_3) & exp(x_3)[1+x_3+x_4] & 5x_1+x_2+sin(x_4) \\ cos(x_1) & 5x_1+2x_2+x_3 & exp(x_3) & (x_3+x_4)cos(x_4)+sin(x_4) \end{bmatrix}.
\end{aligned}
$$

The result is verified.

Chain rule for derivative of matrices w.r.t. a matrix

Consider the matrices $\mathbf{X}, \mathbf{Y}, \mathbf{Z}$ of orders $m \times n$, $p \times q$ and $r \times s$, respectively.

Let $\mathbf{Z}_{r\times s} = \mathbf{Z}(\mathbf{Y}(\mathbf{X}))$. Then $\mathbf{Z} = [z_{ij}(\mathbf{Y}(\mathbf{X}))]_{r\times s}$. To find $\frac{\partial\mathbf{Z}}{\partial\mathbf{X}}$, the chain rule is developed as follows.

For any x_{kl},

$$\frac{\partial\mathbf{Z}}{\partial x_{kl}} = \left[\sum_{v=1}^{q}\sum_{u=1}^{p}\frac{\partial z_{ij}}{\partial y_{uv}}\frac{\partial y_{uv}}{\partial x_{kl}}\right]_{r\times s}.$$

Using the above relation in,

$$
\begin{aligned}
\frac{\partial\mathbf{Z}}{\partial\mathbf{X}} &= \sum_{k,l}\mathbf{P}_{kl}\otimes\frac{\partial\mathbf{Z}}{\partial x_{kl}} \\
&= \sum_{k,l}\mathbf{P}_{kl}\otimes\sum_{i,j}\mathbf{P}_{ij}\sum_{v=1}^{q}\sum_{u=1}^{p}\frac{\partial z_{ij}}{\partial y_{uv}}\frac{\partial y_{uv}}{\partial x_{kl}} \\
&= \sum_{v=1}^{q}\sum_{u=1}^{p}\sum_{k,l}\mathbf{P}_{kl}\frac{\partial y_{uv}}{\partial x_{kl}}\otimes\sum_{i,j}P_{ij}\frac{\partial z_{ij}}{\partial y_{uv}} \\
&= \sum_{v}\sum_{u}\frac{\partial y_{uv}}{\partial\mathbf{X}}I_n\otimes I_r\frac{\partial\mathbf{Z}}{\partial y_{uv}} \\
&= \sum_{v}\sum_{u}\left(\frac{\partial y_{uv}}{\partial\mathbf{X}}\otimes I_r\right)\left(I_n\otimes\frac{\partial\mathbf{Z}}{\partial y_{uv}}\right)
\end{aligned}
$$

Observation.

1. The scalar multiplication with Kronecker product $(\mathbf{P}\otimes\alpha\mathbf{Q}) = \alpha\mathbf{P}\otimes\mathbf{Q}$ is used and the mixed product rule is used in the calculation.

2. The order of $\frac{\partial\mathbf{Z}}{\partial\mathbf{X}}$ is $rm \times ns$.

3. The order of the right side of the equality is

$$(rm \times nr)(nr \times ns) = rm \times ns.$$

Example 7.8.2 *Verify the chain rule for the following matrices* $\mathbf{X} = \begin{bmatrix} x_{11} & x_{12} \\ x_{21} & x_{22} \end{bmatrix}$ $\mathbf{Y} = \begin{bmatrix} y_{11} & y_{12} \end{bmatrix}$ *and* $\mathbf{Z} = \begin{bmatrix} z_{11} & z_{12} \\ z_{21} & z_{22} \end{bmatrix}$ *where* $\mathbf{Z} = \mathbf{Z}(\mathbf{Y})$ *and* $\mathbf{Y} = \mathbf{Y}(\mathbf{X})$, $\mathbf{Z} = \mathbf{Z}(\mathbf{Y}(\mathbf{X}))$, $\mathbf{Z} = \begin{bmatrix} z_{ij}\mathbf{Y}(\mathbf{X}) \end{bmatrix}_{2\times 2}$.

$$\frac{\partial \mathbf{Z}}{\partial \mathbf{X}} = \begin{bmatrix} \frac{\partial z_{11}}{\partial x_{11}} & \frac{\partial z_{12}}{\partial x_{11}} & \frac{\partial z_{11}}{\partial x_{12}} & \frac{\partial z_{12}}{\partial x_{12}} \\ \frac{\partial z_{21}}{\partial x_{11}} & \frac{\partial z_{22}}{\partial x_{11}} & \frac{\partial z_{21}}{\partial x_{12}} & \frac{\partial z_{22}}{\partial x_{12}} \\ \frac{\partial z_{11}}{\partial x_{21}} & \frac{\partial z_{12}}{\partial x_{21}} & \frac{\partial z_{11}}{\partial x_{22}} & \frac{\partial z_{12}}{\partial x_{22}} \\ \frac{\partial z_{21}}{\partial x_{21}} & \frac{\partial z_{22}}{\partial x_{21}} & \frac{\partial z_{21}}{\partial x_{22}} & \frac{\partial z_{22}}{\partial x_{22}} \end{bmatrix}$$

Now since $z_{ij}(\mathbf{Y}(\mathbf{X}))$ *hence*

$$\begin{aligned} \frac{\partial z_{ij}}{\partial x_{kl}} &= \frac{\partial z_{ij}}{\partial y_{11}} \frac{\partial y_{11}}{\partial x_{kl}} + \frac{\partial z_{ij}}{\partial y_{12}} \frac{\partial y_{12}}{\partial x_{kl}} \\ &= \sum_{v=1}^{2} \frac{\partial z_{ij}}{\partial y_{1v}} \frac{\partial y_{1v}}{\partial x_{kl}} \text{ for all } i, j, k, l = 1, 2. \end{aligned}$$

Hence

$$\frac{\partial \mathbf{Z}}{\partial \mathbf{X}} = \begin{bmatrix} \sum \frac{\partial z_{11}}{\partial y_{1v}} \frac{\partial y_{1v}}{\partial x_{11}} & \sum \frac{\partial z_{12}}{\partial y_{1v}} \frac{\partial y_{1v}}{\partial x_{11}} & \sum \frac{\partial z_{11}}{\partial y_{1v}} \frac{\partial y_{1v}}{\partial x_{12}} & \sum \frac{\partial z_{12}}{\partial y_{1v}} \frac{\partial y_{1v}}{\partial x_{12}} \\ \sum \frac{\partial z_{21}}{\partial y_{1v}} \frac{\partial y_{1v}}{\partial x_{11}} & \sum \frac{\partial z_{22}}{\partial y_{1v}} \frac{\partial y_{1v}}{\partial x_{11}} & \sum \frac{\partial z_{21}}{\partial y_{1v}} \frac{\partial y_{1v}}{\partial x_{12}} & \sum \frac{\partial z_{22}}{\partial y_{1v}} \frac{\partial y_{1v}}{\partial x_{12}} \\ \sum \frac{\partial z_{11}}{\partial y_{1v}} \frac{\partial y_{1v}}{\partial x_{21}} & \sum \frac{\partial z_{12}}{\partial y_{1v}} \frac{\partial y_{1v}}{\partial x_{21}} & \sum \frac{\partial z_{11}}{\partial y_{1v}} \frac{\partial y_{1v}}{\partial x_{22}} & \sum \frac{\partial z_{12}}{\partial y_{1v}} \frac{\partial y_{1v}}{\partial x_{22}} \\ \sum \frac{\partial z_{21}}{\partial y_{1v}} \frac{\partial y_{1v}}{\partial x_{21}} & \sum \frac{\partial z_{22}}{\partial y_{1v}} \frac{\partial y_{1v}}{\partial x_{21}} & \sum \frac{\partial z_{21}}{\partial y_{1v}} \frac{\partial y_{1v}}{\partial x_{22}} & \sum \frac{\partial z_{22}}{\partial y_{1v}} \frac{\partial y_{1v}}{\partial x_{22}} \end{bmatrix}$$

Now following the procedure described in this section, it is sufficient to calculate

$$\left(\frac{\partial y_{11}}{\partial \mathbf{X}} \otimes I_2\right)\left(I_2 \otimes \frac{\partial \mathbf{Z}}{\partial y_{11}}\right) + \left(\frac{\partial y_{12}}{\partial \mathbf{X}} \otimes I_2\right)\left(I_2 \otimes \frac{\partial \mathbf{Z}}{\partial y_{12}}\right).$$

Now $\left(\frac{\partial y_{11}}{\partial \mathbf{X}} \otimes I_2\right)\left(I_2 \otimes \frac{\partial \mathbf{Z}}{\partial y_{11}}\right)$

$$= \left(\begin{bmatrix} \frac{\partial y_{11}}{\partial x_{11}} & \frac{\partial y_{11}}{\partial x_{12}} \\ \frac{\partial y_{11}}{\partial x_{21}} & \frac{\partial y_{11}}{\partial x_{22}} \end{bmatrix} \otimes \begin{bmatrix} 1 & 0 \\ 0 & 1 \end{bmatrix}\right)\left(\begin{bmatrix} 1 & 0 \\ 0 & 1 \end{bmatrix} \otimes \begin{bmatrix} \frac{\partial z_{11}}{\partial y_{11}} & \frac{\partial z_{12}}{\partial y_{11}} \\ \frac{\partial z_{21}}{\partial y_{11}} & \frac{\partial z_{22}}{\partial y_{11}} \end{bmatrix}\right)$$

$$= \begin{bmatrix} \frac{\partial y_{11}}{\partial x_{11}} & 0 & \frac{\partial y_{11}}{\partial x_{12}} & 0 \\ 0 & \frac{\partial y_{11}}{\partial x_{11}} & 0 & \frac{\partial y_{11}}{\partial x_{12}} \\ \frac{\partial y_{11}}{\partial x_{21}} & 0 & \frac{\partial y_{11}}{\partial x_{22}} & 0 \\ 0 & \frac{\partial y_{11}}{\partial x_{21}} & 0 & \frac{\partial y_{11}}{\partial x_{22}} \end{bmatrix} \begin{bmatrix} \frac{\partial z_{11}}{\partial y_{11}} & \frac{\partial z_{12}}{\partial y_{11}} & 0 & 0 \\ \frac{\partial z_{21}}{\partial y_{11}} & \frac{\partial z_{22}}{\partial y_{11}} & 0 & 0 \\ 0 & 0 & \frac{\partial z_{11}}{\partial y_{11}} & \frac{\partial z_{12}}{\partial y_{11}} \\ 0 & 0 & \frac{\partial z_{21}}{\partial y_{11}} & \frac{\partial z_{22}}{\partial y_{11}} \end{bmatrix}$$

$$= \begin{bmatrix} \frac{\partial z_{11}}{\partial y_{11}}\frac{\partial y_{11}}{\partial x_{11}} & \frac{\partial z_{12}}{\partial y_{11}}\frac{\partial y_{11}}{\partial x_{11}} & \frac{\partial z_{11}}{\partial y_{11}}\frac{\partial y_{11}}{\partial x_{12}} & \frac{\partial z_{12}}{\partial y_{11}}\frac{\partial y_{11}}{\partial x_{12}} \\ \frac{\partial z_{11}}{\partial y_{11}}\frac{\partial y_{11}}{\partial x_{12}} & \frac{\partial z_{22}}{\partial y_{11}}\frac{\partial y_{11}}{\partial x_{11}} & \frac{\partial z_{12}}{\partial y_{11}}\frac{\partial y_{11}}{\partial x_{12}} & \frac{\partial z_{22}}{\partial y_{11}}\frac{\partial y_{11}}{\partial x_{12}} \\ \frac{\partial z_{11}}{\partial y_{11}}\frac{\partial y_{11}}{\partial x_{21}} & \frac{\partial z_{12}}{\partial y_{11}}\frac{\partial y_{11}}{\partial x_{21}} & \frac{\partial z_{11}}{\partial y_{11}}\frac{\partial y_{11}}{\partial x_{22}} & \frac{\partial z_{12}}{\partial y_{11}}\frac{\partial y_{11}}{\partial x_{22}} \\ \frac{\partial z_{21}}{\partial y_{11}}\frac{\partial y_{11}}{\partial x_{21}} & \frac{\partial z_{22}}{\partial y_{11}}\frac{\partial y_{11}}{\partial x_{21}} & \frac{\partial z_{21}}{\partial y_{11}}\frac{\partial y_{11}}{\partial x_{22}} & \frac{\partial z_{22}}{\partial y_{11}}\frac{\partial y_{11}}{\partial x_{22}} \end{bmatrix}$$

Similarly $\left(\frac{\partial y_{12}}{\partial \mathbf{X}} \otimes I_2\right)\left(I_2 \otimes \frac{\partial \mathbf{Z}}{\partial y_{12}}\right)$

$$= \begin{bmatrix} \frac{\partial z_{11}}{\partial y_{12}}\frac{\partial y_{12}}{\partial x_{11}} & \frac{\partial z_{12}}{\partial y_{12}}\frac{\partial y_{12}}{\partial x_{11}} & \frac{\partial z_{12}}{\partial y_{12}}\frac{\partial y_{12}}{\partial x_{12}} & \frac{\partial z_{12}}{\partial y_{12}}\frac{\partial y_{12}}{\partial x_{12}} \\ \frac{\partial z_{21}}{\partial y_{12}}\frac{\partial y_{12}}{\partial x_{11}} & \frac{\partial z_{22}}{\partial y_{12}}\frac{\partial y_{12}}{\partial x_{11}} & \frac{\partial z_{21}}{\partial y_{12}}\frac{\partial y_{12}}{\partial x_{12}} & \frac{\partial z_{22}}{\partial y_{12}}\frac{\partial y_{12}}{\partial x_{12}} \\ \frac{\partial z_{11}}{\partial y_{12}}\frac{\partial y_{12}}{\partial x_{22}} & \frac{\partial z_{12}}{\partial y_{12}}\frac{\partial y_{12}}{\partial x_{21}} & \frac{\partial z_{11}}{\partial y_{12}}\frac{\partial y_{12}}{\partial x_{22}} & \frac{\partial z_{12}}{\partial y_{12}}\frac{\partial y_{12}}{\partial x_{22}} \\ \frac{\partial z_{21}}{\partial y_{12}}\frac{\partial y_{12}}{\partial x_{21}} & \frac{\partial z_{22}}{\partial y_{12}}\frac{\partial y_{12}}{\partial x_{21}} & \frac{\partial z_{21}}{\partial y_{12}}\frac{\partial y_{12}}{\partial x_{22}} & \frac{\partial z_{22}}{\partial y_{12}}\frac{\partial y_{12}}{\partial x_{22}} \end{bmatrix}$$

Now $\left(\frac{\partial y_{11}}{\partial \mathbf{X}} \otimes I_2\right)\left(I_2 \otimes \frac{\partial \mathbf{Z}}{\partial y_{11}}\right) + \left(\frac{\partial y_{12}}{\partial \mathbf{X}} \otimes I_2\right)\left(I_2 \otimes \frac{\partial \mathbf{Z}}{\partial y_{12}}\right)$

$$= \begin{bmatrix} \sum \frac{\partial z_{11}}{\partial y_{1v}}\frac{\partial y_{1v}}{\partial x_{11}} & \sum \frac{\partial z_{12}}{\partial y_{1v}}\frac{\partial y_{1v}}{\partial x_{11}} & \sum \frac{\partial z_{11}}{\partial y_{1v}}\frac{\partial y_{1v}}{\partial x_{12}} & \sum \frac{\partial z_{12}}{\partial y_{1v}}\frac{\partial y_{1v}}{\partial x_{12}} \\ \sum \frac{\partial z_{21}}{\partial y_{1v}}\frac{\partial y_{1v}}{\partial x_{11}} & \sum \frac{\partial z_{22}}{\partial y_{1v}}\frac{\partial y_{1v}}{\partial x_{11}} & \sum \frac{\partial z_{21}}{\partial y_{1v}}\frac{\partial y_{1v}}{\partial x_{12}} & \sum \frac{\partial z_{22}}{\partial y_{1v}}\frac{\partial y_{1v}}{\partial x_{12}} \\ \sum \frac{\partial z_{11}}{\partial y_{1v}}\frac{\partial y_{1v}}{\partial x_{21}} & \sum \frac{\partial z_{12}}{\partial y_{1v}}\frac{\partial y_{1v}}{\partial x_{21}} & \sum \frac{\partial z_{11}}{\partial y_{1v}}\frac{\partial y_{1v}}{\partial x_{22}} & \sum \frac{\partial z_{12}}{\partial y_{1v}}\frac{\partial y_{1v}}{\partial x_{22}} \\ \sum \frac{\partial z_{21}}{\partial y_{1v}}\frac{\partial y_{1v}}{\partial x_{21}} & \sum \frac{\partial z_{22}}{\partial y_{1v}}\frac{\partial y_{1v}}{\partial x_{21}} & \sum \frac{\partial z_{21}}{\partial y_{1v}}\frac{\partial y_{1v}}{\partial x_{22}} & \sum \frac{\partial z_{22}}{\partial y_{1v}}\frac{\partial y_{1v}}{\partial x_{22}} \end{bmatrix} = \frac{\partial \mathbf{Z}}{\partial \mathbf{X}}$$

and the result is verified.

Example 7.8.3 *Verify the chain rule for differentiation of matrices w.r.t. matrices given* $\mathbf{X} = \begin{bmatrix} x_{11} & x_{12} \\ x_{21} & x_{22} \end{bmatrix}$ $\mathbf{Y} = \begin{bmatrix} y_{11} & y_{12} \end{bmatrix}$ *and* $\mathbf{Z} = \begin{bmatrix} y_{11}^2 & y_{11}y_{12} \\ 2y_{11}^2 y_{12} & y_{12}^2 \end{bmatrix}$ *where* $y_{11} = x_{11}$, $y_{12} = x_{22}^2$.

Solution.

$$\frac{\partial \mathbf{Z}}{\partial \mathbf{X}} = \begin{bmatrix} \frac{\partial(y_{11}^2)}{\partial x_{11}} & \frac{\partial(y_{11}y_{12})}{\partial x_{11}} & \frac{\partial(y_{11}^2)}{\partial x_{12}} & \frac{\partial(y_{11}y_{12})}{\partial x_{12}} \\ \frac{\partial(2y_{11}^2 y_{12})}{\partial x_{11}} & \frac{\partial(y_{12}^2)}{\partial x_{11}} & \frac{\partial(2y_{11}^2 y_{12})}{\partial x_{12}} & \frac{\partial(y_{12}^2)}{\partial x_{12}} \\ \frac{\partial(y_{11}^2)}{\partial x_{21}} & \frac{\partial(y_{11}y_{12})}{\partial x_{21}} & \frac{\partial(y_{11}^2)}{\partial x_{22}} & \frac{\partial(y_{11}y_{12})}{\partial x_{22}} \\ \frac{\partial(2y_{11}^2 y_{12})}{\partial x_{21}} & \frac{\partial(y_{12}^2)}{\partial x_{21}} & \frac{\partial(2y_{11}^2 y_{12})}{\partial x_{22}} & \frac{\partial(y_{12}^2)}{\partial x_{22}} \end{bmatrix}$$

Using chain rule $\frac{\partial z_{ij}}{\partial x_{kl}} = \frac{\partial z_{ij}}{\partial y_{11}} \cdot \frac{\partial y_{11}}{\partial x_{kl}} + \frac{\partial z_{ij}}{\partial y_{12}}\frac{\partial y_{12}}{\partial x_{kl}}$ *for* $i, j = 1, 2$ *and* $k, l = 1, 2$ *gives,*

$$\frac{\partial \mathbf{Z}}{\partial \mathbf{X}} = \begin{bmatrix} 2x_{11} & x_{22}^2 & 0 & 0 \\ 4x_{11}x_{22}^2 & 0 & 0 & 0 \\ 0 & 0 & 0 & 2x_{11}x_{22} \\ 0 & 0 & 4x_{11}^2 x_{22} & 4x_{22}^3 \end{bmatrix}$$

Now, to obtain the result using the chain rule formula, consider

$$\frac{\partial y_{11}}{\partial \mathbf{X}} = \begin{bmatrix} \frac{\partial y_{11}}{\partial x_{11}} & \frac{\partial y_{11}}{\partial x_{12}} \\ \frac{\partial y_{11}}{\partial x_{21}} & \frac{\partial y_{11}}{\partial x_{22}} \end{bmatrix} = \begin{bmatrix} 1 & 0 \\ 0 & 0 \end{bmatrix}, \text{ since } y_{11} = x_{11}$$

Similarly,

$$\frac{\partial y_{12}}{\partial \mathbf{X}} = \begin{bmatrix} 0 & 0 \\ 0 & 2x_{22} \end{bmatrix}, \; since \; y_{12} = x_{22}^2.$$

Now $\dfrac{\partial \mathbf{Z}}{\partial y_{11}} = \begin{bmatrix} 2y_{11} & y_{12} \\ 4y_{11}y_{12} & 0 \end{bmatrix}$ *and* $\dfrac{\partial \mathbf{Z}}{\partial y_{12}} = \begin{bmatrix} 0 & y_{11} \\ 2y_{11}^2 & 2y_{12} \end{bmatrix}$

Using the formula

$$\frac{\partial \mathbf{Z}}{\partial \mathbf{X}} = \sum_{j=1}^{2} \left(\frac{\partial y_{1j}}{\partial \mathbf{X}} \otimes I_2 \right) \left(I_2 \otimes \frac{\partial \mathbf{Z}}{\partial y_{1j}} \right)$$

$$= \left(\frac{\partial y_{11}}{\partial \mathbf{X}} \otimes I_2 \right) \left(I_2 \otimes \frac{\partial \mathbf{Z}}{\partial y_{11}} \right) + \left(\frac{\partial y_{12}}{\partial \mathbf{X}} \otimes I_2 \right) \left(I_2 \otimes \frac{\partial \mathbf{Z}}{\partial y_{12}} \right)$$

$$= \left[\begin{bmatrix} 1 & 0 \\ 0 & 0 \end{bmatrix} \otimes \begin{bmatrix} 1 & 0 \\ 0 & 1 \end{bmatrix} \right] \left[\begin{bmatrix} 1 & 0 \\ 0 & 1 \end{bmatrix} \otimes \begin{bmatrix} 2y_{11} & y_{12} \\ 4y_{11}y_{12} & 0 \end{bmatrix} \right]$$

$$+ \left[\begin{bmatrix} 0 & 0 \\ 0 & 2x_{22} \end{bmatrix} \otimes \begin{bmatrix} 1 & 0 \\ 0 & 1 \end{bmatrix} \right] \left[\begin{bmatrix} 1 & 0 \\ 0 & 1 \end{bmatrix} \otimes \begin{bmatrix} 0 & y_{11} \\ 2y_{11}^2 & 2y_{12} \end{bmatrix} \right]$$

$$= \begin{bmatrix} 1 & 0 & 0 & 0 \\ 0 & 1 & 0 & 0 \\ 0 & 0 & 0 & 0 \\ 0 & 0 & 0 & 0 \end{bmatrix} \begin{bmatrix} 2y_{11} & y_{12} & 0 & 0 \\ 4y_{11}y_{11} & 0 & 0 & 0 \\ 0 & 0 & 2y_{11} & y_{12} \\ 0 & 0 & 4y_{11}y_{12} & 0 \end{bmatrix}$$

$$+ \begin{bmatrix} 0 & 0 & 0 & 0 \\ 0 & 0 & 0 & 0 \\ 0 & 0 & 2x_{22} & 0 \\ 0 & 0 & 0 & 2x_{22} \end{bmatrix} \begin{bmatrix} 0 & y_{11} & 0 & 0 \\ 2y_{11}^2 & 2y_{12} & 0 & 0 \\ 0 & 0 & 0 & y_{11} \\ 0 & 0 & 2y_{11}^2 & 2y_{12} \end{bmatrix}$$

$$= \begin{bmatrix} 2x_{11} & x_{22}^2 & 0 & 0 \\ 4x_{11}x_{22}^2 & 0 & 0 & 0 \\ 0 & 0 & 0 & 2x_{11}x_{22} \\ 0 & 0 & 4x_{11}^2x_{22} & 4x_{22}^3 \end{bmatrix}$$

Verifying the formula.

EXERCISE 7.8

1. Apply for the matrices given below, the product rule for differentiation of matrix w.r.t. a matrix, $\mathbf{X} = \begin{bmatrix} x_{11} & x_{12} \\ x_{21} & x_{22} \end{bmatrix}$

   (i) $\mathbf{Y} = \begin{bmatrix} cos\ (x_{11}) & x_{12}^2 \\ exp(x_{21}) & x_{22}\ x_{11} \end{bmatrix}$ and $\mathbf{Z} = \begin{bmatrix} 2x_{11} \\ cos\ (x_{22}) \end{bmatrix}$

(ii) $\mathbf{Y} = \begin{bmatrix} x_{11}^2 & x_{11}\,x_{12}\,x_{21} \\ x_{12}\,x_{21} & x_{11}^2 \end{bmatrix}$ and $\mathbf{Z} = \begin{bmatrix} cos(x_{11}) & x_{12}^2\,x_{22} \\ x_{21}^2 sin(x_{12}) & x_{22} \end{bmatrix}$

2. Apply for the matrices given below, the chain rule for differentiation of matrices w.r.t. a matrix $\mathbf{X} = \begin{bmatrix} x_{11} & x_{12} \\ x_{21} & x_{22} \end{bmatrix}$ (i) $\mathbf{Y} = \begin{bmatrix} y_{11} & y_{12} \end{bmatrix}$ and $\mathbf{Z} = \begin{bmatrix} y_{11}\,y_{12} & y_{12}^2 \\ y_{11}^2 & y_{11}\,y_{12} \end{bmatrix}$, where $y_{11} = x_{11}\,x_{12}$, $y_{12} = x_{21}\,x_{22}$

(ii) $\mathbf{Y} = \begin{bmatrix} y_{11} & y_{12} \end{bmatrix}$ $\mathbf{Z} = \begin{bmatrix} y_{11}\,y_{12} & sin(y_{11}) \\ cos(y_{22}) & y_{22}^2 \end{bmatrix}$, where $y_{11} = x_{11}^2$, $y_{12} = x_{21}^2$.

7.9 Another Definition for Derivative of a Matrix w.r.t. a Matrix

The derivative of a matrix, with respect to a matrix can be defined in various ways. The following is another definition in this direction.

Let $\mathbf{Z}$ be an $r \times s$ matrix and $\mathbf{X}$ be an $m \times n$ matrix. Assume that $\mathbf{Z}$ is a function of the entries of $\mathbf{X}$.

Definition 7.9.1 *The derivative of the matrix* $\mathbf{Z}$ *w.r.t. the matrix is defined as*

$$\frac{\partial \mathbf{Z}}{\partial \mathbf{X}} = \left[\sum_k \sum_l \frac{\partial z_{ij}}{\partial x_{kl}} \right]_{r \times s}$$

Observation. The definition yields a matrix of the same order as $\mathbf{Z}$.

Example 7.9.1 *Let* $\mathbf{Z} = \begin{bmatrix} x_{11}^2 + x_{12}^2 & x_{12}x_{21} + x_{11}x_{22} \\ x_{11}^2 & x_{22}^2 + x_{21}x_{11} \end{bmatrix}$ *and* $\mathbf{X} = \begin{bmatrix} x_{11} & x_{12} \\ x_{11} & x_{22} \end{bmatrix}$.
Find $\frac{\partial \mathbf{Z}}{\partial \mathbf{X}}$ *using Definition 7.9.1.*
Solution.

$$\frac{\partial \mathbf{Z}}{\partial \mathbf{X}} = \left[\sum_k \sum_l \frac{\partial z_{ij}}{\partial x_{kl}} \right]_{2 \times 2}$$

$$= \begin{bmatrix} \frac{\partial z_{11}}{\partial x_{11}} + \frac{\partial z_{11}}{\partial x_{12}} + \frac{\partial z_{11}}{\partial x_{21}} + \frac{\partial z_{11}}{\partial x_{22}} & \frac{\partial z_{12}}{\partial x_{11}} + \frac{\partial z_{12}}{\partial x_{12}} + \frac{\partial z_{12}}{\partial x_{21}} + \frac{\partial z_{12}}{\partial x_{22}} \\ \frac{\partial z_{21}}{\partial x_{11}} + \frac{\partial z_{21}}{\partial x_{12}} + \frac{\partial z_{21}}{\partial x_{21}} + \frac{\partial z_{21}}{\partial x_{22}} & \frac{\partial z_{22}}{\partial x_{11}} + \frac{\partial z_{22}}{\partial x_{12}} + \frac{\partial z_{22}}{\partial x_{21}} + \frac{\partial z_{22}}{\partial x_{22}} \end{bmatrix}$$

$$= \begin{bmatrix} 2x_{11} + 2x_{12} & x_{21} + x_{22} + x_{12} + x_{11} \\ 2x_{11} & x_{21} + x_{11} + 2x_{22} \end{bmatrix}$$

Example 7.9.2 *Consider* $\mathbf{Z} = \begin{bmatrix} x_{22}^2 + x_{12} & x_{11} + x_{21}^2 \\ x_{12}^2 + x_{22} & x_{11}^2 + x_{11} \end{bmatrix}$. *Find* $\frac{\partial \mathbf{Z}}{\partial \mathbf{X}}$ *using (i) Definition 7.7.1 and (ii) Definition 7.9.1.*

1. $$\frac{\partial \mathbf{Z}}{\partial \mathbf{X}} = \begin{bmatrix} \dfrac{\partial(x_{22}^2 + x_{12})}{\partial \mathbf{X}} & \dfrac{\partial(x_{11} + x_{21}^2)}{\partial \mathbf{X}} \\ \dfrac{\partial(x_{12}^2 + x_{21})}{\partial \mathbf{X}} & \dfrac{\partial(x_{11}^2 + x_{11})}{\partial \mathbf{X}} \end{bmatrix}$$
$$= \begin{bmatrix} 0 & 1 & 1 & 0 \\ 0 & 2x_{22} & 2x_{21} & 0 \\ 0 & 2x_{12} & 1 + 2x_{11} & 0 \\ 0 & 1 & 0 & 0 \end{bmatrix}$$

2. $$\frac{\partial \mathbf{Z}}{\partial \mathbf{X}} = \begin{bmatrix} 1 + 2x_{22} & 1 + 2x_{21} \\ 2x_{12} + 1 & 1 + 2x_{11} \end{bmatrix}$$

EXERCISE 7.9

1. Apply definition 7.9.1 for the problem 1 in Exercise 7.7.
2. Apply definition 7.9.1 for the problem 2 in Exercise 7.7.

7.10 Conclusion

In this chapter, various types of derivatives were introduced. These methods have a rational extension to functions of matrices. Some of the derivatives introduced here are not yet utilized and hence there is a great potential for new types of differential equations. Such a study will enhance the subject in the areas of linear algebra and differential equations. For further reading the readers are suggested to refer to [8], [23] and [24].

Chapter 8

Linear Systems of Differential Equations

8.1 Introduction

The study of linear systems of differential equations (LSDEs) essentially depends on the knowledge of linear algebra included in Chapters 1 to 4 and 6 of the present text. The present chapter includes the derivatives with respect to a scalar. Basic knowledge of the theory of differential equations, particularly, the existence and uniqueness of solutions is useful in the analysis of systems of differential equations and these results are stated without proof.

The study of LSDEs has been found useful in many mathematical models involving coupled effects that exist in real-life situations. The present chapter brings forth a few basic principles of the study of LSDEs.

In Section 8.2, the LSDEs and their variants are described in the n-dimensional space, the aim being as how to solve LSDE. For this purpose, one needs the concept of fundamental matrix (FM) which is formulated by considering eigen-values and eigen-vectors of the involved matrices and is given in Section 8.3.

A general method of successive approximations is included in Section 8.4. The theory of linear homogeneous systems is then extended by considering nonhomogeneous systems for which a variation of parameters formula of Lagrange's type is obtained. This is the content of Section 8.5.

Particularly, it is shown that linear systems of differential equations involving constant coefficients are rather easy to handle by introducing the concept of an exponential matrix function and methods of linear algebra. This has been achieved in Section 8.6.

In addition to the essential features of LSDE's, namely, existence and with uniqueness, one needs to know the nature of solutions of the systems and their path and future behavior involved in the study of stability. This concept is dealt with in Section 8.7. The last section includes applications of LSDEs to some modeling problems. Throughout the chapter adequate number of examples illustrate the theory.

DOI: 10.1201/9781351014953-8

8.2 Linear Systems

Consider a linear system of coupled differential equations of order n

$$\begin{aligned} x_1' &= a_{11}(t)x_1 + a_{12}(t)x_2 + \cdots + a_{1n}(t)x_n \\ x_2' &= a_{21}(t)x_1 + a_{22}(t)x_2 + \cdots + a_{2n}(t)x_n \\ &\vdots \\ x_n' &= a_{n1}(t)x_1 + a_{n2}(t)x_2 + \cdots + a_{nn}(t)x_n \qquad t \in I, \end{aligned} \tag{8.1}$$

where I is a non-empty interval of the real line $\mathbb{R}$.

The system (8.1) has the following vector-matrix representation

$$\mathbf{x}' = \mathbf{A}(t)\mathbf{x}, \quad t \in I, \tag{8.2}$$

where $\mathbf{A}(t) \in \mathbf{C}[I, \mathbb{R}^{n\times n}], \mathbf{x} = [x_1 \ x_2 \ \ldots \ x_n]$ an n-vector, and is a linear system of differential equations. This chapter aims at knowing the unknown function $\mathbf{x}$ for $t \in I$.

Now add to (8.2) the value of the unknown vector $\mathbf{x}(t_0)$, called 'initial data' $t_0 \in I$ and a given n-vector $\mathbf{x}_0 = [x_{10} \ x_{20} \ \ldots \ x_{n0}]^T$. This addition enriches (8.2) as an initial value problem (IVP)

$$\mathbf{x}' = \mathbf{A}(t)\mathbf{x}, \ \mathbf{x}(t_0) = \mathbf{x}_0. \tag{8.3}$$

Here (8.3) is a homogeneous system, for at $\mathbf{x} = \mathbf{0}$, $\mathbf{x}' = \mathbf{0}$. It is further generalized into a nonhomogeneous linear IVP having the form

$$\mathbf{x}' = \mathbf{A}(t)\mathbf{x} + \mathbf{p}(t), \ \mathbf{x}(t_0) = \mathbf{x}_0, \tag{8.4}$$

where $\mathbf{p} \in C[I, \mathbb{R}^n]$, $\mathbf{p}(t) \neq \mathbf{0}, t \in I$, and is an n-vector of the type $[p_1(t) \ p_2(t) \ \ldots \ p_n(t)]^T$. In (8.4) the n-vector $\mathbf{p}(t)$ affects the derivative $\mathbf{x}'$ i.e the rate of change of $\mathbf{x}$. Hence (8.4) is called a 'perturbed system'. Other names for $\mathbf{p}(t)$ are (i) a booster (ii) a control factor depending on the areas of applications of the given system. Here the aim is to solve (8.4) i.e to find $\mathbf{x}(t)$ for $t \in I$ satisfying the relations in equation (8.4).

Definition 8.2.1 $\mathbf{x}(t)$ *is called a solution of the IVP (8.4), if its derivative is equal to the right-hand side for* $t \in I$ *and that* $\mathbf{x}(t_0) = \mathbf{x}_0$.

Remark 8.2.1 *(i) The system (8.4) is nonhomogeneous because the right side of (8.4) is such that* $\mathbf{A}(t)\mathbf{x} + \mathbf{p}(t) \neq \mathbf{0}$ *when* $\mathbf{x} = \mathbf{0}$

(ii) The system (8.4) retains its 'system' structure if all elements of $\mathbf{A}(t)$, *namely* $a_{ij}(t)$ *are not simultaneously zero, for* $t \in I$.

(iii) The system (8.4) involves an $n \times n$ *matrix* $\mathbf{A}(t)$ *and* n*-vectors* $\mathbf{x}, \mathbf{p}, \mathbf{x}_0$ *and therefore is called system in vector–matrix form.*

The following proposition stated without proof guarantees the existence of a unique solution of (8.4) on I.

Proposition 8.2.1 *The IVP (8.4) possesses a solution* $\mathbf{x}(t)$ *on* I *if* $\mathbf{A}(t) \in \mathbf{C}[I, \mathbb{R}^{n\times n}]$ *and* $\mathbf{p} \in C[I, \mathbb{R}^n]$. *Further,* $\mathbf{x}(t)$ *is unique on* I.

In the above proposition the condition that $\mathbf{A}(t)$ is continuous on I is indeed, strong. Under some relaxed conditions, the conclusions still hold. It is claimed that the linear nonhomogeneous differential equation

$$x^{(n)} + a_1(t)\, x^{(n-1)} + ... + a_n(t)\, x = h(t), \tag{8.5}$$

where the scalar functions $a_1, a_2, \ldots, a_n$ and h are all continuous functions on I can be considered as linear system. Let the initial conditions at $t_0 \in I$ be given as

$$x(t_0) = b_1, x'(t_0) = b_2, .., x^{(n-1)}(t_0) = b_n,$$

where $b_1, b_2, \ldots, b_n$ are given constants. The following transformations $x_1 = x$, $x_1' = x_2, \ldots, x_{n-1}' = x_n$ convert (8.5) in to the form of an n-system (8.4), where
$\mathbf{x} = (x_1 \;\; x_2 \;\; \ldots \;\; x_n)^T$ and

$$\mathbf{A}(t) = \begin{bmatrix} 0 & 1 & 0 & \cdots & 0 \\ 0 & 0 & 1 & \cdots & 0 \\ \vdots & \vdots & \vdots & \vdots & \vdots \\ 0 & 0 & 0 & \cdots & 1 \\ -a_n & -a_{n-1} & -a_{n-2} & \cdots & -a_1 \end{bmatrix}.$$

$\mathbf{p}(t) = (0 \;\; 0 \;\; \ldots \;\; h(t))^T$, and $\mathbf{x}_0 = \mathbf{x}(t_0) = (b_1 \;\; b_2 \;\; .... \;\; b_n)^T$.
The following example is illustrative.

Example 8.2.1 *Convert the linear differential equation into a system form.*

$x''' - 8x'' + 15x' - 4x = 0$.
Solution. Let $x_1 = x$, $x_2 = x_1' = x'$, $x_3 = x_2' = x''$ and $x_3' = 8x_3 - 15x_2 + 4x_1$. Hence it follow that

$$\begin{bmatrix} x_1 \\ x_2 \\ x_3 \end{bmatrix}' = \begin{bmatrix} x_1' \\ x_2' \\ x_3' \end{bmatrix} = \begin{bmatrix} 0 & 1 & 0 \\ 0 & 0 & 1 \\ 4 & -15 & 8 \end{bmatrix} \begin{bmatrix} x_1 \\ x_2 \\ x_3 \end{bmatrix}$$

is a system of the form (8.4) with $\mathbf{p}(t) = 0$, $t \in I$.

In case the given equation is strengthened further, i.e. $x''' - 8x'' + 15x' - 4x = h(t),\ t \in \mathbf{I}$. where $h \in C[I, \mathbb{R}]$ then the related system takes the form (8.4) where $\mathbf{p}(t) = [0\ \ 0\ \ h(t)]^T$.

Example 8.2.2 *Convert the following system of 4 differential equations*

$(x'\ \ y'\ \ z'\ \ w') = (4x+5y+6z+7\quad -3x-3z\quad 7y-3z+w+6\quad x+y+z+w+1)$ into a system.

Solution. The system form(8.4) is given by :

$$\begin{bmatrix} x \\ y \\ z \\ w \end{bmatrix}' = \begin{bmatrix} 4 & 5 & 6 & 0 \\ -3 & 0 & -3 & 0 \\ 0 & 7 & -3 & 1 \\ 1 & 1 & 1 & 1 \end{bmatrix} \begin{bmatrix} x \\ y \\ z \\ w \end{bmatrix} + \begin{bmatrix} 7 \\ 0 \\ 6 \\ 1 \end{bmatrix}.$$

EXERCISE 8.2

1. Express the following systems of equations into vector - matrix differential equation,

   (i) $(x'\ \ y'\ \ z') = (3x + 2y \quad 2x + y + z \quad -3z)$;
   (ii) $(x'\ \ y'\ \ z') = (2x + 5y - 9z \quad 7x - 2y + 3z \quad -x + 5y + 7z)$;
   (iii) $(x'\ \ y'\ \ z') = (a_1x + b_1y + c_1z \quad a_2x + b_2y + c_2z \quad a_3x + b_3y + c_3z)$.

2. Express the following scalar equations into vector–matrix form

   (i) $x'' + ax' + w^2x = 0$;
   (ii) $x''' + 7x'' + 9x' - 5x = 0$.
   (iii) $x^{(4)} + 5x'' - 6x' + 9x = 0$;

3. Transform into vector–matrix IVP, the differential equations

   (i) $x'' + 8x' + 25x = 3,\ x(0) = 0, x'(0) = 5$;
   (ii) $x''' + 3x'' + 3x' - 4x = t^2 + 4, x(1) = 10,\ x'(1) = 5, x''(1) = 7$;
   (iii) $x'' = x' + y' - z + t,\ y'' = tx + y' - 2y + t^2 + 1$,
   $z'' = x - y + y' + z,\ x(1) = 3, x'(1) = 10, y(1) = 1, y'(1) = -5, z(1) = 4, z'(1) = 3$.

4. Express into vector–matrix form

   (i) $(x'\ \ y'\ \ z'\ \ w') = (3x + y + w \quad 5z + 3w \quad -x + y + z - w \quad 5w)$,
   $x(1) = y(1) = z(1) = w(1) = 2$;

(ii) $x' = 3x + y + w + t,$
$y' = 5z + 3w - e^t,$
$z' = -x + y + z - w + sin\, t,$
$w' = 5w + te^t.$

(iii) $x^{(5)} + 9x^{(4)} + x = 10e^t, x(0) = x'(0) = x''(0) = x'''(0) = x^{(4)}(0) = 1.$

8.3 Fundamental Matrix

The aim of this section is to develop a technique for solving the system (8.2). The necessary tool is the formation of the fundamental matrix (FM) defined below.

Definition 8.3.1 *The system (8.2) possesses n linearly independent solutions $\phi_1, \phi_2, \ldots, \phi_n$ where each ϕ_i has its components $[\phi_{1i} \;\; \phi_{2i} \;\; \ldots \;\; \phi_{ni}]^T$, for $i = 1, 2, ..., n$. Then solutions written as n columns form an $n \times n$ matrix, called the fundamental matrix $\mathbf{\Phi}(t)$, such that*

$$\mathbf{\Phi}'(t) = \mathbf{A}(t)\mathbf{\Phi}(t), \; t \in \text{I}. \tag{8.6}$$

Firstly, there is a need to determine criteria for a matrix $\mathbf{\Phi}(t)$ to be a FM. The following theorem provides the answer.

Theorem 8.3.1 *Let $\mathbf{A}(t) \in \mathbf{C}[I, \mathbb{R}^{n\times n}]$. Suppose that the relation (8.6) holds. Then the determinant $\Phi(t)$, satisfies the scalar differential equation*

$$[\det \mathbf{\Phi}(t)]' = [\text{tr } \mathbf{A}(t)] \det[\mathbf{\Phi}(t)], \quad t \in I, \tag{8.7}$$

where

$$\text{tr } [\mathbf{A}(t)] = [a_{11}(t) + a_{22}(t) + \cdots + a_{nn}(t)], \quad t \in I. \tag{8.8}$$

Here the solution of (8.7) is given by

$$\det[\mathbf{\Phi}(t)] = \det\mathbf{\Phi}(t_0)\; exp \left[\int_{t_0}^{t} \text{tr } [\mathbf{A}(s)] \right] ds, \quad t_0, t \in \text{I},$$

where t_0 is the initial point.

Proof. Here the column vectors $\phi_1, \phi_2, ..., \phi_n$ are the solutions of the system (8.2). Each $\phi_i (i = 1, 2, \ldots, n)$ has n-elements in it, denoted by $\phi_{1i}, \phi_{2i}, \ldots, \phi_{ni}$. In view of (8.2) it follows that

$$\phi_{ij}'(t) = a_{i1}(t)\, \phi_{1j}(t) + a_{i2}(t)\, \phi_{2j}(t) + \cdots + a_{in}(t)\phi_{nj}(t), \; for\; t \in \text{I}, \tag{8.9}$$

where $i, j = 1, 2, ..., n$. Now consider the determinant function of the matrix $\Phi(t), t \in I$, given by

$$\det \mathbf{\Phi}(t) = \begin{vmatrix} \phi_{11} & \phi_{12} & \cdots & \phi_{1n} \\ \phi_{21} & \phi_{22} & \cdots & \phi_{2n} \\ \vdots & & & \\ \phi_{n1} & \phi_{n2} & \cdots & \phi_{nn} \end{vmatrix} (t), \ t \in I. \tag{8.10}$$

Hence

$$[\det \ \Phi(t)]' = \begin{vmatrix} \phi'_{11} & \phi'_{12} & \cdots & \phi'_{1n} \\ \phi_{21} & \phi_{22} & \cdots & \phi_{2n} \\ \vdots & \vdots & & \vdots \\ \phi_{n1} & \phi_{n2} & \cdots & \phi_{nn} \end{vmatrix} + \begin{vmatrix} \phi_{11} & \phi_{12} & \cdots & \phi_{1n} \\ \phi'_{21} & \phi'_{22} & \cdots & \phi'_{2n} \\ \vdots & \vdots & & \vdots \\ \phi_{n1} & \phi_{n2} & \cdots & \phi_{nn} \end{vmatrix}$$

$$+ \cdots + \begin{vmatrix} \phi_{11} & \phi_{12} & \cdots & \phi_{1n} \\ \phi_{21} & \phi_{22} & \cdots & \phi_{2n} \\ \vdots & \vdots & & \vdots \\ \phi'_{n1} & \phi'_{n2} & \cdots & \phi'_{nn} \end{vmatrix}, \quad t \in \mathbf{I}. \tag{8.11}$$

Hence it follows that, the first determinant in (8.11) is given by

$$\begin{vmatrix} \sum_{k=1}^{n} a_{1k} \ \phi_{k1} & \sum_{k=1}^{n} a_{1k} \ \phi_{k2} & \cdots & \sum_{k=1}^{n} a_{1k} \ \phi_{kn} \\ \phi_{21} & \phi_{22} & \cdots & \phi_{2n} \\ \vdots & \vdots & & \vdots \\ \phi_{n1} & \phi_{n2} & \cdots & \phi_{nn} \end{vmatrix}.$$

The simplification yields that the first determinant contributes, in (8.11), $(a_{11} \ det \ \mathbf{\Phi})$. Continuation of the similar steps, it follows that the contribution of the second determinant, in (8.11), is $(a_{22} \ det \ \mathbf{\Phi})$. Following the same procedure in the third to n determinants, the contribution to (8.11) by the last determinant is the $(a_{nn} \ det \ \mathbf{\Phi})$. Hence

$$\begin{aligned} (\det \ \Phi)' &= (a_{11} + a_{22} + \cdots + a_{nn}) \det \ \Phi \\ &= (\text{tr } \Phi) \det \ \Phi, \ t \in I, \end{aligned}$$

which is (8.7).

Observation. (8.7) is a linear scalar homogeneous, first order differential equation with the initial point at $t_0 \in I$, it has a solution given in Theorem 8.3.1.

Here one query needs to be answered. 'Does the system (8.2) have a unique FM or not? The answer is that the FM in (8.6) is not unique. For, if $\mathbf{\Phi}$ is one FM then $\mathbf{\Phi C}$, where $\mathbf{C}$ is an arbitrary $n \times n$ non-singular constant matrix, is also FM. It is clear that $\mathbf{\Phi C}$ also satisfies (8.6).
The examples given below are useful.

Example 8.3.1 *Show that* $\boldsymbol{\Phi}(t) = \begin{bmatrix} e^{3t} & 2t\, e^{3t} \\ 0 & e^{3t} \end{bmatrix}$ *is the FM of the system* $\mathbf{x}' = \begin{bmatrix} 3 & 2 \\ 0 & 3 \end{bmatrix} \mathbf{x}$.

Solution.

(i) The first criteria is to show that the columns of $\boldsymbol{\Phi}(t)$ are linearly independent. Consider $c_1 \begin{bmatrix} e^{3t} \\ 0 \end{bmatrix} + c_2 \begin{bmatrix} 2te^{3t} \\ e^{3t} \end{bmatrix} = 0$ implies that $e^{3t}(c_1 + 2tc_2] = 0$ and $e^{3t}\, c_2 = 0$, that is, $c_2 = 0$ and hence $c_1 = 0$.

(ii) To prove that $\boldsymbol{\Phi}'(t) = \begin{bmatrix} 3 & 2 \\ 0 & 3 \end{bmatrix} \boldsymbol{\Phi}(t)$:

$$\boldsymbol{\Phi}'(t) = \begin{bmatrix} 3\, e^{3t} & 6t\, e^{3t} + \; 2e^{3t} \\ 0 & 3\, e^{3t} \end{bmatrix}.$$

$$\boldsymbol{\Phi}'(t) = \begin{bmatrix} 3 & 2 \\ 0 & 3 \end{bmatrix} \begin{bmatrix} e^{3t} & 2t\, e^{3t} \\ 0 & e^{3t} \end{bmatrix} = \begin{bmatrix} 3 & 2 \\ 0 & 3 \end{bmatrix} \boldsymbol{\Phi}(t).$$

Hence, $\boldsymbol{\Phi}(t)$ is an FM.

Example 8.3.2 *Consider the linear system* $\mathbf{x}' = \mathbf{A}\mathbf{x}$ *where* $\mathbf{x} = [x_1\ x_2\ x_3]^T$ *and* $\mathbf{A} = \begin{bmatrix} -3 & 1 & 0 \\ 0 & -3 & 1 \\ 0 & 0 & -3 \end{bmatrix}$.

Verify that $\boldsymbol{\Phi}(t) = \begin{bmatrix} e^{-3t} & t\, e^{-3t} & e^{-3t} \dfrac{t^2}{2!} \\ 0 & e^{-3t} & t\, e^{-3t} \\ 0 & 0 & e^{-3t} \end{bmatrix}$ is an FM.

Solution. (i) To show that the 3 columns of $\boldsymbol{\Phi}(t)$ are linearly independent note that

$$e^{-3t} \left\{ c_1 \begin{bmatrix} 1 \\ 0 \\ 0 \end{bmatrix} + c_2 \begin{bmatrix} t \\ 1 \\ 0 \end{bmatrix} + c_3 \begin{bmatrix} t^2/2! \\ t \\ 1 \end{bmatrix} \right\} = 0,$$

implies that $c_1 = c_2 = c_3 = 0$ since $e^{-3t} \neq 0$.
(ii) Further,

$$\boldsymbol{\Phi}'(t) = -3e^{-3t} \begin{bmatrix} 1 & t & t^2/2! \\ 0 & 1 & t \\ 0 & 0 & 1 \end{bmatrix} + e^{-3t} \begin{bmatrix} 0 & 1 & t \\ 0 & 0 & 1 \\ 0 & 0 & 0 \end{bmatrix}$$

$$= e^{-3t} \begin{bmatrix} -3 & 1 - 3t & t - \dfrac{3t^2}{2!} \\ 0 & -3 & 1 - 3t \\ 0 & 0 & -3 \end{bmatrix}$$

$$= \begin{bmatrix} -3 & 1 & 0 \\ 0 & -3 & 1 \\ 0 & 0 & -3 \end{bmatrix} \begin{bmatrix} e^{-3t} & t\,e^{-3t} & e^{-3t}\frac{t^2}{2!} \\ 0 & e^{-3t} & t\,e^{-3t} \\ 0 & 0 & e^{-3t} \end{bmatrix}$$

$$= \mathbf{A}\ \mathbf{\Phi}(t).$$

Hence, $\mathbf{\Phi}(t)$ is an FM.

Example 8.3.3 *Verify relation (8.7) when* $\mathbf{\Phi}(t) = \begin{bmatrix} e^{3t} & 2\,t\,e^{3t} \\ 0 & e^{3t} \end{bmatrix}$ *is the FM in Example 8.3.1.*

Solution. Given,

$$\mathbf{A}\ (t) = \begin{bmatrix} 3 & 2 \\ 0 & 3 \end{bmatrix}.$$

The trace of $\mathbf{A}(t) = 6$. Further, $det\mathbf{\Phi}\ (t) = e^{6t}$ which implies that $[det\ \mathbf{\Phi}\ (t)]' = 6\,e^{6t}$. Clearly, the relation (8.6) holds.

From Theorem 8.3.1, it is noted that there exists a close relationship between the $\mathbf{x}(t)$ of the differential system (8.3) and its FM. Consider the expression $\mathbf{\Phi}(t)\alpha$, where $\boldsymbol{\alpha}$ is an unknown constant n-vector. Assume that

$$\mathbf{x}(t) = \mathbf{\Phi}(t)\alpha, t\ \in\ I, \tag{8.12}$$

is the solution of (8.3). The unknown n-vector α in (8.12) needs to be found. Clearly, $\mathbf{x}'(t) = \mathbf{\Phi}'(t)\alpha = \mathbf{A}(t)\ \mathbf{\Phi}(t)\alpha = \mathbf{A}(t)\ \mathbf{x}(t), t\ \in\ I$ in view of (8.12).

Now, consider IVP (8.3), with initial condition, $\mathbf{x}(t_0) = \mathbf{x}_0$. Hence,

$$\mathbf{x}(t_0) = \mathbf{\Phi}(t_0)\alpha = \mathbf{x}_0,$$

which implies that $\boldsymbol{\alpha} = \mathbf{\Phi}^{-1}\ (t_0)\ \mathbf{x}_0$. Note that $\mathbf{\Phi}^{-1}(t_0)$ exists, it follows that

$$\mathbf{x}(t) = \mathbf{\Phi}(t)\ \alpha = \mathbf{\Phi}(t)\ \mathbf{\Phi}^{-1}(t_0)\ \mathbf{x}_0, \quad t\ \in\ I, \tag{8.13}$$

where $\mathbf{x}(t)$ is the solution of the IVP (8.3). Here (8.13) establishes the relation between the solution $\mathbf{x}(t)$ and the FM $\mathbf{\Phi}(t)$. Thus, if the FM of a system is known, then the solution $\mathbf{x}(t)$ is given by (8.13).

Example 8.3.4 *Find the solution of the system* $\mathbf{x}' = \begin{bmatrix} 3 & 2 \\ 0 & 3 \end{bmatrix} \mathbf{x},\ \mathbf{x}(0) = \begin{bmatrix} 1 \\ 1 \end{bmatrix}$.

Solution. The FM for this problem, as given in Example 8.3.1, is $\mathbf{\Phi}(t) = \begin{bmatrix} e^{3t} & 2\,te^{3t} \\ 0 & e^{3t} \end{bmatrix}$, $t\ \in\ I$. Hence,

$$\mathbf{\Phi}(0) = \begin{bmatrix} 1 & 0 \\ 0 & 1 \end{bmatrix} \quad \text{and} \quad \mathbf{\Phi}^{-1}(0) = \begin{bmatrix} 1 & 0 \\ 0 & 1 \end{bmatrix}.$$

It follows from (8.13) that

$$\begin{aligned}\mathbf{x}(t) &= \mathbf{\Phi}(t)\,\mathbf{\Phi}^{-1}\,(0)\,\mathbf{x}(0)\\ &= \begin{bmatrix} e^{3t} & 2\,te^{3t}\\ 0 & e^{3t}\end{bmatrix}\begin{bmatrix}1 & 0\\ 0 & 1\end{bmatrix}\begin{bmatrix}1\\ 1\end{bmatrix}\\ &= \begin{bmatrix}(1+2t)e^{3t}\\ e^{3t}\end{bmatrix}.\end{aligned}$$

Example 8.3.5 *Find the solution of the IVP*

$$\mathbf{x}' = \mathbf{A}(t)\mathbf{x} = \begin{bmatrix}0 & 1 & 0\\ 0 & -2 & -5\\ 0 & 1 & 2\end{bmatrix}\mathbf{x},\quad \mathbf{x}(0) = \begin{bmatrix}1\\ 0\\ 1\end{bmatrix},\ t\in I.$$

Solution. The FM $\mathbf{\Phi}(t)$ is given by

$$\mathbf{\Phi}(t) = \begin{bmatrix}1 & -2+2\,\cos\,t+\sin\,t & -5+5\,\cos\,t\\ 0 & \cos\,t-2\,\sin\,t & -5\,sin\,t\\ 0 & \sin\,t & \cos\,t+2\,\sin\,t\end{bmatrix}.$$

Verify that $\mathbf{\Phi}(t)$ is, indeed, an FM. Here

$$\mathbf{\Phi}(0) = \begin{bmatrix}1 & 0 & 0\\ 0 & 1 & 0\\ 0 & 0 & 1\end{bmatrix} \quad \text{and} \quad \mathbf{\Phi}^{-1}(0) = \begin{bmatrix}1 & 0 & 0\\ 0 & 1 & 0\\ 0 & 0 & 1\end{bmatrix}.$$

Hence it follows, after substitutions, that

$$\begin{aligned}\mathbf{x}(t) &= \mathbf{\Phi}(t)\,\mathbf{\Phi}^{-1}(0)\,\mathbf{x}_0\\ &= \begin{bmatrix}-4+5\cos\,t\\ -5\sin\,t\\ \cos\,t+2\sin\,t\end{bmatrix}.\end{aligned}$$

EXERCISE 8.3

1. Show that $\mathbf{\Phi}(t)$ given below is an FM of the stated linear systems differential

(i) $\mathbf{\Phi}(t) = \begin{bmatrix}4\,e^{2t}+2\,e^{-4t} & e^{2t}-e^{-4t}\\ 8\,e^{2t}-8\,e^{-4t} & 2\,e^{2t}+e^{-4t}\end{bmatrix},\ \begin{bmatrix}x_1\\ x_2\end{bmatrix}' = \begin{bmatrix}0 & 1\\ 8 & -2\end{bmatrix}\begin{bmatrix}x_1\\ x_2\end{bmatrix},$

(ii) $\mathbf{\Phi}(t) = \begin{bmatrix}e^{-t} & 0 & 2(e^t-e^{-t})\\ 0 & e^{-t} & e^t-e^t\\ 0 & 0 & e^t\end{bmatrix},\ \begin{bmatrix}x_1\\ x_2\\ x_3\end{bmatrix}' = \begin{bmatrix}-1 & 0 & 4\\ 0 & -1 & 2\\ 0 & 0 & 1\end{bmatrix}\begin{bmatrix}x_1\\ x_2\\ x_3\end{bmatrix},$

$$\text{(iii)}\ \mathbf{\Phi}(t) = \begin{bmatrix} e^{2t} & e^{-t} & 0 \\ e^{2t} & 0 & e^{-t} \\ e^{2t} & -e^{-t} & -e^{-t} \end{bmatrix}, \begin{bmatrix} x_1 \\ x_2 \\ x_3 \end{bmatrix}' = \begin{bmatrix} 0 & 1 & 1 \\ 1 & 0 & 1 \\ 1 & 1 & 0 \end{bmatrix} \begin{bmatrix} x_1 \\ x_2 \\ x_3 \end{bmatrix}.$$

2. Suppose that $\mathbf{\Phi}(t)$ and $\mathbf{\Psi}(t)$ are two FMs of the system (8.2) for $t \in I$. Show that there exists a $n \times n$ constant nonsingular matrix $\mathbf{C}$ such that $\Psi = \Phi\, C$. Is $\alpha\, \mathbf{\Phi}(t) + \beta\mathbf{\Psi}(t)$ (α, β are real) an FM?

3. Let $\mathbf{\Phi}(t)$ be FM of (8.2) where $\mathbf{A}(t)$ is a real matrix. Show that

$$\frac{d}{dt}\,(\mathbf{\Phi}^{-1}(t))^T = -\mathbf{A}^T[\mathbf{\Phi}^{-1}(t)]^T.$$

Is $(\mathbf{\Phi}^{-1})^T$ an FM of the system $\mathbf{x}' = -\mathbf{A}^T(t)\mathbf{x}, t \in I$?

8.4 Method of Successive Approximations

The aim of this section is to develop a method for finding the solution of the IVP (8.3).

Definition 8.4.1 *Successive approximations for the IVP (8.3) is a sequence of functions, denoted by* $\{\boldsymbol{\phi_0}, \boldsymbol{\phi_1}, \boldsymbol{\phi_2}, \ldots, \boldsymbol{\phi_n}, \ldots\}$ *and defined on* I *as follows:*

$$\boldsymbol{\phi_0}(t) = \mathbf{x}_0$$

$$\boldsymbol{\phi_1}(t) = \mathbf{x}_0 + \int_{t_0}^{t} \mathbf{A}(s)\boldsymbol{\phi_0}(s)\, ds,$$

$$\boldsymbol{\phi_2}(t) = \mathbf{x}_0 + \int_{t_0}^{t} \mathbf{A}(s)\boldsymbol{\phi_1}(s)\, ds$$

$$\vdots$$

$$\boldsymbol{\phi_n}(t) = \mathbf{x}_0 + \int_{t_0}^{t} \mathbf{A}(s)\boldsymbol{\phi_{n-1}}(s)\, ds \quad n = 1, 2, \ldots \tag{8.14}$$

defined on I.

Observation. $\boldsymbol{\phi_0}(t)$ is the constant vector function $\mathbf{x}_0$ (given in (8.3)) and therefore is known. The next approximation $\phi_1(t)$ depends on $\phi_0(t) = \mathbf{x}_0$ and is well defined. The process thus continues for $n = 2, 3, 4, \cdots$. The conclusion is that all the elements of the sequence $\{\phi_n(t)\}$ exist and are well defined vector functions on I.

This method of successive approximations is explained through the following illustrative example.

Example 8.4.1 *Consider the IVP for* $t \in I$,

$$x'' - 4x' + 4x = 0, \;\; x(0) = 0, \;\; x'(0) = 1, \tag{8.15}$$

where $0 \in I$.
The IVP has the representation

$$\begin{bmatrix} x_1 \\ x_2 \end{bmatrix}' = \begin{bmatrix} 0 & 1 \\ -4 & 4 \end{bmatrix} \begin{bmatrix} x_1 \\ x_2 \end{bmatrix}, \quad \begin{bmatrix} x_{10} \\ x_{20} \end{bmatrix} = \begin{bmatrix} 0 \\ 1 \end{bmatrix}, \quad t \in I. \tag{8.16}$$

Note that the IVP (8.15) is converted into the system form (8.16) by the substitution
$x = x_1$, $x' = x_1' = x_2$ *and* $x'' = x_2' = -4x_1 + 4x_2$. *Now, suppose that* $\boldsymbol{\phi_0} = \begin{bmatrix} 0 \\ 1 \end{bmatrix}$.

Then, $\boldsymbol{\phi_1}(t) = \begin{bmatrix} 0 \\ 1 \end{bmatrix} + \int\limits_0^t \begin{bmatrix} 0 & 1 \\ -4 & 4 \end{bmatrix} \begin{bmatrix} 0 \\ 1 \end{bmatrix} ds = \begin{bmatrix} t \\ 1 + 4t \end{bmatrix}$.

Further, in view of (8.14), $\boldsymbol{\phi_2}(t)$ *is given as*

$$\boldsymbol{\phi_2}(t) = \begin{bmatrix} 0 \\ 1 \end{bmatrix} + \int\limits_0^t \begin{bmatrix} 0 & 1 \\ -4 & 4 \end{bmatrix} \begin{bmatrix} s \\ 1 + 4s \end{bmatrix} ds = \begin{bmatrix} t + 2t^2 \\ 1 + 4t + 6t^2 \end{bmatrix}.$$

Continuation of the process provides

$$\boldsymbol{\phi_3}(t) = \begin{bmatrix} t + 2t^2 + 2t^3 \\ 1 + 4t + 6t^2 + \frac{16}{3}\, t^3 \end{bmatrix} \quad \textit{and}$$

$$\boldsymbol{\phi_4}(t) = \begin{bmatrix} t + 2t^2 + 2t^3 + \frac{4}{3}\, t^4 \\ 1 + 4t + 6t^2 + \frac{16}{3}\, t^3 + \frac{10}{3}\, t^4 \end{bmatrix} \quad \textit{for } \; t \in \mathbf{I}.$$

[*As an exercise, the readers may verify* $\boldsymbol{\phi_3}(t)$ *and* $\boldsymbol{\phi_4}(t)$.]
Taking into account the nature of emerging pattern of polynomials in the vector, $\boldsymbol{\phi_5}(t)$ *and so on, it turns out that*

$$\begin{bmatrix} x_1(t) \\ x_2(t) \end{bmatrix} = \begin{bmatrix} t\, e^{2t} \\ (1 + 2t)\, e^{2t} \end{bmatrix}, \; t \in I. \tag{8.17}$$

The limiting column vector as $n \to \infty$, *is indeed the solution of system (8.15). During the conversion of (8.15) to (8.16), it was assumed that* $x_1 = x$, *hence the component* $x_1(t)$ *of the vector (8.17) is the solution of the IVP (8.15), i.e.,* $x(t) = t\, e^{2t}$. *It is left to the readers to verify that* $x(t)$ *is the solution of differential equation (8.15).*

The method of successive approximations theoretically works well, however, it becomes tedious after a few approximations. With modern developments in computational mathematics, this method yields better results. In the above example, the matrix $\mathbf{A}(t) = \mathbf{A}$ is a constant. Hence the solution $\mathbf{x(t)}$ is obtained smoothly.

The following existence result is stated without proof.

Proposition 8.4.1 *Consider the IVP for the system (8.3) and the scheme of successive approximations given in (8.14), where the matrix $\mathbf{A}(t)$ is a continuous matrix for $t \in I$ and $\mathbf{x}_0$ is a given constant vector. Then, the sequence $\{\phi_0(t), \phi_1(t), \ldots, \phi_n(t), ...\}, t \in I$ converges to a vector function $\mathbf{x}(t), t \in I$ as $n \to \infty$ and $\mathbf{x}(t)$ is the solution of (8.3) existing uniquely for $t \in I$.*

There is yet another theoretical approach to get the solution $\mathbf{x}(t)$ of the IVP (8.3). Integrating (8.3) between the limits t_0 and t in I yields

$$\mathbf{x}(t) = \mathbf{x}_0 + \int_{t_0}^{t} \mathbf{A}(t_1)\, \mathbf{x}(t_1)\, dt_1, \quad t_0, t_1 \in I. \tag{8.18}$$

Since the matrix $\mathbf{A}(t)$ is a continuous function the integral equation remains well defined on I. What has been achieved is that the vector-matrix IVP is converted into a new form called 'integral equation' (8.18). Note that (8.18) is not a solution of the IVP (8.3), since the vector function $\mathbf{x}(t)$ appears on both the sides of (8.18). However, this fact provides an advantage since $\mathbf{x}(t)$ may be replaced in (8.18) under the integral sign providing the relation

$$\begin{aligned} \mathbf{x}(t) &= \mathbf{x}_0 + \int_{t_0}^{t} \mathbf{A}(t_1) \left[\mathbf{x}_0 + \int_{t_0}^{t_1} \mathbf{A}(t_2)\, \mathbf{x}(t_2) dt_2 \right] dt_1, \\ &= \mathbf{x}_0 + \int_{t_0}^{t} \mathbf{A}(t_1)\, \mathbf{x}_0\, dt_1 + \int_{t_0}^{t} \mathbf{A}(t_1) \int_{t_0}^{t_1} \mathbf{A}(t_2)\, \mathbf{x}(t_2)\, dt_2\, dt_1, \end{aligned}$$

which exists and is well defined. Further the double integral contains the term $\mathbf{x}(t_2)$. As the process continues, the existence of each term is guaranteed infinite times, providing the following statement

$$\begin{aligned} \mathbf{x}(t) = \lim_{k \to \infty} [\mathbf{x}_0 &+ \int_{t_0}^{t} \mathbf{A}(t_1)\, \mathbf{x}_0\, dt_1 + \int_{t_0}^{t}\int_{t_0}^{t_1} \mathbf{A}(t_1)\, \mathbf{A}(t_2)\, \mathbf{x}_0\, dt_2\, dt_1 \\ &\cdots + \int_{t_0}^{t} \cdots \int_{t_0}^{t_{k-1}} \mathbf{A}(t_k) \cdots \mathbf{A}(t_1)\, \mathbf{x}_0\, dt_k \cdots dt_1] \\ = \mathbf{\Phi}(t, t_0)\, \mathbf{x}_0. \end{aligned} \tag{8.19}$$

Under the condition of continuity of $\mathbf{A}(t)$ on $\mathbf{I}$, it is logically possible to show that the infinite series occurring in (8.19) converges uniformly to a limit $\mathbf{x}(t)$ which is, indeed, the unique solution of the IVP (8.3). The process of integration in (8.19) is tedious but it is mathematically tenable. If in the IVP (8.3), $\mathbf{A}(t) = \mathbf{A},\ t \in I$ is a constant matrix then the process of infinite times integration in (8.19) works well and provides

$$\mathbf{\Phi}(t, t_0) = e^{(t-t_0)\mathbf{A}} \ and\ \mathbf{x}(t) = e^{(t-t_0)\mathbf{A}}\mathbf{x}_0,\ t \in I.$$

The matrix $\mathbf{\Phi}(t, t_0)$ is called evolution or transition matrix.

Properties of $\mathbf{\Phi}(t, t_0)$

The following are the properties of $\mathbf{\Phi}(t, t_0)$:

(i) $\mathbf{\Phi}(t, t_0)\,\mathbf{\Phi}(t_0, s) = \mathbf{\Phi}(t, s), \quad t, t_0, s \in I;$

(ii) $\mathbf{\Phi}(t, t) = \mathbf{I_n}, \mathbf{I_n}$ being the $n \times n$ identity matrix and $t \in I;$

(iii) $\mathbf{\Phi}^{-1}(t, t_0) = \mathbf{\Phi}(t_0, t);$

(iv) $\frac{\partial}{\partial t}\mathbf{\Phi}(t, t_0) = \mathbf{A}(t)\mathbf{\Phi}(t, t_0)$, or $\mathbf{\Phi}(t, t_0) = \mathbf{I}_n + \int_{t_0}^{t} \mathbf{A}(s)\,\mathbf{\Phi}(s, t_0)\,ds.$

These properties are useful in mathematical modeling problems, particularly when $\mathbf{A}(t) = \mathbf{A}, \mathbf{A}$ being a constant matrix and $\mathbf{\Phi}(t, t_0) = e^{(t-t_0)\mathbf{A}}, t \in I$.

The properties (i) to (iv) of $\mathbf{\Phi}(t, t_0)$ are clearly verified.

Let the n-vectors $\mathbf{x}_i = (x_{i1}\ \ x_{i2}\ \ \cdots\ \ x_{in}),\ i = 1, 2, ...n$ be a set of n-linearly independent solutions with initial conditions $(t_0\ \ x_0)$ where $x_0 = (x_{10}\ \ x_{20}\ \ \ldots\ \ x_{n0})$, then it follows that

$$\mathbf{x}_i(t) = \mathbf{\Phi}(t, t_0)\,x_{i0}, \quad i = 1, 2, \ldots, n$$

which is

$$\mathbf{X}(t) = \mathbf{\Phi}(t, t_0)\mathbf{X}(t_0),$$

where $\mathbf{X}(t)$ and $\mathbf{X}(t_0)$ are $n \times n$ matrices satisfying the matrix differential equation

$$\mathbf{X}'(t) = \ \mathbf{A}(t)\ \ \mathbf{X}(t), \mathbf{X}(t_0) = \mathbf{X}_0.$$

Hence, it follows that

$$\mathbf{\Phi}(t, t_0) = \mathbf{X}(t)\,\mathbf{X}^{-1}(t_0), \quad t_0, t \in I.$$

Note: Matrix differential equations are treated in Chapter 9.

EXERCISE 8.4

1. Find the first three successive approximations for the following systems of equations

(i) $\begin{bmatrix} x_1 \\ x_2 \end{bmatrix}' = \begin{bmatrix} 1 & 0 \\ 0 & -1 \end{bmatrix} \begin{bmatrix} x_1 \\ x_2 \end{bmatrix}, \begin{bmatrix} x_1(0) \\ x_2(0) \end{bmatrix} = \begin{bmatrix} 2 \\ 0 \end{bmatrix}$

(ii) $\begin{bmatrix} x_1 \\ x_2 \end{bmatrix}' = \begin{bmatrix} e^t & 1 \\ t & e^t \end{bmatrix} \begin{bmatrix} x_1 \\ x_2 \end{bmatrix}, \begin{bmatrix} x_1(1) \\ x_2(1) \end{bmatrix} = \begin{bmatrix} 1 \\ 1 \end{bmatrix}.$

(iii) $\begin{bmatrix} x \\ y \end{bmatrix}' = \begin{bmatrix} t & 1 \\ 1 & t \end{bmatrix} \begin{bmatrix} x \\ y \end{bmatrix}, \begin{bmatrix} x(0) \\ y(0) \end{bmatrix} = \begin{bmatrix} 1 \\ 0 \end{bmatrix}$

8.5 Nonhomogeneous Systems

The form of a nonhomogeneous system is given in (8.4) with initial conditions, where $\mathbf{A} \in \mathbf{C}[I, \mathbb{R}^{n \times n}]$, $\mathbf{p}(t) \neq 0$, $t \in I$. Note that $\mathbf{p}(t)$ is defined on I. It has already been mentioned that the n-vector solution $\mathbf{x}(t)$, under the given assumptions, exists on I and is unique.
Consider the system

$$\mathbf{x}' = \mathbf{A}(t)\mathbf{x},\ \mathbf{x}(t_0) = \mathbf{x}_0, \quad t, t_0 \in I,$$

and its perturbed system

$$\mathbf{y}' = \mathbf{A}(t)\mathbf{y} + \mathbf{p}(t),\ \ \mathbf{y}(t_0) = \mathbf{x}_0$$

having unique solutions $\mathbf{x}(t)$ and $\mathbf{y}(t)$ respectively. Note that the two IVP's have identical initial conditions. The aim of this section is to establish a relation between the solutions $\mathbf{x}(t)$ and $\mathbf{y}(t)$, $t \in I$. From (8.13), it is known that

$$\mathbf{x}(t) = \mathbf{\Phi}(t)\, \mathbf{\Phi}^{-1}(t_0)\, \mathbf{x}_0, \quad t \in I.$$

Assume that

$$\mathbf{y}(t) = \mathbf{\Phi}(t)\, \mathbf{\Phi}^{-1}(t_0)\, \mathbf{u}(t),\ \ t \in I, \tag{8.20}$$

where the unknown function $u(t)$ needs to be determined.

Differentiation of (8.20) yields

$$\begin{aligned} \mathbf{y}'(t) &= \mathbf{\Phi}'(t)\mathbf{\Phi}^{-1}(t_0)\, \mathbf{u}(t) + \mathbf{\Phi}(t)\, \mathbf{\Phi}^{-1}(t_0)\mathbf{u}'(t) \\ &= \mathbf{A}(t)\mathbf{\Phi}(t)\, \mathbf{\Phi}^{-1}(t_0)\, \mathbf{u}(t) + \mathbf{\Phi}(t)\, \mathbf{\Phi}^{-1}(t_0)\, \mathbf{u}'(t) \\ &= \mathbf{A}(t)\, \mathbf{y}(t) + \mathbf{\Phi}(t)\, \mathbf{\Phi}^{-1}(t_0)\, \mathbf{u}'(t) = \mathbf{A}(t)\, \mathbf{y}(t) + \mathbf{p}(t). \end{aligned}$$

Thus, it follows that $\mathbf{p}(t) = \mathbf{\Phi}(t)\ \mathbf{\Phi}^{-1}\ (t_0)\ \mathbf{u}'(t)$ or equivalently $u'(t) = \mathbf{\Phi}(t_0)\ \mathbf{\Phi}^{-1}(t)\ \mathbf{p}(t),\quad t \in I$. Integration leads to $\mathbf{u}(t) - \mathbf{u}(t_0) = \mathbf{u}(t) - \mathbf{x}_0 = \int_{t_0}^{t} \mathbf{\Phi}\ (t_0)\ \mathbf{\Phi}^{-1}(s)\ \mathbf{p}(s) ds$.

Substituting the function $\mathbf{u}(t)$ in (8.20) provides

$$\begin{aligned} \mathbf{y}(t) &= \mathbf{\Phi}(t)\ \mathbf{\Phi}^{-1}(t_0) \left[\mathbf{x}_0 + \int_{t_0}^{t} \mathbf{\Phi}(t_0)\ \mathbf{\Phi}^{-1}\ (s)\ \mathbf{p}(s)\ ds \right] \\ &= \mathbf{\Phi}\ (t)\ \mathbf{\Phi}^{-1}(t_0)\ \mathbf{x}_0 + \int_{t_0}^{t} \mathbf{\Phi}(t)\ \mathbf{\Phi}^{-1}(s)\ \mathbf{p}(s)\ ds, \quad t \in I. \end{aligned} \tag{8.21}$$

Hence, $\mathbf{y}(t)$ has been obtained explicitly. Since

$$\mathbf{x}(t) = \mathbf{x}(t, t_0, \mathbf{x}_0) = \mathbf{\Phi}(t)\ \mathbf{\Phi}^{-1}(t_0)\ \mathbf{x}_0,$$

it follows that

$$\mathbf{x}(t, t_0, \mathbf{p}(s)) = \mathbf{\Phi}(t)\ \mathbf{\Phi}^{-1}(t_0)\ \mathbf{p}(s),$$

and

$$\mathbf{y}(t) = \mathbf{x}(t, t_0, \mathbf{x}_0) + \int_{t_0}^{t} \mathbf{x}(t, s, \mathbf{p}(s))\ ds. \tag{8.22}$$

Equation (8.22) clearly establishes the relation between the solutions $\mathbf{x}(t)$ of (8.3) and $\mathbf{y}(t)$ of (8.4).

Observation.

1. The choice of $\mathbf{x}_0 = \mathbf{0}$ in (8.3) implies that $\mathbf{x}(t) \equiv \mathbf{0}$ for $t \in I$. Then, $$\mathbf{y}(t) = \int_{t_0}^{t} \mathbf{\Phi}\ (t)\ \mathbf{\Phi}^{-1}(s)\mathbf{p}(s)\ ds, \quad t \in I.$$

2. If the perturbation $\mathbf{p}(t) = \mathbf{0}$, then $\mathbf{y}(t) = \mathbf{x}(t) = \mathbf{\Phi}(t)\ \mathbf{\Phi}^{-1}\ (t_0)\mathbf{x}_0$.

3. In (8.3) and (8.4) the initial data $((t_0, x_0))$ is identical. The solution of (8.4) depends on the parameters $(t_0\ \ \mathbf{x}_0)$. The parameter $\mathbf{x}_0$ is now replaced by an unknown function $\mathbf{u}(t)$, for $t \in I$. Since the parameter $\mathbf{x}_0$ is replaced, the relations (8.21) and (8.22) are known in the literature as variation of parameters formula (VPF). The VPF formula for scalar linear differential equations was first established by the mathematician Lagrange in the 18th century.

4. In (8.3) and (8.4) the initial conditions are identical. However, they can be different from each other. Then, what is the change in VPF (8.21) in such a situation? The generalized VPF is now available in the

mathematical literature. It is not included here since it is beyond the scope of this book.

5. The VPF (8.21) or (8.22) is important since it has several applications in the theory of DEs. The examples given below are illustrative.

Example 8.5.1 Solve the IVP

$$\begin{bmatrix} x_1 \\ x_2 \end{bmatrix}' = \begin{bmatrix} 1 & 0 \\ 0 & 2 \end{bmatrix} \begin{bmatrix} x_1 \\ x_2 \end{bmatrix} + \begin{bmatrix} 1 \\ 1 \end{bmatrix}, \begin{bmatrix} x_{10}(0) \\ x_{20}(0) \end{bmatrix} = \begin{bmatrix} 1 \\ 0 \end{bmatrix}.$$

Solution. *An FM is given by*

$$\Phi(t) = \begin{bmatrix} e^t & 0 \\ 0 & 2e^{2t} \end{bmatrix}, \Phi^{-1}(t) = \frac{1}{e^{3t}} \begin{bmatrix} e^{2t} & 0 \\ 0 & e^t \end{bmatrix}, \Phi(0) = \begin{bmatrix} 1 & 0 \\ 0 & 2 \end{bmatrix},$$
$$\Phi^{-1}(0) = \begin{bmatrix} 2 & 0 \\ 0 & 1 \end{bmatrix}.$$

[Note: There can be an infinite number of FM's.]

Hence,

$$\begin{aligned} x(t) &= \begin{bmatrix} e^t & 0 \\ 0 & e^{2t} \end{bmatrix} \begin{bmatrix} 1 & 0 \\ 0 & 1 \end{bmatrix} \begin{bmatrix} 1 \\ 0 \end{bmatrix} \\ &\quad + \int_0^t \begin{bmatrix} e^t & 0 \\ 0 & e^{2t} \end{bmatrix} \cdot \begin{bmatrix} e^{t-s} & 0 \\ 0 & e^{2(t-s)} \end{bmatrix} \begin{bmatrix} 1 \\ 1 \end{bmatrix} ds \\ &= \begin{bmatrix} 2e^t - e^{2t} \\ 2e^t - 2e^{2t} \end{bmatrix} + \int_0^t \frac{1}{e^{3s}} \begin{bmatrix} e^t & e^{2t} \\ e^t & 2e^{2t} \end{bmatrix} \cdot \begin{bmatrix} 2e^{2s} - e^{2s} \\ -e^s + e^s \end{bmatrix} ds. \end{aligned}$$

Example 8.5.2 *Consider*

$$\begin{bmatrix} x_1 \\ x_2 \end{bmatrix}' = \begin{bmatrix} 0 & 1 \\ -e^{2t} & 1 \end{bmatrix} \begin{bmatrix} x_1 \\ x_2 \end{bmatrix} + \begin{bmatrix} 0 \\ 1 \end{bmatrix}, \begin{bmatrix} x_{10}(0) \\ x_{20}(0) \end{bmatrix} = \begin{bmatrix} 0 \\ 1 \end{bmatrix}.$$

Find the solution of the system, where

$$\mathbf{\Phi}(t) = \begin{bmatrix} \sin e^t & \cos e^t \\ e^t \cos e^t & -e^t \sin e^t \end{bmatrix}$$

is a fundamental matrix.

Solution.

$$\mathbf{\Phi}^{-1}(t) = \frac{1}{-e^t} \begin{bmatrix} -e^t \sin e^t & -\cos e^t \\ -e^t \cos e^t & \sin e^t \end{bmatrix}, \quad \mathbf{\Phi}^{-1}(0) = -\begin{bmatrix} -\sin 1 & -\cos 1 \\ -\cos 1 & \sin 1 \end{bmatrix}.$$

Also,

$$\mathbf{x}_0 = \begin{bmatrix} 0 \\ 1 \end{bmatrix}. \text{ Hence,}$$

$$\begin{aligned}\mathbf{x}(t, 0, \mathbf{x}_0) &= \begin{bmatrix} sin\ e^t & cos\ e^t \\ e^t cos\ e^t & -e^t \sin e^t \end{bmatrix} \begin{bmatrix} \sin 1 & \cos 1 \\ \cos 1 & -\sin 1 \end{bmatrix} \begin{bmatrix} 0 \\ 1 \end{bmatrix} \\ &= \begin{bmatrix} \sin(e^t - 1) \\ e^t \cos(e^t - 1) \end{bmatrix}.\end{aligned}$$

Observation. $\mathbf{x}(t, 0, \mathbf{x}_0)$ is the solution of the homogeneous part of the IVP. Further, if $\mathbf{p}(t) = \begin{bmatrix} 0 \\ 1 \end{bmatrix}$, then

$$\mathbf{x}(t, s, \mathbf{p}(s)) = \int_0^t \Phi(t)\ \Phi^{-1}(s)\ p(s) ds = \int_0^t \begin{bmatrix} sin(e^t - e^s) \\ e^{(t-s)}\ cos(e^t - e^s) \end{bmatrix} ds.$$

All terms in (8.22) are obtained. Hence,

$$\mathbf{y}(t) = -\begin{bmatrix} cos(e^t - 1) \\ -e^t sin(e^t - 1) \end{bmatrix} + \int_0^t \begin{bmatrix} sin(e^t - e^s) \\ e^{(t-s)} cos(e^t - e^s) \end{bmatrix} ds,$$

is the solution of the nonhomogeneous equation.

EXERCISE 8.5

1. Solve the IVPs

   (i) $x'' + 2x' - 8x = e^t, \;\; x(0) = 1, x'(0) = -4;$

   (ii) $x'' + x = 1, \;\; x(\pi) = 1, \;\; x'(\pi) = 2.$

2. Solve the IVPs

   (i) $\begin{bmatrix} x_1 \\ x_2 \end{bmatrix}' = \begin{bmatrix} 0 & 1 \\ -2 & 3 \end{bmatrix} \begin{bmatrix} x_1 \\ x_2 \end{bmatrix} + \begin{bmatrix} 1 \\ 1 \end{bmatrix}, \begin{bmatrix} x_1(0) \\ x_2(0) \end{bmatrix} = \begin{bmatrix} x_{10} \\ x_{20} \end{bmatrix};$

   (ii) $\begin{bmatrix} x_1 \\ x_2 \\ x_3 \end{bmatrix}' = \begin{bmatrix} -1 & 0 & 4 \\ 0 & -1 & 2 \\ 0 & 0 & 1 \end{bmatrix} \begin{bmatrix} x_1 \\ x_2 \\ x_3 \end{bmatrix} + \begin{bmatrix} e^t \\ e^{-t} \\ 0 \end{bmatrix}; \begin{bmatrix} x_{10} \\ x_{20} \\ x_{30} \end{bmatrix} = \begin{bmatrix} 0 \\ 1 \\ 1 \end{bmatrix}.$

8.6 Linear Systems with Constant Coefficients

Finding the solution of the system (8.3) is difficult when the matrix $\mathbf{A}(t)$ depends on t. The present section aims at developing a method for finding an explicit solution of the system (8.3) when the matrix $\mathbf{A}(t)$ is independent of t, i.e., $\mathbf{A}$ is an $n \times n$ constant matrix. Thus, the system under consideration is

$$\mathbf{x}' = \mathbf{A}\mathbf{x}, \ \mathbf{x}(t_0) = \mathbf{x}_0, \ \ t \in I. \tag{8.23}$$

The method consists of finding eigen values and eigen-vectors of $\mathbf{A}$ which are then used to write $\mathbf{x}(t)$ explicitly.

Definition 8.6.1 $\mathbf{e}^{at}$*, where a is a scalar, $t \in I$, is defined as*

$$e^{at} = 1 + at + \frac{a^2\,t^2}{2!} + \cdots + \frac{a^n\,t^n}{n!} + \ldots = 1 + \sum_{n=1}^{\infty} \frac{a^n\,t^n}{n!}.$$

The series is uniformly convergent on I. This series is now extended by replacing the scalar 'a' by a matrix $\mathbf{A}$, as follows:

$$e^{t\,\mathbf{A}} = \mathbf{I}_n + \sum_{n=1}^{\infty} \frac{t^n\,\mathbf{A}^n}{n!}, \ t \in I, \tag{8.24}$$

where $\mathbf{I}_n$ is the $n \times n$ identity matrix defined in Chapter 1.

The series is uniformly convergent under a matrix norm. Hence, $e^{t\mathbf{A}}$ is well defined. In relation (8.24) $e^{t\mathbf{A}}$ is called a matrix exponential function for $t \in I$.

Example 8.6.1 Find $e^{t\mathbf{A}}$ where $\mathbf{A} = \begin{bmatrix} 0 & 1 \\ -1 & 0 \end{bmatrix}$.

Solution. *Here*

$$\begin{aligned}
e^{t\mathbf{A}} &= \exp\left\{t\begin{bmatrix} 0 & 1 \\ -1 & 0 \end{bmatrix}\right\} \\
&= \begin{bmatrix} 1 & 0 \\ 0 & 1 \end{bmatrix} + \begin{bmatrix} 0 & 1 \\ -1 & 0 \end{bmatrix} t + \begin{bmatrix} 0 & 1 \\ -1 & 0 \end{bmatrix}^2 \frac{t^2}{2!} + \cdots + \begin{bmatrix} 0 & 1 \\ -1 & 0 \end{bmatrix}^n \frac{t^n}{n!} \\
&= \begin{bmatrix} 1 - \frac{t^2}{2!} + \frac{t^4}{4!} + \ldots & t - \frac{t^3}{3!} + \frac{t^5}{5!} - \ldots \\ -(t - \frac{t^3}{3!} + \frac{t^5}{5!} - \ldots) & 1 - \frac{t^2}{2!} + \frac{t^4}{4!} - \ldots \end{bmatrix} \\
&= \begin{bmatrix} \cos t & \sin t \\ -\sin t & \cos t \end{bmatrix}, \ t \in I.
\end{aligned}$$

Observation. The matrix $\begin{bmatrix} 0 & 1 \\ 1 & 0 \end{bmatrix}$ is called a circulant matrix and $\begin{bmatrix} 0 & 1 \\ -1 & 0 \end{bmatrix}$ is anti-circulant.

Example 8.6.2 *Find* $e^{t\mathbf{A}}$ *where* $\mathbf{A} = \begin{bmatrix} 0 & -4 \\ -4 & 0 \end{bmatrix}$.

Solution. *Here*

$$\begin{bmatrix} 0 & -4 \\ -4 & 0 \end{bmatrix}^{2k} = \begin{bmatrix} 4^{2k} & 0 \\ 0 & 4^{2k} \end{bmatrix}, \; k = 1, 2, 3, ..$$

$$\text{and} \quad \begin{bmatrix} 0 & -4 \\ -4 & 0 \end{bmatrix}^{2k+1} = \begin{bmatrix} 0 & -4^{2k+1} \\ -4^{2k+1} & 0 \end{bmatrix}, k = 0, 1, 2, ...$$

$$Hence \; exp\left\{ t \begin{bmatrix} 0 & -4 \\ -4 & 0 \end{bmatrix} \right\} = \frac{1}{2} \begin{bmatrix} e^{4t} + e^{-4t} & e^{-4t} - e^{4t} \\ e^{-4t} - e^{4t} & e^{4t} + e^{4t} \end{bmatrix}.$$

Observation.

(i) Given the initial value problem $\mathbf{x}' = \mathbf{A}\mathbf{x}$, $\mathbf{x}(t_0) = \mathbf{x}_0$, $\mathbf{x}(t) = exp[\mathbf{A}(t - t_0)\; \mathbf{x}_0$ is its unique solution;

(ii) Given a diagonal matrix $\mathbf{A} = \begin{bmatrix} a_1 & 0 \\ 0 & a_2 \end{bmatrix}$, $exp\;(\mathbf{A}t) = \begin{bmatrix} e^{a_1 t} & 0 \\ 0 & e^{a_2 t} \end{bmatrix}$;

(iii) The matrix of the type $\mathbf{A} = \begin{bmatrix} k & 1 \\ 0 & k \end{bmatrix}$, is of the form

$$\mathbf{A} = \begin{bmatrix} k & 0 \\ 0 & k \end{bmatrix} + \begin{bmatrix} 0 & 1 \\ 0 & 0 \end{bmatrix}.$$

Hence,

$$e^{\mathbf{A}\,t} = e^{\begin{bmatrix} k & 0 \\ 0 & k \end{bmatrix} t} e^{\begin{bmatrix} 0 & 1 \\ 0 & 0 \end{bmatrix} t} = \begin{bmatrix} e^{kt} & 0 \\ 0 & e^{kt} \end{bmatrix} \left\{ \begin{bmatrix} 1 & 0 \\ 0 & 1 \end{bmatrix} + \begin{bmatrix} 0 & 1 \\ 0 & 0 \end{bmatrix} t \right\} = e^{kt} \begin{bmatrix} 1 & t \\ 0 & 1 \end{bmatrix};$$

(iv) Let $\mathbf{\Phi}(t) = e^{t\,\mathbf{A}}$, then $\mathbf{\Phi}^{-1}(t) = e^{-t\,\mathbf{A}}$;

(v) Observe that for a nonsingular matrix $\mathbf{P}$, $exp[\mathbf{P}^{-1}\mathbf{X}\;\mathbf{P}] = \mathbf{P}^{-1}[e^{\mathbf{X}}]\mathbf{P}$ where $\mathbf{X}$ and $\mathbf{P}$ are matrices of the order $n \times n$;

(vi) If $\mathbf{X}$ and $\mathbf{Y}$ are square matrices of order $\mathbf{n}$, such that $\mathbf{XY} = \mathbf{YX}$, then $exp(\mathbf{X})exp(\mathbf{Y}) = exp(\mathbf{X} + \mathbf{Y})$ holds.

The system (8.2) when $\mathbf{A}$ is an $n \times n$ constant matrix, has the form $\mathbf{x}' = \mathbf{A}\mathbf{x}$. Assume that, the solution vector is of the form $\mathbf{x}(t) = (e^{t\mathbf{A}}\mathbf{c})$, $\mathbf{c}$ being a constant arbitrary vector. To verify this assumption, differentiate $\mathbf{x}(t)$ to get

$$\mathbf{x}'(t) = (e^{t\mathbf{A}}\mathbf{c})' = \mathbf{A}\; e^{t\mathbf{A}}\mathbf{c} = \mathbf{A}\;\mathbf{x}(t),\; t \in I.$$

The claim holds.

If the initial condition is given as $\mathbf{x}(t_0) = \mathbf{x}_0$, then it follows that the unique solution $\mathbf{x}(t)$ is given by

$$\begin{aligned}\mathbf{x}(t) &= \mathbf{x}(t, t_0, \mathbf{x}_0) = \\ &= \mathbf{\Phi}(t)\, \mathbf{\Phi}^{-1}(t_0)\, \mathbf{x}_0 = e^{t\mathbf{A}}\, e^{-t_0\mathbf{A}}\, \mathbf{x}_0 = e^{\mathbf{A}(t-t_0)}\mathbf{x}_0, \;\; t, \; t_0 \in I.\end{aligned}$$

Further, when

$$t_0 = 0, \;\; \mathbf{x}(t) = e^{\mathbf{A}t}\, \mathbf{x}_0,$$

the solution $\mathbf{x}(t)$ is an n-vector $(x_1(t) \;\; x_2(t) \;\; \ldots \;\; x_n(t))$. Choose the initial conditions $\mathbf{x}_{i0} = (0 \;\; 0 \;\; \ldots \;\; 1 \;\; \ldots \;\; 0) = \mathbf{e}_i$ (1 in ith place), then the fundamental matrix, FM, $\mathbf{\Phi}(t)$ (when $\mathbf{A}$ is a constant matrix) is constructed using n linearly independent solutions in the form of an $n \times n$ matrix. Hence, it follows that

$$\mathbf{X}'(t) = \frac{d}{dt}\, \mathbf{X}(t) = \mathbf{A}\mathbf{X}(t), \;\; \mathbf{X}(0) = \mathbf{I}_n,$$

where the matrix $\mathbf{I}_n = [\mathbf{e}_1 \;\; \mathbf{e}_2 \;\; \ldots \;\; \mathbf{e}_n]$. Now, it is easy to get FM $e^{t\mathbf{A}}$ for a given matrix A.

Methods to find a fundamental matrix

The method of finding a fundamental matrix using eigen-values and eigen-vectors is explained through examples for the various cases of eigen values discussed in Chapter 2. Further, the method is extended to a square matrix of order n, for each case.

(a) Method of eigen-values

Given a square matrix $\mathbf{A}$, the method of finding its eigen-values and eigen-vectors is discussed in Chapter 2. There are three possibilities, namely

(i) The eigen-values are real and distinct.

(ii) The eigen-values are repeated.

(iii) The eigen-values are complex, distinct and/or repeated.

The method is illustrated for the three cases through examples.

Case (i) Eigen-values are real and distinct.

Example 8.6.3 *Consider the system* $\mathbf{x}' = \mathbf{A}\mathbf{X}$ *where* $\mathbf{A} = \begin{bmatrix} 1 & 2 \\ 4 & 3 \end{bmatrix}$ *Find* $e^{\mathbf{A}t}$.

To find the eigen-values, consider the determinant,

$$\begin{vmatrix} 1-\lambda & 2 \\ 4 & 3-\lambda \end{vmatrix} = \lambda^2 - 4\lambda - 5 = (\lambda + 1)(\lambda - 5) = 0.$$

Hence, the roots $\lambda_1 = -1, \lambda_2 = 5$ are the eigen-values. Consider the equation

$$(\mathbf{A} - \lambda I)\mathbf{x} = 0.$$

The corresponding eigen-vectors are $[1 \ \ -1]^T$ and $[1 \ \ 2]^T$. Thus, $\begin{bmatrix} 1 \\ -1 \end{bmatrix} e^{-t}$ and $\begin{bmatrix} 1 \\ 2 \end{bmatrix} e^{5t}$ are the two linearly independent solutions,

and the FM is given by,

$$\mathbf{\Phi}(t) = \begin{bmatrix} e^{-t} & e^{5t} \\ -e^{-t} & 2e^{5t} \end{bmatrix}.$$

To find $e^{t\mathbf{A}}$, we employ the relation $e^{t\mathbf{A}} = \mathbf{\Phi}(t)\ \mathbf{\Phi}^{-1}(0)$. Here, $\mathbf{\Phi}(0) = \begin{bmatrix} 1 & 1 \\ -1 & 2 \end{bmatrix}$ and $\mathbf{\Phi}^{-1}(0) = \frac{1}{3}\begin{bmatrix} 2 & -1 \\ 1 & 1 \end{bmatrix}$. Hence, it follows that

$$e^{t\mathbf{A}} = \frac{1}{3}\begin{bmatrix} 2e^{-t} + e^{5t} & -e^{-t} + e^{5t} \\ -2e^{-t} + 2e^{5t} & e^{-t} + 2e^{5t} \end{bmatrix}.$$

Example 8.6.4 *Find an FM of the system*

$$\mathbf{x}' = \begin{bmatrix} 0 & 1 \\ -2 & 3 \end{bmatrix}\mathbf{x}.$$

Solution. *The eigen-values are* $\lambda_1 = 1$, $\lambda_2 = 2$. *For* $\lambda_1 = 1$, *the eigen-vector is* $[1 \ \ 1]^T$ *and one solution vector is* $[1 \ \ 1]^T\ e^t$.

For $\lambda_2 = 2$, *the eigen-vector is* $[1 \ \ 2]^T$ *and the other solution vector is* $[1 \ \ 2]^T\ e^{2t}$.

Hence, $\mathbf{\Phi}(t) = \begin{bmatrix} e^t & e^{2t} \\ e^t & 2e^{2t} \end{bmatrix}$.

To find e^{tA},

$$\begin{aligned} e^{tA} &= \Phi(t)\ \Phi^{-1}(0) \\ &= \begin{bmatrix} e^t & e^{2t} \\ e^t & 2e^{2t} \end{bmatrix}\begin{bmatrix} 2 & -1 \\ -1 & 1 \end{bmatrix} = \begin{bmatrix} 2e^t - e^{2t} & -e^t + e^{2t} \\ 2e^t - 2e^{2t} & -e^t + 2e^{2t} \end{bmatrix}. \end{aligned}$$

Observation. The eigen-vectors obtained in these two examples are not unique. They have been chosen by inspection. There are infinite ways to choose $\mathbf{\Phi}(t)$.

(a) General method When the eigen-values of a matrix are real and distinct, find for each eigen-value an eigen-vector, Suppose that λ_1 is an eigen value and $[a_1 \ \ a_2 \ \ \ldots \ \ a_n]^T$ is its eigen-vector, then a related solution–vector is $[a_1 \ \ a_2 \ \ \ldots \ \ a_n]^T\ e^{\lambda_1 t}$. Continue the process for the remaining eigen values to get n solution-vectors. Form a matrix with each solution vector as a column. Denote the matrix by $\mathbf{\Phi}(t)$. This is a fundamental matrix, $e^{t\mathbf{A}}$ is also an FM and hence for some nonsingular matrix $\mathbf{C}, e^{t\mathbf{A}} = \mathbf{\Phi}(t)\mathbf{C}$, i.e., for $t = 0, \mathbf{I}_n = \mathbf{\Phi}(0)\mathbf{C}$, which implies that $\mathbf{C} = \mathbf{\Phi}^{-1}(0)$. Hence, $e^{t\mathbf{A}} = \mathbf{\Phi}(t)\mathbf{\Phi}^{-1}(0)$.

Case (ii) Repeated eigen-values.

Example 8.6.5 *Solve the system* $\begin{bmatrix} x_1 \\ x_2 \\ x_3 \end{bmatrix}' = \begin{bmatrix} 1 & 1 & 1 \\ 0 & -2 & 3 \\ 0 & 1 & 0 \end{bmatrix} \begin{bmatrix} x_1 \\ x_2 \\ x_3 \end{bmatrix}$.

Solution. *The matrix* $\mathbf{A}$ *has the eigen values* $-3, 1, 1$ *and* $[1 \ \ -6 \ \ 2]^T \ e^{-3t}$, $[1 \ \ 0 \ \ 0]^T \ e^t$, $[t \ \ 1/2 \ \ 1/2]^T \ e^t$ *are the three linearly independent solution-vectors. (The second linearly independent vector corresponding to the eigen value* $\lambda = 1$ *is obtained as follows.*

Find $\mathbf{w}$ *such that* $(\mathbf{A} - \mathbf{I})\mathbf{w} = [1 \ 0 \ 0]^T$. *then* $\mathbf{u} = te^t[1 \ 0 \ 0]^T + e^t\mathbf{w}$ *is the required vector)*

The FM for the system is then given by

$$\mathbf{\Phi}(t) = \begin{bmatrix} e^{-3t} & e^t & te^t \\ -6e^{-3t} & 0 & 1/2e^t \\ 2e^{-3t} & 0 & 1/2e^t \end{bmatrix},$$

To find $e^{t\mathbf{A}}$, *since*

$$\begin{aligned} e^{t\mathbf{A}} &= \mathbf{\Phi}(t)\ \mathbf{\Phi}^{-1}(0) \\ &= \begin{bmatrix} e^{-3t} & e^t & te^t \\ -6e^{-3t} & 0 & 1/2e^t \\ 2e^{-3t} & 0 & 1/2e^t \end{bmatrix} \begin{bmatrix} 1 & 1 & 0 \\ -6 & 0 & 1/2 \\ 2 & 0 & 1/2 \end{bmatrix}^{-1} \\ &= \frac{1}{8} \begin{bmatrix} e^{-3t} & e^t & te^t \\ -6e^{-3t} & 0 & 1/2e^t \\ 2e^{-3t} & 0 & 1/2e^t \end{bmatrix} \begin{bmatrix} 0 & -1 & 1 \\ 0 & 6 & -6 \\ 0 & -2 & 2 \end{bmatrix} \\ &= \frac{1}{8}e^t \begin{bmatrix} 8 & 4t+1 & 12t-1 \\ 0 & 2 & 6 \\ 0 & 2 & 6 \end{bmatrix}. \end{aligned}$$

(b) General method Find the eigen values of matrix $\mathbf{A}$. Suppose that λ_i is an eigen-value repeated r_i times. Then r_i linearly independent eigen-vectors are required. For this purpose consider the system of equations $(\mathbf{A} - \lambda_i\mathbf{I})x = 0$ to get the first eigen-vector α_1. Then consider the system of equations $(\mathbf{A} - \lambda_i\mathbf{I})\mathbf{x} = \alpha_1$. Choose a second vector α_2 so that $(\mathbf{A} - \lambda_i\mathbf{I})\alpha_2 = \alpha_1$. Continue the procedure until r_i linearly independent vectors are available.

Case (iii) Eigen values are complex.

Example 8.6.6 *Find* $e^{t\mathbf{A}}$ *for* $\mathbf{x}' = \mathbf{A}\mathbf{x}$ *where* $\mathbf{A} = \begin{bmatrix} 2 & -5 \\ 2 & -4 \end{bmatrix}$.

Solution. *The eigen-values are* $-1 \pm i$. *As before, for* $\lambda_1 = -1 \pm i$, *the eigen-vectors are* $[3+i\ 2]^T$ *and* $[3-i\ 2]^T$, *respectively. Hence choose solution vectors as*

$[3+i\ 2]^T\ e^{(-1+i)t}$ *and* $[3-i\ 2]^T\ e^{-(1+i)t}$. *It follows that*

$$\mathbf{\Phi}(t) = \begin{bmatrix} (3+i)e^{(-1+i)t} & (3-i)e^{-(1+i)t} \\ 2e^{(-1+i)t} & 2e^{-(1+i)t} \end{bmatrix}.$$

Here, $\mathbf{\Phi}(0) = \begin{bmatrix} (3+i) & (3-i) \\ 2 & 2 \end{bmatrix}$ *and* $\mathbf{\Phi}^{-1}(0) = \frac{1}{4i}\begin{bmatrix} 2 & -(3-i) \\ -2 & 3+i \end{bmatrix}$,

Now,

$$\begin{aligned} e^{t\mathbf{A}} &= \mathbf{\Phi}(t)\mathbf{\Phi}^{-1}(0) \\ &= e^{-t}\begin{bmatrix} 3\,sin\,t + cos\,t & -5\,sin\,t \\ 2\,sin\,t & cos\,t - 3\,sin\,t \end{bmatrix} \end{aligned}$$

Example 8.6.7 *Find* $e^{t\mathbf{A}}$ *for* $\mathbf{x}' = \mathbf{A}\mathbf{x}$ *where* $\mathbf{A} = \begin{bmatrix} 1 & 2 & -1 \\ 0 & 1 & 1 \\ 0 & -1 & 1 \end{bmatrix}$

Solution. *Eigen-values are* $1, 1+i, 1-i$. *The solution vector for* $\lambda_1 = 1$ *can be chosen as* $\begin{bmatrix} 1 & 0 & 0 \end{bmatrix}^T e^t$. *For* $\lambda_2 = 1+i$ *the solution vector is* $\begin{bmatrix} -2+i & -i & 1 \end{bmatrix}^T e^{(1+i)t}$ *and for* $\lambda_3 = (1-i)$ *it is* $\begin{bmatrix} -2-i & i & 1 \end{bmatrix}^T\ e^{(1-i)t}$. *Hence*

$$\mathbf{\Phi}(t) = \begin{bmatrix} e^t & (-2+i)e^{(1+i)t} & (-2-i)e^{(1-i)t} \\ 0 & -ie^{(1+i)t} & ie^{(1-i)t} \\ 0 & e^{(1+i)t} & e^{(1-i)t} \end{bmatrix},\ \mathbf{\Phi}(0) = \begin{bmatrix} 1 & -2+i & -2-i \\ 0 & -i & i \\ 0 & 1 & 1 \end{bmatrix}.$$

$$\mathbf{\Phi}^{-1}(0) = \begin{bmatrix} 1 & 1 & 2 \\ 0 & \frac{i}{2} & \frac{1}{2} \\ 0 & \frac{-i}{2} & \frac{1}{2} \end{bmatrix}$$

and

$$e^{t\mathbf{A}} = e^t \begin{bmatrix} 1 & 1 + 2\,sin\,t - cos\,t & 2 - sin\,t - 2\,cos\,t \\ 0 & cos\,t & sin\,t \\ 0 & -sin\,t & cos\,t \end{bmatrix},$$

which is obtained on simplifying the entries using the fact that $e^{it} = cos\,t + i\,sin\,t$

(c) General method Find the eigen-values of $\mathbf{A}$ and follow the method in finding solution vectors for each eigen-value real or complex, distinct or repeated. Note that when eigen-values are complex, the exponential expression can also be expressed in terms of trigonometric ratios, since $e^{it} = cost + i\ sint$. If the complex eigen-values are distinct then apply the procedure in (a) to obtain the solution. If the complex values are repeated then apply the procedure in (b) to obtain the solution.

Aliter method to find $e^{t\mathbf{A}}$

From the Cayley-Hamilton Theorem, if $p(x)$ is characteristic equation for matrix A, then $p(A) = 0$. Also any scalar function $f(x)$ can be expressed in terms of a remainder and quotient as $f(x) = b(x)p(x) + r(x)$, where $p(x)$ is a nth degree polynomial and $r(x)$ is a polynomial of degree less than n. Set $r(x) = r_{n-1}x^{n-1} + r_{n-2}x^{n-2} + ... + r_0$. This result is also true when x is replaced by a matrix, $\mathbf{A}$, that is, $f(\mathbf{A}) = b(\mathbf{A})p(\mathbf{A}) + r(\mathbf{A})$. From this result it follows that $e^{t\mathbf{A}}$ can be reduced to a matrix polynomial of degree $n-1$. Thus $e^{t\mathbf{A}} = b(\mathbf{A})p(\mathbf{A}) + r(\mathbf{A}) = r(\mathbf{A})$ where $r(\mathbf{A}) = r_{n-1}\mathbf{A}^{n-1} + r_{n-2}\mathbf{A}^{n-2} + ... + r_0\mathbf{I}_{n\times n}$. The r_i's can be obtained from the scalar case by

evaluating $e^{t\lambda}$ at the roots of $p(\lambda)$, which are the eigen-values of the matrix $\mathbf{A}$. Let the eigen-values be given as $\{\lambda_1, \lambda_2, ..., \lambda_n\}$ then (i) if the n eigen-values are distinct, consider the n simultaneous equations $e^{t\lambda_i} = r(\lambda_i), i = 1, ..., n$ and solving them yield the values r_i's.

(ii) If one or more of the eigen-values is repeated then for the repeated value, say $\lambda_j = \lambda_i$, the equation is $\frac{de^{tx}}{dx}|_{x=\lambda_j} = te^t\lambda_j = \frac{dr(x)}{dx}_{x=\lambda_j}$. If $\lambda_k = \lambda_j = \lambda_i$ the equation is $\frac{d^2e^{tx}dx^2}{x=\lambda_k} = t^2e^t\lambda_k = \frac{d^2r(x)}{dx^2}_{x=\lambda_k}$. This procedure continues on for higher derivatives if the eigen-values are repeated more number of times. The following examples are illustrative

Example 8.6.8 *Find* $e^{t\mathbf{A}}$ *when* $\mathbf{A} = \begin{bmatrix} 1 & 2 \\ 4 & 3 \end{bmatrix}$.

Solution. Using the method given above, $e^{t\mathbf{A}} = \alpha_0\ \mathbf{I}_2 + \alpha_1\mathbf{A}t, r(\lambda) = \alpha_0 + \alpha_1\lambda$.

For the given matrix $\mathbf{A}$ the eigen values are $\lambda_1 = -1, \lambda_2 = 5$.

Hence $e^{-t} = \alpha_0 - \alpha_1, e^{5t} = \alpha_0 + 5\alpha_1$ which yield

$$\alpha_0 = \frac{1}{6}(e^{5t} + 5e^{-t}) \text{ and } \alpha_2 = \frac{1}{6}(e^{5t} - e^t), \text{ providing}$$

$$e^{t\mathbf{A}} = \frac{1}{6}\begin{bmatrix} 2e^{5t} + 4e^{-t} & 2e^{5t} - 2e^{-t} \\ 4e^{5t} - 4e^{-t} & 4e^{5t} + 2e^{-t} \end{bmatrix}.$$

The following example contains repeated roots of $\mathbf{A}$.

Example 8.6.9 *Find* $e^{t\mathbf{A}}$ *when* $\mathbf{A} = \begin{bmatrix} 0 & 1 \\ -4 & 4 \end{bmatrix}$.

Solution. *The eigen values of* $\mathbf{A}$ *are* 2, 2. *Further,*

$$\begin{aligned}
e^{t\mathbf{A}} &= \alpha_0\mathbf{I}_2 + \alpha_1\mathbf{A}t, r(\lambda) = \alpha_0 + \alpha_1\lambda; \\
e^{2t} &= r(2) = \alpha_0 + 2\alpha_1 \textit{ and } te^{2t} = \alpha_1 \textit{ yielding} \\
\alpha_0 &= (1-2t)e^{2t} \textit{ and } \alpha_1 = te^{2t}, \textit{ providing} \\
e^{t\mathbf{A}} &= \begin{bmatrix} (1-2t)e^{2t} & te^{2t} \\ -4te^{2t} & (1+2t)e^{2t} \end{bmatrix}.
\end{aligned}$$

Example 8.6.10 *Find* $e^{t\mathbf{A}}$, *where* $\mathbf{A} = \begin{bmatrix} 0 & 1 & 0 \\ 0 & -i & 0 \\ 0 & 1 & i \end{bmatrix}$.

The matrix $\mathbf{A}t$ has the eigen values $0, it, -it$. Further, let

$$e^{t\mathbf{A}} = \alpha_0 I + \alpha_1\mathbf{A}t + \alpha_2\mathbf{A}^2t^2 \textit{ and } r(\lambda) = \alpha_0 + \alpha_1\lambda + \alpha_2\lambda^2.$$

Hence,

$$\begin{aligned}
e^0 &= r(0) = \alpha_0 + \alpha_1(0) + \alpha_2(0) \\
e^{it} &= r(i) = \alpha_0 + \alpha_1(i) + \alpha_2(i)^2 \\
e^{-it} &= r(-i) = \alpha_0 + \alpha_1(-i) + \alpha_2(-i)^2.
\end{aligned}$$

Solving these simultaneous equations for α_0, α_1 and α_2 it follows that

$$\alpha_0 = 1,\ \alpha_1 = \frac{e^{it} - e^{-it}}{2i} = 1\ sint,\ \alpha_2 = \frac{e^{it} + e^{-it} - 2}{-2} = 1 - cos\ t.$$

Hence,

$$\begin{aligned}
e^{t\mathbf{A}} &= 1\begin{bmatrix} 1 & 0 & 0 \\ 0 & 1 & 0 \\ 0 & 0 & 1 \end{bmatrix} + sint\begin{bmatrix} 0 & 1 & 0 \\ 0 & -i & 0 \\ 0 & 1 & i \end{bmatrix} + 1 - cost\begin{bmatrix} 0 & -i & 0 \\ 0 & -1 & 0 \\ 0 & 0 & -1 \end{bmatrix} \\
&= \begin{bmatrix} 1 & \sin t - i(1-\cos t) & 0 \\ 0 & \cos t - isint & 0 \\ 0 & \sin t & \cos t + i\sin t \end{bmatrix}.
\end{aligned}$$

In respect of repeated roots, employ the idea involved in the method of variation of parameters. The following example illustrates the method. Here again, consider example 8.6.9.

Example 8.6.11 *Find* $e^{t\mathbf{A}}$ *when* $\mathbf{A} = \begin{bmatrix} 0 & 1 \\ -4 & 4 \end{bmatrix}$

The eigen values of $\mathbf{A}$ are 2 repeated twice.
For $\lambda = 2$, by the usual method the solution vector is

$$\mathbf{x}_1(t) = [1\ \ 2]^T\ e^{2t}.$$

To find $\mathbf{x}_2(t)$, assume that $\mathbf{x}_2(t) = \begin{bmatrix} c_1(t) \\ c_2(t) \end{bmatrix} e^{2t}$,
where c_1 and c_2 are unknown functions of t.
To determine c_1 and c_2, verify the conditions under which $\mathbf{x}_2(t)$ is a solution. Consider

$$\mathbf{x}_2'(t) \quad = \quad \begin{bmatrix} c_1(t) \\ c_2(t) \end{bmatrix} 2e^{2t} + \begin{bmatrix} c_1'(t) \\ c_2'(t) \end{bmatrix} e^{2t} = \begin{bmatrix} 0 & 1 \\ -4 & 4 \end{bmatrix} \begin{bmatrix} c_1(t) \\ c_2(t) \end{bmatrix} e^{2t}.$$

Comparing the coefficients of e^{2t},

$$\begin{aligned} c_1'(t) &= -2c_1(t) + c_2(t) \\ c_2'(t) &= -4c_1(t) + 2c_2(t) = 2[-2c_1(t) + c_2(t)] \end{aligned}$$

which implies that $c_2'(t) = 2c_1'(t)$ or $c_2(t) = 2c_1(t) + \alpha$. Substituting for $c_2(t)$ gives $c_1'(t) = \alpha$. Hence,

$$c_1(t) = \alpha t + \beta \textit{ and } c_2(t) = 2\alpha t + 2\beta + \alpha.$$

Since α and β are arbitrary constants of integration, choose $\alpha = 1$ and $\beta = 0$. Hence

$$x_2(t) = \begin{bmatrix} c_1(t) \\ c_2(t) \end{bmatrix} e^{2t} = \begin{bmatrix} t \\ 2t+1 \end{bmatrix} e^{2t} = \begin{bmatrix} 1 \\ 2 \end{bmatrix} te^{2t} + \begin{bmatrix} 0 \\ 1 \end{bmatrix} e^{2t}.$$

It follows that, $x(t) = \bar{c}_1 x_1(t) + \bar{c}_2 x_2(t) = \bar{c}_1 [1\ 2]^T e^{2t} +$
$\bar{c}_2 \{[1\ 2]^T\ te^{2t} + [0\ 1]^T e^{2t}\}$.

Hence, $\mathbf{\Phi}(t) = \begin{bmatrix} 1 & t \\ 2 & 2t+1 \end{bmatrix} e^{2t}$, $\mathbf{\Phi}(0) = \begin{bmatrix} 1 & 0 \\ 2 & 1 \end{bmatrix}$, $\mathbf{\Phi}^{-1}(0) = \begin{bmatrix} 1 & 0 \\ -2 & 1 \end{bmatrix}$ and

$$e^{t\mathbf{A}} = \mathbf{\Phi}(t)\mathbf{\Phi}^{-1}(0) = \begin{bmatrix} 1 & t \\ 2 & 2t+1 \end{bmatrix} \begin{bmatrix} 1 & 0 \\ -2 & 1 \end{bmatrix} e^{2t} = \begin{bmatrix} 1-2t & t \\ -4t & 2t+1 \end{bmatrix} e^{2t}.$$

EXERCISE 8.6

1. Show that the matrix $\mathbf{A} = \begin{bmatrix} i & 1 & 0 \\ 0 & i & 1 \\ 0 & 0 & i \end{bmatrix}$ has eigen-value i repeated three times. Find $e^{t\mathbf{A}}$.

2. Show that the solutions of $\begin{bmatrix} x \\ y \\ z \end{bmatrix}' = \begin{bmatrix} 0 & 2 & 0 \\ -2 & 0 & 0 \\ 0 & 0 & 0 \end{bmatrix} \begin{bmatrix} x \\ y \\ z \end{bmatrix}$ are given by

$$\begin{bmatrix} x(t) \\ y(t) \\ z(t) \end{bmatrix} = c_1 \begin{bmatrix} \sin 2t \\ \cos 2t \\ 0 \end{bmatrix} + c_2 \begin{bmatrix} \cos 2t \\ -\sin 2t \\ 0 \end{bmatrix} + \begin{bmatrix} 0 \\ 0 \\ 1 \end{bmatrix}.$$

3. Write the equation $x'' + 3x' - 4x = 0$ into system form and show that the system, in general, is unbounded. Hint: Find $e^{t\mathbf{A}}$.

4. Solve $\begin{bmatrix} x_1 \\ x_2 \end{bmatrix}' = \begin{bmatrix} 8 & -1 \\ 4 & 12 \end{bmatrix} \begin{bmatrix} x_1 \\ x_2 \end{bmatrix}.$

5. Solve the IVP

$$\begin{bmatrix} x \\ y \end{bmatrix}' = \begin{bmatrix} 1 & 2 \\ 0 & 3 \end{bmatrix} \begin{bmatrix} x \\ y \end{bmatrix} + \begin{bmatrix} t \\ t \end{bmatrix}, \mathbf{x}(0) = \begin{bmatrix} 8/9 \\ -1/9 \end{bmatrix}.$$

6. Find $e^{t\mathbf{A}}$ where

$$(i)\ \mathbf{A} = \begin{bmatrix} 1 & 1 & 0 \\ 0 & 1 & 0 \\ 0 & 0 & 1, \end{bmatrix}, \quad (ii)\ \mathbf{A} = \begin{bmatrix} 0 & 1 \\ 8 & 2, \end{bmatrix}, \quad (iii)\ \mathbf{A} = \begin{bmatrix} 0 & 1 \\ -9 & 6, \end{bmatrix}.$$

8.7 Stability Analysis of a System

The term stability is commonly employed in several life situations. The dictionary meaning of this term is: continued existence, permanence, standing firmly in place, not easily moved, shaken or overthrown etc. In a narrative form of language this term is in use on many occasions, such as economic stability, stability of health, political stability or stability of a government. These kinds of experiences take a qualitative or quantitative form when real-life situations are treated in a mathematical sense. Mathematical treatment of nonlinear or complex situations is not discussed.

Note that the solution of (8.3) is of the form $\mathbf{x}(t) = \mathbf{\Phi}(t)\mathbf{\Phi}^{-1}(t_0)\mathbf{x}_0$, t, $t_0 \in I$, where $\mathbf{\Phi}(t)$ is a fundamental matrix. When $\mathbf{x_0} = \mathbf{0}$, clearly $\mathbf{x(t)} \equiv \mathbf{0}$, $t \in I$, $\mathbf{x}(t) \equiv 0$ is called as the trivial or zero solution and stability of the system (8.3) is defined in terms of the the trivial solution. When $\mathbf{x_0} \neq \mathbf{0}$, the solution $\mathbf{x}(t) \not\equiv \mathbf{0}$.

Definition 8.7.1 *The trivial solution of (8.3), is said to be stable if any* $\mathbf{x}(t)$ *with* $\mathbf{x}(t_0)$ *remains close to the trivial solution, as* $t \to \infty$ *or* $\lim_{t\to\infty} \mathbf{x}(t) = 0$.

The present section, deals with the stability of solutions of the systems described in terms of

(i) polynomial form differential equations, with constant coefficients,

(ii) the systems of the form (8.3) where the matrix $\mathbf{A}$ is constant,

(iii) the matrix $\mathbf{A}$ in (8.3) depends on t, for $t \in I = [0, \infty)$

The criteria for stability are not for the cases listed above are gien below without proof.

(i) Consider the nth order homogeneous linear differential equation

$$x^{(n)} + a_1\, x^{(n-1)} + \cdots + a_n\, x = 0, \quad t \in [0, \infty), \tag{8.25}$$

where $a_1, a_2, ..., a_n$ are constants. Under these assumptions, it is known that the solution exists uniquely on $[0, \infty)$. It is known that the stability of trivial solution $x(t) \equiv 0$ is related to the roots is also the characteristic equation (CE) given by

$$r^n + a_1\, r^{n-1} + \cdots + a_n = 0, \tag{8.26}$$

The following statement provides the criteria for the stability of $x(t) \equiv 0$.

Criteria (i) Let all the roots of the CE (8.29) have negative real parts and let ρ be a real number such that

$$-\rho > max\ real\ (\lambda_i)$$
$$1 \leq i \leq n, \tag{8.27}$$

where $\lambda_1, \lambda_2, \ldots, \lambda_m \quad (m < n)$ are the distinct roots of (8.29), then there exists a number $k > 0$ having the property

$$||x(t)|| \leq\ k\ \ exp\,(-\rho\, t), \ \ t\ \in\ I = [0, \infty)$$

where ρ is given in (8.30) and $||x|| = sup\{|x(t)| : t \in I\}$

Observation. The right side of the above inequality suggests that $\lim_{t\to\infty} |x(t)| = 0$ indicating that the solution $x(t)$ remains close to the ideal stability position.

The nature of the roots of the CE depends on the nature of the co-efficients $a_1, a_2, \ldots, a_n$. This fact is included in the stability criteria given by E. Routh and A. Hurwitz and is stated below.

Routh-Hurwitz Criteria

Assume that the coefficients $a_1, a_2, ..., a_n$ in the CE (8.29) are real numbers. Define the determinants

Let $D_1 = a_1$ and $D_2 = \begin{vmatrix} a_1 & a_3 \\ 1 & a_2 \end{vmatrix}$, $D_3 = \begin{vmatrix} a_1 & a_3 & a_5 \\ 1 & a_2 & a_4 \\ 0 & a_1 & a_3 \end{vmatrix}$, and

$$...D_k = \begin{vmatrix} a_1 & a_3 & a_5 & \cdots & a_{2k-1} \\ 1 & a_2 & a_4 & \cdots & a_{2k-2} \\ 0 & a_1 & a_3 & \cdots & a_{2k-3} \\ 0 & 0 & a_4 & \cdots & a_{2k-4} \\ \vdots & \vdots & \vdots & \vdots & \vdots \\ 0 & 0 & 0 & \cdots & a_k \end{vmatrix}, k = 1, 2, \ldots, n \text{ and } a_j = 0 \text{ when } j > n.$$

If the det $D_k > 0$ for $k = 1, 2, \ldots, n$, then the roots of the CE (8.29) have negative real parts and the solution $x(t)$ of (8.28) is such that $\lim_{t\to\infty} |x(t)| = 0$. The following examples illustrate the forementioned result.

Example 8.7.1 *Consider the DE* $x'' + 9x' + 20\,x = 0.$

Solution. *The CE is given by*
$\lambda^2 + 9\lambda + 20 = (\lambda + 4)(\lambda + 5) = 0$. *The roots are* $\lambda_1 = -4$ *and* $\lambda_2 = -5$ *are negative. Hence, the solution* $x(t)$ *is stable. The Routh - Hurwitz criteria may also be applied. With* $a_1 = 9, a_2 = 20, a_3 = 0.$
$D_1 = 9 > 0 \;\; D_2 = \begin{vmatrix} 9 & 0 \\ 1 & 20 \end{vmatrix} = 180 > 0.$ *Hence,* $\lim_{n\to\infty} ||x(t)|| = 0.$

Observation. For a general second order equation $x'' + a_1 x' + a_2\,x = 0$, $D_1 = a_1 > 0$ and $D_2 = \begin{vmatrix} a_1 & 0 \\ 1 & a_2 \end{vmatrix} = a_1\,a_2 > 0$ are the conditions required for stability. This is possible if $a_1 > 0, a_2 > 0$. Hence, any second order differential equation having $a_1 > 0,\; a_2 > 0$ is always stable.

Example 8.7.2 *For the DE,* $x''' + a_1\,x'' + a_2 x' + a_3 x = 0$

$$D_1 = a_1,\; D_2 = \begin{vmatrix} a_1 & a_3 \\ 1 & a_2 \end{vmatrix}, D_3 = \begin{vmatrix} a_1 & a_3 & 0 \\ 1 & a_2 & 0 \\ 0 & a_1 & a_3 \end{vmatrix}$$

Here D_1, D_2 *and* D_3 need to be positive. Assume that $a_1 > 0,\; a_2 > 0$ and $a_3 > 0$, then the required condition holds for the given DE.

Example 8.7.3 *Consider the DE,* $z''' + a_1\,z'' + (8 - a_1)\,z' + 7z = 0.$

Determine the value of a_1 *so that the given DE remains stable.*

Solution. $D_1 = a_1, D_2 = \begin{vmatrix} a_1 & 7 \\ 1 & 8 - a_1 \end{vmatrix}, D_3 = \begin{vmatrix} a_1 & 7 & 0 \\ 1 & 8 - a_1 & 0 \\ 0 & a_1 & 7 \end{vmatrix}$

Here $D_1 > 0$ implies $a_1 > 0$. Further,
$D_2 = 8\,a_1 - a_1^2 - 7 > 0$ implies $1 < a_1 < 7$. With this choice of a_1 it follows that $D_3 > 0$.

Case (ii) Consider the system (8.3) where the matrix **A** is a constant. The following theorem gives the stability criteria of (8.3).

Theorem 8.7.1 *Suppose that the eigen-values of* **A** *are* $\lambda_1, \lambda_2, ..., \lambda_m$ *and let* λ_j, $j = 1, 2, ..., m$ *be repeated* n_j *times, such that* $n_1 + n_2 + \cdots + n_m = n$. *Let the eigen values have a representation*

$\lambda_j = \alpha_j + i\,\beta_j,\ \ j = 1, 2, \ldots, m$, and that $\alpha_j < \eta < 0\ for\ \ j = 1, 2, \ldots, m$. Then there exists a constant $M > 0$, such that
$||e^{t\mathbf{A}}|| \leq M\ exp(\eta\, t), t\ \in\ I = [0, \infty)$, and the system (8.3) is said to be stable.

Consider the following examples.

Example 8.7.4 *Consider the system*

$$\begin{aligned} x_1' &= -4x_1 - 6x_2 \\ x_2' &= 5x_1 - 10\,x_2. \end{aligned}$$

Solution. Given $\mathbf{A} = \begin{bmatrix} -4 & -6 \\ 5 & -10 \end{bmatrix}$ has eigen-values $-7 \pm \sqrt{21}\ i$.
The real parts of the eigen-values are negative. Here $\alpha_j = -7$. Hence η can be chosen a negative number < -7. Therefore the conclusion is $\lim\limits_{t\to\infty} ||x(t)|| = 0$ and that the system is stable.

Example 8.7.5 *Let* $\mathbf{x}' = \mathbf{A}\mathbf{x} = \begin{bmatrix} 1 & 1 & 1 \\ 0 & -2 & 3 \\ 0 & 1 & 0 \end{bmatrix} \mathbf{x}$

The matrix $\mathbf{A}$ has eigen-values $\lambda_1 = 1, \lambda_2 = 1, \lambda_3 = -3$. The first two eigen values have positive real part. Hence $\lim\limits_{t\to\infty} ||x(t)|| = \infty$; i.e. the solution $x(t)$ is unstable.

Case (iii) Consider $\mathbf{x}' = \mathbf{A}(t)\mathbf{x}, t\ \in\ I = [0, \infty)$ where the matrix $\mathbf{A}$ depends on t, is real valued and continuous in $t\ \in\ I$. The stability of the given system depends on the nature of the eigen values of the symmetric matrix $\mathbf{A}(t) + \mathbf{A}^T(t)$, $\mathbf{A}^T$ is the transpose of $\mathbf{A}$ and the eigen values are functions of t.

The following are the stability criteria.

Theorem 8.7.2 *Let $H(t)$ be the largest eigen-value of the matrix $\mathbf{A}(t)+\mathbf{A}^T(t)$ and let t_0 be a fixed real number. If $\lim\limits_{t\to\infty} \int_{t_0}^{t} H(s)ds = -\infty, t\ \in\ I$, then every solution $x(t)$ of the given linear system is such that $\lim\limits_{t\to\infty} x(t) = 0$.*
On the other hand, let $m(t)$ be the smallest eigen-value of the matrix $\mathbf{A}(t) + \mathbf{A}^T(t)$ and is such that $\lim\limits_{t\to\infty} \int_{t_0}^{t} m(s)ds = +\infty$.
Then every nonzero solution $x(t)$ of the given linear system is unbounded, i.e. $\lim\limits_{t\to\infty} ||x(t)|| = \infty$.

Example 8.7.6 *Consider the system* $\mathbf{x}' = \begin{bmatrix} \frac{2}{t^3} & t^3 \\ -t^3 & -2 \end{bmatrix} \mathbf{x}$

Here

$$\mathbf{A}(t) + \mathbf{A}^T(t) = \begin{bmatrix} \frac{2}{t^3} & t^3 \\ -t^3 & -2 \end{bmatrix} + \begin{bmatrix} \frac{2}{t^3} & -t^3 \\ t^3 & -2 \end{bmatrix} = \begin{bmatrix} \frac{4}{t^3} & 0 \\ 0 & -4 \end{bmatrix}$$

has two eigen-values $\frac{4}{t^3}$ and -4. Clearly, H(t) $= \frac{4}{t^3}$ and m(t) $= -4$. Applying the stability criteria,

$$\begin{aligned}\int_{t_0}^{t} H(s)ds &= \int_{t_0}^{t} \frac{4}{s^3} ds \\ &= \left[\frac{-2}{t^2} + \frac{2}{t_0^2}\right] \to \frac{2}{t_0^2} \text{ as } t \to \infty.\end{aligned}$$

Hence solution $x(t)$ remains bounded.

EXERCISE 8.7

1. Consider the equation
   $z''' + (a - \alpha_1)\, z'' - 3(\alpha_2 - 1)\, z' + 9z = 0.$

   Apply Routh-Hurwitz's criteria for stability and determine values of α_1, α_2 so that the system remains stable.

2. Show that all solutions of the following systems are bounded on $(0, \infty)$.

   (i) $x'' + e^t\, x = 0.$

   (ii) $x'' + (t - 1)\, x = 0.$

3. Using the following matrices show that the system (8.3) is stable

   (i) $\begin{bmatrix} -2 & 1 & 1 \\ 0 & -2 & 4 \\ 0 & 0 & -1 \end{bmatrix}$ (ii) $\mathbf{A} = \begin{bmatrix} -1 & 5 & 6 \\ 0 & -4 & -2 \\ 0 & 1 & -1 \end{bmatrix}.$

4. For the linear system

   $$(x_1 \;\; x_2 \;\; x_3 \;\; x_4 \;\; x_5)' = (-3x_1 \quad -3x_2 \quad -3x_1 \quad -x_4 - x_5 \;\; 6x_2 - x_5)$$

   Show that the eigen-values are $-3, 0, -1$ with multiplicity $2, 2, 1$, respectively and that $x(t)$ is bounded.

5. Show that the system $x' = A(t)\, x$

(i) is bounded as $t \to \infty$ when $A(t) = \begin{bmatrix} \frac{1}{t^2} & t^2 \\ -t^2 & -1 \end{bmatrix}$,

(ii) is stable when $A(t) = \begin{bmatrix} -\frac{1}{t} & t^2+1 \\ -(t^2+1) & -2 \end{bmatrix}$.

(Hint: Find the eigen-values of $A(t) + A^T(t)$.)

8.8 Election Mathematics

Many countries in the world have adopted democratic set up for administrating the welfare activities of villages, cities, states and the country. The method of selecting the administrators is, generally, by way of 'elections' i.e. adult franchise. Naturally there exists a tough competition among the members of the voting community, and also between the candidates contesting the elections. Some experts, through their experience, evaluate the abilities of the contesting members and opine about 'who will win the elections'. For many years the evaluation process was 'abstract' in nature. Slowly experts decided the factors involved in a competition and finally the election process took the shape of a mathematical modeling problem which is now called as 'election mathematics'. The aim of this section is to present such a simple model involving a system of differential equations.

Suppose that in a city two members A and B wish to contest an election to become a corporator in its administration, possessing variable (winning) potential x and y, respectively depending on an independent variable t. In a simplistic and pure competitive atmosphere the rate of change of potential x of A depends on the preparation of the potential of y and vice versa. In a mathematical sense

$$\begin{aligned} \frac{dx}{dt} &= x'(t) \propto y(t) \\ \frac{dy}{dt} &= y'(t) \propto x(t), \\ x' &= ay \text{ and } y' = bx. \end{aligned} \tag{8.28}$$

The constants of proportion 'a' and 'b' can be time dependent but for the sake of simplicity, let 'a' and 'b' be assumed to be constants. However, the model (8.28) is not realistic since many factors retard or affect the potential of the candidates in real-life situations. For example, the rate of change of the potential of the candidate will depend on various factors like (i) shortage of funds (ii) nonavailability of manpower (iii) rules and regulations of election offices. Taking into account the affect of retardation the refined model takes the form

$$x' = ay - px, \quad y' = bx - qy, \tag{8.29}$$

where p and q are assumed to be constants.

The model (8.29) needs one more refinement since local disputes, false promises and rumors in the election competition, also affect x' and y'. The refined model is then given by

$$x' = ay - px + \alpha, \quad y' = bx - qy + \beta, \tag{8.30}$$

where again, it is assumed that α and β are constants and take into account the above mentioned negative factors. These constants act like perturbation terms.

The effectiveness of (8.30) may increase if (i) the constants $a, p, \alpha, b, q, \beta$ are chosen realistically and (ii) these are made available as functions of t. The second factor enhances the complexity in (8.30).

The system (8.30) is linear in its composition and hence matrix methods are available to find its solution.

It is experienced that some expert election-agencies collect useful data during the process of voting and predict the outcome of the elections as soon as the voting process is completed. If the findings of these agencies quite often differ from each other it is because of some presumptions about the parameters appearing in the system (8.30).

The model (8.30) includes some special cases of interest. These are stated below:

(i) Let $\alpha = 0$ and $\beta = 0$.

This situation suggests that the local disputes among voters are absent, false promises or rurmors are ineffective. This leads to a fair-practice election, a spirit of pure competition exists between the two candidates.

(ii) The case, $x' = 0, y' = 0$, is an indication that increase or decrease in the potential of two competing candidates is zero. It suggests that the election process is indeed peaceful. In particular in (8.33), in addition to $\alpha = 0, \beta = 0$, suppose that $ay - px = 0, \quad bx - qy = 0$. This means that $x = 0$ and $y = 0$ (when $aq \neq bp$) suggesting that an equilibrium stage is reached. In elections, this is a rare phenomena. In this case contesting candidates must be close friends or are saints.

(iii) Consider the situation when $ay - px = 0$ and $bx - qy = 0$. Suppose that local disputes and rumors are prevailing in the community of the voters. Then in view of (8.30), the system takes the form $x' = \alpha$ and $y' = \beta$. Assume that $\alpha > 0, \beta > 0$. The system suggests that there is competition between the two candidates.

(iv) Suppose that the constants a and b are large numbers and dominate the value of p, α, q, β, then the system (8.30) becomes

$$x' = ay \text{ and } y' = bx$$

which means $x'' = ay' = abx$. Solving this equation results in solving the system (8.30) in this case, and the solution is $x(t) = A\ e^{\sqrt{ab}t} + B\ e^{\sqrt{-ab}t}$ (A and B are constants). In case $A > 0$, then $x(t) \to \infty$ as $t \to \infty$. This fact suggests that the elections are fought vigorously.

(v) Suppose that one of the contesting candidates, say B takes a lean position i.e, $y = 0$ then the system (8.30) takes the form

$$\frac{dx}{dt} = -px + \alpha \text{ and } \frac{dy}{dt} = bx + \beta.$$

Here, if $b > 0$ or $x > 0$ it implies that y increases as time t increases implying there by that the lean position of candidate B may not last long enough. This fact suggests that elections need to be fought, in the right spirit.

Observation.

1. The mathematical model presented in (8.30) involves two candidates in an election. In many elections there is a multi-cornered contest for one seat. Then the model becomes complex.

2. The arms race between two conflicting nations represents a similar model. Mathematicians have studied the available data for World War II between two conflicting nations. It was found that the model of the type (8.30) works well.

3. In an ocean, big fish rely, for their food, on small fish. There exists a conflicting situation which is described by the model (8.30). This model is also known as prey–predator model in mathematics.

4. In an industry conflicting situations may arise between management and labourers about the raise in salaries, bonuses or economic problems. Model (8.30) works in such a situation also.

5. In agricultural sciences, farmers plant trees at a distance from each other following scientific guidelines between two plantations. It has been established that the roots of the trees face conflicting phenomena for food and water in the soil.

   In all these cases (1) to (5), the mathematical model is similar, however the meanings of the terms involved are different.

8.9 Conclusion

This chapter considered the study of vector-matrix differential equations which includes the existence and uniqueness of solutions as a basic feature. Methods of solving linear vector-matrix differential equations are explained through second and third-order matrices. This approach can be extended to higher order matrices using computational techniques. The examples are simple and illustrate the technique. The application given in Section 8.8 helps to widen the scope of understanding of the subject. For further reading the readers are referred to [3], [6], [11] and [16].

Chapter 9

Linear Matrix Differential Equations

9.1 Introduction

During the last four to five decades the study of linear matrix differential equations (LMDEs) and their properties have witnessed a rapid growth because of the demands in modern engineering and in technology. The emergence of areas such as control systems, theory of communications and optimization, demanded extensive study of matrix analysis. This chapter deals with the methods of obtaining explicit solutions of LMDEs. In Section 9.2 an initial-value-problem for an LMDE and other different types of LMDEs are introduced.

In Sections 9.3 to 9.5, the methods of obtaining explicit solutions of LMDEs of the first order occurring in various forms are presented. The linear variation of parameters formula(VPF) has been developed to evaluate the effect of a perturbation, a booster or a control term on the system. In Section 9.6, an LMDE of higher order is treated and solved explicitly to get a solution matrix.

The methods of solving LMDEs are then employed (i) to solve matrix boundary value problem and (ii) for the development of matrix trigonometry and matrix hyperbolic functions and extended trigonometric functions. The examples illustrate the methods developed in the Sections 9.3 to 9.7.

9.2 Initial-value-problems of LMDEs

The simplest form of a linear matrix differential equation (LMDE) is given by

$$\mathbf{X}' = \mathbf{A}(\mathbf{t})\mathbf{X}, \quad t \in I, \tag{9.1}$$

where $\mathbf{X}$ is an $n \times n$ matrix defined on an interval $I \subset \mathbb{R}$, the elements of which are differentiable functions on I, $\mathbf{A}(t)$ is an $n \times n$ matrix whose elements are continuous functions of $t \in I$. Here (9.1) is a homogeneous first order LMDE.

Definition 9.2.1 *A solution of* (9.1) *is defined as an* $n \times n$ *matrix function*

DOI: 10.1201/9781351014953-9

$\mathbf{\Phi}(\mathbf{t})$ *defined on* I*, which is differentiable on* I *and is such that*

$$\mathbf{\Phi}'(\mathbf{t}) = \mathbf{A}(\mathbf{t})\,\mathbf{\Phi}(\mathbf{t}), \quad t \in I. \tag{9.2}$$

Observation. To get a specific solution $\mathbf{X}(t)$ of (9.1), it is necessary to set an initial data $\mathbf{X}(\mathbf{t_0}) = \mathbf{X_0}$, $t_0 \in I$, which is a given initial matrix belonging to $\mathbb{R}^{n\times n}$. Now (9.1) with its initial data (t_0 $\mathbf{X}_0$) is an IVP

$$\mathbf{X}' = \mathbf{A}(\mathbf{t})\mathbf{X}, \quad \mathbf{X}(\mathbf{t_0}) = \mathbf{X_0}, \quad t_0, t \in I. \tag{9.3}$$

It is known that the solution of this IVP (9.3) exists on I and is unique.

Types of LMDEs

1. The IVP (9.3) may be further strengthened by adding a perturbation term $\mathbf{H}(t)$ to get a linear perturbed matrix differential equation

$$\mathbf{Y}' = \mathbf{A}(\mathbf{t})\mathbf{Y} + \mathbf{H}(\mathbf{t}), \quad \mathbf{Y}(\mathbf{t_0}) = \mathbf{Y_0}, \tag{9.4}$$

where $\mathbf{H}(t)$ is an $n \times n$ matrix, continuous on I. The existence and uniqueness property of (9.4), as in (9.3), continues to hold for all $t \in I$.

Observation. $\mathbf{H}(\mathbf{t})$ depends only on the scalar $t \in I$. In case, the matrix $\mathbf{H}(\mathbf{t})$ is replaced by $\mathbf{H}(\mathbf{t}, \mathbf{y})$ then the LMDE(9.4) becomes nonlinear. In such a situation a solution of (9.4) may exist uniquely under some conditions but it becomes difficult to obtain explicit solutions. Such nonlinear situations are avoided in the chapter.

2. There is yet another possibility to strengthen the linearity in (9.1) by considering a linear matrix differential equation of the form

$$\mathbf{X}' = \mathbf{A}(\mathbf{t})\mathbf{X} + \mathbf{X}\,\mathbf{B}(\mathbf{t}), \quad t \in I, \tag{9.5}$$

where $\mathbf{A}$ and $\mathbf{B} \in C[I, \mathbb{R}^{n\times n}]$ and its perturbed format is

$$\mathbf{Y}' = \mathbf{A}(\mathbf{t})\,\mathbf{Y} + \mathbf{Y}\,\mathbf{B}(\mathbf{t}) + \mathbf{H}(\mathbf{t}), \tag{9.6}$$

where $\mathbf{H} \in C[I, \mathbb{R}^{n\times n}]$. Under these hypotheses, (9.6) possesses a unique solution on I. Generally, the equations (9.5) and (9.6) need the additional condition

$$\mathbf{AB} \neq \mathbf{BA} \quad \text{for} \quad t \in I. \tag{9.7}$$

3. One more LMDE format that invites attention is

$$\mathbf{X}' = \mathbf{A}(\mathbf{t})\,\mathbf{X}\,\mathbf{B}(\mathbf{t}), \quad t \in I. \tag{9.8}$$

and its perturbed equation format is

$$\mathbf{Y}' = \mathbf{A}(\mathbf{t})\,\mathbf{Y}\,\mathbf{B}(\mathbf{t}) + \mathbf{H}(\mathbf{t}), \tag{9.9}$$

where $\mathbf{H}(t) \in C\,[I,\ \mathbb{R}^{n\times n}]$.

The matrix systems noted in (9.1) to (9.9) are linear and are of the first order. In a subsequent section, linear systems of higher order are also considered.

9.3 LMDE $\mathbf{X}' = \mathbf{A(t)X}$

Given a linear system of the vector- matrix form $\mathbf{x}' = \mathbf{A(t)x}$ (discussed in Chapter 8) it has been noted that a fundamental matrix(FM) $\mathbf{\Phi}(t)$ needs to be developed. Further, the FM satisfies

$$\mathbf{\Phi}'(\mathbf{t}) = \mathbf{A}(\mathbf{t})\,\mathbf{\Phi}(\mathbf{t}), \quad t \in I.$$

Observe that $\mathbf{\Phi}(\mathbf{t})$ is an $n \times n$ matrix. Clearly, it is a solution of the matrix differential equation (9.1). Further, given the FM, $\mathbf{\Phi}(t)$, matrices of the form $\mathbf{\Psi}(\mathbf{t}) = \mathbf{\Phi}(\mathbf{t})\mathbf{C}$, where $\mathbf{C}$ is an $n\times n$ constant, nonsingular matrix, are all FMs. Hence, it follows that

$$\begin{aligned} \mathbf{\Phi}'(\mathbf{t})\,\mathbf{C} &= \mathbf{A}(\mathbf{t})\,\mathbf{\Phi}(\mathbf{t})\,\mathbf{C}\,, \quad t \in I, \text{ or} \\ \mathbf{\Psi}'(\mathbf{t}) &= \mathbf{A}(\mathbf{t})\,\mathbf{\Psi}(\mathbf{t}). \end{aligned}$$

Hence, to solve LMDE (9.1), it is necessary to find an FM $\mathbf{\Phi}(t)$. It can be easily verified that

$$\mathbf{X}(\mathbf{t}) = \mathbf{\Phi}(\mathbf{t})\,\mathbf{\Phi}^{-1}(\mathbf{t_0})\,\mathbf{X_0}, \text{ (since } \mathbf{X}(\mathbf{t_0}) = \mathbf{X_0}) \tag{9.10}$$

is a solution of the IVP (9.1).

The following examples are illustrative.

Example 9.3.1 *Solve the LMDE* $\mathbf{X}' = \begin{bmatrix} 1 & 2 \\ 4 & 3 \end{bmatrix}\mathbf{X}$.

Solution. *In Example 8.6.3, it was shown that an FM of* $\mathbf{x}' = \begin{bmatrix} 1 & 2 \\ 4 & 3 \end{bmatrix}\mathbf{x}$ *is given by*

$$exp\left[\begin{bmatrix} 1 & 2 \\ 4 & 3 \end{bmatrix} t\right] = \frac{1}{3}\begin{bmatrix} 2\,e^{-t} + e^{5t} & -e^{-t} + e^{5t} \\ -2\,e^{-t} + 2\,e^{5t} & e^{-t} + 2\,e^{5t} \end{bmatrix} = \mathbf{\Phi}(t),$$

for $t \in I$. *It follows that* $\mathbf{\Phi}(\mathbf{t})$ *is also a solution of the given LMDE,* $\mathbf{\Phi}'(\mathbf{t}) = \begin{bmatrix} 1 & 2 \\ 4 & 3 \end{bmatrix}\mathbf{\Phi}(\mathbf{t}),\ t \in I.$

Example 9.3.2 *For the IVP* $\mathbf{X}' = \begin{bmatrix} 0 & 1 \\ -4 & 4 \end{bmatrix}\mathbf{X},\quad \mathbf{X}(0) = \begin{bmatrix} 1 & 0 \\ 0 & 1 \end{bmatrix}$, *obtain the* *solution of the given IVP when a FM is given by* $\mathbf{\Phi}(t) = \begin{bmatrix} 1 & t \\ 2 & 2t+1 \end{bmatrix} e^{2t}$.

Solution. *From the given information,* $\mathbf{\Phi}(0) = \begin{bmatrix} 1 & 0 \\ 2 & 1 \end{bmatrix}$ *and* $\mathbf{\Phi}^{-1}(0) = \begin{bmatrix} 1 & 0 \\ -2 & 1 \end{bmatrix}$.
By (9.10), $\mathbf{X}(t) = \mathbf{\Phi}(t)\ \mathbf{\Phi}^{-1}(0)\ \mathbf{X}(0)$

$$= e^2t \begin{bmatrix} 1-2t & t \\ -4t & 2t+1 \end{bmatrix} \begin{bmatrix} 1 & 0 \\ 0 & 1 \end{bmatrix} = e^{2t} \begin{bmatrix} 1-2t & t \\ -4t & 2t+1 \end{bmatrix}$$

is a solution to the considered problem.

Example 9.3.3 *Solve the IVP* $\mathbf{X}' = \mathbf{A}(t)\mathbf{X}$,

$\mathbf{X}(\pi) = \begin{bmatrix} 0 & 1 & 0 \\ 1 & 0 & 0 \\ 0 & 0 & 1 \end{bmatrix}$, *where* $\mathbf{A}(t) = \begin{bmatrix} i & 1 & 0 \\ 0 & i & 1 \\ 0 & 0 & i \end{bmatrix}$ *given that*
$\mathbf{\Phi}(t) = \begin{bmatrix} \sin 2t & \cos 2t & 0 \\ \cos 2t & -\sin 2t & 0 \\ 0 & 0 & 1 \end{bmatrix}$ *is a fundamental matrix.*

Solution. *By* (9.10), *it follows that*

$$\mathbf{X}(t) = \mathbf{\Phi}(t)\ \mathbf{\Phi}^{-1}(\pi)\mathbf{X}(\pi)$$

$$= \begin{bmatrix} \sin 2t & \cos 2t & 0 \\ \cos 2t & -\sin 2t & 0 \\ 0 & 0 & 1 \end{bmatrix} \begin{bmatrix} 0 & 1 & 0 \\ 1 & 0 & 0 \\ 0 & 0 & 1 \end{bmatrix} \begin{bmatrix} 0 & 1 & 0 \\ 1 & 0 & 0 \\ 0 & 0 & 1 \end{bmatrix} = \begin{bmatrix} \sin 2t & \cos 2t & 0 \\ \cos 2t & -\sin 2t & 0 \\ 0 & 0 & 1 \end{bmatrix}$$

The method of solving a perturbed LMDE of the type (9.4)

The method of solving vector-matrix differential equation was already discussed in Section 8.5. Therefore, it follows that given $\mathbf{\Phi}(\mathbf{t})$ satisfying (9.1) the solution of (9.4) is given by

$$\mathbf{Y}(t) = \mathbf{\Phi}(t)\ \mathbf{\Phi}^{-1}(t_0)\mathbf{X_0} + \int_{t_0}^{t} \mathbf{\Phi}(t)\ \mathbf{\Phi}^{-1}(s)\ \mathbf{H}(s)ds, \quad t_0, t \in I, \tag{9.11}$$

which is equivalent to

$$\mathbf{Y}(t) = \mathbf{Y}(t, t_0, \mathbf{X}_0) = \mathbf{X}(t, t_0, \mathbf{X}_0) + \int_{t_0}^{t} \mathbf{X}(t, s,\ \mathbf{H}(s))ds, \quad t_0, t \in I. \tag{9.12}$$

Here $\mathbf{X}(t, t_0, \mathbf{X}_0)$ is the solution of (9.3), which has the form $\mathbf{X}(t) = \mathbf{\Phi}(t)\mathbf{\Phi}^{-1}(t_0)\mathbf{X}_0$. The steps in obtaining (9.11) are similar to those given in

Section 8.5 and hence are omitted. As a particular case, let $\mathbf{A}(t) = \mathbf{A}$, a constant matrix, then

$$\mathbf{Y}(t) = \mathbf{Y}(t, t_0, \mathbf{X}_0) = e^{\mathbf{A}(t-t_0)}\mathbf{X_0} + \int_{t_0}^{t} e^{\mathbf{A}(t-s)}\mathbf{H}(s)ds, \quad t, t_0 \in I. \tag{9.13}$$

The relations (9.11), (9.12) and (9.13) are all called as 'variation of parameters formula(VPF)'.

Example 9.3.4 *Solve the LMDE* $\mathbf{Y}'(t) = \begin{bmatrix} 1 & -1 \\ 0 & 2 \end{bmatrix} \mathbf{Y}(t) + \begin{bmatrix} -t & 1 \\ 0 & 0 \end{bmatrix}$
$\mathbf{Y}(t_0) = \mathbf{X}_0, \ t_0, t \in I.$

Solution. $\mathbf{A} = \begin{bmatrix} 1 & -1 \\ 0 & 2 \end{bmatrix}$ *and* $\mathbf{H}(t) = \begin{bmatrix} -t & 1 \\ 0 & 0 \end{bmatrix}$. *By the VPF*

$$\mathbf{Y}(t) = \begin{bmatrix} e^{t-t_0} & e^{2(t-t_0)} \\ 0 & -2e^{2(t-t_0)} \end{bmatrix}$$
$$\mathbf{X_0} + \int_{t_0}^{t} \begin{bmatrix} e^{t-s} & e^{2(t-s)} \\ 0 & -2e^{2(t-s)} \end{bmatrix} \begin{bmatrix} -s & 1 \\ 0 & 0 \end{bmatrix} ds.$$

All the terms under the integral sign are integrable. The simplification is left to the reader. Choose $\mathbf{X}_0 = \begin{bmatrix} 1 & 1 \\ 0 & 1 \end{bmatrix}$ and complete the calculation.

EXERCISE 9.3

1. Solve the LMDEs

   (i) $\mathbf{X}' = \begin{bmatrix} 0 & 1 \\ -4 & 4 \end{bmatrix} \mathbf{X},$ (ii) $\mathbf{X}' = \begin{bmatrix} -1 & 1 & 1 \\ 0 & -2 & 3 \\ 0 & 1 & 0 \end{bmatrix} \mathbf{X},$

2. Solve the IVPs

   (i) $\mathbf{X}' = \begin{bmatrix} 0 & 1 & 0 \\ 0 & -2 & -5 \\ 0 & 1 & 2 \end{bmatrix} \mathbf{X}, \ \mathbf{X}(0) = \begin{bmatrix} 1 & 0 & 0 \\ 1 & 2 & 1 \\ 0 & 0 & 1 \end{bmatrix},$

   (ii) $\mathbf{X}' = \begin{bmatrix} 3 & 2 \\ 0 & 3 \end{bmatrix} \mathbf{X}, \ \mathbf{X}(0) = \begin{bmatrix} 1 & 0 \\ 0 & 3 \end{bmatrix}.$

3. Solve the perturbed LMDEs

   (i) $\mathbf{X}' = \begin{bmatrix} 1 & -1 \\ 0 & 2 \end{bmatrix} \mathbf{X} + \begin{bmatrix} -t & 1 \\ 0 & 0 \end{bmatrix}, \ \mathbf{X}(0) = \begin{bmatrix} 1 & 0 \\ 0 & 1 \end{bmatrix},$

   (ii) $\mathbf{X}' = \begin{bmatrix} 1 & 0 \\ 0 & -1 \end{bmatrix} \mathbf{X} + \begin{bmatrix} 0 & t \\ 0 & 1 \end{bmatrix}, \ \mathbf{X}(0) = \begin{bmatrix} 1 & 0 \\ 2 & 0 \end{bmatrix}.$

9.4 The LMDE $\mathbf{X}' = \mathbf{AXB}$

In the present section, the properties of Kronecker products have been used extensively. The $(n \times n)$ solution matrix $\mathbf{X}(t)$ of the LMDE (9.14) (given) is converted into a vector-matrix form using properties in Chapter 6. Some of the properties of interest in the subsequent sections of this chapter are the following:

1. Vec $(\mathbf{A}\,\mathbf{B}) = (\mathbf{B}^T \otimes \mathbf{I_n})\, Vec\ \mathbf{A}$
2. $exp\ (\mathbf{B}^T\, t) = \exp\ (\mathbf{B}\, t)$
3. $[(exp\ \mathbf{B}^T\, t) \otimes \mathbf{I_n}]$ Vec $\mathbf{C} =$ Vec $[\mathbf{C}\ \exp\ (\mathbf{B}\, t)]$
4. Vec $(\mathbf{AXB}) = (\mathbf{B}^T \otimes \mathbf{A})$ Vec $(\mathbf{X})$.

Here $\mathbf{I_n}$ denotes the $n \times n$ identity matrix and $\mathbf{A}, \mathbf{B}$ are $n \times n$ constant matrices. Consider an LMDE of the form

$$\mathbf{X}' = \mathbf{AXB},\ \mathbf{X}(0) = \mathbf{X}_0,\ t \in \mathbb{R}, \tag{9.14}$$

where $\mathbf{A}$ and $\mathbf{B}$ are constant matrices of appropriate order, $\mathbf{AB} \neq \mathbf{BA}$ and $\mathbf{X_0}$ is a given constant matrix. By the property (iv) mentioned above, it follows that

$$Vec(\mathbf{AXB}) = (\mathbf{B}^T \otimes \mathbf{A})\ Vec(\mathbf{X}). \tag{9.15}$$

Now, the equation (9.14) together with (9.15) takes the form

$$Vec(\mathbf{X}') = Vec(\mathbf{A}\ \mathbf{X}\ \mathbf{B}) = (\mathbf{B}^T \otimes \mathbf{A})\ Vec(\mathbf{X}). \tag{9.16}$$

Here (9.16) is a linear vector-matrix equation of the type $\mathbf{X}' = \mathbf{AX}$ with constant coefficients and has, therefore, a unique solution

$$Vec(\mathbf{X}(t)) = [\exp\ ((\mathbf{B}^T \otimes \mathbf{A})t)]\ Vec(\mathbf{X}(0)),\ t \in \mathbb{R}. \tag{9.17}$$

Observe that the RHS of (9.17) is completely known and that $\exp\ [\mathbf{B}^T \otimes \mathbf{A}\ t]$ needs a resolution. The following example explains the situation.

Example 9.4.1 *Solve the LMDE IVP*

$$\mathbf{X}' = \begin{bmatrix} 3 & -1 \\ 4 & -2 \end{bmatrix} \mathbf{X} \begin{bmatrix} 2 & 2 \\ 1 & 3 \end{bmatrix},\ \mathbf{X}(0) = \begin{bmatrix} 0 & 0 \\ 0 & 1 \end{bmatrix}.$$

Solution. *By* (9.17), *it follows that*

$$Vec(\mathbf{X}(t)) = \exp\left(\begin{bmatrix} 2 & 2 \\ 1 & 3 \end{bmatrix}^T \otimes \begin{bmatrix} 3 & -1 \\ 4 & -2 \end{bmatrix} t \right) Vec\left(\begin{bmatrix} 0 & 0 \\ 0 & 1 \end{bmatrix} \right).$$

Here, exp $((\mathbf{B}^T \otimes \mathbf{A})t)$ is a 4×4 matrix. To resolve this matrix, find the eigen-values and eigen-vectors of the matrices $\mathbf{A}$ and $\mathbf{B}$.

For matrix $\mathbf{A}$, the eigen-values are $\mu_1 = -1$ and $\mu_2 = 2$ and the eigen-vectors respectively are $\begin{bmatrix}1\\4\end{bmatrix}, \begin{bmatrix}1\\1\end{bmatrix}$.

For the matrix $\mathbf{B}$ the eigen-values are $\lambda_1 = 1$, $\lambda_2 = 4$, and the respective eigen-vectors are $\begin{bmatrix}-1\\1\end{bmatrix}, \begin{bmatrix}1\\2\end{bmatrix}$. Further, the matrix

$$\mathbf{B}^T \otimes \mathbf{A} = \begin{bmatrix} 6 & -2 & 3 & -1 \\ 8 & -4 & 4 & -2 \\ 6 & -2 & 9 & -3 \\ 8 & -4 & 12 & -6 \end{bmatrix}.$$

The eigen-values of $(\mathbf{B}^T \otimes \mathbf{A})$ are of the form $\lambda_i \, \mu_j (i, j = 1, 2)$, i.e. -4, -1, 2, 8. The respective eigen-vectors of $(\mathbf{B}^T \otimes \mathbf{A}) = \begin{bmatrix}-1 & 1\\1 & 2\end{bmatrix} \otimes \begin{bmatrix}1 & 1\\4 & 1\end{bmatrix}$ are

$$\begin{bmatrix}1\\4\\2\\8\end{bmatrix}, \begin{bmatrix}-1\\-4\\1\\4\end{bmatrix}, \begin{bmatrix}-1\\-1\\1\\1\end{bmatrix}, \begin{bmatrix}1\\1\\2\\2\end{bmatrix}$$

leading to solution vectors

$$\begin{bmatrix}-1\\-4\\1\\4\end{bmatrix} e^{-t}, \begin{bmatrix}1\\4\\2\\8\end{bmatrix} e^{-4t}, \begin{bmatrix}-1\\-1\\1\\1\end{bmatrix} e^{2t}, \begin{bmatrix}1\\1\\2\\2\end{bmatrix} e^{8t}, \quad t \in \mathbb{R}.$$

Hence the FM is given by

$$\begin{aligned} \mathbf{\Phi}(t) &= \begin{bmatrix} -e^{-t} & -e^{2t} & e^{-4t} & e^{8t} \\ -4e^{-t} & -e^{2t} & 4e^{-4t} & e^{8t} \\ e^{-t} & e^{2t} & 2\,e^{-4t} & 2e^{8t} \\ 4\,e^{-t} & e^{2t} & 8\,e^{-4t} & 2\,e^{8t} \end{bmatrix}, \textit{ leading to} \\ \mathbf{\Phi}(0) &= \begin{bmatrix} -1 & -1 & 1 & 1 \\ -4 & -1 & 4 & 1 \\ 1 & 1 & 2 & 2 \\ 4 & 1 & 8 & 2 \end{bmatrix} = \begin{bmatrix}-1 & 1\\1 & 2\end{bmatrix} \otimes \begin{bmatrix}1 & 1\\4 & 1\end{bmatrix}. \end{aligned}$$

Observe the nature of $\mathbf{\Phi}(0)$ in the Kronecker product. Hence

$$e^{(\mathbf{B}^T \otimes \mathbf{A})t} = \mathbf{\Phi}(t)\, \mathbf{\Phi}^{-1}(0) = \mathbf{\Phi}(t) \left(\begin{bmatrix}-1 & 1\\1 & 2\end{bmatrix} \otimes \begin{bmatrix}1 & 1\\4 & 1\end{bmatrix} \right)^{-1}.$$

Noting that $(\mathbf{P} \otimes \mathbf{Q})^{-1} = \mathbf{P}^{-1} \otimes \mathbf{Q}^{-1}$, it follows that

$$\begin{bmatrix}1 & 1\\4 & 1\end{bmatrix}^{-1} = \frac{-1}{3} \begin{bmatrix}1 & -1\\-4 & 1\end{bmatrix} \text{ and } \begin{bmatrix}-1 & 1\\1 & 2\end{bmatrix}^{-1} = \frac{-1}{3} \begin{bmatrix}2 & -1\\-1 & -1\end{bmatrix}.$$

Hence it follows that

$$\mathbf{\Phi}^{-1}(0) = \frac{1}{9}\begin{bmatrix} 2 & -1 \\ -1 & -1 \end{bmatrix} \otimes \begin{bmatrix} 1 & -1 \\ -4 & 1 \end{bmatrix}.$$

Hence

$$\begin{aligned} Vec(\mathbf{X}(t)) &= \exp\{(\mathbf{B}^T \otimes \mathbf{A})t\}.\ Vec(\mathbf{X}(0)). \\ &= \frac{1}{9}\begin{bmatrix} -e^{-t} & -e^{2t} & e^{-4t} & e^{8t} \\ -4e^{-t} & -e^{2t} & 4e^{-4t} & e^{8t} \\ e^{-t} & e^{2t} & 2e^{-4t} & 2e^{8t} \\ 4e^{-t} & e^{2t} & 8e^{-4t} & 2e^{8t} \end{bmatrix} \begin{bmatrix} 2 & -2 & -1 & 1 \\ -8 & 2 & 4 & -1 \\ -1 & 1 & -1 & 1 \\ 4 & -1 & 4 & -1 \end{bmatrix} \begin{bmatrix} 0 \\ 0 \\ 0 \\ 1 \end{bmatrix} \end{aligned}$$

which leads to

$$\mathbf{X(t)} = \frac{1}{9}\begin{bmatrix} -e^{-t} + e^{-2t} + e^{-4t} - e^{8t} & e^{-t} - e^{2t} + 2e^{-4t} - 2\,e^{8t} \\ -4\,e^{-t} + 4\,e^{-4t} + e^{2t} - e^{8t} & 4\,e^{-t} + 8\,e^{-4t} - e^{2t} - 2\,e^{8t} \end{bmatrix}.$$

Variation of parameters formula (VPF)

The LMDE (9.8) or (9.14), after converting into the vector-matrix form provides its solution

$$Vec(\mathbf{X(t)}) = \exp((\mathbf{B}^T \otimes \mathbf{A})\,t)\,Vec(\mathbf{X_0}),\quad t\ \in\ \mathbb{R}.$$

To get the solution of the perturbed equation (9.9) assume that the solution $Vec(\mathbf{Y}(t))$ is such that

$$Vec\,(\mathbf{Y}(t)) = \exp((\mathbf{B}^T \otimes \mathbf{A})\,t)\ Vec(\mathbf{U}(t)),\quad t\ \in\ \mathbb{R},$$

where $\mathbf{U}(t)\ \in\ C[I, \mathbb{R}^{n\times n}]$ is an unknown matrix which needs to be found. Following the steps in the VPF of vector-matrix equation studied in Section 8.5, it follows that

$$Vec(\mathbf{Y}(t)) = \exp\,((\mathbf{B}^T \otimes \mathbf{A})\,t)\ Vec(\mathbf{X_0}) + \int_{t_0}^{t} \exp\,\{(\mathbf{B}^T \otimes \mathbf{A})(t-s)\}\,Vec(\mathbf{H}(s))ds$$

which is the same as

$$Vec(\mathbf{Y}(t) = Vec(\mathbf{X}(t, t_0\ \mathbf{X}_0)) + \int_{t_0}^{t} Vec(\mathbf{X}(t, s,\ \mathbf{H}(s)))ds.$$

Here, $Vec(\mathbf{Y}(t))$ is explicitly obtained. After setting the $n^2 \times 1$ column matrix into an $n \times n$ matrix form, $\mathbf{Y}(t)$, the solution of the perturbed equation (9.9) is obtained.

EXERCISE 9.4

1. Consider the illustrative example in Section 9.4. Replace $\mathbf{X}(0)$ by $\mathbf{X}(0) = \begin{bmatrix} 1 & 0 \\ 0 & 1 \end{bmatrix}$. Find the solution to the given problem with the new initial condition.

2. Let and $\mathbf{A} = \begin{bmatrix} 1 & 0 \\ 1 & 2 \end{bmatrix}$ and $\mathbf{B} = \begin{bmatrix} 0 & 8 \\ 1 & -2 \end{bmatrix}$.

   (i) Find the eigen-values and eigen-vectors of $\mathbf{A}$ and $\mathbf{B}$;

   (ii) Find the eigen-values and eigen-vectors of $\mathbf{B}^T \otimes \mathbf{A}$;

   (iii) Determine the matrix $\mathbf{\Phi}(t)$ and $\mathbf{\Phi}(0)$;

   (iv) Use the Kronecker product method to find $\mathbf{\Phi}^{-1}(0)$;

   (v) Determine exp $(\mathbf{B}^T \otimes \mathbf{A})\, t$.

3. Solve the IVPs

   (i) $\mathbf{X}' = \begin{bmatrix} 1 & 0 \\ 1 & 2 \end{bmatrix} \mathbf{X} \begin{bmatrix} 0 & 8 \\ 1 & -2 \end{bmatrix}, \ \mathbf{X}(0) = \begin{bmatrix} 0 & 1 \\ 0 & 1 \end{bmatrix};$

   (ii) $\mathbf{X}' = \begin{bmatrix} 1 & 0 \\ 1 & 2 \end{bmatrix} \mathbf{X} \begin{bmatrix} 0 & 8 \\ 1 & -2 \end{bmatrix} + \begin{bmatrix} 0 & 1 \\ 1 & 0 \end{bmatrix}, \ \mathbf{X}(0) = \begin{bmatrix} 0 & 1 \\ 0 & 1 \end{bmatrix};$

9.5 More General LMDE

Consider the LMDE

$$\mathbf{X}' = \mathbf{A}(t)\mathbf{X} + \mathbf{X}\,\mathbf{B}(t), \ \ \mathbf{X}(t_0) = \mathbf{C}, \ \ t, t_0 \in \mathbb{R}. \tag{9.18}$$

Assume that the $\mathbf{A}(t)$ and $\mathbf{B}(t)$ are continuous noncommutative matrices of order $n \times n$ for $t \in \mathbb{R}$ and $\mathbf{C}$ is an $n \times n$ constant matrix. Under the continuity condition the existence and uniqueness of the solution of the LMDE (9.18) are guaranteed.

To solve the LMDE (9.18), assume that $\mathbf{Y}$ is the solution of the IVP for LMDE

$$\mathbf{Y}' = \mathbf{A}(t)\mathbf{Y}, \ \ \mathbf{Y}(t_0) = \mathbf{I_n}, \ \ t, t_0 \in \mathbb{R},$$

where $\mathbf{I_n}$ denotes $n \times n$ identity matrix. Further, assume that $\mathbf{Z}(t)$ satisfies

$$\mathbf{Z}' = \mathbf{B}^T(t)\mathbf{Z}, \ \ \mathbf{Z}(t_0) = \mathbf{I_n}.$$

It is clear that $\mathbf{Z}^T(t)$ satisfies the LMDE

$$[\mathbf{Z}^T(t)]' = \mathbf{Z}^T(t)\mathbf{B}(t), \ \mathbf{Z}^T(t_0) = \mathbf{I_n}.$$

The solution of LMDE (9.18) is then given by

$$\mathbf{X}(t) = \mathbf{Y}(t)\mathbf{C}\mathbf{Z}^T(t), \quad t \in \mathbb{R}. \tag{9.19}$$

Suppose, $A(t) = A$ and $B(t) = B$ are constant matrices, then

$$\mathbf{Y}(t) = e^{\mathbf{A}(t-t_0)} \text{ and } \mathbf{Z}(t) = e^{\mathbf{B}^T(t-t_0)}.$$

are solutions of $\mathbf{Y}' = \mathbf{A}\mathbf{Y}$, $\mathbf{Y}(t_0) = \mathbf{I_n}$ and $\mathbf{Z}' = \mathbf{B}^T\mathbf{Z}$, $\mathbf{Z}(t_0) = \mathbf{I_n}$, respectively. Then the solution of (9.18) is then given by

$$\mathbf{X}(t) = e^{\mathbf{A}(t-t_0)}\,\mathbf{C}[e^{\mathbf{B}^T(t-t_0)}]^T = e^{\mathbf{A}(t-t_0)}\,\mathbf{C}\,e^{\mathbf{B}(t-t_0)}. \tag{9.20}$$

Variation of parameters formula

Consider the nonhomogeneous LMDE,

$$\mathbf{U}' = \mathbf{A}(t)\mathbf{U} + \mathbf{U}\,\mathbf{B}(t) + \mathbf{H}(t), \ \mathbf{U}(t_0) = \mathbf{C}, \tag{9.21}$$

where $\mathbf{H}$ is an $n \times n$ continuous matrix on I. The solution of the linear part of (9.21) is given by (9.19). To apply the technique of variation of parameters formula, let matrix $\mathbf{C}$ in (9.21) depend on t, i.e.

$$\mathbf{U}(t) = \mathbf{Y}(t)\mathbf{C}(t)\mathbf{Z}^T(t), \quad t \in I. \tag{9.22}$$

Clearly, $\mathbf{C}(t_0) = \mathbf{C}$. Determine the matrix $\mathbf{C}(t)$ such that (9.22) satisfies (9.21). This gives,

$$\begin{aligned}
\mathbf{U}'(t) &= \mathbf{Y}'(t)\mathbf{C}(t)\mathbf{Z}^T(t) + \mathbf{Y}(t)\mathbf{C}'(t)\mathbf{Z}^T(t) + \mathbf{Y}(t)\mathbf{C}(t)[\mathbf{Z}^T(t)]' \\
&= \mathbf{A}(t)\mathbf{Y}(t)\mathbf{C}(t)\mathbf{Z}^T(t) + \mathbf{Y}(t)\mathbf{C}'(t)\mathbf{Z}^T(t) + \mathbf{Y}(t)\mathbf{C}(t)\mathbf{Z}^T(t)\mathbf{B}(t). \\
\text{Also, } \mathbf{U}'(t) &= \mathbf{A}(t)\mathbf{U}(t) + \mathbf{U}(t)\mathbf{B}(t) + \mathbf{H}(t) \\
&= \mathbf{A}(t)\mathbf{Y}(t)\mathbf{C}(t)\mathbf{Z}^T(t) + \mathbf{Y}(t)\mathbf{C}(t)\mathbf{Z}^T(t)\mathbf{B}(t) + \mathbf{H}(t).
\end{aligned}$$

Now equating the above two expressions of $\mathbf{U}'(t)$, gives,

$$\mathbf{Y}(t)\mathbf{C}'(t)\mathbf{Z}^T(t) = \mathbf{H}(t)$$

or $\mathbf{C}'(t) = \mathbf{Y}^{-1}(t)\,\mathbf{H}(t)[\mathbf{Z}^T(t)]^{-1}$. Since, $\mathbf{C}(t_0) = \mathbf{C}$, integrating between t_0 to t

$$\mathbf{C}(t) = \mathbf{C} + \int_{t_0}^{t} \mathbf{Y}^{-1}(s)\mathbf{H}(s)[\mathbf{Z}^T(s)]^{-1}ds$$

and substituting in (9.22),

$$\begin{aligned}\mathbf{U}(t) &= \mathbf{Y}(t)[\mathbf{C} + \int_{t_0}^{t} \mathbf{Y}^{-1}(s)\mathbf{H}(s)[\mathbf{Z}^T(s)]^{-1}ds]\mathbf{Z}^T(t) \\ &= \mathbf{Y}(t)\mathbf{C}\mathbf{Z}^T(t) + \int_{t_0}^{t} \mathbf{Y}(t)\mathbf{Y}^{-1}(s)\mathbf{H}(s)[\mathbf{Z}^T(s)]^{-1}\,\mathbf{Z}^T\,(t)ds\end{aligned}$$

$$= \mathbf{Y}(\mathbf{t})\mathbf{C}\mathbf{Z}^T(t) + \int_{t_0}^{t} \mathbf{Y}(t,s)\mathbf{H}(s)\mathbf{Z}^T(t,s)ds, \tag{9.23}$$

where $\mathbf{Y}(t,s) = \mathbf{Y}(t)\mathbf{Y}^{-1}(s)$ and $\mathbf{Z}^T(t,s) = [\mathbf{Z}^T(s)]^{-1}\,\mathbf{Z}^T(t)$.

In the terminology of control theory $\mathbf{Y}(t,s)$ is called a transition matrix of the equation $\mathbf{X}' = \mathbf{A}(t)\mathbf{X}$. The relation (9.23) is the desired variation of parameters formula and it is the solution of the IVP (9.21).

According to the logic followed in relation (8-22) of Chapter 8, the variation of parameters formula has the form

$$\mathbf{U}(t,t_0,\mathbf{C}) = \mathbf{X}(t,t_0,\mathbf{C}) + \int_{t_0}^{t} \mathbf{X}(t,s,\mathbf{H}(s))ds, \tag{9.24}$$

where $\mathbf{X}(t,t_0,\mathbf{C}) = \mathbf{Y}(t)\,\mathbf{C}\,\mathbf{Z}^T(t) \quad t,t_0 \in I$.

Observation.

(i) VPF (9.11) is a particular case of (9.24)

(ii) Let $\mathbf{A}(\mathbf{t}) = \mathbf{A}$ and $\mathbf{B}(\mathbf{t}) = \mathbf{B}$ be constant matrices in (9.5). Then $\mathbf{Y}(t,s) = e^{\mathbf{A}(t-s)}$ and $\mathbf{Z}^T(t,s) = e^{\mathbf{B}(t-s)}$. Now the relation (9.24) takes the form

$$\mathbf{U}(t) = e^{\mathbf{A}(t-t_0)}\mathbf{C}\,e^{\mathbf{B}(t-t_0)} + \int_{t_0}^{t} e^{\mathbf{A}(t-s)}\mathbf{H}(s)e^{\mathbf{B}(t-s)}ds. \tag{9.25}$$

(ii) The formula (9.24) and its particular form (9.25) have several applications in stability theory, control and optimization theory, and other engineering and technology problems.

Solution using Kronecker Products

To solve (9.5) when $\mathbf{A}(t)$ and $\mathbf{B}(t)$ are constant matrices. Using Kronecker products, consider the equation

$$\mathbf{X}' = \mathbf{AX} + \mathbf{XB}, \ \ \mathbf{X}(0) = \mathbf{C}. \tag{9.26}$$

Then $\begin{cases} Vec(\mathbf{X}') = Vec(\mathbf{AX} + \mathbf{XB}) & = (\mathbf{I_n} \otimes \mathbf{A} + \mathbf{B}^T \otimes \mathbf{I_n})\, Vec(\mathbf{X}), \\ Vec(\mathbf{X}(0)) = Vec(\mathbf{C}). \end{cases}$

This is a vector-matrix equation with initial condition, whose solution is given by

$$\begin{aligned} Vec(\mathbf{X}) &= exp[(\mathbf{I_n} \otimes \mathbf{A})t + (\mathbf{B}^T \otimes \mathbf{I_n})t]\, Vec(\mathbf{C}) \\ &= [exp(\mathbf{I_n} \otimes \mathbf{A})t][exp(\mathbf{B}^T \otimes \mathbf{I_n})t]\, Vec(\mathbf{C}) \\ &= Vec[exp(\mathbf{A}t)]\, Vec[\mathbf{C}\ exp(\mathbf{B}^T\, t)]. \end{aligned}$$

$$\text{Hence, } \mathbf{X}(t) = exp(\mathbf{A}t)\mathbf{C}\ exp(\mathbf{B}^T\ t), \tag{9.27}$$

using the properties of Kronecker products in Section 9.4. Let

$$\mathbf{Y}(t, t_0) = \mathbf{Y}(t)\ \mathbf{Y}^{-1}(t_0), \ \ \mathbf{Z}(t, t_0) = \mathbf{Z}(t)\ \mathbf{Z}^{-1}(t_0).$$

Define the matrix $\mathbf{\Psi}(t, t_0) = \mathbf{Z}(t, t_0) \otimes \mathbf{Y}(t, t_0)$. Then,

$$\begin{aligned} \mathbf{\Psi}'(t, t_0) &= \mathbf{Z}'(t, t_0) \otimes \mathbf{Y}(t, t_0) + \mathbf{Z}(t, t_0) \otimes \mathbf{Y}'(t, t_0) \\ &= [\mathbf{B}^T(t)\mathbf{Z}(t, t_0)]\mathbf{Y}(t, t_0) + \mathbf{Z}(t, t_0) \otimes [\mathbf{A}(t)\mathbf{Y}(t, t_0) \\ &= [\mathbf{B}^T(t)\mathbf{Z}(t, t_0)] \otimes \mathbf{I_n}\ \mathbf{Y}(t, t_0) + \mathbf{I_n}\mathbf{Z}(t, t_0) \otimes [\mathbf{A}(t)\mathbf{Y}(t, t_0)] \\ &= [\mathbf{B}^T \otimes \mathbf{I_n} + \mathbf{I_n} \otimes \mathbf{A}][\mathbf{Z}(t, t_0) \otimes \mathbf{Y}(t, t_0)] \\ &= [\mathbf{B}^T \otimes \mathbf{I_n} + \mathbf{I_n} \otimes \mathbf{A}]\mathbf{\Psi}(t, t_0). \end{aligned}$$

Hence, $\psi(t, t_0)$ satisfies the equation,

$$Vec(\mathbf{X}') = [\mathbf{B}^T \otimes \mathbf{I_n} + \mathbf{I_n} \otimes \mathbf{A}]\, Vec(\mathbf{X}).$$

Further,

$$\mathbf{\Psi}(t_0, t_0) = \mathbf{Z}(t_0, t_0) \otimes \mathbf{Y}(t_0, t_0) = \mathbf{I_n} \otimes \mathbf{I_n} = \mathbf{I_n}^2.$$

Hence, $\mathbf{\Psi}(t, t_0)$ is the transition matrix for the LMDE in (9.18).

Example 9.5.1 *Solve the matrix differential equation*

$$\mathbf{X}' = \begin{bmatrix} 3 & 2 \\ 0 & 3 \end{bmatrix} \mathbf{X} + \mathbf{X} \begin{bmatrix} 0 & 1 \\ -1 & 0 \end{bmatrix}, \ \mathbf{X}(0) = \mathbf{C},$$

where $\mathbf{C}$ *is a* 2×2 *constant matrix* $\begin{bmatrix} c_{11} & c_{12} \\ c_{21} & c_{22} \end{bmatrix}$.

Solution. *For the IVP*

$$\mathbf{X}' = \begin{bmatrix} 3 & 2 \\ 0 & 3 \end{bmatrix} \mathbf{X}, \qquad \mathbf{X}(0) = \mathbf{I_2},$$

the solution is $\begin{bmatrix} e^{3t} & 2t\, e^{3t} \\ 0 & e^{3t} \end{bmatrix}$

and for the IVP $\mathbf{Z}' = \begin{bmatrix} 0 & -1 \\ 1 & 0 \end{bmatrix} \mathbf{Z}$, $\mathbf{X}(0) = \mathbf{I_2}$, *it is* $\begin{bmatrix} \cos\, t & -\sin\, t \\ \sin\, t & \cos\, t \end{bmatrix}$
Hence, from (9.19) *the solution of the given problem is*

$$\mathbf{X}(t) = \begin{bmatrix} e^{3t} & 2t\, e^{3t} \\ 0 & e^{3t} \end{bmatrix} \begin{bmatrix} c_{11} & c_{12} \\ c_{21} & c_{22} \end{bmatrix} \begin{bmatrix} \cos\, t & -\sin\, t \\ \sin\, t & \cos\, t \end{bmatrix}.$$

Example 9.5.2 *Solve the IVP*

$$\mathbf{X}' = \begin{bmatrix} 1 & 2 \\ 4 & 3 \end{bmatrix} \mathbf{X} + \mathbf{X} \begin{bmatrix} 0 & 1 \\ -4 & 4 \end{bmatrix} + \begin{bmatrix} 1 & 1 \\ 1 & 1 \end{bmatrix}, \ \mathbf{X}(0) = \begin{bmatrix} 2 & 1 \\ 1 & 2 \end{bmatrix}.$$

Solution. *For the IVP* $\mathbf{X}' = \begin{bmatrix} 1 & 2 \\ 4 & 3 \end{bmatrix} \mathbf{X}$, $\mathbf{X}(0) = \mathbf{I_2}$ *the solution is*

$$e^{\begin{bmatrix} 1 & 2 \\ 4 & 3 \end{bmatrix} t} = \frac{1}{3} \begin{bmatrix} 2e^{-t} + e^{5t} & -e^{-t} + e^{5t} \\ -2e^{-t} + 2e^{5t} & e^{-t} + 2e^{5t} \end{bmatrix}$$

and for the IVP $\mathbf{X}' = \mathbf{X} \begin{bmatrix} 0 & 1 \\ -4 & 4 \end{bmatrix}$, $\mathbf{X}(0) = \mathbf{I_2}$, *the solution is*

$$e^{\begin{bmatrix} 0 & 1 \\ -4 & 4 \end{bmatrix} t} = \begin{bmatrix} e^{2t} - 2t\, e^{2t} & te^{2t} \\ -4te^{2t} & (1+2t)e^{2t} \end{bmatrix}.$$

By following the VPF (9.25), one gets

$$\mathbf{X}(t) = e^{\begin{bmatrix} 1 & 2 \\ 4 & 3 \end{bmatrix} t} \begin{bmatrix} 2 & 1 \\ 1 & 2 \end{bmatrix} e^{\begin{bmatrix} 0 & 1 \\ -4 & 4 \end{bmatrix} t}$$

$$+ \int_0^t e^{\begin{bmatrix} 1 & 2 \\ 4 & 3 \end{bmatrix} (t-s)} \begin{bmatrix} 1 & 1 \\ 1 & 1 \end{bmatrix} e^{\begin{bmatrix} 0 & 1 \\ -4 & 4 \end{bmatrix} (s-t)} ds$$

$$= \frac{1}{3} \int_0^t \begin{bmatrix} 2e^{-(t-s)} + e^{5(t-s)} & -e^{-(t-s)} + e^{5(t-s)} \\ -2e^{-(t-s)} + 2e^{5(t-s)} & e^{-(t-s)} + 2e^{5(t-s)} \end{bmatrix} \begin{bmatrix} 1 & 1 \\ 1 & 1 \end{bmatrix}$$

$$\begin{bmatrix} e^{2(t-s)} - 2(t-s)\, e^{2(t-s)} & (t-s)e^{2(t-s)} \\ -4(t-s)e^{2(t-s)} & (1+2(t-s))e^{2(t-s)} \end{bmatrix} ds \ (simplify!)$$

Example 9.5.3 *Find the transition matrix* $\mathbf{\Psi}(t, t_0)$ *for the LMDE*

$$\mathbf{X}' = \begin{bmatrix} 3 & 2 \\ 0 & 3 \end{bmatrix} \mathbf{X} + \mathbf{X} \begin{bmatrix} 0 & 1 \\ -1 & 0 \end{bmatrix}.$$

Solution. *From the Example 9.5.1,*

$$\begin{aligned} \mathbf{Y}(t, t_0) &= \mathbf{Y}(t)\mathbf{Y}^{-1}(t_0) = e^{3(t-t_0)} \begin{bmatrix} 1 & 2(t-t_0) \\ 0 & 1 \end{bmatrix} \quad \textit{and} \\ \mathbf{Z}(t, t_0) &= \mathbf{Z}(t)\mathbf{Z}^{-1}(t_0) = \begin{bmatrix} \cos(t-t_0) & -\sin(t-t_0) \\ \sin(t-t_0) & \cos(t-t_0) \end{bmatrix}. \end{aligned}$$

Hence,

$$\mathbf{\Psi}(t, t_0) = \begin{bmatrix} \cos(t-t_0) & -\sin(t-t_0) \\ \sin(t-t_0) & \cos(t-t_0) \end{bmatrix} \otimes e^{3(t-t_0)} \begin{bmatrix} 1 & 2(t-t_0) \\ 0 & 1 \end{bmatrix}.$$

EXERCISE 9.5

1. Solve $\mathbf{X}' = \begin{bmatrix} 1 & -1 \\ 0 & 2 \end{bmatrix} \mathbf{X} + \mathbf{X} \begin{bmatrix} 1 & 0 \\ 0 & -1 \end{bmatrix}$, $\mathbf{X}(0) = \begin{bmatrix} -2 & 0 \\ 1 & 1 \end{bmatrix}$.

2. Solve $\mathbf{X}' = \begin{bmatrix} 1 & -1 \\ 0 & 2 \end{bmatrix} \mathbf{X} + \mathbf{X} \begin{bmatrix} 1 & 0 \\ 0 & -1 \end{bmatrix} + \begin{bmatrix} t & 0 \\ 0 & -1 \end{bmatrix}$, $\mathbf{X}(0) = \begin{bmatrix} -2 & 0 \\ 1 & 1 \end{bmatrix}$.

3. Find the transition matrix for the LMDE

$$\mathbf{X}' = \begin{bmatrix} 1 & -1 \\ 0 & 2 \end{bmatrix} \mathbf{X} + \mathbf{X} \begin{bmatrix} 1 & 2 \\ 0 & 3 \end{bmatrix}.$$

9.6 A Class of LMDE of Higher Order

From results in Section 9.5, it can be observed that there exists a close relationship between the representation of solutions of the scalar differential equation $\mathbf{x}' = \mathbf{ax}$, $\mathbf{x}(0) = \mathbf{c}$ and the LMDE $\mathbf{X}' = \mathbf{AX} + \mathbf{XB}$, $\mathbf{X}(0) = \mathbf{C}$.

This approach can be extended for studying LMDE of higher orders. Consider an nth order homogeneous scalar differential equation with constant coefficients,

$$y^{(n)} + a_1 y^{(n-1)} + \cdots + a_{n-1} y' + a_n y = 0, \tag{9.28}$$

whose corresponding characteristic equation is

$$\lambda^n + a_1 \lambda^{n-1} + \cdots + a_{n-1} \lambda + a_n = 0, \tag{9.29}$$

where $a_1, a_2, \ldots, a_n$ are real numbers.

Let $\lambda_1, \lambda_2, \ldots, \lambda_n$ be the roots of the characteristic equation. The n linearly independent solutions of the scalar differential equation (9.28) are described in the following cases (refer to Section 2.7 and Section 8.6).

(a) $\lambda_i, i = 1, 2, \ldots, n$ are real and distinct.

The n solutions of (9.28) are of the form $x_i(t) = c_i e^{\lambda_i t}$, where c_i is a constant for $i = 1, 2, \ldots, n$.

(b) $\lambda_1 = \lambda$ is real and repeated r times $\lambda_i, i = n - r + 1, \ldots, n$ are real and distinct. Then the r solutions of (9.28) are given by

$$x_1(t) = c_1 e^{\lambda t},\ x_2(t) = c_2 t e^{\lambda t}, \ldots, x_r(t) = c_r t^{r-1} e^{\lambda t}$$

the remaining $(n - r)$ solutions of (9.28) are as in case (a).

(c) Let $\lambda_1 = a + ib$ be a complex root. Then $\lambda_2 = a - ib$ is also a root of the considered characteristic equation and the two linearly independent solutions are given by

$$\lambda_1(t) = c_1 e^{at} cos\, bt,\ \lambda_2(t) = c_2 e^{at} sin\, bt$$

and the other $(n - 2)$ linearly independent solutions are obtained from cases (a) and (b).

Now consider a class of LMDE related to (9.26) and is given by

$$\begin{aligned}
&\mathbf{X}^{(n)} + a_1[\mathbf{A}\mathbf{X}^{(n-1)} + \mathbf{X}^{(n-1)}\mathbf{B}] \\
&\quad + a_2[\mathbf{A}^2\mathbf{X}^{(n-2)} + 2\mathbf{A}\mathbf{X}^{(n-2)}\mathbf{B} + \mathbf{X}^{(n-2)}\mathbf{B}^2] + \ldots \\
&\quad + a_n[\mathbf{A}^n\mathbf{X} + n\mathbf{A}^{n-1}\mathbf{X}\mathbf{B} + \frac{n(n-1)}{2}\mathbf{A}^{n-2}\mathbf{X}\mathbf{B}^2 + \cdots + \mathbf{X}\mathbf{B}^n] = 0,
\end{aligned} \tag{9.30}$$

where $\mathbf{A}$ and $\mathbf{B}$ are constant matrices of order $n \times n$.

From equation (9.30), some special cases are as follows.

$$\begin{cases}
\text{For } n = 1, & \mathbf{X}' + a_1(\mathbf{AX} + \mathbf{XB}) = 0; \\
\text{For } n = 2, & \mathbf{X}'' + a_1(\mathbf{AX}' + \mathbf{X}'\mathbf{B}) + a_2(\mathbf{A^2X} + \mathbf{2AXB} + \mathbf{XB^2}) = 0; \\
\text{For } n = 3, & \mathbf{X}''' + a_1(\mathbf{AX}'' + \mathbf{X}''\mathbf{B}) + a_2(\mathbf{A^2X}' + \mathbf{2AX'B} + \mathbf{X'B^2}) \\
& +a_3(\mathbf{A}^3X + 3\mathbf{A}^2\mathbf{XB} + 3\mathbf{AXB}^2 + \mathbf{XB}^3) = 0.
\end{cases}$$

All the LMDEs of the above type have explicit solutions. Listed below are the respective nature of solutions related to (a), (b) and (c). The theoretical details are omitted.

(a_1) Relative to (a) of (9.28), the solution set for equation (9.30) is given by

$$\mathbf{X}_1(t) = e^{\lambda_1 \mathbf{A}t}\mathbf{C}_1 e^{\lambda_1 \mathbf{B}t},\ \ \mathbf{X}_2(t) = e^{\lambda_2 \mathbf{A}t}\mathbf{C}_2 e^{\lambda_2 \mathbf{B}t}, \ldots$$
$$\mathbf{X}_m(t) = e^{\lambda_m \mathbf{A}t}\mathbf{C}_m e^{\lambda_m \mathbf{B}t},$$

where $\mathbf{C_1}, \mathbf{C_2}, \ldots, \mathbf{C_m}$ are constant matrices of appropriate order. The sum of these solutions is also a solution.

(b_1) Relative to (b) of (9.28), the solution set for equation (9.30) is given by

$$e^{\lambda_{m+1}\mathbf{A}t}\mathbf{D}_1\, e^{\lambda_{m+1}\mathbf{B}t},\ te^{\lambda_{m+1}\mathbf{A}t}\,\mathbf{D}_2\, e^{\lambda_{m+1}\mathbf{B}t}, \ldots$$
$$t^{r-1}e^{\lambda_{m+1}\mathbf{A}t}\mathbf{D}_r\, e^{\lambda_{m+1}\mathbf{B}t}.$$

The sum of these solutions

$$e^{\lambda_{m+1}\mathbf{A}t}(\mathbf{D}_1 + t\mathbf{D}_2 + \cdots + t^{r-1}\mathbf{D}_r)e^{\lambda_{m+1}\mathbf{B}t}$$

is also a solution. Here $\mathbf{D_1}, \mathbf{D_2}, ..., \mathbf{D_r}$ are constant matrices of appropriate order.

(c_1) Relative to (c), the solution set is given by $e^{\mathbf{A}t} cos\ bt$ and $e^{\mathbf{A}t} sin\ bt$ where the root $\alpha = a + ib$. Here $e^{\mathbf{A}t}$ $(\mathbf{C}_1 cos\ bt + \mathbf{C}_3 sin\ bt)$ where $\mathbf{C}_1$ and $\mathbf{C}_2$ are arbitrary constant matrices of appropriate order is the set of solutions of (9.30).

The procedure for solving LMDE (9.30) is as follows. Given LMDE (9.30), determine its related scalar form (9.28). Solve it by using algebraic method and follow the scheme of writing solutions as given in (a_1), (b_1) and (c_1). The following examples, illustrate the method.

Example 9.6.1 *Solve the matrix differential equation*

$$\mathbf{X}'' + 5(\mathbf{AX}' + \mathbf{X}'\mathbf{B}) + 6(\mathbf{A}^2\mathbf{X} + 2\mathbf{AXB} + \mathbf{XB}^2) = 0.$$

Solution. *From* (9.30), *the related scalar equation* (9.28) *is given by* $y'' + 5y' + 6y = 0$ *and the characteristic equation is* $r^2 + 5r + 6 = 0$ *having characteristic roots* -3 *and* -2. *The solutions of the corresponding scalar differential equation are then given by* e^{-3t} *and* e^{-2t}. *Now following the scheme of solutions given in* (a_1), *the solution of the given LMDE is*

$$\mathbf{X}_1(t) = e^{-3\mathbf{A}t}\mathbf{C}_1 e^{-3\mathbf{B}t},\ \mathbf{X}_2(t) = e^{-2\mathbf{A}t}\mathbf{C}_2 e^{-2\mathbf{B}t},$$

where $\mathbf{C}_1$ *and* $\mathbf{C}_2$ *are arbitrary constant matrices. Hence* $\mathbf{H}(t) = \mathbf{X}_1(t) + \mathbf{X}_2(t)$ *is also a solution.*

Observation. The solution of the above LMDE is valid for any choice of matrices A and B of appropriate order.

Example 9.6.2 *Solve the LMDE*

$$\mathbf{X}'' - 2(\mathbf{AX}' + \mathbf{X}'\mathbf{B}) + (\mathbf{A}^2\mathbf{X} + 2\mathbf{AXB} + \mathbf{XB}^2) = 0. \tag{9.31}$$

Solution. *The related scalar differential equation is* $y'' - 2y' + y = 0$. *Its* *characteristic equation has* $r = 1$ *as a repeated root. The solutions of the corresponding scalar differential equation are given by* e^t *and* te^t. *Hence the solution of* (9.31) *following the scheme in* (b_1) *above, is given by*

$$\mathbf{X}(t) = e^{\mathbf{A}t}(\mathbf{C}_1 + \mathbf{C}_2 t)e^{\mathbf{B}t}. \tag{9.32}$$

In particular, let $\mathbf{A} = \begin{bmatrix} 0 & 1 \\ -1 & 0 \end{bmatrix}$ *and* $\mathbf{B} = \begin{bmatrix} 0 & 1 \\ 8 & 2 \end{bmatrix}$.

Observe that $\mathbf{AB} \neq \mathbf{BA}$. *Further,*

$$\begin{aligned} e^{\mathbf{A}t} &= \begin{bmatrix} \cos t & \sin t \\ -\sin t & \cos t \end{bmatrix} \text{ and} \\ e^{\mathbf{B}t} &= \frac{1}{6}\begin{bmatrix} 4e^{-2t} + 2e^{4t} & e^{4t} \\ 8e^{4t} - 8e^{-2t} & 4e^{4t} + 2e^{-2t} \end{bmatrix} \end{aligned}$$

Now, in view of (9.32)

$$\mathbf{X}(t) = e^{\begin{bmatrix} 0 & 1 \\ -1 & 0 \end{bmatrix}t}(\mathbf{C}_1 + \mathbf{C}_2 t)e^{\begin{bmatrix} 0 & 1 \\ 8 & -2 \end{bmatrix}t}$$

$$= \begin{bmatrix} \cos t & \sin t \\ -\sin t & \cos t \end{bmatrix}(\mathbf{C}_1 + \mathbf{C}_2\, t)\frac{1}{6}\begin{bmatrix} 4e^{-2t} + 2e^{4t} & e^{4t} - 2e^{-2t} \\ 8e^{4t} - 8e^{-2t} & 4e^{4t} - 2e^{-2t} \end{bmatrix} \tag{9.33}$$

Further, suppose that equation (9.31) *has the initial conditions at* $t = 0$ *given by*

$$\mathbf{X}(0) = \begin{bmatrix} 0 & 0 \\ 1 & 2 \end{bmatrix} \text{ and } \mathbf{X}'(0) = \begin{bmatrix} 1 & 0 \\ 0 & 1 \end{bmatrix}. \tag{9.34}$$

From (9.32) *and* (9.33), *for* $t = 0$

$$\mathbf{X}(0) = \mathbf{C}_1 = \begin{bmatrix} 0 & 0 \\ 1 & 2 \end{bmatrix} \text{ and } \mathbf{X}'(0) = \mathbf{AC}_1 + \mathbf{C}_2 + \mathbf{C}_1\mathbf{B} = \begin{bmatrix} 1 & 0 \\ 0 & 1 \end{bmatrix}.$$

Substituting for $\mathbf{A}$, $\mathbf{C}_1$ *and* $\mathbf{B}$, *and simplifying gives* $\mathbf{C}_2 = \begin{bmatrix} -8 & 0 \\ 1 & 2 \end{bmatrix}$.

Hence,

$$\mathbf{X}(t) = \frac{1}{6}\begin{bmatrix} \cos t & \sin t \\ -\sin t & \cos t \end{bmatrix}\begin{bmatrix} -8t & 0 \\ 1 & 2 + 2t \end{bmatrix}\begin{bmatrix} 4e^{-2t} + 2e^{4t} & e^{4t} - e^{-2t} \\ 8e^{4t} - 8e^{-2t} & 4e^{4t} + 2e^{-2t} \end{bmatrix}$$

is the required solution.

Example 9.6.3 *Solve the equation*

$$\mathbf{X}''' - 3(\mathbf{A}^2\mathbf{X}' + 2\mathbf{A}\mathbf{X}'\mathbf{B} + \mathbf{X}'\mathbf{B}^2) + 2(\mathbf{A}^3\mathbf{X} + 3\mathbf{A}^2\mathbf{X}\mathbf{B} + 3\mathbf{A}\mathbf{X}\mathbf{B}^2 + \mathbf{X}\mathbf{B}^3) = 0.$$

Solution. *The related scalar equation is* $y''' - 3y' + 2y = 0$ *whose characteristic equation has roots* $-2, 1$ *and 1. Following the scheme of solutions given in* (a_1) *and* (b_1) *we have*

for $\lambda = -2, \mathbf{X}_1(t) = e^{-2\mathbf{A}t}\mathbf{C}_0 e^{-2\mathbf{B}t}$;
for $\lambda = 1$, *a repeated root* $\mathbf{X}_2(t) = e^{\mathbf{A}t}(\mathbf{C}_1 + \mathbf{C}_2 t)e^{\mathbf{B}t}$.

The general solution is given by

$$\mathbf{X}(t) = e^{-2\mathbf{A}t}\mathbf{C}_0 e^{-2\mathbf{B}t} + e^{\mathbf{A}t}(\mathbf{C}_1 + \mathbf{C}_2 t)e^{\mathbf{B}t}.$$

EXERCISE 9.6

1. Solve the following LMDE
   (a) $\mathbf{X}''' - 3(\mathbf{A}\mathbf{X}'' + \mathbf{X}''\mathbf{B}) + 3(\mathbf{A}^2\mathbf{X}' + 2\mathbf{A}\mathbf{X}'\mathbf{B} + \mathbf{X}'\mathbf{B}^2) - (\mathbf{A}^3\mathbf{X} + 3\mathbf{A}^2\mathbf{X}\mathbf{B} + 3\mathbf{A}\mathbf{X}\mathbf{B}^2 + \mathbf{X}\mathbf{B}^3) = 0$;

   (b) $\mathbf{X}''' - (\mathbf{A}\mathbf{X}'' + \mathbf{X}''\mathbf{B}) - 12(\mathbf{A}^2\mathbf{X}' + 2\mathbf{A}\mathbf{X}'\mathbf{B} + \mathbf{X}'\mathbf{B}^2) = 0$

   $$\text{where } \mathbf{A} = \begin{bmatrix} 1 & -1 \\ 0 & 2 \end{bmatrix} \text{ and } \mathbf{B} = \begin{bmatrix} 1 & 2 \\ 0 & 3 \end{bmatrix}.$$

2. Solve the IVP
   $$\mathbf{X}'' - 2(\mathbf{A}\mathbf{X}' + \mathbf{X}'\mathbf{B}) + (\mathbf{A}^2\mathbf{X} + 2\mathbf{A}\mathbf{X}\mathbf{B} + \mathbf{X}\mathbf{B}^2) = 0,$$
   $$\mathbf{X}(0) = \begin{bmatrix} 0 & 0 \\ 2 & 2 \end{bmatrix} \text{ and } \mathbf{X}'(0) = \begin{bmatrix} 0 & 1 \\ 1 & 0 \end{bmatrix},$$
   $$\text{where } \mathbf{A} = \begin{bmatrix} 0 & 1 \\ -1 & 0 \end{bmatrix} \text{ and } \mathbf{B} = \begin{bmatrix} 1 & -1 \\ 0 & 2 \end{bmatrix}.$$

9.7 Boundary Value Problem of LMDE

In this section an attempt will be made to apply the material developed in the earlier sections. In several real-life situations boundary value problems (BVP) arise naturally in applications in control engineering, dynamical systems and ecological systems. For example, BVP can be used to model the following:

(i) The trajectory of a satellite to be placed in a geocentric orbit.

(ii) The trajectory formed by shooting a flying plane with a missile can be modeled as a BVP.

When the parameters are many in any physical situation, then it can be formulated as BVP of LMDE.
Consider a BVP of an LMDE given by,

$$\mathbf{X}' = \mathbf{A}(t)\mathbf{X} + \mathbf{X}\mathbf{B}(t), +\mathbf{H}(t), \tag{9.35}$$

where $\mathbf{A}(t), \mathbf{B}(t), \mathbf{H}(t), \mathbf{X} \in \mathbf{C}[I, \mathbb{R}^{n\times n}]$ with $I = [a, b]$, a≤b, satisfying the boundary condition

$$\mathbf{M}\mathbf{X}(a) + \mathbf{N}\mathbf{X}(b) = \mathbf{K}, \tag{9.36}$$

where $\mathbf{M}, \mathbf{N}$ and $\mathbf{K}$ are given $n \times n$ constant matrices.

Definition 9.7.1 *A solution of the BVP of the LMDE* (9.35) *and* (9.36) *is a function* $\mathbf{X} \in \mathbf{C}[I, \mathbb{R}^{n\times n}]$ *satisfying both the relations* (9.35) *and* (9.36).

Procedure to find the solution of the BVP (9.35) and (9.36)

By the VPF (9.25), the explicit solution of (9.35) is given by

$$\mathbf{X}(t) = \mathbf{Y}(t)\mathbf{C}\mathbf{Z}^T(t) + \int_a^t \mathbf{Y}(t)\mathbf{Y}^{-1}(s)\mathbf{H}(s)[\mathbf{Z}(t)\mathbf{Z}^{-1}(s)]^T\, ds, \tag{9.37}$$

where $\mathbf{C}$ is an arbitrary $n \times n$ constant matrix. Substituting (9.37) in (9.36) gives

$$\mathbf{M}\mathbf{Y}(a)\mathbf{C}\mathbf{Z}^T(a)+\mathbf{N}\mathbf{Y}(b)\mathbf{C}\mathbf{Z}^T(b)+\mathbf{N}\int_a^b \mathbf{Y}(b)\mathbf{Y}^{-1}(s)\mathbf{H}(s)[\mathbf{Z}(b)\mathbf{Z}^{-1}(s)]^T ds = \mathbf{K}. \tag{9.38}$$

To find the unknown matrix $\mathbf{C}$, set $\mathbf{A}_1 = \mathbf{M}\mathbf{Y}(a)$, $\mathbf{A}_2 = \mathbf{N}\mathbf{Y}(b)$, $\mathbf{B}_1 = \mathbf{Z}^T(a)$, $\mathbf{B}_2 = \mathbf{Z}^T(b)$ and $\mathbf{P} = \mathbf{K} - \mathbf{N}\int_a^b \mathbf{Y}(b)\mathbf{Y}^{-1}(s)\, \mathbf{H}(s)[\mathbf{Z}(b)\mathbf{Z}^{-1}(s)]^T ds$.

With this notation it follows that

$$\mathbf{A}_1\mathbf{C}\mathbf{B}_1 + \mathbf{A}_2\mathbf{C}\mathbf{B}_2 = \mathbf{P}, \tag{9.39}$$

where $\mathbf{C}$ is an unknown matrix to be determined. Using the following properties of the Kronecker Product (see relation 9.15),
$Vec\ (\mathbf{A}_1\mathbf{C}\mathbf{B}_1) = (\mathbf{B}_1^T \otimes \mathbf{A}_1)\ Vec\ \mathbf{C}$, and
$Vec\ (\mathbf{A}_2\mathbf{C}\mathbf{B}_2) = (\mathbf{B}_2^T \otimes \mathbf{A}_2)\ Vec\ \mathbf{C}$,
relation (9.39) becomes

$$[(\mathbf{B}_1^T \otimes \mathbf{A}_1) + (\mathbf{B}_2^T \otimes \mathbf{A}_2)]\ Vec\ \mathbf{C} = Vec\ \mathbf{P}$$

Thus

$$Vec\ \mathbf{C} = [(\mathbf{B}_1^T \otimes \mathbf{A}_1) + (\mathbf{B}_2^T \otimes \mathbf{A}_2)]^{-1}\ Vec\ \mathbf{P}, \tag{9.40}$$

whenever the inverse in (9.40) exists, the RHS is known therefore the LHS, Vec $\mathbf{C}$ is known and hence $\mathbf{C}$ can be expressed. Replacing the value of $\mathbf{C}$ in (9.37) gives the explicit solution $\mathbf{X}(t)$ satisfying (9.35) and (9.36).

The following example illustrates the methodology for solving BVPs.

Example 9.7.1 *Solve the BVP* $\mathbf{X}' = \begin{bmatrix} 1 & 0 \\ 0 & 1 \end{bmatrix} \mathbf{X} + \mathbf{X} \begin{bmatrix} 1 & 0 \\ 0 & -1 \end{bmatrix}$,

$$\begin{bmatrix} 1 & 1 \\ 0 & 1 \end{bmatrix} \mathbf{X}(0) + \begin{bmatrix} 0 & 1 \\ 1 & 0 \end{bmatrix} \mathbf{X}(1) = \begin{bmatrix} 1 & 0 \\ 0 & -1 \end{bmatrix}, \quad t \in [0, 1].$$

Solution. *Comparing the given problem with the BVP of LMDE (9.35) and (9.36),* $a = 0, b = 1, \mathbf{A} = \begin{bmatrix} 1 & 0 \\ 0 & 1 \end{bmatrix}$, $\mathbf{B} = \begin{bmatrix} 1 & 0 \\ 0 & -1 \end{bmatrix}$, $\mathbf{M} = \begin{bmatrix} 1 & 1 \\ 0 & 1 \end{bmatrix}$, $\mathbf{N} = \begin{bmatrix} 0 & 1 \\ 1 & 0 \end{bmatrix}$

$$\begin{aligned} \mathbf{K} &= \begin{bmatrix} 1 & 0 \\ 0 & -1 \end{bmatrix}, \mathbf{H} = \begin{bmatrix} 0 & 0 \\ 0 & 0 \end{bmatrix}. \text{ Clearly,} \\ \mathbf{Y}(t) &= \begin{bmatrix} e^t & 0 \\ 0 & e^t \end{bmatrix}, \ \mathbf{Z}(t) = \begin{bmatrix} e^t & 0 \\ 0 & e^{-t} \end{bmatrix}, \ \mathbf{P} = \mathbf{K} = \begin{bmatrix} 1 & 0 \\ 0 & -1 \end{bmatrix}. \\ \mathbf{A}_1 &= \mathbf{M}\,\mathbf{Y}(0) = \begin{bmatrix} 1 & 1 \\ 0 & 1 \end{bmatrix}, \ \mathbf{A}_2 = \begin{bmatrix} 0 & e \\ e & 0 \end{bmatrix}, \ \mathbf{B}_1 = \begin{bmatrix} 1 & 0 \\ 0 & 1 \end{bmatrix}. \\ \mathbf{B}_2 &= \begin{bmatrix} e & 0 \\ 0 & e^{-1} \end{bmatrix} \text{ by (9.40).} \end{aligned}$$

Further,

$$\begin{aligned} Vec\,\mathbf{C} &= \left\{ \begin{bmatrix} 1 & 0 \\ 0 & 1 \end{bmatrix} \otimes \begin{bmatrix} 1 & 1 \\ 0 & 1 \end{bmatrix} + \begin{bmatrix} e & 0 \\ 0 & e^{-1} \end{bmatrix} \otimes \begin{bmatrix} 0 & e \\ e & 0 \end{bmatrix} \right\} Vec\,\mathbf{P} \\ &= \begin{bmatrix} 1 & 1+e^2 & 0 & 0 \\ e^2 & 1 & 0 & 0 \\ 0 & 0 & 1 & 2 \\ 0 & 0 & 1 & 1 \end{bmatrix} \begin{bmatrix} 1 \\ 0 \\ 0 \\ -1 \end{bmatrix}. \end{aligned}$$

Hence,

$$\mathbf{C} = \begin{bmatrix} 1 & -2 \\ e^2 & -1 \end{bmatrix}.$$

In view of (9.37) *the solution of the given BVP is given by*

$$\mathbf{X}(t) = \begin{bmatrix} e^t & te^t \\ e^t & te^t + e^t \end{bmatrix} \begin{bmatrix} 1 & -2 \\ e^2 & -1 \end{bmatrix} \begin{bmatrix} e^t & 0 \\ 0 & e^{-t} \end{bmatrix}.$$

Observation. (i) The boundary condition (9.36) may occur in several types. Only one type is given here.
(ii) In most of the cases, finding the matrix **C** from the relation (9.40) manually is very difficult. Using programming tools the problems can be resolved.

9.8 Trigonometric and Hyperbolic Matrix Functions

In this section an LMDE is considered and the solutions are obtained in terms of matrix trigonometric functions.

Following a similar approach as in Section 9.6, a scalar linear differential equation is considered and its solutions are obtained in terms of trigonometric and hyperbolic functions.

Consider the IVP of the linear scalar differential equation

$$x'' + x = 0,\ x(0) = 1,\ x'(0) = 1, \quad t \in \mathbb{R}. \tag{9.41}$$

Clearly, $\cos t$ and $\sin t$ are solutions of the IVP (9.41).

Transforming the IVP (9.41) to a LSDE or a linear system of differential equations gives,

$$\begin{bmatrix} x_1 \\ x_2 \end{bmatrix}' = \begin{bmatrix} 0 & -1 \\ 1 & 0 \end{bmatrix} \begin{bmatrix} x_1 \\ x_2 \end{bmatrix}, \ \begin{bmatrix} x_1(0) \\ x_2(0) \end{bmatrix} = \begin{bmatrix} 1 \\ 0 \end{bmatrix}. \tag{9.42}$$

Properties of solutions of the LSDE (9.42)

Suppose $x_1(t)$ and $x_2(t)$ are solutions of the LSDE (9.42).

(i) Consider $\frac{d}{dt}(x_1^2 + x_2^2) = 2x_1x_1' + 2x_2x_2' = 0$.
Integrating w.r.t. t between 0 to t, yields

$$x_1^2\ (t) + x_2^2\ (t) = 1, \tag{9.43}$$

on using (9.42) along with the initial conditions. Then the basic trigonometrical relation,

$$x_1^2(t) + x_2^2(t) = 1, \tag{9.44}$$

holds for all $t \in \mathbb{R}$

Observations.

1. The solution $cos\, t$ and $sin\, t$ of (9.41) are also the solutions of the LSDE (9.42), which can be observed by writing $x_1(t) = \cos t$ and $x_2(t) = \sin t$.
2. By considering $x_1(t)$ and $x_2(t)$ as two linearly independent solutions of the LSDE (9.41) all properties of $cos\, t$ and $sin\, t$ can be proved.

This leads to the generalization of trigonometric functions by setting $M_{20}(t) = x_1(t) = cos\ t$ and $M_{21}(t) = x_2(t) = sin\ t$.

Matrix Trigonometric Functions

For any given n × n matrix $\mathbf{T}$, the matrix trigonometric functions $cos\,\mathbf{T}$ and $sin\,\mathbf{T}$ are defined in terms of the power series,

$$cos\,\mathbf{T} = \sum_{n=0}^{\infty}(-1)^n \frac{\mathbf{T}^{2n}}{(2n)!}$$

and

$$sin\,\mathbf{T} = \sum_{n=0}^{\infty}(-1)^n \frac{\mathbf{T}^{2n+1}}{(2n+1)!}$$

with the assumption that $\|\mathbf{T}\|$ is well defined and $\|\mathbf{T}\| < \infty$.

Further, cos $\mathbf{T}$ and sin $\mathbf{T}$ are two linearly independent solutions of the matrix IVP,

$$\mathbf{X}'' + \mathbf{X} = 0, \;\; \mathbf{X}(0) = \mathbf{I}_n, \;\; \mathbf{X}'(0) = \mathbf{0}. \tag{9.45}$$

Corresponding to the LMDE (9.45) its related scalar differential equation is (9.27).

Also, the following properties hold for any two $\mathbf{T}_{n\times n}$ and $\mathbf{R}_{n\times n}$ matrices.

(ii) $cos(\mathbf{T}+\mathbf{R}) = cos(\mathbf{T})\,cos(\mathbf{R}) - sin(\mathbf{T})\,sin(\mathbf{R})$, and

$sin(\mathbf{T}+\mathbf{R}) = sin(\mathbf{T})\,cos(\mathbf{R}) + sin(\mathbf{R})\,cos(\mathbf{T})$

(iii) $e^{i\,\mathbf{T}} = cos\,\mathbf{T} + i\,sin\,\mathbf{T}$ and $e^{-i\mathbf{T}} = cos\;\mathbf{T} - i\,sin\;\mathbf{T}$;
(the above relations are Euler's formulae for matrix trigonometry)

(iv) $cos\,\mathbf{T} = \dfrac{e^{i\,\mathbf{T}} + e^{-i\,\mathbf{T}}}{2}, \; sin\,\mathbf{T} = \dfrac{e^{i\,\mathbf{T}} - e^{-i\,\mathbf{T}}}{2i}$;

(v) For any rational number n DeMoivre's Theorem for matrix trigonometry is given by,

$$(e^{i\mathbf{T}})^n = cos\,n\mathbf{T} + i\,sin\,n\mathbf{T} \tag{9.46}$$

The properties listed in (i) to (v) are well defined and well-established for matrix trigonometry.

Example 9.8.1 *Let* $\mathbf{T} = \begin{bmatrix} 0 & 1 & 0 \\ 0 & 0 & 1 \\ 0 & 0 & 0 \end{bmatrix}$. *Find* $cos\mathbf{T}$, $sin\mathbf{T}$ *and show that* $cos^2\mathbf{T} + sin^2\mathbf{T} = I_3$.

Solution. *Observe that* $\mathbf{T}^3 = 0$. *Hence* $\mathbf{T}^n = 0$ *for all* $n \geq 3$.
Thus, $cos\mathbf{T} = I_3 - \frac{\mathbf{T}^2}{2!}$, $sin\mathbf{T} = \mathbf{T}$
and

$cos^2(\mathbf{T}) + sin^2(\mathbf{T}) = \left(\boldsymbol{I_3} - \frac{\mathbf{T}^2}{2!}\right)^2 + \mathbf{T}^2 = \boldsymbol{I_3} - \mathbf{T}^2 + \frac{\mathbf{T}^4}{4} + \mathbf{T}^2 = \boldsymbol{I_3}$, *since* $T^4 = 0$.

Example 9.8.2 *Prove relation* (9.46), *for* $n = 2$, *using the* **T** *in Example 9.8.1.*

Solution. *Consider*

$$\begin{aligned}(cos\,(\mathbf{T}) + isin\,(\mathbf{T}))^2 &= \left(\boldsymbol{I_2} + i\,\mathbf{T} - \frac{\mathbf{T}^2}{2!}\right)^2 = (\boldsymbol{I_2} - 2\mathbf{T}^2) + i2\,\mathbf{T}. \\ &= cos\,(2\mathbf{T}) + i\,sin\,(2\mathbf{T}).\end{aligned}$$

Example 9.8.3 *For* $\mathbf{A} = \begin{bmatrix} 0 & 1 & 0 \\ 0 & 0 & 1 \\ 0 & 0 & 0 \end{bmatrix}$ and $\mathbf{B} = \begin{bmatrix} 1 & 1 & 3 \\ 5 & 2 & 6 \\ -2 & -1 & -3 \end{bmatrix}$

Verify that $\mathbf{AB} \neq \mathbf{BA}$ and

$$\begin{aligned} M_{20}\,(\mathbf{A}+\mathbf{B}) &= M_{20}\,(\mathbf{A})\,M_{20}\,(\mathbf{B}) - M_{21}(\mathbf{A})\,M_{21}(\mathbf{B}) \\ M_{21}\,(\mathbf{A}+\mathbf{B}) &= M_{21}(\mathbf{A})\,M_{20}(\mathbf{B}) + M_{21}(\mathbf{B})\,M_{20}\,(\mathbf{A}).\end{aligned}$$

Solution. Both **A** and **B** are nilpotent matrices $\mathbf{A}^n = 0,\ \mathbf{B}^n = 0$ for $n \geq 3$. Hence, it follows that

$$cos\,\mathbf{A} = \mathbf{I}_3 - \frac{\mathbf{A}^2}{2!},\ sin\,\mathbf{A} = \mathbf{A} \text{ and}$$
$$cos\,\mathbf{B} = \mathbf{I}_3 - \frac{\mathbf{B}^2}{2!},\ sin\,\mathbf{B} = \mathbf{B}.$$

$$\begin{aligned}\text{The LHS } &= M_{20}(\mathbf{A}+\mathbf{B}) = cos(\mathbf{A}+\mathbf{B}) = \mathbf{I}_3 - \frac{(\mathbf{A}+\mathbf{B})^2}{2!} + \frac{(\mathbf{A}+\mathbf{B})^4}{4!} \\ &= \mathbf{I}_3 - \frac{\mathbf{A}^2 + \mathbf{AB} + \mathbf{BA} + \mathbf{B}^2}{2} + \frac{\mathbf{A}^2\,\mathbf{B}^2 + \mathbf{B}^2\,\mathbf{A}^2}{\mathbf{8}}\end{aligned}$$

Here,

$$\begin{aligned}\text{RHS } &= M_{20}(\mathbf{A})M_{20}(\mathbf{B}) + M_{21}(\mathbf{A})M_{21}(\mathbf{B}) \\ &= \left(\mathbf{I}_3 - \frac{\mathbf{A}^2}{2}\right)\left(I_3 - \frac{\mathbf{B}^2}{2}\right) - \frac{\mathbf{AB}}{2} - \frac{\mathbf{BA}}{2} \\ &= \mathbf{I}_3 - \frac{\mathbf{A}^2}{2} - \frac{\mathbf{B}^2}{2} + \frac{\mathbf{A}^2\mathbf{B}^2}{\mathbf{8}} + \frac{\mathbf{B}^2\,\mathbf{A}^2}{8} - \frac{\mathbf{AB}}{2} - \frac{\mathbf{BA}}{2}.\end{aligned}$$

Hence, the first trigonometrical addition formula is proved. The second follows on similar lines.

Extended trigonometric functions

Consider a third-order IVP of the form

$$x''' + x = 0,\ x(0) = 1,\ x'(0) = 0,\ x''(0) = 0. \tag{9.47}$$

Then the three linearly independent solutions of the IVP are given by

$$M_{30}(t) = 1 - \frac{t^3}{3!} + \frac{t^6}{6!} - \frac{t^9}{9!} + \cdots = \sum_{n=0}^{\infty}(-1)^n \frac{t^{3n}}{(3n)!}$$

$$M_{31}(t) = t - \frac{t^4}{4!} + \frac{t^7}{7!} - \frac{t^{10}}{10!} + \cdots = \sum_{n=0}^{\infty}(-1)^n \frac{t^{3n+1}}{(3n+1)!}$$

$$M_{32}(t) = \frac{t^2}{2} - \frac{t^5}{5!} + \frac{t^8}{8!} - \frac{t^{11}}{11!} + \cdots = \sum_{n=0}^{\infty}(-1)^n \frac{t^{3n+2}}{(3n+2)!}$$

The system representation of the IVP (9.47) is obtained by writing

$$x_1' = -x_3,\ x_2' = x_1,\ x_3' = x_2$$

with $x_1(0) = 1,\ x_2(0) = x_3(0) = 0$ and is

$$\begin{bmatrix} x_1 \\ x_2 \\ x_3 \end{bmatrix}' = \begin{bmatrix} 0 & 0 & -1 \\ 1 & 0 & 0 \\ 0 & 1 & 0 \end{bmatrix} \begin{bmatrix} x_1 \\ x_2 \\ x_3 \end{bmatrix} \begin{bmatrix} x_1(0) \\ x_2(0) \\ x_3(0) \end{bmatrix} = \begin{bmatrix} 1 \\ 0 \\ 0 \end{bmatrix}.$$

The solutions are given by

$$x_1(t) = M_{30}(t),\ x_2(t) = M_{31}(t) \text{ and } x_3(t) = M_{32}(t).$$

Following are some of the properties satisfied by these three functions.

$$\begin{aligned}
(i)\ W(M_{30}, M_{31}, M_{32})(t) &= \begin{vmatrix} M_{30}(t) & M_{31}(t) & M_{32}(t) \\ M_{30}'(t) & M_{31}'(t) & M_{32}'(t) \\ M_{30}''(t) & M_{31}''(t) & M_{32}''(t) \end{vmatrix} \\
&= \begin{vmatrix} M_{30}(t) & M_{31}(t) & M_{32}(t) \\ -M_{32}(t) & M_{30}(t) & M_{31}(t) \\ -M_{31}(t) & -M_{32}(t) & M_{30}(t) \end{vmatrix} \\
&= M_{30}^3(t) - M_{31}^3(t) + M_{32}^3 + 3M_{30}M_{31}M_{32} \\
&= 1 \text{ for each } t.
\end{aligned}$$

This is a basic relation [similar to (9.43)] for the third order. This fact suggests that one can obtain similar basic relations for differential equations of higher orders $n = 4, 5, ..$

(ii) By following the method discussed earlier, it can be proved that addition formulae for any $t \in \mathbb{R}$,

$$\begin{aligned}
M_{30}(t+\alpha) &= M_{30}(t)\,M_{30}(\alpha) - M_{31}(t)\,M_{32}(\alpha) - M_{32}(t)\,M_{31}(\alpha)\\
M_{31}(t+\alpha) &= M_{30}(t)\,M_{31}(\alpha) + M_{31}(t)\,M_{30}(\alpha) - M_{32}(t)\,M_{32}(\alpha)\\
M_{32}(t+\alpha) &= M_{30}(t)\,M_{32}(\alpha) + M_{31}(t)\,M_{31}(\alpha) + M_{32}(t)\,M_{30}(\alpha) \quad (9.48)
\end{aligned}$$

Substituting $\alpha = t$, gives formulae for

$$M_{30}(2t), M_{31}(2t) \text{ and } M_{32}(2t).$$

(iii) The characteristic equation of the IVP (9.47) is $r^3 + 1 = 0$. The characteristic roots are $-1, \frac{1}{2}(1 - i\sqrt{3}), \frac{1}{2}(1 + i\sqrt{3})$, which are denoted by $-1, w$ and $-w^2$, respectively. Then,

$$\begin{aligned}
e^{-t} &= M_{30}(t) - M_{31}(t) + M_{32}(t)\\
e^{wt} &= M_{30}(t) + wM_{31}(t) + w^2M_{32}(t)\\
e^{-w^2t} &= -M_{30}(t) - wM_{31}(t) - w^2M_{22}(t) \quad (9.49)
\end{aligned}$$

Observation. These equations can be solved to get M_{30}, M_{31} and M_{32} in terms of e^{-t}, e^{wt} and e^{-w^2t}.

Extended matrix trigonometric functions for any $n \times n$ matrix $\mathbf{T}$ can now be defined as follows.

$$\begin{aligned}
\boldsymbol{M}_{30}(\mathbf{T}) &= \sum_{n=0}^{\infty}(-1)^n \frac{\mathbf{T}^{3n}}{(3n)!},\\
\boldsymbol{M}_{31}(\mathbf{T}) &= \sum_{n=0}^{\infty}(-1)^n \frac{\mathbf{T}^{3n+1}}{(3n+1)!},\\
\boldsymbol{M}_{32}(\mathbf{T}) &= \sum_{n=0}^{\infty}(-1)^n \frac{\mathbf{T}^{3n+2}}{(3n+2)!}. \quad (9.50)
\end{aligned}$$

These infinite series are convergent for all $\mathbf{T}$, $||\mathbf{T}|| < 1$. All the forementioned formulae can be obtained by replacing both t and α by $\mathbf{T}$, an $n \times n$ matrix.

Example 9.8.4 *For the matrix* $\mathbf{T} = \begin{bmatrix} 0 & 1 & 0 & 0\\ 0 & 0 & 1 & 0\\ 0 & 0 & 0 & 1\\ 0 & 0 & 0 & 0 \end{bmatrix}$ *find* $\boldsymbol{M}_{30}(\mathbf{T}), \boldsymbol{M}_{31}(\mathbf{T}), \boldsymbol{M}_{32}(\mathbf{T})$ *and verify the relation* (9.48).

Solution. *Here, it is observed that* $\mathbf{T}^n = 0$ *for* $n \geq 4$. *Hence from the definition,* $M_{30}(\mathbf{T}) = \mathbf{I}_4 - \frac{\mathbf{T}^3}{3!} = \begin{bmatrix} 1 & 0 & 0 & -\frac{1}{3} \\ 0 & 1 & 0 & 0 \\ 0 & 0 & 1 & 0 \\ 0 & 0 & 0 & 1 \end{bmatrix}$,

$$M_{31}(\mathbf{T}) = \mathbf{T} = \begin{bmatrix} 0 & 1 & 0 & 0 \\ 0 & 0 & 1 & 0 \\ 0 & 0 & 0 & 1 \\ 0 & 0 & 0 & 0 \end{bmatrix},$$

$$M_{32}(\mathbf{T}) = \frac{\mathbf{T}^2}{2!} = \begin{bmatrix} 0 & 0 & \frac{1}{2} & 0 \\ 0 & 0 & 0 & \frac{1}{2} \\ 0 & 0 & 0 & 0 \\ 0 & 0 & 0 & 0 \end{bmatrix}.$$

To verify the basic relation in (i) for the matrix $\mathbf{T}$,

$$\begin{aligned} \textit{Consider} \quad & M_{30}^3 - M_{31}^3 + M_{32}^3 + 3M_{30}M_{31}M_{32} \\ &= \left(\mathbf{I}_4 - \frac{\mathbf{T}^3}{3!}\right) - \mathbf{T}^3 + \frac{\mathbf{T}^6}{(2!)^3} + 3\left(\mathbf{I}_4 - \frac{\mathbf{T}^3}{3!}\right)(\mathbf{T})\left(\frac{\mathbf{T}^2}{(2!)}\right) \\ &= \mathbf{I}_4, \quad \textit{proving the result in (i)} \end{aligned}$$

Example 9.8.5 *For the matrix* $\mathbf{T}$ *as given in Example* (9.8.3), *find*

$$M_{30}(2\mathbf{T}), M_{31}(2\mathbf{T}), M_{32}(2\mathbf{T}).$$

Solution. *Replacing* t *by* $\mathbf{T}$ *and* α *by* $\mathbf{T}$ *in the relations (9.48) gives,*

$$\begin{aligned} M_{30}(2\mathbf{T}) &= M_{30}^2(\mathbf{T}) - 2M_{31}(\mathbf{T})M_{32}(\mathbf{T}) \\ &= \left(\mathbf{I}_4 - \frac{\mathbf{T}^3}{3!}\right)^2 - 2\mathbf{T}\left(\frac{\mathbf{T}^2}{2!}\right) \\ &= \mathbf{I}_4 - 2\frac{\mathbf{T}^3}{3!} + \frac{\mathbf{T}^6}{(3!)^2} - \mathbf{T}^3 = \mathbf{I}_4 - \frac{4\mathbf{T}^3}{3} = \begin{bmatrix} 1 & 0 & 0 & -\frac{4}{3} \\ 0 & 1 & 0 & 0 \\ 0 & 0 & 1 & 0 \\ 0 & 0 & 0 & 1 \end{bmatrix}; \\ M_{31}(2\mathbf{T}) &= 2M_{30}(\mathbf{T})M_{31}(\mathbf{T}) - [M_{32}(\mathbf{T})]^2 \\ &= 2\left(\mathbf{I}_4 - \frac{\mathbf{T}^3}{3!}\right)\mathbf{T} - \frac{\mathbf{T}^4}{4} = 2\mathbf{T} = \begin{bmatrix} 0 & 2 & 0 & 0 \\ 0 & 0 & 2 & 0 \\ 0 & 0 & 0 & 2 \\ 0 & 0 & 0 & 0 \end{bmatrix}; \\ M_{32}(2\mathbf{T}) &= 2M_{30}(\mathbf{T})M_{32}(\mathbf{T}) + [M_{32}(\mathbf{T})]^2 \\ &= 2\left(\mathbf{I}_4 - \frac{\mathbf{T}^3}{3!}\right)\frac{\mathbf{T}^2}{2} + \frac{\mathbf{T}^4}{4} = \mathbf{T}^2 = \begin{bmatrix} 0 & 0 & 1 & 0 \\ 0 & 0 & 0 & 1 \\ 0 & 0 & 0 & 0 \\ 0 & 0 & 0 & 0 \end{bmatrix}. \end{aligned}$$

Hyperbolic and extended hyperbolic functions

The solutions of the IVP

$$x'' - x = 0, \quad \text{with initial conditions} \;\; x(0) = 1, \; x'(0) = 0 \tag{9.51}$$

are the two linearly independent functions $\cosh t$ and $\sinh t$

$$\text{writing} \;\; N_{20}(t) = \sum_{n=0}^{\infty} \frac{t^{2n}}{(2n)!}$$

$$\text{and} \;\; N_{21}(t) = \sum_{n=0}^{\infty} \frac{t^{2n+1}}{(2n+1)!}$$

These functions can be extended as solutions of higher-order differential equations parallel to those of extended trigonometric functions.

Now replacing t with $\mathbf{T}$, an $n \times n$ matrix, the matrix hyperbolic functions are defined as follows

$$\boldsymbol{N}_{20}(\mathbf{T}) = \cosh \mathbf{T} = \sum_{n=0}^{\infty} \frac{\mathbf{T}^{2n}}{(2n)!}$$

$$\boldsymbol{N}_{21}(\mathbf{T}) = \sinh \mathbf{T} = \sum_{n=0}^{\infty} \frac{\mathbf{T}^{2n+1}}{(2n+1)!}.$$

In a similar fashion extended matrix hyperbolic functions can be obtained.

EXERCISE 9.8

1. Find $\boldsymbol{M}_{30}(\mathbf{T}), \boldsymbol{M}_{31}(\mathbf{T}), \boldsymbol{M}_{32}(\mathbf{T}), \boldsymbol{M}_{30}(2\mathbf{T}), \;\; \boldsymbol{M}_{31}(2\mathbf{T}), \boldsymbol{M}_{32}(2\mathbf{T})$ and verify the relation when

$$\mathbf{T} = \begin{bmatrix} 1 & 1 & 3 \\ 5 & 2 & 6 \\ -2 & -1 & -3 \end{bmatrix}.$$

2. Show that the matrix differential equation

$$\mathbf{X}''' + (\mathbf{A}^3\mathbf{X} + 3\mathbf{A}^2\mathbf{X}\mathbf{B} + 3\mathbf{A}\mathbf{X}\mathbf{B}^2 + \mathbf{X}\mathbf{B}^3) = 0$$

has solution set $\{\boldsymbol{M}_{30}(\mathbf{A}t)\mathbf{C}_1\boldsymbol{M}_{30}(\mathbf{B}t), \boldsymbol{M}_{31}(\mathbf{A}t)\mathbf{C}_2\boldsymbol{M}_{31}(\mathbf{B}t), \boldsymbol{M}_{31}(\mathbf{A}t)$ $\mathbf{C}_3\boldsymbol{M}_{32}(\mathbf{B}t)\}$, where $\mathbf{C}_1, \mathbf{C}_2, \mathbf{C}_3$ are arbitrary constant matrices of appropriate order.

3. Let $\mathbf{T}$ be an $n \times n$ matrix. Define

$$\begin{aligned} \boldsymbol{N}_{30}(\mathbf{T}) &= \sum_{n=0}^{\infty} \frac{\mathbf{T}^{3n}}{(3n)!}, \boldsymbol{N}_{31}(\mathbf{T}) = \sum_{n=0}^{\infty} \frac{\mathbf{T}^{3n+1}}{(3n+1)!} \text{ and} \\ \boldsymbol{N}_{32}(\mathbf{T}) &= \sum_{n=0}^{\infty} \frac{\mathbf{T}^{3n+2}}{(3n+2)!}. \end{aligned}$$

These are extended hyperbolic functions. Show that
$\boldsymbol{N}'_{30}(\mathbf{T}) = \boldsymbol{N}_{32}(\mathbf{T}), \boldsymbol{N}'_{31}(\mathbf{T}) = \boldsymbol{N}_{30}(\mathbf{T}), \boldsymbol{N}'_{32}(\mathbf{T}) = \boldsymbol{N}_{31}(\mathbf{T})$.

For $A = \begin{bmatrix} 0 & 1 & 0 & 0 \\ 0 & 0 & 1 & 0 \\ 0 & 0 & 0 & 1 \\ 0 & 0 & 0 & 0 \end{bmatrix}$,

find $\boldsymbol{N}_{30}, \boldsymbol{N}_{31}, \boldsymbol{N}_{32}$ and show that

$$\boldsymbol{N}_{30}^3 + \boldsymbol{N}_{31}^3 + \boldsymbol{N}_{32}^3 - 3\boldsymbol{N}_{30}\boldsymbol{N}_{31}\boldsymbol{N}_{32} = \boldsymbol{I}_4.$$

4. For any $n \times n$ matrices $\mathbf{T}$ and $\mathbf{R}$ such that $(\mathbf{TR} = \mathbf{RT})$, show that

$$\begin{aligned} \boldsymbol{N}_{30}(\mathbf{T}+\mathbf{R}) &= \boldsymbol{N}_{30}(\mathbf{T})\boldsymbol{N}_{30}(\mathbf{R}) + \boldsymbol{N}_{31}(\mathbf{T})\boldsymbol{N}_{32}(\mathbf{R}) + \boldsymbol{N}_{32}(\mathbf{T})\boldsymbol{N}_{31}(\mathbf{R}); \\ \boldsymbol{N}_{31}(\mathbf{T}+\mathbf{R}) &= \boldsymbol{N}_{30}(\mathbf{T})\boldsymbol{N}_{31}(\mathbf{R}) + \boldsymbol{N}_{31}(\mathbf{T})\boldsymbol{N}_{30}(\mathbf{R}) + \boldsymbol{N}_{32}(\mathbf{T})\boldsymbol{N}_{32}(\mathbf{R}); \\ \boldsymbol{N}_{32}(\mathbf{T}+\mathbf{R}) &= \boldsymbol{N}_{30}(\mathbf{T})\boldsymbol{N}_{32}(\mathbf{R}) + \boldsymbol{N}_{31}(\mathbf{T})\boldsymbol{N}_{31}(\mathbf{R}) + \boldsymbol{N}_{32}(\mathbf{T})\boldsymbol{N}_{30}(\mathbf{R}); \end{aligned}$$

Find $\boldsymbol{N}_{30}(2\mathbf{T}), \boldsymbol{N}_{31}(2\mathbf{T})$ and $\boldsymbol{N}_{32}(2\mathbf{T})$.

5. Show that the matrix differential equation

$$\begin{gathered} \mathbf{X}''' + (\mathbf{A}^3\mathbf{X} + 3\mathbf{A}^2\mathbf{X}\mathbf{B} + 3AXB^2 + \mathbf{X}\mathbf{B}^3) = 0; \\ X(0) = 1,\ X'(0) = 0, X''(0) = 0 \end{gathered}$$

has a solution set $\{N_{30}(\mathbf{A}t)\mathbf{C}_1 N_{30}(\mathbf{B}t), N_{31}(\mathbf{A}t)\mathbf{C}_2 N_{31}(\mathbf{B}t), N_{32}(\mathbf{A}t)$ $\mathbf{C}_3 N_{32}(\mathbf{B}t)\}$, where $\mathbf{C}_2, \mathbf{C}_2, \mathbf{C}_3$ are arbitrary $n \times n$ constant matrices.

9.9 Conclusion

The contents of this chapter are natural extensions of the results obtained in Chapter 8. The theory is based on the study of vectors and matrices included in Chapters 1 and 3 and Kronecker product results in Chapter 6 and techniques in calculus of matrices given in Chapter 7. The methods of explicitly solving the LMDEs are elegant, smooth and help to accommodate the wide applications in linear algebra. Applications are provided to new topics such as boundary value problems involving matrices and matrix trigonometry. For further reading the readers are refered to [2], [4], [7], [11] and [22].

Bibliography

[1] H. Anton, C. Rorres, Elementary Linear Algebra with Applications, (10th Edition), Wiley Publication, USA.

[2] S. Barnet, Matrix Differential Equations and Kronecker Products, SIAM J. Appl. Math. 24, No 1(1973) pp1–5.

[3] Richard Bellman, Modern Elementary Differential Equations, Addison Wesley, Reading, Massachusetts (1968).

[4] Richard Bellman, Introduction to Matrix Analysis, SIAM, PA, USA (1997).

[5] K. B. Datta, Matrix and Linear Algebra aided with MATLAB, (2nd Edition), PHI Learning PVT Ltd., New Delhi, India. (2008).

[6] S. G. Deo, V. Raghavendra, Rasmita Kar and V. Lakshmikantham, Text Book of Ordinary Differential Equations (3rd Edition), McGraw-Hill Education, India , New Delhi (2015).

[7] J. Foreman, Data Smart: Using Data Science to Transform Information into Insight, Wiley New York (2013).

[8] A. Graham, Kronecker Products and Matrix Calculus: With Applications (1st Edition), Ellis Horwood Limited, West Sussex, England (1981).

[9] F. S. Hill, Computer Graphics using Open GL, Pearson, New York (2000).

[10] K. Hoffman and R. Kunze, Linear Algebra (2nd Edition), Prentice Hall Inc. NJ, USA (1971).

[11] V. Lakshmikantham and S. G. Deo, Variation of parameters Methods in Dynamic System, Gordon & Breach Science Publishers, Amsterdam (1998).

[12] A. Lesk, Introduction to Bioinformatics (5th Edition), OUP Oxford (2019).

[13] J. Liesen and Z. D. Strakos, Krylov Subspace Methods – Principles and Analysis, Oxford University Press, UK (2013).

[14] B. Noble, Applied Linear Algebra, Prentice-Hall, Inc. New Jersey, USA (1969).

[15] S. K. Saha, Introduction to Robotics, McGraw-Hill Education, India (2014).

[16] D. A. Sanchez, Ordinary Differential Equations and Stability Theory: An Introduction, W. H Freeman and Company, San Francisco (1968).

[17] W. F. Stallings, Cryptography and Network Security, Pearson-Prentice Hall, NJ, USA (2005).

[18] R. G. Stanton, Numerical Methods for Science and Engineering, Prentice Hall Inc., NJ, USA (1964).

[19] G. Strang, Introduction to Linear Algebra (5th Edition), Wellesley Cambridge Press, MA, USA (2016).

[20] R. Schilling, Fundamentals of Robotics Analysis and Control, Prentice Hall, India (1996).

[21] F. E. Udwadia and R. E. Kalaba, Analytical Dynamics: A New Approach, Cambridge University Press, UK (1996).

[22] J. Vasundhara Devi, S. N. R. G. Bharat Iragavarapu and S. Srinivasa Rao, Quasilinearization Technique Periodic Boundary Value Problem for Graph Differential Equations and Its Associated Matrix Differential Equations with Dynamics of Continuous, Discrete and Impulsive Systems, Series B: Applications & Algorithms 23 (2016) pp. 287–300.

[23] W. J. Vetter, Derivative Operations on Matrices, IEEE Trans. Auto. Control. AC-15, 241–244 (1970).

[24] W. J. Vetter, Matrix Calculus Operations and Taylor Expansions, SIAM Rev. 2, 352–369 (1973).

Answers

Exercise 1.2

1. (i) $(4\ -4\ 3)$ (ii)$(5\ 0\ 10)$ (iii) $(11\ -12\ 7)$ (iv)$(9\ -12\ 3)\ -4$
2. (i) $(-1, 1, 8), 9$ (ii) $(1, 11, 3), 26$ 3. $\frac{-1}{\sqrt{41}\sqrt{105}}$

4. $\frac{(4\ 5\ 2)}{\sqrt{45}}$ 10. $\frac{7\bar{\mathbf{i}}-6\bar{\mathbf{j}}-10\bar{\mathbf{k}}}{\sqrt{185}}$ 12. 15.

Exercise 1.3.

1. (i) 4×4; (ii) 4×1; (iii) 2×3, (iv) 3×4, (v) 2×5.

2. 3 , 2 (1×11 and 11×1) 3. (i) $\begin{bmatrix} 3 & 5 & 7 \\ 4 & 6 & 8 \\ 5 & 7 & 9 \\ 6 & 8 & 10 \end{bmatrix}$ (ii) $\begin{bmatrix} 2 & -1 & -6 \\ 5 & 2 & -3 \\ 8 & 5 & 0 \\ 11 & 8 & 3 \end{bmatrix}$

(iii) $\begin{bmatrix} 0 & \frac{1}{3} & \frac{4}{3} \\ \frac{1}{3} & 0 & \frac{1}{3} \\ \frac{4}{3} & \frac{1}{3} & 0 \\ 3 & \frac{4}{3} & \frac{1}{3} \end{bmatrix}$

4. x=1, y=2, z=3 5. $\alpha = 2,\ \beta = 3,\ \gamma = 4,\ \delta = 1$

6. (i) $\begin{bmatrix} -12 & 0 & -3 \\ -3 & -9 & 6 \\ -6 & -6 & -9 \end{bmatrix}$, $\begin{bmatrix} 12 & 16 & 8 \\ -4 & 8 & 0 \\ 8 & 0 & 12 \end{bmatrix}$ (ii) $\begin{bmatrix} -19 & 21 & -4 \\ -4 & -2 & 7 \\ -23 & -15 & -5 \end{bmatrix}$

(iii) $\begin{bmatrix} 2 & -2 & -4 \\ 7 & 1 & 0 \\ 1 & 1 & 1 \end{bmatrix}$

9. $\begin{bmatrix} 3 & 8 \\ -12 & -1 \end{bmatrix}$

13. (i) $\frac{1}{5}\begin{bmatrix} -1 & -11 & -7 \\ -9 & 4 & -9 \end{bmatrix}$ (ii) $\frac{1}{\gamma}\begin{bmatrix} 3\alpha + 4\beta & 5\alpha + 2\beta & -\alpha - 5\beta \\ \alpha - 3\beta & 2\alpha + 5\beta & 5\alpha + 3\beta \end{bmatrix}$

Exercise 1.4.

7. (i) a+b−c = 0 (ii) a+b−c = 1.

Exercise 1.6.

2. 486 5. $\begin{bmatrix} x^2+y^2 & 0 & 0 & ax+by \\ 0 & x^2+a^2 & xy+ab & 0 \\ 0 & xy+ab & y^2+b^2 & 0 \\ ax+by & 0 & 0 & a^2+b^2 \end{bmatrix}$

Exercise 1.7.

2. Take $\mathbf{A} = \mathbf{O}$ 3. (i) $\begin{bmatrix} 2 & -3 \\ -3 & 5 \end{bmatrix}$ (ii) $\frac{1}{2}\begin{bmatrix} 2 & 0 \\ 1 & 1 \end{bmatrix}$ (iii) $\frac{1}{2}\begin{bmatrix} -4 & 2 \\ 3 & -1 \end{bmatrix}$

(iv) $\begin{bmatrix} 1 & 2 & -1 \\ -2 & -3 & 2 \\ 0 & -3 & 1 \end{bmatrix}$ (v). $\begin{bmatrix} 1 & -1 & -2 \\ -2 & 3 & 4 \\ -5 & 5 & 11 \end{bmatrix}$ 4. (i) $\begin{bmatrix} 9 & 5 & -1 & -2 \\ 18 & 12 & -3 & -4 \\ -2 & -1 & 1 & 0 \\ -3 & -2 & 0 & 1 \end{bmatrix}$

(ii) $\begin{bmatrix} 3 & -15 & 5 \\ -1 & 6 & -2 \\ 1 & -5 & 2 \end{bmatrix}$

5. (i) $\frac{1}{50}\begin{bmatrix} -1 & 7 \\ -7 & -1 \end{bmatrix}$ (ii) $\begin{bmatrix} \frac{3}{2} & \frac{-1}{4} & \frac{-9}{4} \\ -1 & \frac{1}{2} & \frac{3}{2} \\ \frac{1}{2} & \frac{-1}{4} & \frac{-1}{4} \end{bmatrix}$ (iii) $\begin{bmatrix} 1 & 0 & 0 \\ -3 & 1 & 1 \\ -4 & 1 & 2 \end{bmatrix}$

Exercise 1.8.

1. $\left[\begin{array}{cc|c} 3 & 0 & 8 \\ 1+a & 1+b & 1+c \\ \hline -6 & 14 & 13 \\ 1 & 9 & 12 \end{array}\right]$, $\left[\begin{array}{cc|c} 1 & 0 & -2 \\ 1-a & 1-b & 1-c \\ \hline -6 & 10 & 3 \\ 1 & -7 & -6 \end{array}\right]$

2. $\begin{bmatrix} 15 & 29 & 19 \\ 14 & 29 & 21 \\ 12 & 23 & 22 \\ 21 & 42 & 28 \\ 13 & 26 & 13 \end{bmatrix}_{5\times 3}$ 3. $\frac{1}{5}\begin{bmatrix} 8 & -7 & 2 & -4 \\ 1 & 1 & -1 & -3 \\ -13 & 12 & -2 & 9 \\ -1 & 4 & -4 & -2 \end{bmatrix}$ 4. $\begin{bmatrix} 1 \\ 2 \\ 2 \\ 1 \end{bmatrix}$

Exercise 1.9

4. $\begin{bmatrix} 2 & 12 \\ 15 & 1 \end{bmatrix}$ 5. $\begin{bmatrix} 3 & -1 \\ 12 & -8 \end{bmatrix}$

Exercise 2.3

2. (i) 2 (ii) 3 (iii) 3 (iv) 2 3. 3

Exercise 2.4

1. (a) 4 (b) 4 2. 3 3 (a) $\begin{bmatrix} 1 & 0 & 0 & 0 \\ -1 & 1 & 0 & 0 \\ -1 & 1 & 1 & 0 \\ -3 & -9 & -15 & 6 \end{bmatrix}$, $\begin{bmatrix} 1 & 0 & 2 & -1 \\ 0 & 1 & -8/6 & -1 \\ 0 & 0 & 0 & 0 \\ 0 & 0 & -1/6 & 0 \end{bmatrix}$

(b) $\begin{bmatrix} 1 & 0 & 0 \\ -2/5 & 1/5 & 0 \\ -1/3 & 0 & 1/3 \end{bmatrix}, \quad \begin{bmatrix} 1 & 1 & 1 \\ 0 & 1 & -1 \\ 0 & 0 & 1 \end{bmatrix}$

Exercise 2.5

1. $k = -1$ $x = \frac{3}{5}(z+1),\ y = \frac{-4}{5}(z+1);\quad k = 2,\ x = \frac{3}{5}z,\ y = \frac{-4}{5}z + 1$
2. $a = -2,\ b = 0$, Infinitely many solutions;
   $a \neq -2$, unique solution; $\quad a = -2,\ b \neq 0$, inconsistent.
3. (a) $x = \frac{79}{18},\ y = \frac{-997}{162},\ z = \frac{-485}{243}$. $\qquad$ (b) $x = \frac{251}{35},\ y = \frac{-69}{35},\ z = \frac{-8}{35}$.
4. $a = \frac{5}{2},\ b = \frac{51}{2}$, Infinitely many no. of solutions;
   $a \neq \frac{5}{2}$, unique solution; $\quad a = \frac{5}{2},\ b \neq \frac{51}{2}$, no solution
5. $\mathbf{x} = 0$
6. (a) $x = 3,\ y = 3,\ z = 1$ $\quad$ (b) $x = -1,\ y = -1,\ z = 1$
   (c) $x = 1,\ y = 1,\ z = 1$

Exercise 2.6

1. (i) $\begin{bmatrix} \frac{-3}{5} & \frac{4}{5} \\ \frac{2}{5} & \frac{-1}{5} \end{bmatrix}$ $\quad$ (ii) $\begin{bmatrix} -1 & \frac{-2}{3} & \frac{2}{3} \\ 2 & \frac{5}{3} & \frac{-2}{3} \\ -2 & \frac{-2}{3} & \frac{5}{3} \end{bmatrix}$ $\quad$ (iii) $\begin{bmatrix} \frac{2}{3} & \frac{-1}{3} & \frac{-1}{3} \\ 0 & 1 & 0 \\ \frac{-1}{3} & \frac{-1}{3} & \frac{2}{3} \end{bmatrix}$

   (iv) $\begin{bmatrix} 0 & \frac{1}{3} & \frac{1}{3} \\ 1 & \frac{-4}{3} & \frac{-1}{3} \\ 1 & \frac{-7}{3} & \frac{-1}{3} \end{bmatrix}$

Exercise 2.7

1.(i) $\lambda = 1,\ -2,\ 3;\quad \begin{bmatrix} 2 \\ -1 \\ 1 \end{bmatrix}, \begin{bmatrix} 1 \\ -1 \\ 1 \end{bmatrix}, \begin{bmatrix} -1 \\ 1 \\ 0 \end{bmatrix}$ $\quad$ (ii) $\lambda = 3,\ 2;\quad \begin{bmatrix} 1 \\ 3 \end{bmatrix}, \begin{bmatrix} 1 \\ 2 \end{bmatrix}$

(iii) $\lambda = 2,\ 4,\ -5;\quad \begin{bmatrix} 1 \\ 0 \\ 0 \end{bmatrix}, \begin{bmatrix} 19 \\ 2 \\ 6 \end{bmatrix}, \begin{bmatrix} -1 \\ 7 \\ 0 \end{bmatrix}$ $\quad$ (iv) $\lambda = 4,\ -1;\quad \begin{bmatrix} 3 \\ 2 \end{bmatrix}, \begin{bmatrix} -1 \\ 1 \end{bmatrix}$

(v) $\lambda = 9,\ 2;\quad \begin{bmatrix} 1 \\ 2 \end{bmatrix}, \begin{bmatrix} -1 \\ 5 \end{bmatrix},$

(vi) $\lambda = 1,\ 4,\ 6;\quad \begin{bmatrix} 1 \\ 0 \\ 0 \end{bmatrix}, \begin{bmatrix} 2 \\ 3 \\ 0 \end{bmatrix}, \begin{bmatrix} 16 \\ 25 \\ 10 \end{bmatrix}$

2. (i) $\begin{bmatrix} -1 & -1 & 1 \\ 1 & 0 & 1 \\ 0 & 1 & 1 \end{bmatrix}, \begin{bmatrix} 0 & 0 & 0 \\ 0 & 0 & 0 \\ 0 & 0 & 6 \end{bmatrix}$ $\quad$ (ii) $\begin{bmatrix} -1 & -2 & 3 \\ 3 & 1 & 0 \\ 1 & 0 & 1 \end{bmatrix}, \begin{bmatrix} 0 & 0 & 0 \\ 0 & 2 & 0 \\ 0 & 0 & 2 \end{bmatrix}$

   (iii) $\begin{bmatrix} 2 & -2 & 1 \\ 1 & 0 & -2 \\ 0 & 1 & 2 \end{bmatrix}, \begin{bmatrix} 9 & 0 & 0 \\ 0 & 9 & 0 \\ 0 & 0 & -9 \end{bmatrix}$ $\quad$ (iv) $\begin{bmatrix} 7 & 0 & 1 \\ 15 & 1 & 0 \\ 35 & 0 & 0 \end{bmatrix}, \begin{bmatrix} 6 & 0 & 0 \\ 0 & -1 & 0 \\ 0 & 0 & 1 \end{bmatrix}$

Exercise 2.8

1. (i) $\frac{3}{\sqrt{2}}$; 0, $\frac{1}{\sqrt{2}}\begin{bmatrix}1\\1\end{bmatrix}$, $\frac{1}{3}\begin{bmatrix}2\\1\\2\end{bmatrix}$ and $\frac{1}{\sqrt{2}}\begin{bmatrix}-1\\1\end{bmatrix}$, $\frac{1}{\sqrt{5}}\begin{bmatrix}-1\\2\\0\end{bmatrix}$

   (ii) $\sqrt{2}$, 0; $\begin{bmatrix}0\\\frac{1}{\sqrt{2}}\\0\\\frac{1}{\sqrt{2}}\end{bmatrix}$, $\begin{bmatrix}1\\0\end{bmatrix}$ and $\begin{bmatrix}\frac{-1}{\sqrt{2}}\\0\\\frac{1}{\sqrt{2}}\\0\end{bmatrix}$, $\begin{bmatrix}0\\1\end{bmatrix}$

Exercise 2.9

1. normal form	index	signature	nature
(i) $2u^2 + 2v^2$	2	0	positive semi definite.
(ii) $u^2 + 6v^2$	2	0	positive definite.
(iii) $-9u^2 + 9v^2 + 9w^2$	2	1	indefinite.

Exercise 3.2

4. $\begin{bmatrix}-a & -b\\-c & -d\end{bmatrix} \notin S$. S is not a vector space.

11. (i) $\mathbf{H} = \{(x,0)/x \in \mathbb{R}\}$ and $\mathbf{K} = \{(0,y)/y \in \mathbb{R}\}$

    (ii) $S_1 = \{(x,0,0)/\ x \in \mathbb{R}\}$ & $S_2 = \{(0,y,z)/\ y,z \in \mathbb{R}\}$

13. **H** and **K** are not subspaces of $\mathbb{R}^2$

15. (iii) $\mathbf{W}_1 = \{(x,0)/x \in \mathbb{R}\}$, $\mathbf{W}_2 = \{(0,y)/y \in \mathbb{R}\}$

Exercise 3.3

1. $\left\{\begin{bmatrix}2\\0\\1\end{bmatrix}, \begin{bmatrix}0\\1\\0\end{bmatrix}, \begin{bmatrix}0\\0\\1\end{bmatrix}\right\}$ 4. $\{x\}$

5. (i) $\mathbf{M}_1 = \left\{\begin{bmatrix}a&0&0\\0&0&0\\0&0&0\end{bmatrix}, \begin{bmatrix}0&0&0\\0&b&0\\0&0&0\end{bmatrix}, \begin{bmatrix}0&0&0\\0&0&0\\0&0&c\end{bmatrix} /a,b,c \in \mathbb{R}\right\}$

   (ii) & (iii) $\mathbf{M}_2 = \mathbf{M}_1 \cup \left\{\begin{bmatrix}0&d&0\\0&0&0\\0&0&0\end{bmatrix}, \begin{bmatrix}0&0&e\\0&0&0\\0&0&0\end{bmatrix}, \begin{bmatrix}0&0&0\\0&0&f\\0&0&0\end{bmatrix} /d,e,f \in \mathbb{R}\right\}$

6. (i) 2 (ii) 2 7. (a) 4 (b) 3 (c) 3 8. 4

Exercise 3.4

1. $\begin{bmatrix} \frac{1}{2} & \frac{2}{3} \\ \frac{-1}{4} & \frac{-1}{6} \end{bmatrix}$ 2. $\begin{bmatrix} 14 \\ -5 \end{bmatrix}$ 3. $\begin{bmatrix} \frac{3}{5} & \frac{-1}{5} & \frac{-1}{5} \\ \frac{1}{10} & \frac{4}{5} & \frac{3}{10} \\ \frac{-3}{10} & \frac{-7}{5} & \frac{1}{10} \end{bmatrix}$ 4. $\begin{bmatrix} 0 & 0 & -1 & 1 \\ 0 & 1 & -1 & 0 \\ 1 & 0 & 1 & -1 \\ -1 & 0 & 0 & 1 \end{bmatrix}$

5. $\begin{bmatrix} 0 & 1 & 0 & -1 \\ 2 & 0 & -2 & -1 \\ 2 & -2 & 0 & 3 \\ -2 & 3 & 2 & -1 \end{bmatrix}$

Exercise 3.5

1. 3 2. x^2 is not linear 5. rank T=3

6. (a), (b) are linear transformations, (c) is not a linear transformation. 7. $\begin{bmatrix} -4 & -3 & -30//11 & 9 & 17 \end{bmatrix}$ 8 (a)$\begin{bmatrix} -1 & 4//0 & 3 \end{bmatrix}$ (b) $\begin{bmatrix} 1 & 9//-2 & -5 \end{bmatrix}$

Exercise 3.6

1. $\begin{bmatrix} -1 & 0 \\ 0 & 1 \end{bmatrix}$ 2. $\begin{bmatrix} 1 & 0 \\ k & 1 \end{bmatrix}$ 4. $\begin{bmatrix} 3 & 2 & 0 \\ -3 & -3 & 1 \\ 3 & 3 & 3 \end{bmatrix}$ 5. $\begin{bmatrix} 0 & -1 & 1 \\ 2 & 1 & 0 \end{bmatrix}$

(7) $\begin{bmatrix} -4 & -3 & -30 \\ 11 & 9 & 17 \end{bmatrix}$

(8) (a) $\begin{bmatrix} -1 & 0 \\ 4 & 3 \end{bmatrix}$ (b) $\begin{bmatrix} 5 & 7 \\ -6 & -3 \end{bmatrix}$

Exercise 3.7

2. (i) $6||x||^2 + 36||y||^2 + 35\langle x, y\rangle$

(ii) $40x - 30y + 12\|x\|^2 + 16\langle x, y\rangle + i\left(30x + 40y - 16\|x\|^2 - 12\langle x, y\rangle\right)$

6. $\mathbf{U} = \begin{bmatrix} 1 & 0 \\ 0 & 1 \end{bmatrix}$, $\mathbf{V} = \begin{bmatrix} 0.707 & 0.707 \\ 0.707 & -0.707 \end{bmatrix}$

Exercise 3.8

(2) $A = \begin{bmatrix} 1 & -1 & 1 \\ 1 & 1 & 0 \\ -1 & 0 & 1 \end{bmatrix}$

$\frac{1}{\sqrt{3}}\begin{pmatrix} 1 \\ 1 \\ -1 \end{pmatrix}, \frac{1}{\sqrt{2}}\begin{pmatrix} -1 \\ 1 \\ 0 \end{pmatrix}, \frac{1}{\sqrt{6}}\begin{pmatrix} \frac{1}{2} \\ \frac{1}{2} \\ 1 \end{pmatrix}$

3. $x = \frac{22}{\sqrt{5}}(1 \quad -2) + \frac{14}{\sqrt{5}}(2 \ 1)$

Exercise 4.2

3. (i)$k(\mathbf{A}) = 4.2652$, (ii) $k(\mathbf{A}) = 5.795$ 4. $\begin{bmatrix} 2 \\ 1 \end{bmatrix}$, $\begin{bmatrix} 2.0046 \\ 0.99899 \end{bmatrix}$, system is well conditioned.

5. $\begin{bmatrix} -998.3333 \\ 1000 \end{bmatrix}$, $\begin{bmatrix} -2247250 \\ 2250250 \end{bmatrix}$, ill conditioned. 6. $\begin{bmatrix} 1.975 \\ 1.015 \end{bmatrix}$, well-conditioned.

Exercise 4.3

1. $x = -\frac{1}{2}$, $y = 3$, $z = \frac{-3}{4}$ 2. $\begin{bmatrix} \frac{-1}{2} & \frac{-5}{3} & \frac{7}{6} \\ 0 & \frac{7}{3} & \frac{-4}{3} \\ \frac{1}{2} & -1 & \frac{1}{2} \end{bmatrix}$

3. $\begin{bmatrix}1&0&0\\4&3&0\\5&2&4\end{bmatrix}$, $\mathbf{U}=\mathbf{L}^T$. 4. $\begin{bmatrix}1&0&0\\3&1&0\\-1&\frac{-5}{7}&1\end{bmatrix}$, $\begin{bmatrix}1&4&3\\0&-7&-8\\0&0&\frac{2}{7}\end{bmatrix}$

Exercise 4.4

1. (i) (a) (1 0.5 0.5); (b) (1 0.5 0.5) (ii) (a) (5 1 2); (b) (5 1 2).
2. (i) $z=0.5,\ y=1,\ x=1$

(ii) & (iii) The iteration scheme diverges as the equations are not strictly dominant.

3. $\begin{pmatrix}2\\3\\-1\end{pmatrix}$

Exercise 4.5

1. $\begin{bmatrix}-2.4495&-3.6742&-4.0825\\0&1.8708&1.669\\0&0&1.0911\end{bmatrix}$ 2. $\mathbf{x}=\begin{bmatrix}2\\1\\1\end{bmatrix}$

Exercise 4.6

2. $\begin{bmatrix}1&2\sqrt{2}&0&0\\2\sqrt{2}&0&\sqrt{2}&0\\0&\sqrt{2}&\frac{-1}{\sqrt{2}}&\frac{5}{2}\\0&0&\frac{5}{2}&\frac{-5}{2}\end{bmatrix}$

Exercise 4.7

1. $\begin{bmatrix}-0.3714&-0.2682&\frac{8}{9}\\-0.5571&-0.7015&\frac{-4}{9}\\-0.7428&0.6603&\frac{-1}{9}\end{bmatrix}$ 2. $\begin{bmatrix}1&0.88&2.05\\0.88&-0.54&0\\2.05&0&5.5\end{bmatrix}$

Exercise 4.8

1. (i) $9,7$; (ii) $9,7$ 2. (i) $(-8,\ 7),(-7,\ 4)$ 3. (i) $\lambda=6.1$, $[0.86\ \ 1\ \ 0.66]^T$.

(ii) $\lambda=5.98$, $\begin{bmatrix}1\\0.52\\0.44\\0.74\end{bmatrix}$

Exercise 5.2.

1. $x^2+\frac{7}{6}y^2+2xy-2x-\frac{13}{6}y=0$ 2. $x^2+y^2+z^2-2y=0$
3. $9x-3y+3z-9=0$

Exercise 5.3.

1. $\begin{bmatrix}0.55\\0.45\end{bmatrix}$, $\begin{bmatrix}0.4825\\0.5175\end{bmatrix}$, $\begin{bmatrix}0.472375\\0.527625\end{bmatrix}$, $\begin{bmatrix}0.47085625\\0.52914375\end{bmatrix}$
2. $\begin{bmatrix}0.5\\0.4\\0.1\end{bmatrix}$, $\begin{bmatrix}0.57\\0.27\\0.16\end{bmatrix}$, $\begin{bmatrix}0.589\\0.238\\0.173\end{bmatrix}$ 3. (i) $\begin{bmatrix}\frac{1}{3}\\\frac{2}{3}\end{bmatrix}$ (ii) $\begin{bmatrix}\frac{67}{89}\\\frac{22}{89}\end{bmatrix}$ 4. $\frac{5}{11}$

5. $x = 16.67\%, y = 29.17\%$ and $z = 54.16\%$.

Exercise 5.4.

1. $x = \frac{6100}{90},\ y = \frac{103400}{273},\ z = \frac{8500}{39}$ 2. $r = 1,\ w = 2,\ i = 5$
3. $x = 1000,\ y = 2200,\ z = 1800$ 4. 2%, 4%, 11% 5. $\frac{20}{3},\ 0,\ \frac{20}{3}$

Exercise 5.5.

1. encrypted message $\begin{bmatrix} 5 \\ 13 \end{bmatrix}$ 2. encrypted message $\begin{bmatrix} 0 \\ 5 \\ 12 \end{bmatrix}, \begin{bmatrix} 11 \\ 17 \\ 16 \end{bmatrix}, \begin{bmatrix} 3 \\ 10 \\ 8 \end{bmatrix}$

Exercise 5.6.

1. (3 5), (4 5), (3 2), (1 2) 2. (3 5 4), (5 4 3), (7 3 5)

3. $\begin{bmatrix} a_{11} & -a_{12} & 0 & 0 \\ a_{21} & -a_{22} & 0 & 0 \\ a_{31} & -a_{32} & 0 & 0 \\ 0 & 0 & 0 & 1 \end{bmatrix}$ 4. $\begin{bmatrix} 2 & -\frac{1}{\sqrt{2}} & \frac{3}{\sqrt{2}} & 1 \\ 1 & -\frac{1}{\sqrt{2}} & \frac{7}{\sqrt{2}} & 1 \\ 5 & \frac{3}{\sqrt{2}} & \frac{5}{\sqrt{2}} & 1 \end{bmatrix}$ 5. $\begin{bmatrix} x_1 & y_1 & 0 & 0 \\ x_2 & y_2 & 0 & 0 \\ x_3 & y_3 & 0 & 0 \end{bmatrix}$

Exercise 5.7.

1. (i) $\begin{bmatrix} 2 \\ 3.036 \\ 6.33 \\ 1 \end{bmatrix}$; (ii) $\begin{bmatrix} 3.732 \\ 0 \\ 2.464 \\ 1 \end{bmatrix}$ 2. $\begin{bmatrix} 1 & 0 & 0 & 3 \\ 0 & 1 & 0 & -6 \\ 0 & 0 & 1 & -1 \\ 0 & 0 & 1 & 1 \end{bmatrix}$ 3. $\begin{bmatrix} \frac{1}{2} & \frac{5}{2} & 2 \\ \frac{3}{2} & \frac{3}{2} & -1 \\ -\frac{1}{2} & \frac{1}{2} & 2 \end{bmatrix}, \begin{bmatrix} \frac{1}{2} \\ \frac{1}{2} \\ \frac{7}{2} \end{bmatrix}, \begin{bmatrix} 13 \\ -8 \\ 7 \end{bmatrix},$

4. $\begin{bmatrix} 1 & 0 & 0 \\ 0 & 0 & -1 \\ 0 & 1 & 0 \end{bmatrix}$ 5. $30°$ *or* $\frac{\pi}{6}$

Exercise 5.8.

1.

A	M	P	-	S
A	-	P	E	S

2.

S	P	A	-	M
-	P	A	A	M

Exercise 5.9.

1. $\begin{bmatrix} -245.51929243 & -1743.16677514 \\ -420.98861985 & -2976.93615271 \\ -99.93679855 & -708.55908509 \\ -169.13971335 & -1206.7523688 \end{bmatrix}$ 2. $\begin{bmatrix} -0.61732479 & 1.50020466 \\ -3.31444678 & -8.36360509 \\ -2.69712199 & 6.8340043 \end{bmatrix}$

Exercise 5.10.

1. $\begin{bmatrix} 77.62148014 & -3.76833911 \\ -3.49809178 & 0.20034556 \\ 0.82372516 & -0.27093975 \\ -0.82372516 & 0.27093945 \end{bmatrix}$ 2. (i) $\begin{bmatrix} -3.57366847e^0 & 4.31847612e^0 \\ -8.48478342e^0 & 9.71681260e^0 \\ 7.84596841e^0 & -9.41043273e^0 \\ -2.64511790e^3 & 3.10073903e^3 \\ 2.9348546e^0 & -4.01209624e^0 \end{bmatrix}$

Exercise 6.2.

1. $(\mathbf{P}_{1!})_{(3\times1)}(\mathbf{P}_{!1})_{(1\times3)},\ (\mathbf{P}_{1!})_{(3\times1)}(\mathbf{P}_{!2})_{(1\times3)},\ (\mathbf{P}_{3!})_{(3\times1)}(\mathbf{P}_{!1})_{(1\times3)}$.
2. $(\mathbf{P}_{11})_{(3\times3)} + (\mathbf{P}_{22})_{(3\times3)} + (\mathbf{P}_{33})_{(3\times3)}$
4. (i) $\Sigma_1^3 \mathbf{A}_{*l}\mathbf{P}_{!l}$, (ii) $\Sigma_1^3 \mathbf{P}_{k!}\mathbf{A}_{k*}^T$, (iii) $\Sigma_{i=1}^3 \Sigma_{j=1}^3 a_{ij}\mathbf{P}_{ij}$, (iv) $\mathbf{P}_{!3}\mathbf{A}\mathbf{P}_{2!}$

Exercise 6.3.

2. (i) $\begin{bmatrix} -12 & 12 & -48 & 48 \\ 24 & 0 & 96 & 0 \\ -36 & 36 & -24 & 24 \\ 72 & 0 & 48 & 0 \end{bmatrix}$ 3. (i) $\begin{bmatrix} 2 & 1 & 8 & 4 \\ 3 & 2 & 12 & 8 \\ 4 & 2 & 6 & 3 \\ 6 & 4 & 9 & 6 \end{bmatrix}$,
(ii) $\begin{bmatrix} 2 & 8 & 1 & 4 \\ 4 & 6 & 2 & 3 \\ 3 & 12 & 2 & 8 \\ 6 & 9 & 4 & 6 \end{bmatrix}$ 4. (i) $\begin{bmatrix} -2 & 2 & -1 & 1 \\ 4 & 0 & 2 & 0 \\ 0 & 0 & -1 & 1 \\ 0 & 0 & 2 & 0 \end{bmatrix}$, (ii) $\begin{bmatrix} -2 & -1 & 2 & 1 \\ 0 & -1 & 0 & 1 \\ 4 & 2 & 0 & 0 \\ 0 & 2 & 0 & 0 \end{bmatrix}$

Exercise 6.4. 1. (i) $\begin{bmatrix} -2 & 4 & 0 & 0 \\ 2 & 0 & 0 & 0 \\ -1 & 2 & -1 & 2 \\ 1 & 0 & 1 & 0 \end{bmatrix}$, (ii) $\frac{1}{4}\begin{bmatrix} 0 & 1 & 0 & -1 \\ 2 & 1 & -2 & -1 \\ 0 & 0 & 0 & 2 \\ 0 & 0 & 4 & 2 \end{bmatrix}$

2. $\begin{bmatrix} 147 & 42 & 154 & 44 \\ -126 & -21 & -132 & -22 \\ 77 & 22 & 84 & 24 \\ -66 & -11 & -72 & -12 \end{bmatrix}$ 3. $\begin{bmatrix} 2 & 3 & 6 & 9 \\ -1 & 5 & -3 & 15 \\ 8 & 12 & 4 & 6 \\ -4 & 20 & -2 & 10 \end{bmatrix}$

4. $\begin{bmatrix} 6 & -3 & -10 & 5 \\ -9 & 6 & 15 & 10 \\ -2 & 1 & 4 & -2 \\ 3 & -2 & -6 & 4 \end{bmatrix}$, 5. 400, 6. 2, $\begin{bmatrix} 0 \\ 1 \\ -1 \\ 0 \end{bmatrix}$, 3, $\begin{bmatrix} 1 \\ 1 \\ -1 \\ -1 \end{bmatrix}$, 4, $\begin{bmatrix} 0 \\ 0 \\ 1 \\ 0 \end{bmatrix}$, 6, $\begin{bmatrix} 0 \\ 0 \\ 1 \\ 1 \end{bmatrix}$

Exercise 6.5.

1.
$$\begin{bmatrix} x_1z_1 \\ x_1z_2 \\ x_2z_1 \\ x_2z_2 \end{bmatrix} = \begin{bmatrix} a_{11}b_{11} & a_{11}b_{12} & a_{12}b_{11} & a_{12}b_{12} \\ a_{11}b_{21} & a_{11}b_{22} & a_{12}b_{21} & a_{12}b_{22} \\ a_{21}b_{11} & a_{21}b_{12} & a_{22}b_{11} & a_{22}b_{12} \\ a_{21}b_{21} & a_{21}b_{22} & a_{22}b_{21} & a_{22}b_{22} \end{bmatrix} \begin{bmatrix} y_1w_1 \\ y_1w_2 \\ y_2w_1 \\ y_2w_2 \end{bmatrix}$$

2. $\begin{bmatrix} x_1z_1 \\ x_1z_2 \\ x_2z_1 \\ x_2z_2 \end{bmatrix} = \begin{bmatrix} -1 & 0 & 0 & 0 \\ 2 & -1 & 0 & 0 \\ 0 & 0 & -1 & 0 \\ 0 & 0 & 2 & -1 \end{bmatrix} \begin{bmatrix} y_1w_1 \\ y_1w_2 \\ y_2w_1 \\ y_2w_2 \end{bmatrix}$

3. $\begin{bmatrix} x_1z_1 \\ x_1z_2 \\ x_2z_1 \\ x_2z_2 \end{bmatrix} = \begin{bmatrix} 2 & 1 & 8 & 4 \\ 3 & 2 & 12 & 8 \\ 4 & 2 & 6 & 3 \\ 6 & 4 & 9 & 6 \end{bmatrix} \begin{bmatrix} y_1w_1 \\ y_1w_2 \\ y_2w_1 \\ y_2w_2 \end{bmatrix}$

Exercise 6.6. 1. $\begin{bmatrix} a_1y_1 + a_3y_2 \\ a_2y_1 + a_4y_2 \\ a_1y_3 + a_2y_4 \\ a_3y_3 + a_4y_4 \end{bmatrix} = \begin{bmatrix} b_1 \\ b_2 \\ b_3 \\ b_4 \end{bmatrix}$, 3. $\begin{bmatrix} 6y_1 & -3y_2 & 4y_3 & -2y_4 \\ 9y_1 & 12y_2 & 6y_3 & 8y_4 \\ -8y_1 & 4y_2 & 2y_3 & -y_4 \\ -12y_1 & -16y_2 & 3y_3 & 4y_4 \end{bmatrix}$,

4. $\begin{bmatrix} 2z_1 & 2z_2 & 11z_3 & 11z_4 \\ -4z_1 & 6z_2 & -22z_3 & 33z_4 \\ 7z_1 & 7z_2 & 9z_3 & 9z_4 \\ -14z_1 & 21z_2 & -18z_3 & 27z_4 \end{bmatrix}$

Exercise 6.7. 1. **U** is a 9×9 matrix, for problems 2. & 3., $\mathbf{U} = \begin{bmatrix} 1 & 0 & 0 & 0 \\ 0 & 0 & 1 & 0 \\ 0 & 1 & 0 & 0 \\ 0 & 0 & 0 & 1 \end{bmatrix}$

Exercise 6.8.

1. (i) $\frac{1}{3}\begin{bmatrix} 2e^2+e^5 & -2e^2+2e^5 \\ -e^2+e^5 & e^2+2e^5 \end{bmatrix}$ (ii) $\begin{bmatrix} \frac{e^5+2e^2}{3} & 0 & \frac{2e^5-2e^2}{3} & 0 \\ 0 & \frac{e^5+2e^2}{3} & 0 & \frac{2e^5-2e^2}{3} \\ \frac{e^5-e^2}{3} & 0 & \frac{2e^5+e^2}{3} & 0 \\ 0 & \frac{e^5-e^2}{3} & 0 & \frac{2e^5+e^2}{3} \end{bmatrix}$

2. (i) $\lambda = 3, 2$ $\begin{bmatrix} 2 \\ 1 \end{bmatrix}, \begin{bmatrix} 1 \\ 1 \end{bmatrix}$; $\mu = 6, 1$ $\begin{bmatrix} 4 \\ 1 \end{bmatrix}, \begin{bmatrix} -1 \\ 1 \end{bmatrix}$ (ii) $\begin{bmatrix} -e^3+2e^2 & e^3-e^2 \\ -2e^3+2e^2 & 2e^3-e^2 \end{bmatrix}$, $\begin{bmatrix} \frac{4e^6+e}{5} & \frac{4e^6-4e}{5} \\ \frac{e^6-e}{5} & \frac{e^6+4e}{5} \end{bmatrix}$

Exercise 6.9.

1. $\begin{bmatrix} 10 & 1 & 4 & 0 \\ 5 & 8 & 0 & 4 \\ 1 & 0 & 11 & 1 \\ 0 & 1 & 5 & 9 \end{bmatrix}$ 2. $\lambda = -2,\ 3,\ 5,\ 10;$ $\begin{pmatrix} 3 \\ -3 \\ -2 \\ 2 \end{pmatrix}, \begin{pmatrix} -1 \\ 1 \\ -1 \\ 1 \end{pmatrix}, \begin{pmatrix} -9 \\ -12 \\ 6 \\ 8 \end{pmatrix}, \begin{pmatrix} 3 \\ 4 \\ 3 \\ 4 \end{pmatrix}$

Exercise 6.10.

1. $\begin{bmatrix} 0 & 2 \\ 1 & -1 \end{bmatrix}$, 2. $\begin{bmatrix} \frac{7}{6} & \frac{1}{6} \\ \frac{-13}{6} & \frac{-1}{6} \end{bmatrix}$, 3. $\begin{bmatrix} 0 & 0 \\ 0 & 0 \end{bmatrix}$

Exercise 7.2.

1. $[3t^2 \quad -cost \quad 1]^T$, $\left[\frac{t^4}{4} \quad cost \quad \frac{t^2}{2}\right]^T + [c_1 \quad c_2 \; c_3]^T$ 2. $[e^t \quad -2sin2t \quad 5]$

3. $[te^t \quad -t\,cost + sint \quad sint]^T + [c_1 \quad c_2 \quad c_3]^T$

4. (i) $\begin{bmatrix} 0 & 1 \\ 2t & 3t^2 \end{bmatrix}$, (ii) $\begin{bmatrix} cost - t\,sint & 1 & cost \\ 0 & sint + t\,cost & e^t \\ 2t & e^t + te^t & tant + t\,sec\,2t \end{bmatrix}$

5. (i) $\begin{bmatrix} sint & \frac{t^2}{2} \\ \frac{t^3}{3} & -cost \end{bmatrix} + \begin{bmatrix} c_1 & c_3 \\ c_2 & c_4 \end{bmatrix}$

(ii) $\begin{bmatrix} t & \frac{t^2}{2} & \frac{t^3}{6} \\ sint & -cost & 0 \\ t & \ln sect & \ln(tant + sect) \end{bmatrix} + \begin{bmatrix} c_1 & c_4 & c_7 \\ c_2 & c_5 & c_8 \\ c_3 & c_6 & c_9 \end{bmatrix}$

6. $\frac{\partial f}{\partial x} = \begin{bmatrix} y & 2x & yz \end{bmatrix}$, $\frac{\partial f}{\partial y} = \begin{bmatrix} x & 2 & xz \end{bmatrix}$, $\frac{\partial f}{\partial z} = \begin{bmatrix} 0 & 2z & xy \end{bmatrix}$

7. (i) 0 (ii) 0 (iii) $cost[e^t - t\ln t[te^t + e^t]] + t^2e^t(te^t + e^t)sint + te^t[cost(\ln t + 1) - t^3e^{2t}(2sint + cost) + t^2e^tlnt(-sint + cost) - 3t^2e^{2t}sint + te^tcost(2lnt + 1) + e^t(sint - cost) + sint + cost$

Exercise 7.3.

1. $\begin{bmatrix} 3x_1^2 - 3x_2x_3 & cosx_1 \\ -3x_1x_3 + 2x_2 & -e^{x_2} \\ -3x_1x_2 & \frac{1}{x_3} \end{bmatrix}$ 2. $\begin{bmatrix} 2 & 27x_1^2 + 1 \\ 4x_2 & 2x_2 \end{bmatrix}$ 4. $r^2\ cos\phi$

Exercise 7.4.

1. (i) $\begin{bmatrix} \frac{||\mathbf{X}||x_{22} - |\mathbf{X}|}{||\mathbf{X}||^2} & \frac{-||\mathbf{X}||x_{21} - |\mathbf{X}|}{||\mathbf{X}||^2} \\ \frac{-||\mathbf{X}||x_{12} - |\mathbf{X}|}{||\mathbf{X}||^2} & \frac{||\mathbf{X}||x_{11} - |\mathbf{X}|}{||\mathbf{X}||^2} \end{bmatrix}$ (ii) $\begin{bmatrix} x_{22}||\mathbf{X}|| + |\mathbf{X}| & -x_{21}||\mathbf{X}|| + |\mathbf{X}| \\ -x_{12}||\mathbf{X}|| + |\mathbf{X}| & x_{11}||\mathbf{X}|| + |\mathbf{X}| \end{bmatrix}$

2. $\frac{\partial \mathbf{y}}{\partial \mathbf{x}} = \begin{bmatrix} x_{22}x_{33} - x_{23}x_{32} & -x_{21}x_{33} + x_{23}x_{31} & x_{21}x_{32} - x_{22}x_{31} \\ -x_{12}x_{33} + x_{13}x_{32} & -x_{13}x_{31} + x_{11}x_{33} & -x_{11}x_{32} + x_{12}x_{31} \\ -x_{13}x_{22} + x_{12}x_{23} & -x_{11}x_{23} + x_{21}x_{13} & -x_{12}x_{21} + x_{11}x_{22} \end{bmatrix}$

3. $\frac{\partial \mathbf{y}}{\partial \mathbf{x}} = \begin{bmatrix} x_{22}x_{33} - x_{23}x_{32} & -x_{21}x_{33} + x_{23}x_{31} & x_{21}x_{32} - x_{22}x_{31} \\ -x_{12}x_{33} + x_{13}x_{32} & -x_{13}x_{31} + x_{11}x_{33} & -x_{11}x_{32} + x_{12}x_{31} \\ -x_{13}x_{22} + x_{12}x_{23} & -x_{11}x_{23} + x_{21}x_{13} & -x_{12}x_{21} + x_{11}x_{22} \end{bmatrix}$

Exercise 7.5.

1. $\begin{bmatrix} 2 & 6 \\ 5 & 15 \end{bmatrix}, \begin{bmatrix} 4 & 12 \\ 6 & 18 \end{bmatrix}, \begin{bmatrix} 6 & 8 \\ 12 & 16 \end{bmatrix}$ 2. $\begin{bmatrix} x_2 & x_1 + x_4 \\ 0 & x_2 \end{bmatrix}, \begin{bmatrix} 0 & x_3 \\ x_2 & 2x_4 \end{bmatrix}$

3. $\begin{bmatrix} y_1 + x_1\frac{\partial y_1}{\partial x_1} + x_3\frac{\partial y_2}{\partial x_1} & y_3 + x_1\frac{\partial y_3}{\partial x_1} + x_3\frac{\partial y_4}{\partial x_1} \\ x_2\frac{\partial y_1}{\partial x_1} + x_4\frac{\partial y_2}{\partial x_1} & x_2\frac{\partial y_3}{\partial x_1} + x_4\frac{\partial y_4}{\partial x_1} \end{bmatrix},$

$\begin{bmatrix} x_1\frac{\partial y_1}{\partial x_2} + x_3\frac{\partial y_2}{\partial x_2} & x_1\frac{\partial y_3}{\partial x_2} + x_3\frac{\partial y_4}{\partial x_2} \\ y_1 + x_2\frac{\partial y_1}{\partial x_2} + x_4\frac{\partial y_2}{\partial x_2} & y_3 + x_2\frac{\partial y_3}{\partial x_2} + x_4\frac{\partial y_4}{\partial x_2} \end{bmatrix}$

$\begin{bmatrix} x_1\frac{\partial y_1}{\partial x_3} + y_2 + x_3\frac{\partial y_2}{\partial x_3} & x_1\frac{\partial y_3}{\partial x_3} + y_4 + x_3\frac{\partial y_4}{\partial x_3} \\ x_2\frac{\partial y_1}{\partial x_3} + x_4\frac{\partial y_2}{\partial x_3} & x_2\frac{\partial y_3}{\partial x_3} + x_4\frac{\partial y_4}{\partial x_3} \end{bmatrix},$

$\begin{bmatrix} x_1\frac{\partial y_1}{\partial x_4} + x_3\frac{\partial y_2}{\partial x_4} & x_1\frac{\partial y_3}{\partial x_4} + x_3\frac{\partial y_4}{\partial x_4} \\ x_2\frac{\partial y_1}{\partial x_4} + y_2 + x_4\frac{\partial y_2}{\partial x_4} & x_2\frac{\partial y_3}{\partial x_4} + y_4 + x_4\frac{\partial y_4}{\partial x_4} \end{bmatrix}$

Exercise 7.6.

1. (i) $\begin{bmatrix} 6x\,dx & 6y\,dy \\ 3y\,dx + 3x\,dy & 2xy\,dx + x^2\,dy \end{bmatrix}$

(ii) $\begin{bmatrix} 2x\,dx & 2y\,dy + 8\,dx \\ (y + 8x)\,dx + x\,dy & 2xy\,dx + (x^2 + 8y)\,dy \end{bmatrix}$

2. $\begin{bmatrix} (18x^2y+4)\,dx+6x^3dy & (36x^3+12x^2)\,dx \\ 32xy^2\,dx+(32x^2y+12y)dy & (72x^2y+12xy^2)\,dx+(24x^3+12x^2y)\,dy \end{bmatrix}$

Exercise 7.7.

1. $\begin{bmatrix} x_{12}+2x_{11} & 0 & x_{11} & 0 \\ sin(x_{12}+x_{11}) & 2x_{11} & sin(x_{12}+x_{11}) & 0 \\ 0 & sin(x_{21}+x_{22}) & 0 & sin(x_{21}+x_{22}) \\ 0 & x_{22} & 0 & x_{21} \end{bmatrix}$

2. $\begin{bmatrix} 1 & 0 & x_{12} & x_{11} & x_{21} & 0 \\ 0 & 0 & 0 & 0 & x_{11} & 0 \\ x_{21} & 0 & 0 & x_{22}^2 & x_{22} & 0 \\ x_{11} & 0 & 0 & 2x_{22}x_{12} & 0 & x_{11} \end{bmatrix}$

3. $\begin{bmatrix} [exp(x_{11}^2+x_{22}^2)]2x_{11} & 0 & 0 & \frac{2x_{12}}{x_{22}^2+x_{12}^2} \\ 2x_{11} & 0 & 0 & 1 \\ 0 & 0 & 2x_{22}\,exp(x_{11}^2+x_{22}^2) & \frac{2x_{22}}{x_{22}^2+x_{12}^2} \\ 0 & 0 & 2x_{22} & 2x_{22} \end{bmatrix}$

4. $\begin{bmatrix} -sin(x_{11}) & 0 & 0 & -sin(x_{12}) \\ 2x_{11}cos(x_{12}) & 2x_{11}\,sin(x_{22}) & -x_{11}^2\,sin(x_{12}) & 0 \\ 0 & cos(x_{21}) & cos(x_{22}) & 0 \\ 0 & 0 & 0 & cos(x_{22})[x_{11}^2+x_{22}]+sin(x_{22}) \end{bmatrix}$

Exercise 7.8.

1. (i) $\begin{bmatrix} -2x_{11}sin(x_{11})+2cos(x_{11}) & 2x_{12}cos(x_{22}) \\ x_{22}cos(x_{21})+2exp(x_{21}) & 0 \\ 0 & -x_{12}^2sin(x_{22}) \\ 2x_{11}exp(x_{21}) & x_{11}cos(x_{22})-x_{22}x_{11}sin(x_{22}) \end{bmatrix}$, (ii)

$\begin{bmatrix} 2x_{11}cos(x_{11})-x_{11}^2sin(x_{11}) & 2x_{11}x_{12}^2x_{21} & x_{21}sin(x_{12})+x_{12}x_{21}cos(x_{12}) & 2x_{11}^2x_{12}x_{22}+x_{21}^2 \\ -x_{12}x_{22}sin(x_{11}) & 0 & x_{22}cos(x_{11})+x_{21}^2cos(x_{12}) & 3x_{12}^2x_{22}^2 \\ x_{12}sin(x_{12}) & 2x_{12}x_{21} & 0 & x_{11}^2x_{12}^2 \\ 2x_{21}sin(x_{12}) & 3x_{21}^2 & x_{12}cos(x_{11}) & 2x_{12}^3x_{22} \end{bmatrix}$

2. (i) $\begin{bmatrix} x_{21}x_{12}x_{22} & 0 & x_{22}x_{11}x_{21} & 2x_{12}x_{22}^2 \\ 2x_{11}x_{21}^2 & x_{12}x_{21}x_{22} & 0 & x_{11}x_{21}x_{22} \\ x_{11}x_{12}x_{22} & 0 & x_{11}x_{12}x_{21} & 2x_{12}^2x_{22} \\ 2x_{11}^2x_{21} & x_{11}x_{12}x_{22} & 0 & x_{11}x_{12}x_{21} \end{bmatrix}$

(ii) $\begin{bmatrix} 2x_{11}x_{22}^2 & 2x_{11}cos(x_{11}^2) & 0 & 0 \\ 0 & 0 & 0 & 0 \\ 0 & 0 & 2x_{22}x_{11}^2 & 0 \\ 0 & 0 & -2x_{22}sin(x_{22}^2) & 4x_{22}^3 \end{bmatrix}$

Exercise 7.9.

1. $\begin{bmatrix} x_{12}+3x_{11} & 2cos(x_{21}+x_{22}) \\ 2cos(x_{12}+x_{11}) & 2x_{11}+x_{21}+x_{22}) \end{bmatrix}$

2. $\begin{bmatrix} 1 & x_{11}+x_{12} & x_{11}+x_{21}) \\ x_{11}+x_{21} & x_{22}^2+2x_{12}x_{22} & x_{11}+x_{22} \end{bmatrix}$

Exercise 8.2.

1. $\mathbf{x}' = \mathbf{A}\mathbf{x}$ and $\mathbf{x} = \begin{bmatrix} x & y & z \end{bmatrix}^T$ and $\mathbf{A}$=

(i) $\begin{bmatrix} 3 & 2 & 0 \\ 2 & 1 & 1 \\ 0 & 0 & -3 \end{bmatrix}$ (ii) $\begin{bmatrix} 2 & 5 & -9 \\ 7 & -2 & 3 \\ -1 & 5 & 7 \end{bmatrix}$ (iii) $\begin{bmatrix} a_1 & b_1 & c_1 \\ a_2 & b_2 & c_2 \\ a_3 & b_3 & c_3 \end{bmatrix}$

2. (i) $\begin{bmatrix} x_1 \\ x_2 \end{bmatrix}' = \begin{bmatrix} 0 & 1 \\ -w^2 & -a \end{bmatrix} \begin{bmatrix} x_1 \\ x_2 \end{bmatrix}$, (ii) $\begin{bmatrix} x_1 \\ x_2 \\ x_3 \end{bmatrix}' = \begin{bmatrix} 0 & 1 & 0 \\ 0 & 0 & 1 \\ 5 & -9 & -7 \end{bmatrix} \begin{bmatrix} x_1 \\ x_2 \\ x_3 \end{bmatrix}$

(iii) $\begin{bmatrix} x_1 \\ x_2 \\ x_3 \\ x_4 \end{bmatrix}' = \begin{bmatrix} 0 & 1 & 0 & 0 \\ 0 & 0 & 1 & 0 \\ 0 & 0 & 0 & 1 \\ -9 & 6 & -5 & 0 \end{bmatrix} \begin{bmatrix} x_1 \\ x_2 \\ x_3 \\ x_4 \end{bmatrix}$ 3. (i) $\begin{bmatrix} 0 & 1 \\ -25 & -8 \end{bmatrix} \begin{bmatrix} x_1 \\ x_2 \end{bmatrix} + \begin{bmatrix} 0 \\ 3 \end{bmatrix}$, $\mathbf{x}_0 = \begin{bmatrix} 0 \\ 5 \end{bmatrix}$

(ii) $\begin{bmatrix} 0 & 1 & 0 \\ 0 & 0 & 1 \\ 4 & -3 & -3 \end{bmatrix} \begin{bmatrix} x_1 \\ x_2 \\ x_3 \end{bmatrix} + \begin{bmatrix} 0 \\ 0 \\ t^2+4 \end{bmatrix}$, $\mathbf{x}_0 = \begin{bmatrix} 10 \\ 5 \\ 7 \end{bmatrix}$

(iii) $\begin{bmatrix} 0 & 1 & 0 & 0 & 0 & 0 \\ 0 & 1 & 0 & 1 & -1 & 0 \\ 0 & 0 & 0 & 1 & 0 & 0 \\ t & 0 & -2 & 1 & 0 & 0 \\ 0 & 0 & 0 & 0 & 0 & 1 \\ 1 & 0 & -1 & 1 & 1 & 0 \end{bmatrix} \begin{bmatrix} x_1 \\ x_2 \\ y_1 \\ y_2 \\ z_1 \\ z_2 \end{bmatrix} + \begin{bmatrix} 0 \\ t \\ 0 \\ t^2+1 \\ 0 \\ 0 \end{bmatrix}$ $\mathbf{x}(1) = \begin{bmatrix} 3 & 10 & 1 & -5 & 4 & 3 \end{bmatrix}^T$

4. (i) $\mathbf{A} = \begin{bmatrix} 3 & 1 & 0 & 1 \\ 0 & 0 & 5 & 3 \\ -1 & 1 & 1 & -1 \\ 0 & 0 & 0 & 5 \end{bmatrix}$, $\mathbf{x}(1) = \begin{bmatrix} 2 \\ 2 \\ 2 \\ 2 \end{bmatrix}$ (ii) $\mathbf{A} = \begin{bmatrix} 3 & 1 & 0 & 1 \\ 0 & 0 & 5 & 3 \\ -1 & 1 & 1 & -1 \\ 0 & 0 & 0 & 5 \end{bmatrix}$,

$\mathbf{b} = \begin{bmatrix} t \\ -e^t \\ sint \\ te^t \end{bmatrix}$ (iii) $\mathbf{A} \begin{bmatrix} 0 & 1 & 0 & 0 & 0 \\ 0 & 0 & 1 & 0 & 0 \\ 0 & 0 & 0 & 0 & 1 \\ -1 & 0 & 0 & 0 & 9 \end{bmatrix}$, $\mathbf{b} = \begin{bmatrix} 0 \\ 0 \\ 0 \\ 0 \\ 10e^t \end{bmatrix}$; $\mathbf{x}(0) = \begin{bmatrix} -1 \\ 1 \\ 1 \\ 1 \\ 1 \\ 1 \end{bmatrix}$

Exercise 8.4.

1. (i) $\begin{bmatrix} 2 \\ 0 \end{bmatrix}$, $\begin{bmatrix} 2+2t \\ 0 \end{bmatrix}$, $\begin{bmatrix} 2+2t+t^2 \\ 0 \end{bmatrix}$

(ii) $\begin{bmatrix} 1 \\ 1 \end{bmatrix}$, $\begin{bmatrix} 1+e^t+t \\ 1-e^{-t}+\frac{t^2}{2} \end{bmatrix}$, $\begin{bmatrix} \frac{1}{2}e^{2t}+te^t+\frac{t^3}{6}+e^{-t}+t+1 \\ 1+\frac{t^2}{2}+(t-1)e^t+\frac{t^3}{3}-\frac{1}{2}(t^2+2t+2)e^{-t}+\frac{e^{-2t}}{2} \end{bmatrix}$

(iii) $\begin{bmatrix} 1 \\ 0 \end{bmatrix}, \begin{bmatrix} 1+t+\frac{t^2}{2} \\ 1 \end{bmatrix}, \begin{bmatrix} 1+\frac{1}{2}t^2+\frac{t^4}{8} \\ t+\frac{t^3}{2} \end{bmatrix}$

Exercise 8.5.

1. (i) $\begin{bmatrix} -\frac{1}{5}e^t+\frac{1}{6}e^{2t}+\frac{31}{30}e^{-4t} \\ -\frac{1}{5}e^t+\frac{1}{3}e^{2t}-\frac{62}{15}e^{-4t} \end{bmatrix}$ (ii) $\begin{bmatrix} 1+2sin(t+\pi) \\ 2cos(t-\pi) \end{bmatrix}$

2. (i) $\begin{bmatrix} e^{2t}(-x_{10}+x_{20})+e^t[2x_{10}-x_{20}-1]+1 \\ 2e^{2t}[-x_{10}+x_{20}]+e^t[2x_{10}-x_{20}-1]+1 \end{bmatrix}$, (ii) $\begin{bmatrix} \frac{5}{2}e^t-\frac{5}{2}e^{-t} \\ e^t+te^{-t} \\ e^t \end{bmatrix}$

Exercise 8.6.

1. $\begin{bmatrix} e^{it} & te^{it} & \frac{1}{2}t^2e^{it} \\ 0 & e^{it} & te^{it} \\ 0 & 0 & e^{it} \end{bmatrix}$ 3. (i) $\frac{1}{5}\begin{bmatrix} e^{-4t}+4e^t & -e^{-4t}+e^t \\ -4e^{-4t}+4e^t & 4e^{-4t}+e^t \end{bmatrix}$

4. $\bar{c_1}\begin{bmatrix} 1 \\ -2 \end{bmatrix}e^{10t}+\bar{c_2}\begin{bmatrix} -t \\ 2t+1 \end{bmatrix}e^{10t}$ 5. $\begin{bmatrix} e^t-t/3-1/9 \\ -\frac{t}{3}-\frac{1}{9} \end{bmatrix}$

6. (i) $\begin{bmatrix} e^t & te^t & 0 \\ 0 & e^t & 0 \\ 0 & 0 & e^t \end{bmatrix}$ (ii) $\frac{1}{6}\begin{bmatrix} 2e^{4t}+4e^{-2t} & e^{4t}-e^{-2t} \\ 8e^{4t}-8e^{2t} & 4e^{4t}+2e^{-2t} \end{bmatrix}$

(iii) $\begin{bmatrix} e^{3t}-3te^{3t} & te^{3t} \\ -9te^{3t} & e^{3t}+3te^{3t} \end{bmatrix}$

Exercise 8.7.

1. $\alpha_1 > -a$; $\alpha_2 < 1-\frac{3}{a+\alpha_1}$
3. (i) $\lambda = -2,\ -2,\ -1$; stable (ii) $\lambda = -1,\ -2,\ -3$; stable
5. (i) $\lambda = \frac{2}{s^2}$; bounded (ii) $\lambda = -4$; stable

Exercise 9.3.

1. (i) $\begin{bmatrix} e^{2t}-2te^{2t} & te^{2t} \\ -4te^{2t} & (1+2t)e^{2t} \end{bmatrix}$, (ii) $\frac{1}{4}\begin{bmatrix} 4e^{-t} & e^t-e^{-3t} & 3e^t-4e^{-t}+e^{-3t} \\ 0 & e^t+3e^{-3t} & 3e^t-3e^{-3t} \\ 0 & e^t-e^{-3t} & 3e^t+e^{-3t} \end{bmatrix}$

2. (i) $\begin{bmatrix} -1+2\cos t+\sin t & -4+4\cos t+2\sin t & -7+7\cos t+sin\,t \\ \cos t-2\,sin\,t & 2\,cos\,t-4\sin t & \cos t-7sin\,t \\ \sin t & 2\sin t & \cos t+3\sin t \end{bmatrix}$,

(ii) $\begin{bmatrix} e^{3t} & 6te^{3t} \\ 0 & 3e^{3t} \end{bmatrix}$. 3 (i) $\begin{bmatrix} t+1 & 2e^t-e^{2t}-1 \\ 0 & e^{2t} \end{bmatrix}$,

(ii) $\begin{bmatrix} e^t & e^t-t-1 \\ 2e^{-t} & 2(t+1-e^{-t}) \end{bmatrix}$

Exercise 9.4.

1. $\frac{1}{9}\begin{bmatrix} -3e^{-t}+9e^{2t}+3e^{8t} & 3e^{-t}-9e^{2t}+6e^{8t} \\ -12e^{-t}+9e^{2t}+3e^{8t} & 12e^{-t}-9e^{2t}+6e^{8t} \end{bmatrix}$ 2. (i) $\lambda = 2,\ -4$; $\begin{bmatrix} 1 \\ 2 \end{bmatrix}, \begin{bmatrix} -1 \\ 4 \end{bmatrix}$,
and $\mu = 1,\ 2$; $\begin{bmatrix} 1 \\ 0 \end{bmatrix}, \begin{bmatrix} 1 \\ 1 \end{bmatrix}$

(ii) $\lambda = 2,\ -4,\ 4,\ -8;\quad \begin{bmatrix}1\\0\\2\\0\end{bmatrix}, \begin{bmatrix}-1\\0\\4\\0\end{bmatrix}, \begin{bmatrix}1\\1\\2\\2\end{bmatrix}, \begin{bmatrix}-1\\-1\\4\\4\end{bmatrix};$

(iii) $$\begin{bmatrix} \frac{e^{-4t}+2e^{2t}}{3} & \frac{-2e^{2t}-e^{-4t}+2e^{4t}+e^{-8t}}{3} & \frac{-e^{-4t}+e^{2t}}{6} & \frac{e^{2t}+e^{-4t}+e^{4t}-e^{-8t}}{6} \\ 0 & \frac{e^{-8t}+2e^{4t}}{3} & 0 & \frac{e^{4t}-e^{-8t}}{6} \\ \frac{-4e^{-4t}+4e^{2t}}{3} & \frac{-4e^{2t}+4e^{-4t}+4e^{4t}-4e^{-8t}}{3} & \frac{2e^{-4t}+e^{2t}}{3} & \frac{-e^{2t}-2e^{-4t}+e^{4t}+2e^{-8t}}{3} \\ 0 & \frac{-4e^{-8t}+4e^{4t}}{3} & 0 & \frac{e^{4t}+2e^{-8t}}{3} \end{bmatrix}$$

(iv) $\mathbf{I}_4$, (v) same as (iii).

3. (a) $$\begin{bmatrix} \frac{-e^{-4t}+e^{2t}}{6} & \frac{e^{2t}+e^{-4t}+e^{4t}-e^{-8t}}{6} \\ 0 & \frac{e^{4t}-e^{-8t}}{6} \\ \frac{-4e^{-4t}+4e^{2t}}{3} & \frac{-e^{-2t}-2e^{-4t}+e^{4t}+2e^{-8t}}{3} \\ 0 & \frac{e^{4t}+2e^{-8t}}{3} \end{bmatrix}$$

(b) $$\begin{bmatrix} \frac{6e^{2t}-e^{4t}-e^{-4t}-3}{12} & \frac{12e^{2t}+6e^{-4t}+10e^{4t}-7e^{-8t}-5}{48} \\ 0 & \frac{10e^{4t}-7e^{-8t}-3}{48} \\ \frac{24e^{2t}-15e^{-4t}-9}{12} & \frac{-6e^{2t}-10e^{-4t}+5e^{4t}+7e^{-8t}}{12} \\ 0 & \frac{10e^{4t}+15e^{-8t}-1}{24} \end{bmatrix}$$

Exercise 9.5.

1. $\begin{bmatrix} -e^{2t}-e^{3t} & 1-e^t \\ e^{3t} & e^t \end{bmatrix}$ 2. $\begin{bmatrix} -t/2+1/4-5/4e^{2t}-e^{3t} & 2(1-e^t) \\ e^{3t} & 2e^t-1 \end{bmatrix}$

3. $\begin{bmatrix} e^t & e^{2t} \\ 0 & -2e^{2t} \end{bmatrix} \begin{bmatrix} c_1 & c_3 \\ c_2 & c_4 \end{bmatrix} \begin{bmatrix} e^t & -e^{3t} \\ 0 & e^{3t} \end{bmatrix}$

Exercise 9.6.

1. (i) $e^{\mathbf{A}t}[c_0 + c_1 t + c_2 t^2]e^{\mathbf{B}t}$

(ii) $C_0 + e^{4\mathbf{A}t}C_1 e^{4\mathbf{B}t} + e^{-3\mathbf{A}t}C_3 e^{-3\mathbf{B}t}$

2. $e^{\mathbf{A}t}\begin{bmatrix} -2t & -t \\ 2-t & 2-2t \end{bmatrix} e^{\mathbf{B}t}$

Exercise 9.8.

1. $\begin{bmatrix} 1 & 0 & 0 \\ 0 & 1 & 0 \\ 0 & 0 & 1 \end{bmatrix}$, $\begin{bmatrix} 1 & 1 & 3 \\ 5 & 2 & 6 \\ -2 & -1 & -3 \end{bmatrix}$, $\frac{1}{2}\begin{bmatrix} 0 & 0 & 0 \\ 3 & 3 & 9 \\ -1 & -1 & -3 \end{bmatrix}$;

$\mathbf{I}_3$, $2\begin{bmatrix} 1 & 1 & 3 \\ 5 & 2 & 6 \\ -2 & -1 & -3 \end{bmatrix}$, $\begin{bmatrix} 0 & 0 & 0 \\ 3 & 3 & 9 \\ -1 & -1 & -3 \end{bmatrix}$

Index